Computer Modeling of Geologic Surfaces and Volumes

Edited by
David E. Hamilton
and
Thomas A. Jones

AAPG Computer Applications in Geology, No. 1

Published by
The American Association of Petroleum Geologists
Tulsa, Oklahoma, U.S.A.

Copyright © 1992
The American Association of Petroleum Geologists
All Rights Reserved
Published 1992

ISBN: 0-89181-700-X

AAPG grants permission for a single photocopy of an item from this publication for personal use. Authorization for additional copies of items from this publication for personal or internal use is granted by AAPG provided that the base fee of $3.00 per copy is paid directly to the Copyright Clearance Center, 27 Congress Street, Salem, Massachusetts 01970. Fees are subject to change. Any form of electronic or digital scanning or other digital transformation of portions of this publication into computer-readable and/or transmittable form for personal or corporate use requires special permission from, and is subject to fee charges by, the AAPG.

Association Editor: Susan A. Longacre
Science Director: Gary D. Howell
Publications Manager: Cathleen P. Williams
Special Projects Editor: Anne H. Thomas

AAPG
Wishes to thank the following
for their generous contributions
to

Computer Modeling
of
Geologic Surfaces
and
Volumes

Amerada Hess Limited

❖

Exxon Production Research Company

❖

Marathon Oil Company

❖

Integrated Software Technology, Morrison Knudsen
Corporation

❖

Placer Dome Inc.

Contributions are applied against the production costs of the
publication, thus directly reducing the book's purchase price
and thereby increasing the availability of the volume to a
greater audience.

ACKNOWLEDGMENTS

We would like to thank all the authors for their efforts, and for tolerating our many edits. Without their work, this book would not have come into being. In addition, we appreciate the efforts of other individuals who submitted papers that, because of their style, subject, or audience, were not appropriate for this volume. We want to thank them for the hours they contributed and encourage them to publish those papers in appropriate journals or books.

Masera Corporation, Morrison Knudsen Incorporated, and Petroleum Information Incorporated contributed data used in several papers and published in Appendix B. We appreciate this contribution and feel the data will be useful to many for analyzing algorithms and possibly for developing new ones.

Dick Banks, John Deck, Dale Davis, John Harbaugh, Inda Immega, Doug Palkowsky, Kay Plavidal, and Steve Zoraster read some or all of the papers and made many useful suggestions.

Our employers, Landmark/Zycor Incorporated and Exxon Production Research Company, were very tolerant of our less-than-total attention to our jobs during the time this book was being put together. We appreciate that tolerance and the permission to present material contained in papers we contributed to this book.

Thanks go to Cathleen Williams and others at AAPG for coaching us through the publication process.

David E. Hamilton
Thomas A. Jones

ABOUT THE EDITORS

David E. Hamilton was born in 1952 in Winthrop, Iowa. He earned B.S. and M.S. degrees in geology from Iowa State University, with emphasis on structure and geomorphology and a thesis entitled "Statistical Definition of Geomorphic Regions of Iowa." He joined Exxon Production Research Company in 1979, was trained by Carlton Johnson in the area of computer mapping, and worked 2-D and 3-D modeling projects involving minerals, coal, petroleum, and groundwater chemistry. The job also involved developing new ways to solve geologic modeling problems, and training others in the modeling procedures. Dave joined Zycor Incorporated, in Austin, Texas, in 1987 and has done consulting, training, and program specification in the area of surface mapping. Dave is currently working on 3-D modeling techniques. He has written articles about mapping methods, 3-D modeling projects, the user interface for mapping systems, and with Tom Jones and Carlton Johnson he co-authored the book *Contouring Geologic Surfaces With The Computer*.

Thomas A. Jones was born in 1942 in Delano, California. He earned B.S. and M.S. degrees in mathematical statistics at Colorado State University, and then attended Northwestern University and was granted M.S. and Ph.D. degrees in geology. Upon completion of university training in 1969, he joined Exxon Production Research Company, where he now holds the position of Research Advisor. The bulk of his work at Exxon has been in the development and application of 3-D modeling systems, although a related part of this effort involved methods used in computer-based mapping. During 1983-84, Tom was Exxon's expert witness on computer mapping for arbitration of equity in the Prudhoe Bay field. He is currently working in assessment and basin analysis. Tom has published approximately 45 papers on geologic application of statistics and computers. With Dave Hamilton and Carlton Johnson, he is co-author of a book on computer mapping. He has served as Editor-in-Chief, Associate Editor, and Book Review Editor of the journal *Mathematical Geology*.

Table of Contents

Contributions Acknowledgment iii

Acknowledgments iv

About the Editors v

Preface ix
       David E. Hamilton and Thomas A. Jones

Chapter 1
A Philosophy of Computer Mapping with the Computer 1
       Thomas A. Jones and David E. Hamilton

Chapter 2
Graphical Techniques for Locating Anomalies 9
       William R. Green

Chapter 3
The Shape-Assist Technique: Incorporating Stream Channel Interpretations
into Computer-Generated Surface Models: Clark County, Kansas 27
       John E. Fierstien and Arthur V. Brewster

Chapter 4
Computer-Generated Surfaces and Grids of the Geometry, Rock Type,
and Chemistry of a Bedded Mineral Deposit 37
       Walter J. Bawiec and Wilbur D. Grundy

Chapter 5
Computer Mapping of Pinnacle Reefs, Evaporites, and Carbonates:
       Northern Trend, Michigan Basin 47
       David E. Hamilton and Skye K. Henize

Chapter 6
Techniques for Modeling Compaction Applied to the Pinnacle Reef/
Carbonate/Evaporite Environment 61
       David E. Hamilton and William G. Riehl

Chapter 7
Modeling Anisotropy in Computer Mapping of Geologic Data 75
       Gregory Kushnir and Jeffrey M. Yarus

Chapter 8
Surface Modeling for Sedimentary Basin Simulation 93
       John C. Tipper

Chapter 9
Testing Hydrocarbon Saturation Models for Use in Original Oil-In-Place
Estimation: South Dome of Oregon Basin Field, Park County, Wyoming 105
       Michael J. Heymans, Douglas A. Steed, David E. Hamilton, and Barbara A. Pavlov

Chapter 10
Fault Representation in Automated Modeling of Geologic Structures and Geologic Units.............123
Steven Zoraster

Chapter 11
The Gridded Fault Surface..141
Don Clarke

Chapter 12
Computer Modeling of Multiple Surfaces with Faults: The Ivanhoe Field,
Outer Moray Firth Basin, U.K. North Sea..161
N.J. Hooper, J.G.M. Raven, and M.J. Kilpatrick

Chapter 13
Extensions to Three Dimensions: Introduction to the Section on 3-D
Geologic Block Modeling...175
Thomas A. Jones

Chapter 14
Three-Dimensional Geologic Block Model of a Polar Fan-Delta Complex,
Canning River, North Slope, Alaska...183
John E. Lindsay

Chapter 15
Three-Dimensional Geologic Block Modeling of the Kutcho Creek
Massive Sulfide Deposit, British Columbia...203
David E. Hamilton and Robert S. Didur

Chapter 16
The Impact of Vertical Averaging on Hydrocarbon Volumetric
Calculations—A Case Study..219
Larry E. Denver and Danny C. Phillips

Chapter 17
Application of Variable Zone Modeling to Modeling and Mapping
of Canadian Oil Sands..235
Khosrow Badiozamani, Foad Roghani, and George Hawes

Chapter 18
Three-Dimensional Modeling Techniques in the Analysis of a
Mature Steam Drive..251
Ramsay A. Barrett and Jeffery Bailey

Chapter 19
Three-Dimensional Modeling of Complex Geological Structures:
New Development Tools for Creating 3-D Volumes...261
Raphael Mayoraz, Carol E. Mann, and Aurele Parriaux

Appendix A
Algorithm Comparison with Cross Sections..273

Appendix B
Five Geologic Datasets 279

Index 293

PREFACE

In 1986, with Carlton Johnson, we published the book, *Contouring Geologic Surfaces With The Computer*. That book generally has been well received. There have been comments, however, that examples that apply the methods to real situations are needed. Based on that feedback and discussions with mapping software users, we realized that the geologic mapping community needed more than a second edition with a few added examples. We decided that a new mapping book, covering a wide variety of geologic settings and a large number of applications, would be more valuable.

To expand the 1986 volume beyond our experience base, we chose to make it a collection of contributed papers on computer-mapping case studies. By offering this range of examples, we hope to provide useful "go-bys" for both beginners and advanced users of computer-mapping software. For the most part, the papers concentrate on the geologic features of significance to mapping, the methods used and their justification, and results obtained. Mathematical equations and details of algorithms are kept to a minimum, although obviously some details of this nature must be presented.

The papers are independent and can be read in any order, although we have tried to place introductory papers near the beginning and group related papers together, and have arranged them in order to build the reader's understanding of the subject. Some papers use one dataset, and others use many, to demonstrate mapping procedures. Other papers present complete studies, showing project flow and areas where the programs were most useful. A few of the papers provide general background information useful for understanding an area of computer modeling.

The book is separated into two parts. Part 1 consists of 12 papers and deals with data and surface modeling. Applications range from geology, to mining, to petroleum engineering. Part 1 begins with our philosophy of solving geologic mapping problems, and the subsequent papers demonstrate variations of that philosophy in many application areas.

The papers in Part 1 represent our original goal for the book. However, we recognized that the geoscience and computing communities are moving from mapping problems in two-dimensions to mapping them in three-dimensions. In addition, it is clear that no simple boundary exists between 2-D and 3-D analysis. Accordingly, we expanded the scope of the book.

Part 2 consists of 7 papers dealing with three-dimensional geologic block modeling. It begins with a paper that discusses problems encountered in 2-D computer mapping that are solved by 3-D modeling, and presents some of the techniques used to solve them. Other papers in this part present a variety of methods currently used to store and build 3-D models. The applications range from minerals, to heavy oil, to planning tunnels in mountainous terrain. Because this is a new area of computer applications to geology, several of the papers describe in detail the methods used and the advantages and disadvantages of various approaches.

Two appendices have been provided. Appendix A lists a dataset and demonstrates the application of those data to the evaluation of mapping algorithms in cross section. Appendix B provides five geological datasets that may be used to test mapping algorithms and methods. Some of the papers refer to these appendices and use data presented in them.

Programs specifically written to address the mapping and modeling of geologic data are required to solve the problems presented in these papers. Most of these programs, or their equivalents, are available from vendors of computer mapping and modeling software. Many of the papers in Part 1 could have been done using programs available for personal computers. All of the papers in Part 1 could have been done interactively; however, some procedures involve several hundred steps and others require many minutes to hours of computer time to complete. For these more complex problems, batch functionality is used to advantage.

Papers in Part 2 involve 3-D geologic block-modeling programs that, in the past, required many computational hours on large main-frame computers. More recently, high-powered workstations have allowed these programs to be run quickly on the geologist's desktop. The 3-D graphics capabilities of these workstations have sparked the addition of 3-D visualization to the evaluation of 3-D models. All of these 3-D visualization programs currently are efficient only with "high-end" workstations.

David E. Hamilton
Thomas A. Jones

Chapter 1

A Philosophy of Contour Mapping with the Computer

Thomas A. Jones
Exxon Production Research Company
Houston, Texas, U.S.A.

David E. Hamilton
Landmark / Zycor, Inc.
Austin, Texas, U.S.A.

ABSTRACT

Geologic interpretation is a critical requirement for mapping geological variables. Several acceptable though different maps could be drawn with a set of observations, so the map that best honors geologic concepts and interpretation is most desirable. We propose a philosophy of contouring with the computer that involves combining three factors: geologic interpretation, data, and the program or algorithm to be used. Each of these is important and all must be considered jointly to combine good procedures with appropriate software. In fact, correctly combining these three factors places the required interpretation into the map.

INTRODUCTION

Several methods are available for contouring by hand. Tearpock and Bischke (1991) describe four of these: mechanical contouring, parallel contouring, equal-spaced contouring, and interpretive contouring. The first three of these are usually inadequate because they allow for no geologic interpretation or because they produce geologically unrealistic maps. Interpretive contouring allows use of geologic license, with no inherent assumptions of constant slope, constant spacing, etc. This method allows the incorporation of the geologist's interpretation and experience, plus regional information. The goal of interpretive contouring is to develop the most reasonable and realistic interpretation of the subsurface.

An analogous range of methods is found when using the computer to map. Many programs or algorithms are available for mapping; some make mechanically sound maps that are geologically ridiculous, whereas others are designed to work with certain types of data or surface forms. A contour map that merely honors data points reasonably well is no longer adequate, as computer mapping has expanded to encompass modeling geologic surfaces, as shown by papers in this volume. Because several acceptable though different maps could be drawn, the map now must also honor geologic concepts and interpretation, as well as be consistent with maps of other variables. Just as interpretive contouring blends geologic interpretation and experience to properly contour data, so too must computer algorithms use interpretation to produce an acceptable map.

In many projects, data are passed to the machine without special instructions—and the resulting maps are accepted without question, even though they may violate geologic principles, knowledge, or interpretation. Modern programs have the capability to incorporate geologic interpretation, and geologists are negligent if that information is not used. A computer-mapping program that is provided only measured data will not produce good maps, perhaps leading the geologist to believe that computer mapping is not useful.

Jones et al. (1986) discuss the use of standard software to force geological concepts into computer maps. Most modern programs are flexible enough that a general, three-step approach to mapping can be followed: (1) interpret the geology in the area, both spatially and temporally; (2) process the input data according to this interpretation; and (3) operate the mapping program such that the geologic interpretation is honored. Our philosophy is to follow these three steps when contouring with the computer. In a sense, this approach sequentially reconstructs geologic events with the mapping program.

A second way to look at computer contouring is in terms of three factors: interpretation, data, and the computer program or algorithm. All three of these equally important aspects must be taken into account for success. A non-geologic variable may be contoured if only data and algorithms are taken into account, but interpretation is the single absolute requirement for a geologic map. Some papers in the literature concentrate on one or two of these factors, but rarely does the discussion encompass all three. Combining these three factors according to the needs of the project corresponds to the third step in our mapping methodology.

INTERPRETATION

Modeling geologic features demands conceptual information and interpretation obtained from practical knowledge and experience, as well as measurements obtained in the field. A map may honor the data faithfully, but it may neither portray geology accurately nor predict reliably without valid interpretation to extend the data. Relevant information typically includes the sequence of events that led to the present geologic configuration in an area. This encompasses such things as depositional environments, erosion, relationships between intersecting horizons, mechanisms and timing of folding and faulting, and correlation. This interpretive work requires an understanding of the local and regional geology.

For mapping a single variable, interpretation may be as simple as postulating trends. If a map of sand thickness does not show known directional trends, the map has been made incorrectly; the algorithms used or the treatment of the data did not reflect the interpretation. Although our interpretation of trend direction and strength cannot always be made with confidence, we must use our best estimate, because omitting the trend entirely is also an interpretation, albeit unstated.

Another example relates to the form of a surface. For instance, we may interpret that the surface was first created as a simple planar form, and that later erosion dissected it into several plateaus. The data and map should then show the effect of two events: formation of the original surface, and erosion. Similarly, crests of pinnacle reefs will form a population distinct from inter-reef locations. Mapping such complex surfaces depends on the data and algorithm used, but the first step is recognition of their existence. Of course, interpretation can also be a key to effective sampling.

Multiple surfaces are common, and interpretations must be made regarding these. Several geologic concepts are involved here, including conformability, unconformities, baselap, truncation, and sequences (c.f., Jones et al., 1986). Independently mapping conformable surfaces ignores this additional information, and problems (e.g., impossible crossings of surfaces, or changes in structural complexity due to decreased data density with depth) can arise in the maps (Walters, 1969; Fontaine, 1985). Ignoring this information can create maps that are not consistent geologically.

In practice, the recognition of significant geologic features and interpretation of their implications involve standard methods. The computer may aid in understanding the observed data and geologic relationships. For instance, Green (1985, 1992) presents methods and displays that help highlight significant features.

DATA CONSIDERATIONS

The observed measurements are the second requirement for constructing a contour map. Many possible measurements or observations, across many different specialties, are available in geology. Elements of interpretation are involved in data collection and manipulation, and problems can result if measured values are not handled correctly. Two common errors occur during mapping: one is mixing data that represent different surface forms or variables, and the second is mixing data from different sources that purport to represent the same variable.

It might seem obvious to avoid mixing different variables when mapping, but mixtures can be subtle. An example involves mapping pinnacle reefs; the mapped surface will have different properties on the reef top, reef flank, and inter-reef areas, that is, we have three potential populations. Hamilton and Henize (1989, 1992) show that reefs may not have a realistic appearance if all data are combined and mapped simply (Figure 1). Data from different portions of the reef must be separated and mapped individually, and then the individual portions combined into a final map of the reefs, to obtain realistic results (Figure 1B). Data from this region are supplied in Appendix B of this volume.

A second example is mixing rock tops with stratigraphic tops. A stratigraphic pick is the intersection of a borehole with a litho-, bio-, or chronostratigraphic horizon. If a rock unit is partially truncated, its upper boundary is represented by both stratigraphic and

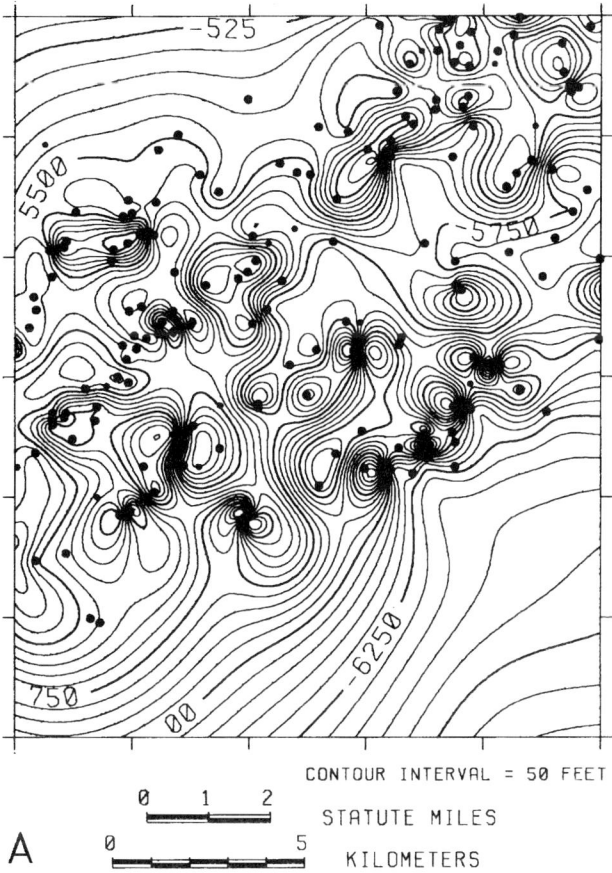
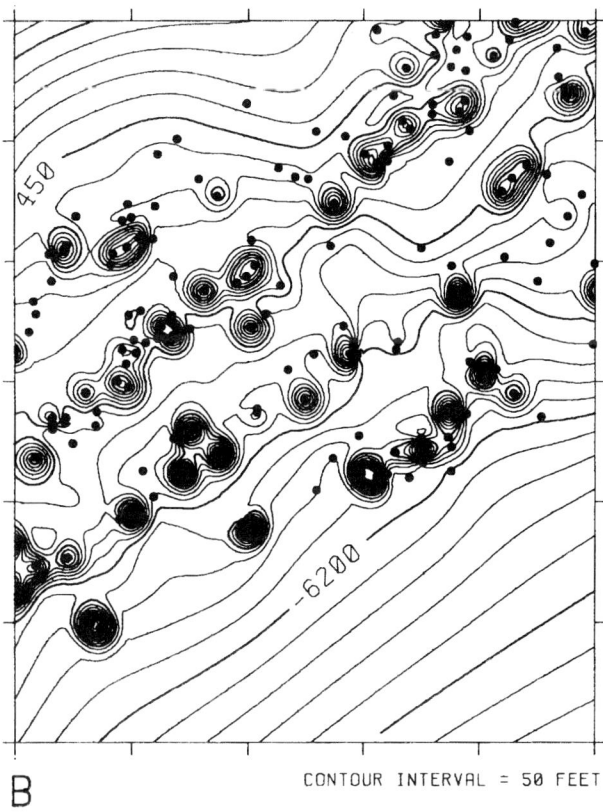
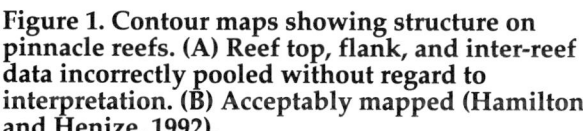

Figure 1. Contour maps showing structure on pinnacle reefs. (A) Reef top, flank, and inter-reef data incorrectly pooled without regard to interpretation. (B) Acceptably mapped (Hamilton and Henize, 1992).

unconformity surfaces. When mapping the top of this truncated unit, use only stratigraphic tops; rock tops include both the stratigraphic and unconformity tops and therefore do not represent the same geologic horizon at all locations (c.f., Iglehart, 1970), as shown in Figure 2A. A separate map should be constructed from the unconformity picks for this unit and all other affected units. The two maps which represent the stratigraphic and unconformity portions of the unit's top should be combined to form the complete top of the rock unit (c.f., Jones et al., 1986), as shown in Figure 2B.

A third example of mixing variables occurs in isochore mapping. If we are mapping the thickness of a unit that pinches out, our data will consist of positive values where the unit exists and zero values where it does not. If zero values are mixed with observed positive thicknesses, a contour map will be unacceptable because the zero contour will erratically try to honor each zero value (Figure 3; Jones et al., 1986, figure 9.7). Further, the surface is essentially uncontrolled away from the positive values. Although one well may have been drilled exactly on the edge of the unit, it is too great a coincidence to believe all wells were so drilled. Such unacceptable results are caused by mixing quantitative positive thickness values with qualitative zero indicators. In this case, "zero" indicates absence of the unit, and is not a contourable thickness. A solution to this problem is modification of the data, discussed below. This problem also is discussed by Fierstein and Brewster (1992). The delta-thickness data in Figure 3 are provided in Appendix B of this volume.

Most projects involve data from several sources (e.g., seismic, well, and outcrop) or from different periods or methods of data collection (e.g., different vintages of seismic data). A mixture of well tops and seismic tops (converted from travel time to elevation) is a common problem. The seismic data are closely spaced along the seismic lines, so details of structure are present. On the other hand, widely spaced wells show only broad features when contoured. In addition to having data with different degrees of detail, time-depth conversion errors can make the two kinds of tops disagree with each other. Combining the datasets into a common pool can produce maps with serious anomalies. Methods are available for merging different datasets into a consistent set (c.f., Jones et al., 1986).

COMPUTER ALGORITHMS

The third major factor for computer contouring is the program or algorithm to be used. It is, of course, important to produce a reasonable representation of an individual surface or spatial variable. However, this representation is only part of mapping. It is also important that the algorithm's properties be consistent with the geologic interpretation and data characteristics.

An irregular data distribution usually is more difficult to contour by hand than an evenly spaced distribution, and similar considerations come into effect with the computer. In order to draw a contour map, most computer programs first convert scattered data into an ordered form, and then use the result for contouring. Two forms are used widely—a rectangular grid, or a triangular network. The grid is implemented with most commercial contouring packages, but triangular methods are now being used more frequently.

Rectangular Grid

A rectangular grid consists of a set of rows (horizontal lines) and columns (vertical lines). Crossings between rows and columns define grid intersections or nodes, each of which represents a geographic (X, Y) location in the area to be contoured. A number (Z-value) assigned to each node represents an estimate of the mapped variable at that location. The Z-values are calculated for the nodes from the data in such a way as to represent the variable (e.g., elevation). Contours are generated from the grid and not from the original observation points.

Demands by geologists for ways to control mapping have led to many different algorithms for assigning values to grid nodes. The selection of various algorithms and controls by the geologist is often done in response to geologic interpretation, although perhaps not thought of as such. Discussions of several basic methods (and variations) to calculate node values are given by Walters (1969), Peucker (1980), and Jones et al. (1986), among many others. A summary of four algorithms is presented in Appendix A of this volume.

All gridding methods select data points near (and ideally surrounding) the node to be calculated. Given the set of points, the calculation may be done in several ways. Simple methods use weighted averages of data values, with weights based on distance. Somewhat more complex ways calculate values by fitting simple, low-order functions, commonly planes or quadratics. Point kriging and spline-fitting are two other methods. Special algorithms have been devised for handling measurements taken along lines (e.g., seismic-line or aeromagnetic data) and digitized contour lines.

Triangular Network

The second general approach to surface replication is the triangular network, sometimes called a triangulated irregular network (TIN). Rather than a square or rectangular pattern, a network is made up of a series of triangles. These triangles are usually defined by the data distribution, a vertex being placed at each data

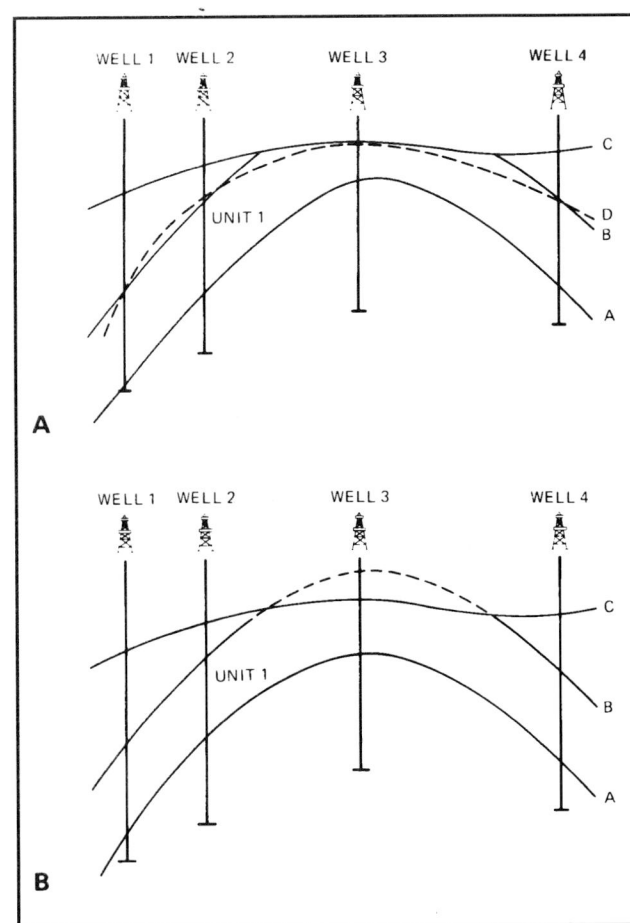

Figure 2. Cross section showing surface B truncated by unconformity C. (A) Surface B does not exist in well 3, but has been picked as the top of rock unit 1; the resulting surface D is incorrect. (B) The pick for surface B in well 3 is set to null (missing data) and correctly produces surface B (Jones et al., 1986, figure 5.1).

point. Vertices thus are analogous to grid intersections, and the number assigned to each vertex is that value observed at the data point.

The triangular method has several advantages: calculation routines are not needed to assign values to the vertices; editing can be done easily; if several observations are made at each data location, the triangular framework only needs to be made once and several values can be assigned to each vertex; and when the map is drawn, all data values will be honored, that is, contours will pass on the correct sides of all points.

The map built from a triangular network will be limited to the area within the data locations. If contours are desired along the map boundary, additional points may have to be created manually. These boundary defining points must be assigned Z-values based on extrapolation from nearby data, as with a grid. Another disadvantage is that the method has no flexibility in how a single-surface map is generated. Unlike grid-

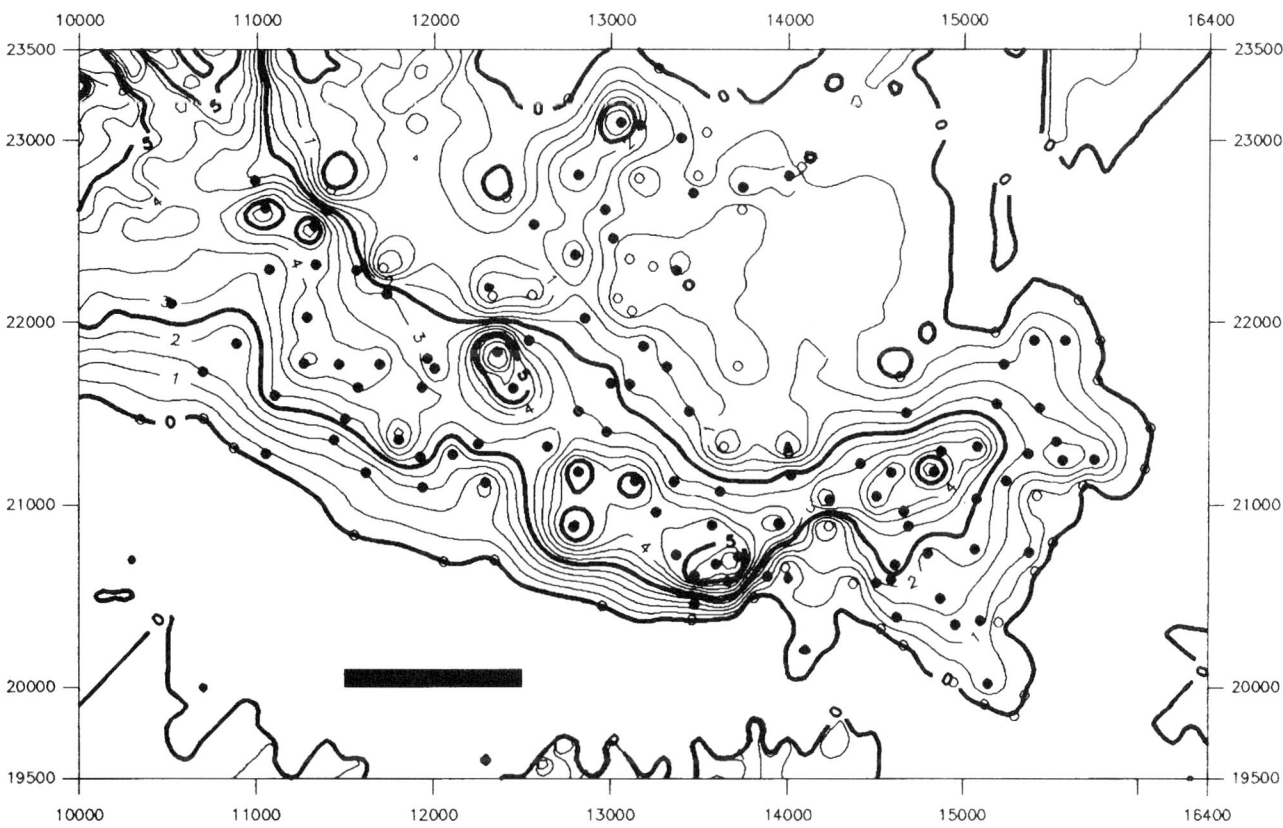

Figure 3. Map showing contours of delta thickness; the zero values were incorrectly included in the calculations, leading to erroneous projections and zero contour. Contour interval is 0.5 ft (0.15 m). Scale bar represents 1000 ft (305 m). Solid dots represent positive thickness values. Open dots represent zero thickness values. Data interpolated from maps in Donaldson et al., 1970, used with permission of SEPM.

ding, the TIN does not allow the user to introduce interpretive features (e.g., tight versus broad closures).

Many ways exist to connect a set of points into triangles, potentially leading to a variety of maps. Delaunay triangles produce a unique set with good properties. Green and Sibson (1978), McCullagh and Ross (1980), Watson (1981, 1982), Philip and Watson (1982), and Auerbach and Schaeben (1990) describe efficient methods that form Delaunay triangles. Banks (1991) and Tipper (1992) also discuss triangulation networks.

BLENDING DATA, INTERPRETATION, AND ALGORITHM

Incorporating even a simple interpretation into a map is affected strongly by properties of available algorithms and how data are organized. This section points out examples of situations in which algorithms or data handling methods are influenced by geologic interpretation.

An important consideration for realistic mapping is the ability to project Z-values. The actual limits of the variable or horizon being mapped are unlikely to occur at sample locations. Yet, gridding algorithms that use averaging techniques imply this by not generating Z-values that exceed the observed maximum or that are less than the observed minimum value; peaks and ridges are lowered, basins are raised, and extrapolations tend to the local data average (Figure 4). This characteristic of a gridding algorithm may produce a result as incorrect as an algorithm that allows wild extrapolations based on data gradients.

Triangular networks can have the same difficulty, but from a more fundamental basis. The only nodes are typically found at data locations, so no Z-values exist in the network that are not observed data values. If values are assigned at "no-data" or "boundary-defining" nodes, the calculations should be done in such a way as to allow projection of trends.

Another type of projection problem, usually thought to be restricted to grids but also possible at boundary-defining network nodes, can occur with extrapolation away from data control. When nodes beyond edges of control are calculated, algorithms may project existing trends from a few edge data points (Figure 4). Although a linear trend is reasonable near the data, extending the same trend without limit is not reasonable. Some algorithms will even

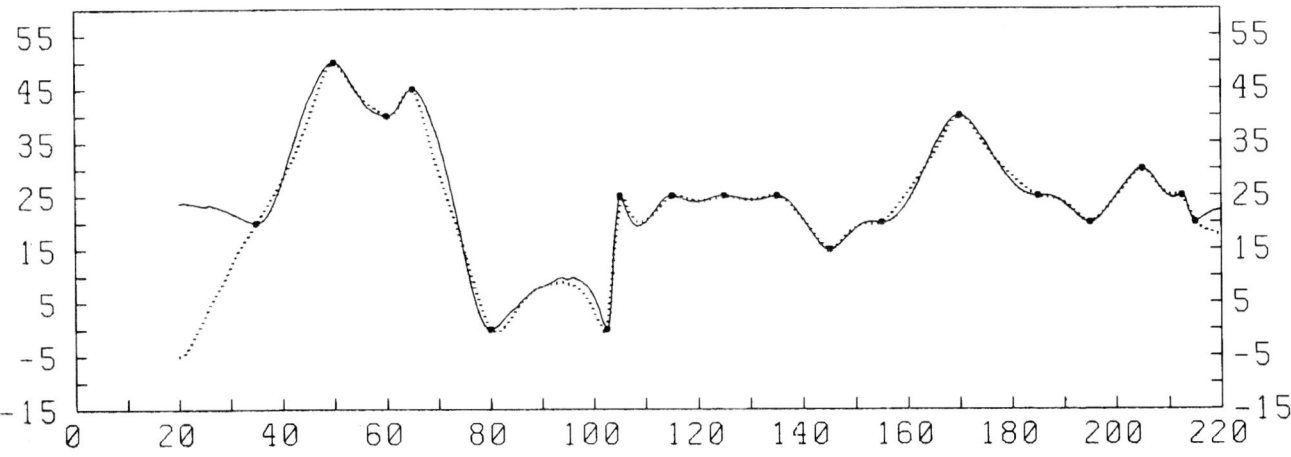

Figure 4. Diagrammatic cross section showing least-squares grid (dotted line) and weighted-average grid (solid line). Dots represent data points.

steepen projected gradients, giving an ever-worsening projection. On the other hand, averaging methods (e.g., distance weighted averages and kriging) may converge incorrectly to a regional mean value.

A related aspect of projection occurs when dealing with thickness of a unit which pinches out. Jones et al. (1986, chapter 9) point out examples (e.g., drawing the zero-contour or integrating volumes) in which negative values should be projected where the unit is absent. If zero values are mixed with observed positive thicknesses and mapped, the map will attempt to honor the zero value at each well in the pinch-out region, implying falsely that the unit goes to zero thickness precisely at each such well. This problem was pointed out above in the discussion of mixtures (Figure 3).

A solution to this isochore-mapping problem, for use both with rectangular grids and triangular networks, uses a modification of the data. Here all observed values that equal zero are changed to "no data" (so these wells will not be used) or to negative values (to ensure a correct zero-line projection). All points could be assigned the same negative value (e.g., –2 in Figure 5), although some algorithms calculate individual values for each point according to local trends and data distribution.

Another example is incorporation of such stratigraphic relationships as baselap and truncation into a contour map. The primary tool for such operations is manipulations between grids or triangular networks which are analogous to cross contouring (Jones et al., 1986). The combining process uses several grids or networks, each created to stress certain geologic properties. Operations normally are done on a node-by-node basis for corresponding nodes or vertices which represent the same X-Y location; calculations are made using Z-values at matching nodes. The grids are combined according to the interpretation to create a final map.

In triangular networks, simple operations between surfaces can become complicated if each triangular network does not have the same set of (X,Y) vertices (e.g., if deeper horizons have fewer data points than do shallow horizons, or if deviated wells are involved). A program that implements a triangular system must have the ability to create and interpolate values at extra or no-data vertices, such as boundary-defining points, short wells, or deviated wells intersecting horizons at varying X-Y locations. This interpolation of values at no-data vertices must use surrounding points to calculate values, somewhat the same as when generating node values in a grid. Similarly, if nodes of different grids do not occupy the same positions, the program must be able to interpolate node values using surrounding nodes.

The inclusion of geologic interpretation thus places requirements on implementation of the algorithm to be used. A program that can not perform operations between grids or networks, or that can not interpolate or project Z-values, would not be useful for many geologic applications. Not only the mathematical properties of a program, but also its application, should be considered.

CONCLUDING REMARKS

The first step in modeling geologic surfaces requires geologic knowledge, concepts, and interpretation for the area being studied. With this background, the data then are manipulated and algorithms selected to be consistent with the geology. Finally, calculations on data and surfaces are done in such a way as to incorporate the geologic knowledge.

The main point stressed in this philosophy of mapping is to think like a geologist; don't ignore training and experience just because a computer is involved. Neither data nor algorithms should rule how mapping is done without other considerations. When all issues have been considered, perhaps the best approach is to let one or another aspect dominate processing. However, this decision should be the geolo-

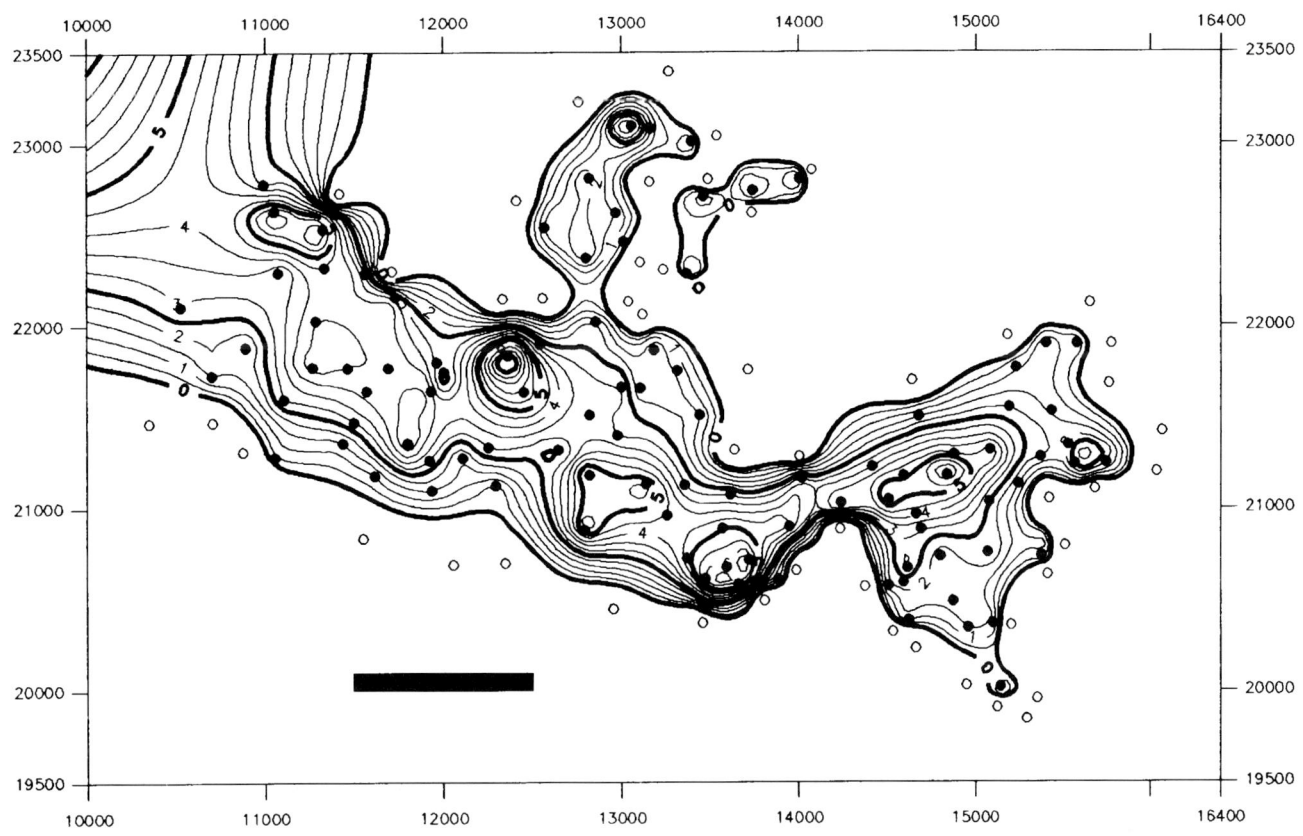

Figure 5. Map showing contours of delta thickness; zero values were changed to –2 to induce correct projections. Contour interval is 0.5 ft (0.15 m). Scale bar represents 1000 ft (305 m). Solid dots represent positive thickness values. Open dots represent zero values changed to –2. Data interpolated from maps in Donaldson et al., 1970, used with permission of SEPM.

gist's, based on geologic common sense and knowledge. The methods used should not be dictated by the way a particular program was designed.

Regardless of advancements in computer hardware and software, placing geological interpretation into contour maps must be done through geological reasoning, a problem-solving approach or philosophy to mapping, and a set of mapping procedures. A good grasp of these factors, plus access to virtually any modern mapping system, allows consistent, realistic maps to be generated.

REFERENCES CITED

Auerbach, S., and H. Schaeben, 1990, Surface representations reproducing given digitized contour lines: Mathematical Geology, v. 22, p. 723–742.

Banks, R., 1991, Contouring algorithms: Geobyte, v. 6, October, p. 15–23.

Donaldson, A. C., R. H. Martin, and W. H. Kanes, 1970, Holocene Guadalupe delta of Texas Gulf Coast: *in* J. P. Morgan, ed., Deltaic sedimentation: modern and ancient, SEPM Special Publication 15, p. 107–137.

Fierstien, J. F., and A. V. Brewster, 1992, The shape-assist technique: incorporating steam channel interpretation into computer generated surface models: Clark County, Kansas, (this volume).

Fontaine, D. A., 1985, Mapping techniques that pay: Oil and Gas Journal, v. 83, no. 12, p. 146–147.

Green, P. J., and R. Sibson, 1978, Computing Dirichlet tessellations in the plane: The Computer Journal, v. 21, p. 168–173.

Green, W. R., 1985, Computer-aided data analysis: New York, John Wiley & Sons, 268 p.

Green, W. R., 1992, Graphical techniques for locating anomalies, (this volume).

Hamilton, D. E., and S. K. Henize, 1989, Modeling pinnacle reefs and their associated surfaces (abs.): AAPG Bulletin, v. 73, p. 360.

Hamilton, D. E., and S. K. Henize, 1992, Computer mapping of pinnacle reefs, evaporites, and carbonates: northern trend, Michigan basin, (this volume).

Iglehart, C. F., 1970, Descriptive classification of subsurface correlative tops: AAPG Bulletin, v. 54, p. 1697–1705.

Jones, T. A., D. E. Hamilton, and C. R. Johnson, 1986, Contouring geologic surfaces with the computer: New York, Van Nostrand Reinhold, 314 p.

McCullagh, M. J., and C. G. Ross, 1980, Delaunay triangulation of a random data set for isarithmic mapping: The Cartographic Journal, v. 17, p. 93–99.

Peucker, T. K., 1980, The impact of different mathematical approaches to contouring: Cartographics, v. 17, p. 73–95.

Philip, G. M., and D. F. Watson, 1982, A precise method for determining contoured surfaces: Journal Australian Petroleum Exploration Association, v. 22, p. 205–212.

Tearpock, D. J., and R. E. Bischke, 1991, Applied subsurface geological mapping: Englewood Cliffs, Prentice-Hall, 648 p.

Walters, R. F., 1969, Contouring by machine: a user's guide: AAPG Bulletin v. 53, p. 2324–2340.

Watson, D. F., 1981, Computing the n-dimensional Delaunay tessellation with application to Voronoi polytopes: Computer Journal, v. 24, p. 161–172.

Watson, D. F., 1982, ACORD: Automatic contouring of raw data: Computers & Geosciences, v. 8, p. 97–101.

Chapter 2

Graphical Techniques for Locating Anomalies

William R. Green
*Placer Dome Inc.
Vancouver, British Columbia
Canada*

ABSTRACT

A primary objective of many exploration programs is to find structural features or anomalous areas as targets for more detailed exploration. Defining what is anomalous may be subjective, so that rigorous mathematical methods are not always available. A variety of graphical display techniques can be applied to this problem.

The essence of the graphical approach is to display data in such a way that unusual patterns may be easily detected visually. Since many types of data are acquired in exploration, different methods are used in different situations. It is advantageous to display a set of data in more than one way, as distinct patterns may be visible in each.

The graphical approach is particularly useful in regional surveys, where large areas and large volumes of data are involved. The main requirement in regional surveys is to quickly scan large geophysical and/or geochemical data files to define smaller areas for more detailed exploration.

INTRODUCTION

In early and intermediate stages of exploration for minerals and petroleum resources, the essence of exploration is to find anomalous areas. The nature of the anomaly may be quite variable, depending on the type of exploration, the characteristics of the target, and the types of data available. An anomaly may be a local area with abnormal characteristics, or a structural feature which may indicate the presence of smaller targets. Many common aspects to the search for anomalies allow the same techniques to be applied in different situations.

The most important common feature is that exploration involves spatial data; that is, data which are at least partly dependent on their location. The problem of locating anomalies then may be considered as defining smaller areas within a larger region, based on characteristics of the exploration data.

A wide variety of powerful statistical and analytical techniques can be used to detect anomalies. The discussion here deals instead with empirical methods, relying on graphical display to show anomalies visually. This generally means plotting data on maps, so that spatial relationships are immediately apparent. For overviews of more statistical approaches, see Green (1985, 1991), Rock (1988), and Agterberg (1989).

The graphical approach implies that anomalies are chosen by a human interpreter of the data, rather than by a computer algorithm. The display methods to be discussed below are often useful adjuncts to the mathematical approaches, however. Statistical measures of anomalous behavior are seldom completely unambiguous, leaving the interpreter to make some selection from statistical rankings. While the examples below deal primarily with raw data, it is a simple matter to use the same forms for derived quantities.

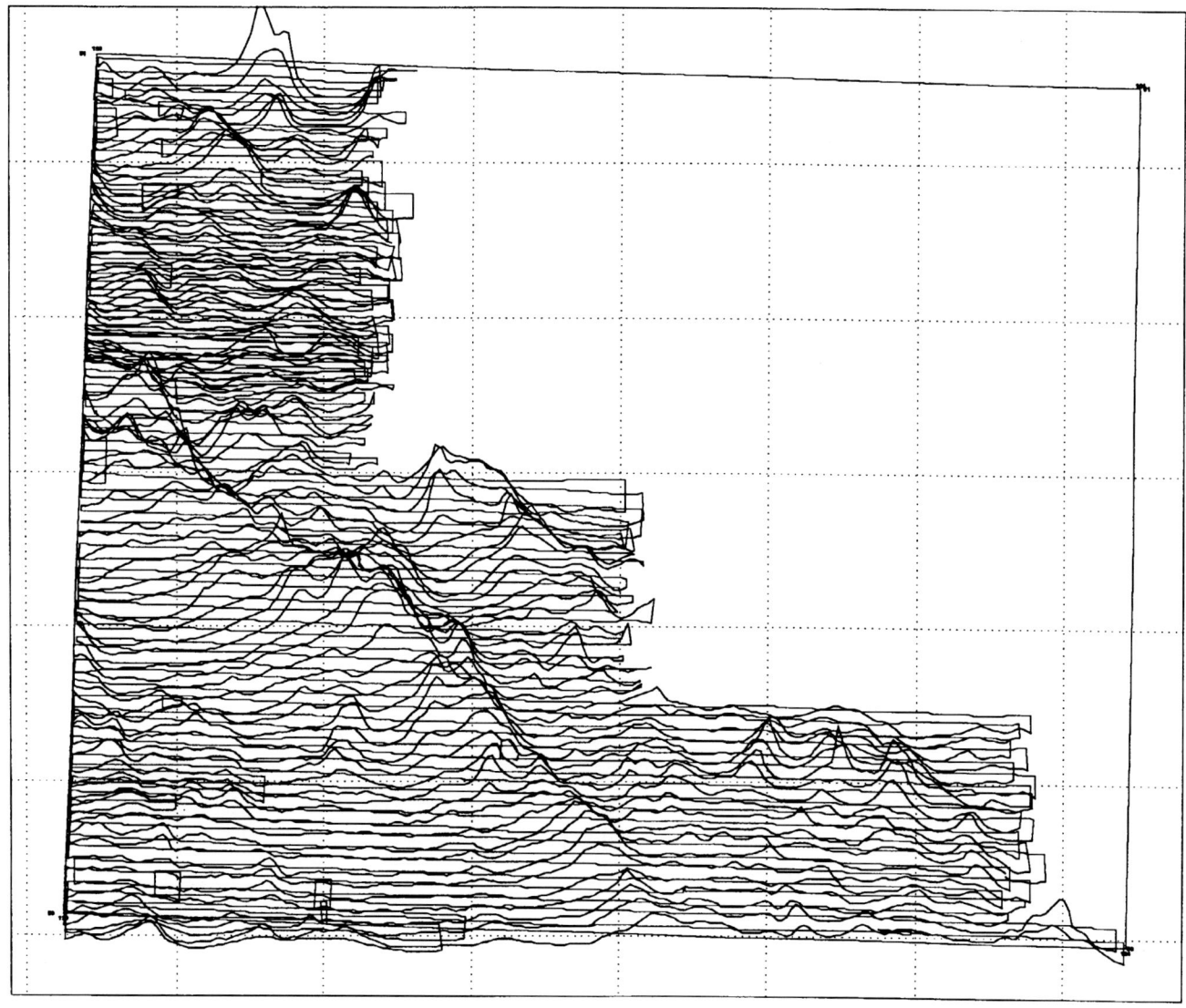

Figure 1. A stacked profile map. Data from the Geological Survey of Canada, for an aeromagnetic survey of central Vancouver Island. (A) Simple profiles.

Computer Mapping Methods

The use of a computer for producing exploration maps started simply as a means of automating the labor-intensive process of drafting numerical data onto maps. Early on, computer algorithms for drawing contours were developed, and they continue to be refined. In many exploration projects, the great benefit of computer mapping is to allow a much greater variety of map formats to be used than would be practical manually. In many instances, the volume of data is so great that it would be virtually impossible to manually draw even a single map (for example, airborne magnetic and radiometric surveys commonly involve hundreds of thousands of readings).

Computers also allow new types of maps that cannot be generated manually. Satellite images, transformations of geophysical surveys, and seismic tomography are among the powerful techniques that are rapidly becoming nearly routine due to the increasing power of computer hardware and sophistication of software.

This paper consists largely of examples of various display formats for mapping exploration data. A basic understanding of mapping methods and computer applications is assumed in the discussion. For more background on these topics, see Green (1985, 1991) for a review of the underlying principles of computer plotting, and guidelines for effective software systems. Jones et al. (1986) provide an in-depth review of techniques for computer contouring.

Note that in many of the examples, additional supporting information is not shown on the plot. This is to emphasize the form of the display: in normal use coordinate labels, color bars, symbol legends and other features are needed to make the information on the map accessible.

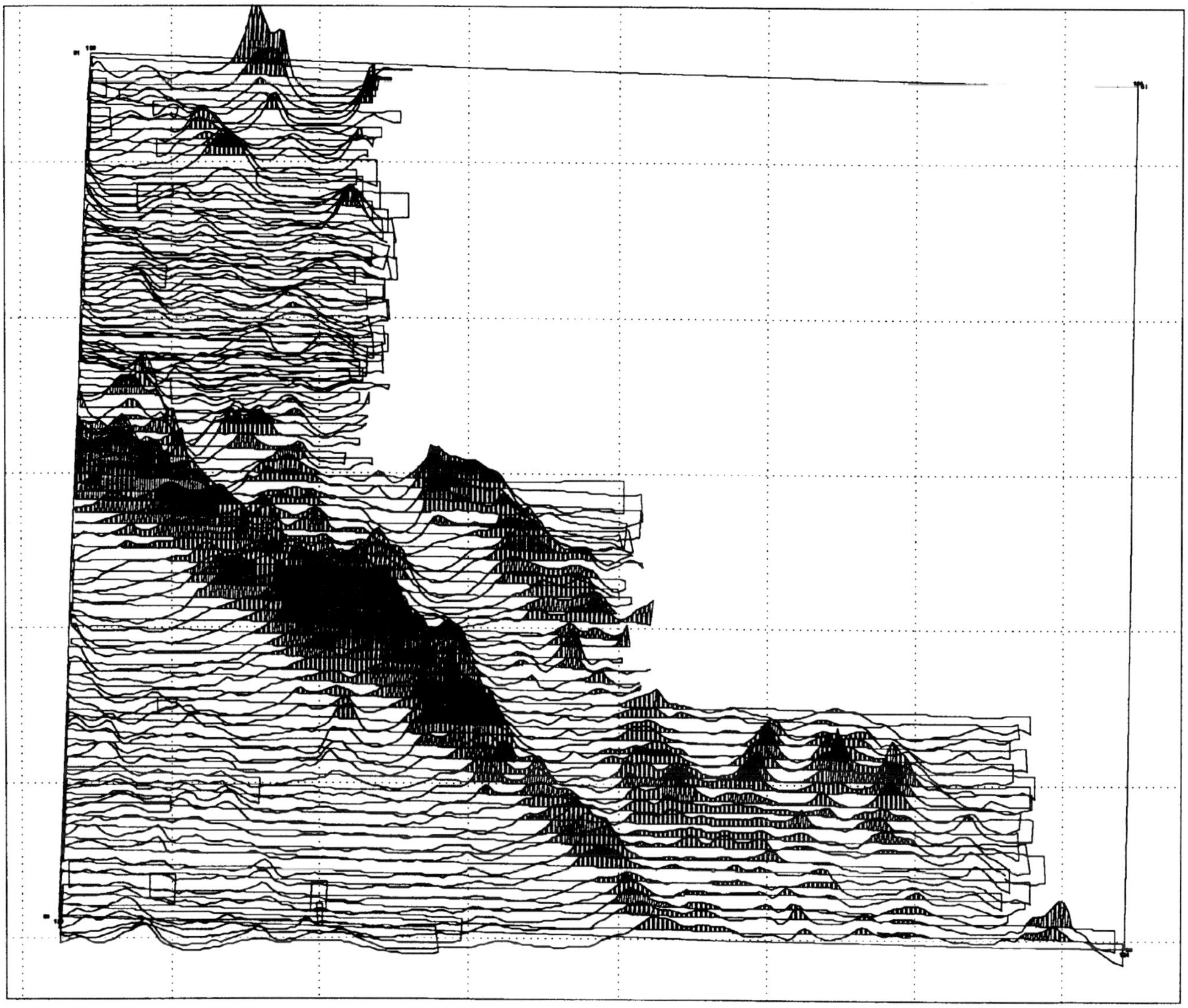

Figure 1. (B) Profiles with shading of values above a base level.

Regional Surveys and Public Domain Data

As noted above, the volume of data is in itself a problem which calls for computer treatment. In creating maps, additional measures may be required to reduce the amount of data being shown. Excessive clutter on a map may obscure features of interest in the data. In addition, the time to plot a single map may become large, in part removing the benefits derived from showing data in a variety of forms.

The problem of volume can arise in all stages of exploration, but is especially prevalent in regional surveys, simply due to the area covered by the survey. The availability of low-cost data from government agencies in many parts of the world also means that large volumes of data may be acquired by even the smallest organizations. Airborne geophysical surveys, regional geochemical sampling, and satellite imagery are among the types of data in this category.

METHODS FOR PLOTTING GEOPHYSICAL MAPS

Airborne geophysical surveys (magnetics, electromagnetics, radiometrics) are a primary tool in early stages of regional exploration. The airborne methods allow large areas to be surveyed in considerable detail, without the expensive logistics of covering the area on the ground. Where group participation or public domain data are available, the costs of acquiring data are minimal.

The basic requirements in mapping regional geophysics are two-fold. Regional structural trends may be present, to provide information on the gross geological characteristics of the region. In addition, local anomalies may provide leads to mineralized areas. The methods for displaying these features are different, and in general mutually exclusive. That is, the optimum display for interpreting regional structure

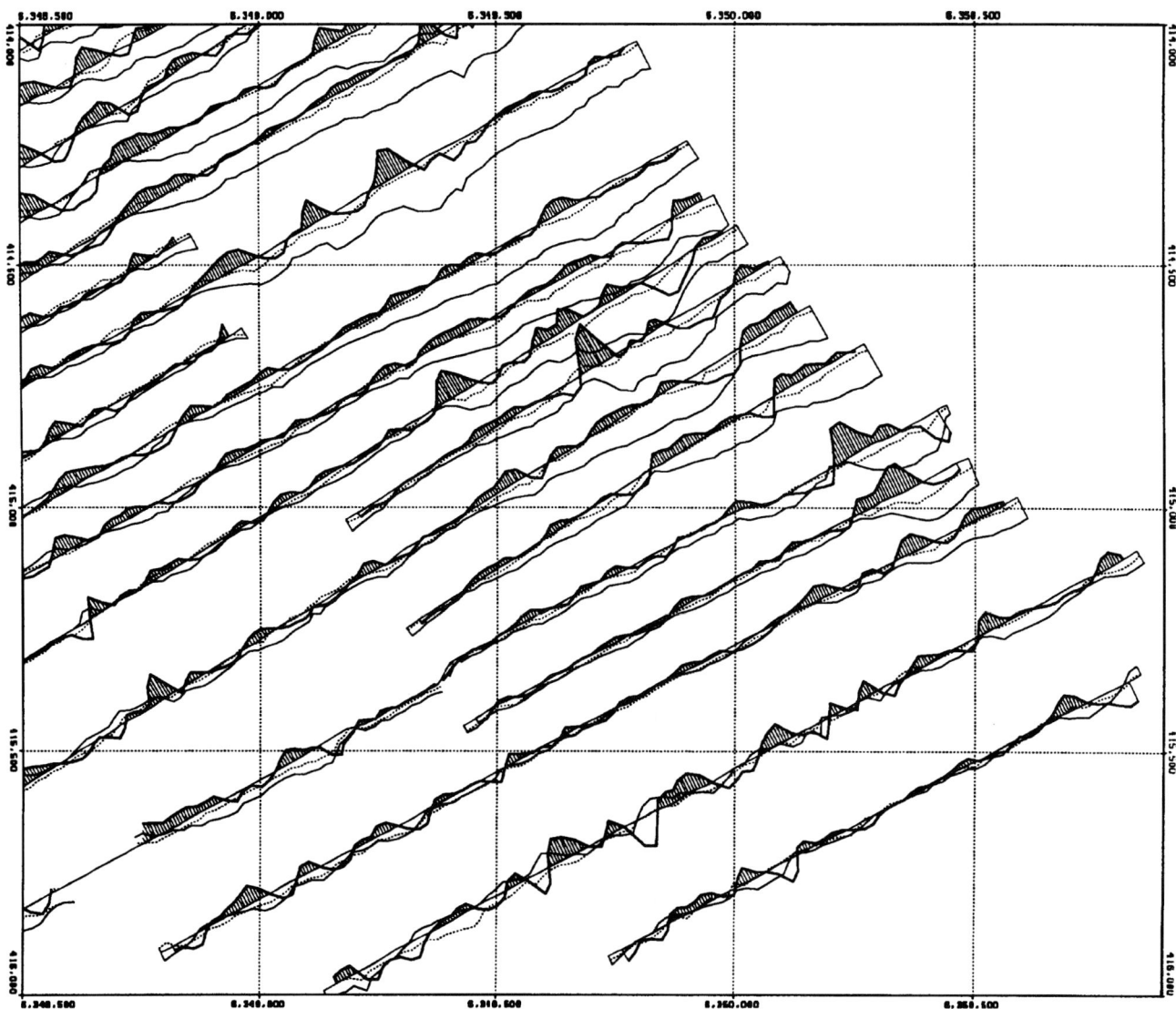

Figure 2. Stacked profiles for multiple data types. Data from a ground VLF survey, showing the in-phase and quadrature components. The Fraser filter of the in-phase component is also shown, with shading of the positive portion.

generally obscures anomalies, while effective display of anomalies often requires suppression of larger trends.

Figure 1A shows a stacked profile map of aeromagnetic data. The values along each survey line are represented by the perpendicular offset of the data lines from the survey locations. With shading as shown in 1B, this format is similar to plotting seismic sections, except for the variable position of each line. Because the shading is not fully black, overlapping traces from adjacent lines can be seen through the shading. This compact form of data display allows the full detail of the magnetic readings to be seen along each line. Trends which cross lines can also be seen, although it can be difficult to visualize the actual shape of the feature. This example shows raw data, although the same form can easily be used for derived quantities such as vertical derivatives, continuation, ratios, and so on.

In Figure 2, the same form of display is used to simultaneously plot more than one geophysical measurement (in this case, the in-phase and quadrature components from a Very Low Frequency (VLF) survey, plus the Fraser filter of the in-phase). Creating the combined display takes some care in choosing scales and line styles (thickness, color etc.), so that the different variables can be readily distinguished.

The most common form for mapping regional features is the contour map. With large surveys, the procedure for plotting contours almost inevitably involves some averaging of the original data. This may be necessary simply to make a more readable map, or may be imposed by hardware and software

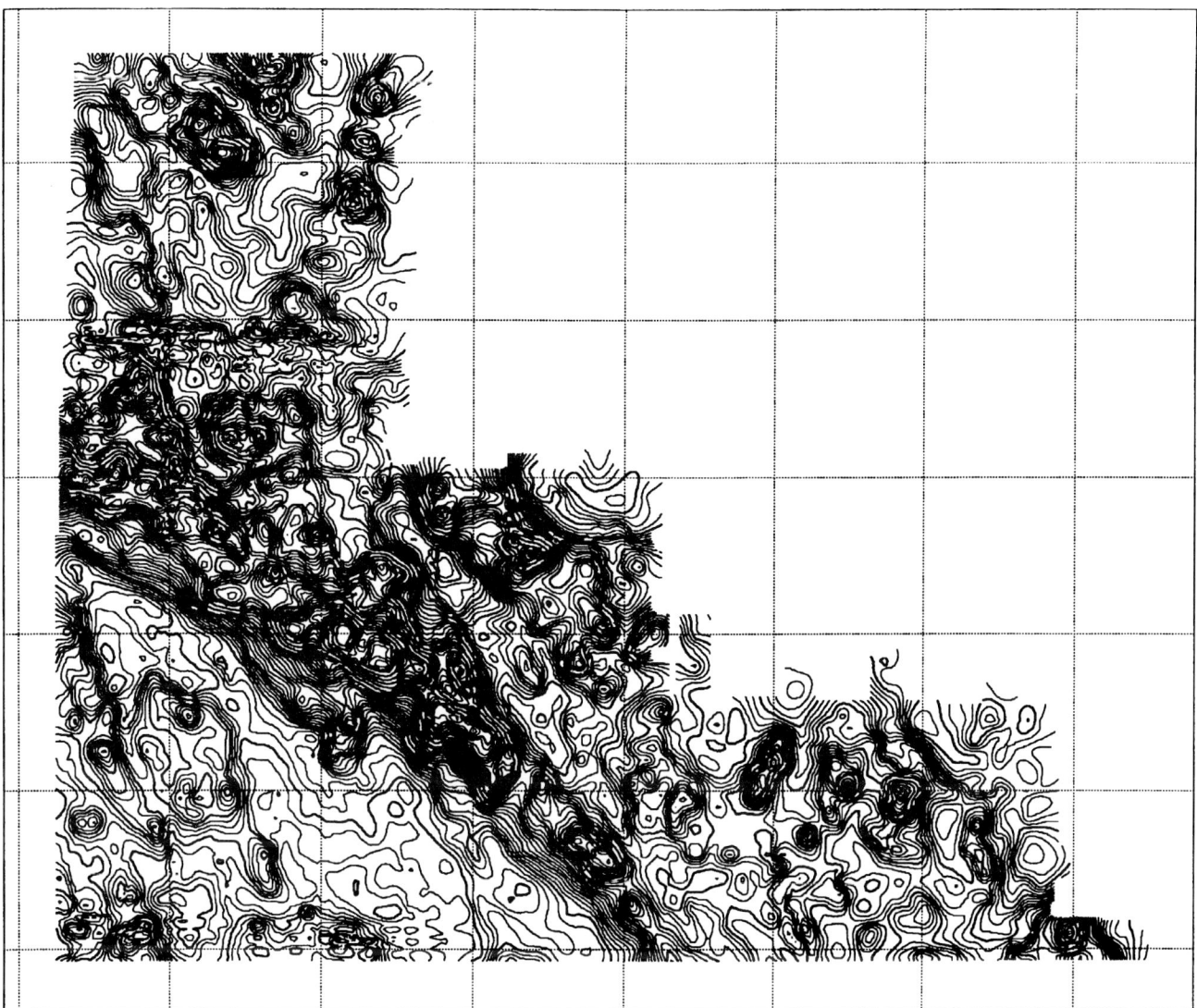

Figure 3. Contour display of aeromagnetic data for the same area as Figure 1. (A) Detailed contours, from a 500 meter grid cell size. (Figure 3B and 3C on following pages)

limitations. In Figure 3, different degrees of averaging are shown by using progressively larger grid cells for interpolation of original data locations. In 3A, the grid cell size is 500 meters. In 3B, 9 cells were averaged into one, to give a grid size of 1500 m. In 3C, the averaging was repeated, to yield a grid of 4500 m. Note the contours become much smoother, and the trends are more obvious. This comes at the expense of losing many smaller features, however. More than one contour map will typically be plotted, to allow both the large scale structures and more localized features to be studied. To emphasize smaller features, the difference of the original grid from the averaged version could be contoured.

An alternate form is a perspective view (Figure 4). While traditionally used for topographic surfaces, it can be effective for illustrating the relative size of structural features. It is more difficult to relate the position of each feature to a ground location, however.

A variation on the normal contour map is a color map (Figure 5). The method for plotting the map is similar, and limitations of the contour method usually still apply (averaging, use of interpolation schemes, etc.). The advantage of color is that highs and lows in the data are more apparent. It is also much faster to experiment with different contouring levels, by varying the color assignments.

One disadvantage of color is that it is relatively more difficult to produce hard-copy maps, particularly when a large size format is used. This is primarily a matter of cost—large color plotters are much more expensive than pen plotters. (Here a color plotter means one that can fill areas with solid colors, as opposed to multi-pen plotters, which can use several colors but are impractical for full coloring.)

The use of color also allows more effective combi-

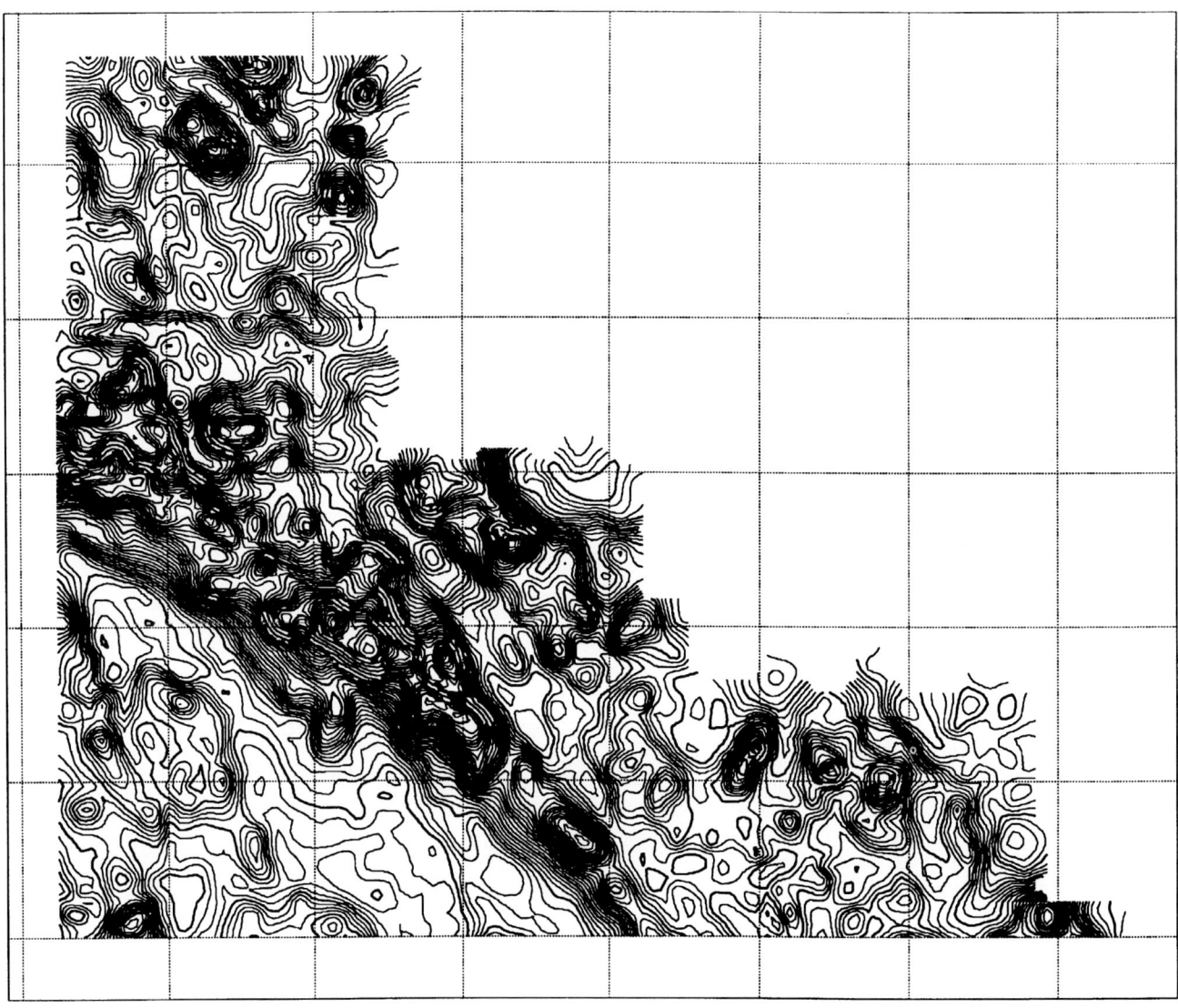

Figure 3. (B) Contours from a smoothed grid, averaged to 1500 meter grid cells.

nations of data display. A growing trend is to use image analysis techniques to create displays of different types of data. For example, regional aeromagnetics may be shown on top of a satellite image (Figure 6). The first image (6A) is a standard false-color display from the Landsat 7-channel Thematic Mapper (TM) scanner, for an area in southeastern British Columbia. A contour map of the magnetic data for the same area is shown in 6B. In 6B, the TM data are shown, with the color assignment controlled by the magnetics. The surface structural details from the satellite data can be immediately compared to the magnetic anomalies generated by hidden crustal features. The method is to use an intensity-hue-saturation (IHS) transformation (Kruse, 1984; Kruse and Raines, 1984), with intensity scaled by the TM data, and hue by the magnetics (saturation is constant). The selection of the particular transform method is dependent on the nature of the data. See Sabins (1987, pp. 251-252) for a discussion of the use of IHS transforms for satellite imagery.

The magnetic data shown in the example can be enhanced to show more structural details. As noted above, conventional processing methods for derivatives, vertical continuation and so on can be combined with the same formats of maps already shown. Alternative enhancement techniques from image processing can be adapted for geophysical data as well. See Mather (1987) for details on image analysis and enhancement techniques, and Kowalik and Glenn (1987) for geophysical applications of these techniques.

Density slicing is a simple way to scan for hidden structural features. This involves viewing the image with colors assigned only to a narrow range of data values. The image is then redisplayed repeatedly, by moving this range across the full range of the data values (e.g., if the data range is 1000 to 5000, colors might be assigned first only in the range 1000 to 1200, then

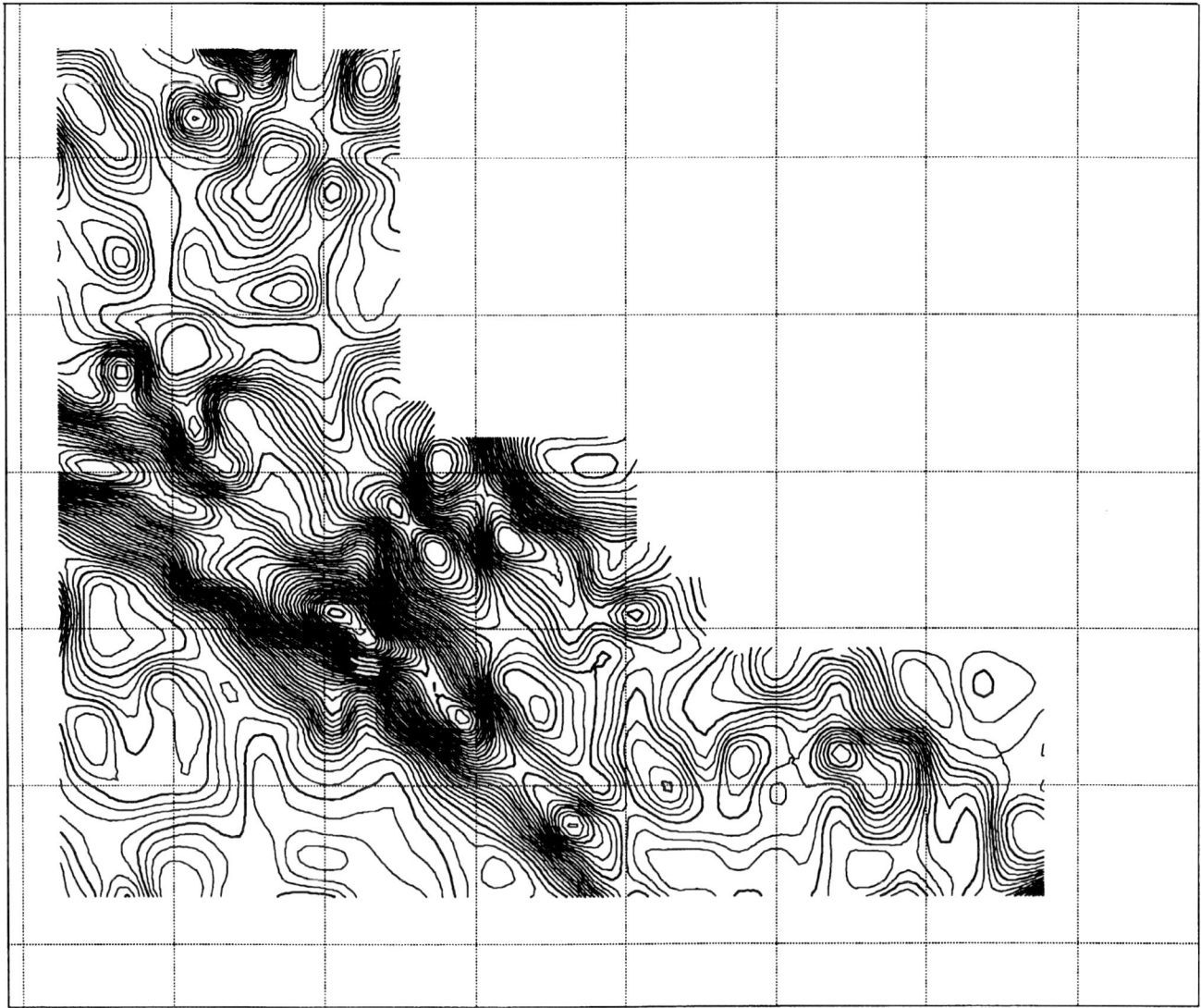

Figure 3. (C) Contours after heavy smoothing, averaged to 4500 meter grid cells. (Note: normal contour labels omitted here for simplicity)

1200 to 1400, and so on). Typically this is done interactively by pointing to the desired slice in a color bar displayed beside the map. Sliding the slice along the color bar allows the full data range to be examined in a very short time.

The technique of sun-shading provides a way of enhancing directional trends in the data. The procedure treats any type of data to be displayed as if it were a topographic surface, and adjusts the intensity for each pixel according to an assumed illumination from a given direction. Portions of the data surface which slope away from the false sun have reduced intensity relative to those which face the false sun. In Figure 7, the effect of changing the illumination direction is illustrated, for the same area shown in Figures 3-5. As for density slicing, the use of shading is most effective on an interactive system which continuously adjusts the display according to the current position of a pointer. For example, pointing to the top of the screen means illuminating from the north (which would emphasize east-west trends). Figure 8 shows a combined color display of total field magnetics and shadow enhanced magnetic data, using the same procedure as the Landsat-magnetics combination of Figure 6C.

Like directional filtering, shading may emphasize noise in the data as well as real structural features. Note the horizontal striping pattern in Figure 8, which is due to minor gridding artifacts along the original survey lines.

DISPLAY OF GEOCHEMICAL DATA

A basic problem in plotting geochemical data is the great range of values. This means that standard con-

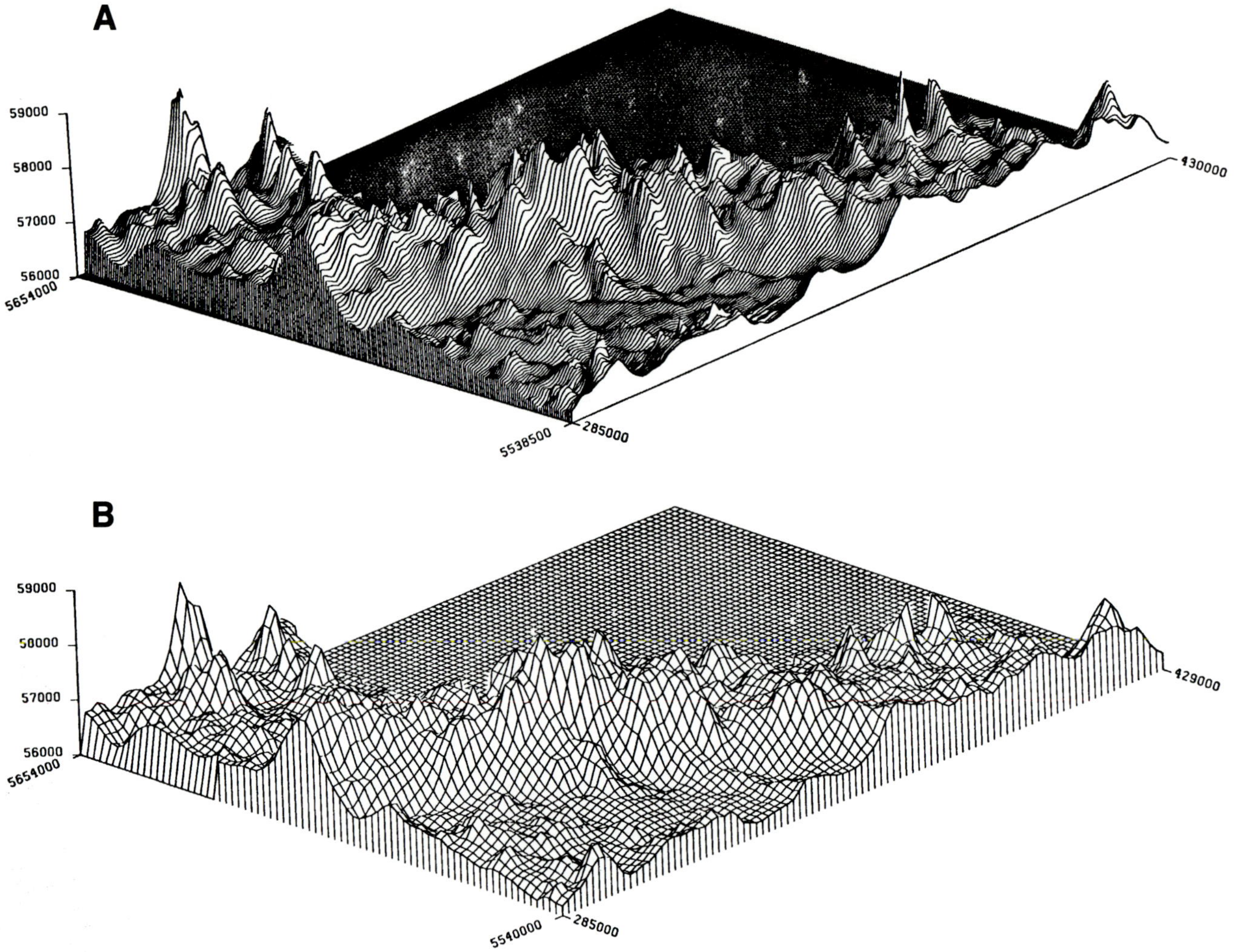

Figure 4. Isometric views of the aeromagnetic data. (A) Data from the 1500 m grid as in Figure 3B. (B) Data from the 4500 m grid as in Figure 3C.

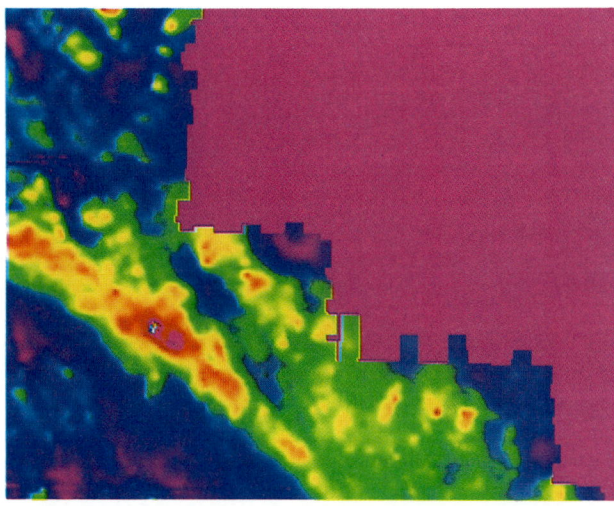

Figure 5. Color map of the aeromagnetic data, using the detailed data as in Figure 3A.

touring methods (or other displays using linear scaling) often fail, by putting too many contours in some areas and too few in others. This is not just a matter of choosing which contours to plot (programs normally allow using a list of levels to provide variable increments between contours). The data interpolation schemes used in contouring are based on implicit assumptions of spatial continuity. Geochemical samples may show extremely high readings adjacent to low (background) ones, making this assumption at best questionable, if not completely invalid. A further complication is that geochemical anomalies may be due to multiple sources, which can be considered as statistically independent populations that should not be mixed in data analysis.

The nature of a contour map itself gives the impression of closed areas of similar values: a single large value may appear to represent a large region, when it is in fact a local feature. This may be seen in Figure 9, which represents copper concentrations in a regional stream sediment survey for an area in southwestern

Figure 6. Combined display of satellite imagery and aeromagnetics data from an area in southeastern British Columbia. (A) False color image of LANDSAT Thematic Data.

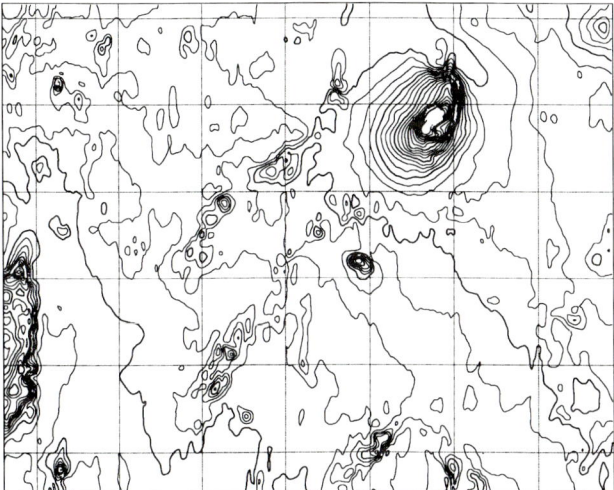

Figure 6. (B) Contour map of aeromagnetics over the same area.

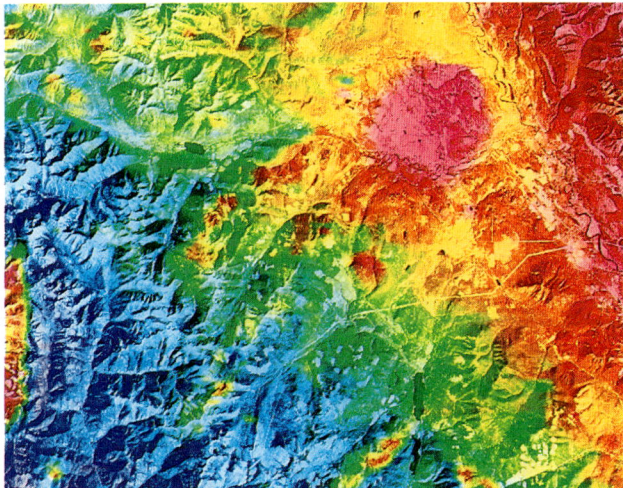

Figure 6. (C) Combined color display. Color hue is assigned according to magnetic data values, while intensity is controlled by the LANDSAT data.

Figure 7. Synthetic shading of aeromagnetic data. (A) Illumination from the northeast emphasizes the major NW-SE trend.

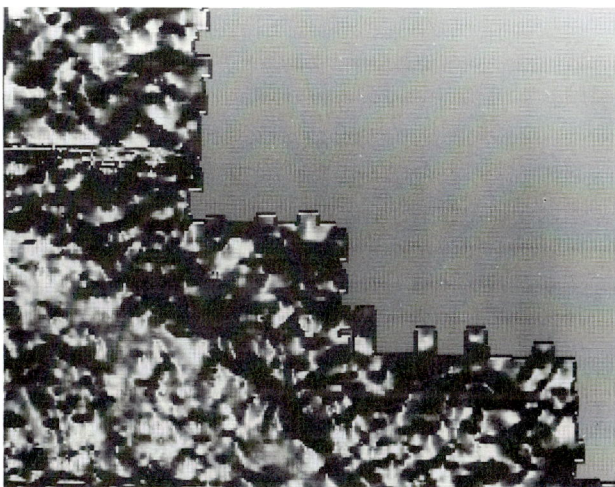

Figure 7. (B) Illumination from the south emphasizes features that cut across the main trend. Note the spiky horizontal features in the SW corner and halfway up the western edge. These are misties in the original data (three different aeromagnetic surveys were combined into one gridded file).

British Columbia. In 9A, the copper values are indicated by the size of the circle plotted at each location. In 9B, the values were contoured. Two areas in the lower central part of the map have pronounced closures in the contours which are due to single samples.

The problem of misrepresenting zones of influence can be even greater when plotting soil geochemical surveys. Like ground geophysics, the procedure is to obtain data at fixed intervals along lines. For economic and logistical reasons, the spacing between lines is usually much greater than along lines. Gridding algorithms frequently have problems with such non-random data distributions. The result may be that anomalies are artificially extended to adjacent lines. In

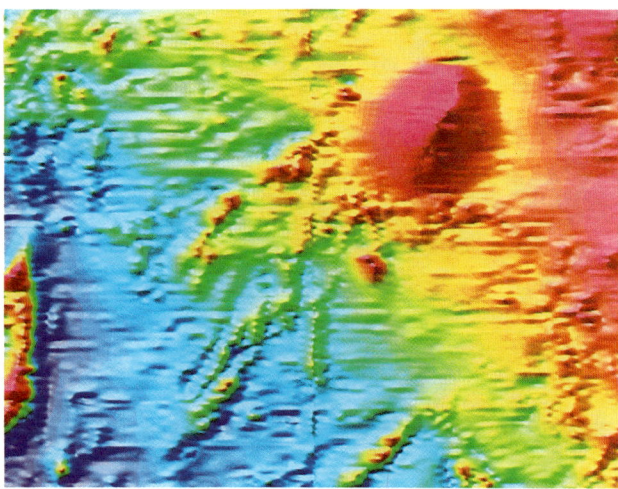

Figure 8. Combined display of total field magnetics and sun-shaded magnetics data for the same area as in Figure 6. Note horizontal gridding artifacts emphasized by shading.

extreme cases, the grid interpolation may fail completely, so that the gridded values (and corresponding contours) no longer even honor the data. The use of symbol or size coding provides an effective way to avoid these difficulties, just as it did in the stream sediment example discussed previously. Figure 10 shows the lead concentrations in soil samples, from an exploration project in Indonesia. In 10A, larger

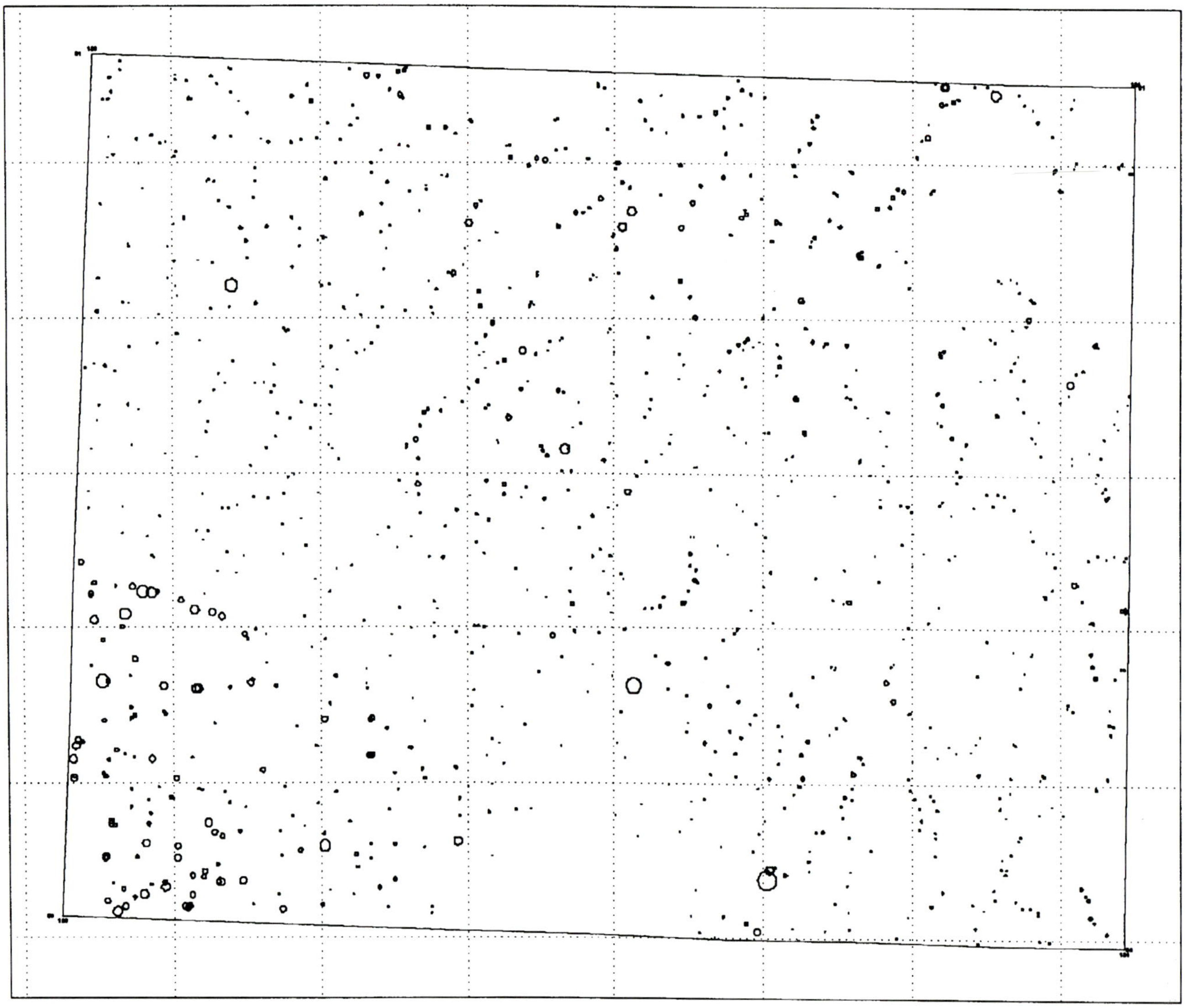

Figure 9. Geochemical data maps. Data from the Geological Survey Branch of the British Columbia government, for central Vancouver Island. (A) Copper concentration represented by symbol size (radius of each circle is proportional to the copper concentration).

symbols are used to show the samples with high lead values. Note that the size of symbols has been chosen so that there is some overlap for high values: this helps emphasize continuity along the lines. In 10B, the same data are contoured. Note the contour closure in the middle of the map, which is due to a single high sample.

Another problem when plotting geochemical maps is that there are often many variables to be considered. Finding joint anomalies in several elements may be the objective in plotting the maps; this requires a compact form of data presentation to combine the data onto one map. The following section describes some general methods for combining different datasets on the same map. These can be used for completely independent sets of data (i.e., measurements at different locations), as well as for the special case when multiple values are known at the same location.

The most effective ways of showing the multivariate behavior of geochemical surveys is to devise special symbols whose shape can vary according to the data values. Chambers et al. (1983) describe many different forms that might be used, including trees, star diagrams, profile plots, and faces. Visual grouping of distinct shapes may be useful as a classification technique.

A simple example is shown in Figure 11. Each sample point is marked by a cross, whose four arms have variable lengths controlled by the amounts of copper, lead, zinc, and mercury in the sample. Samples that are high in all four elements thus are shown by a large regular cross, while those which are high in only some elements appear as distorted shapes.

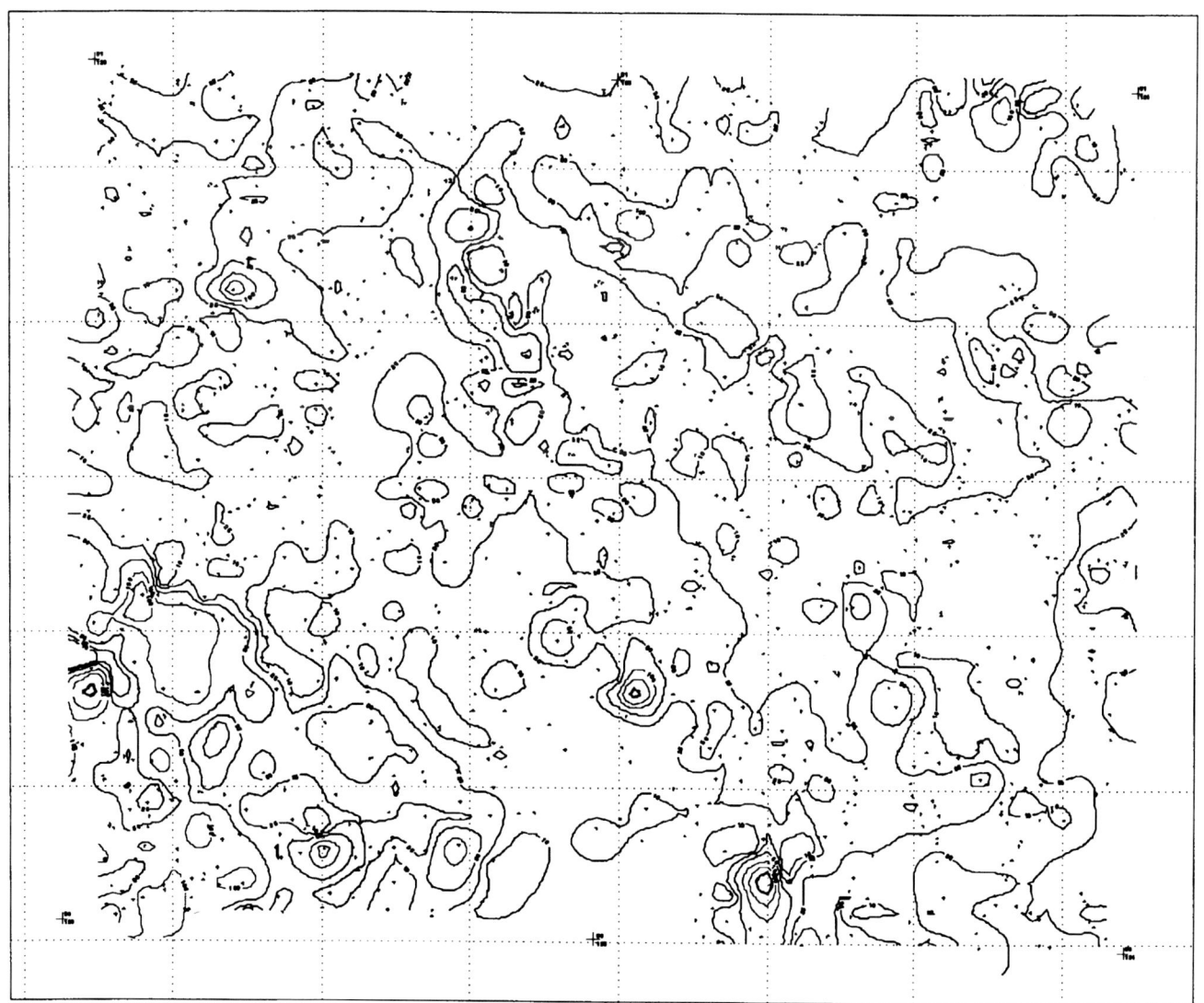

Figure 9. (B) Contour map of copper concentrations (Note: normal contour labels and legends omitted here for simplicity).

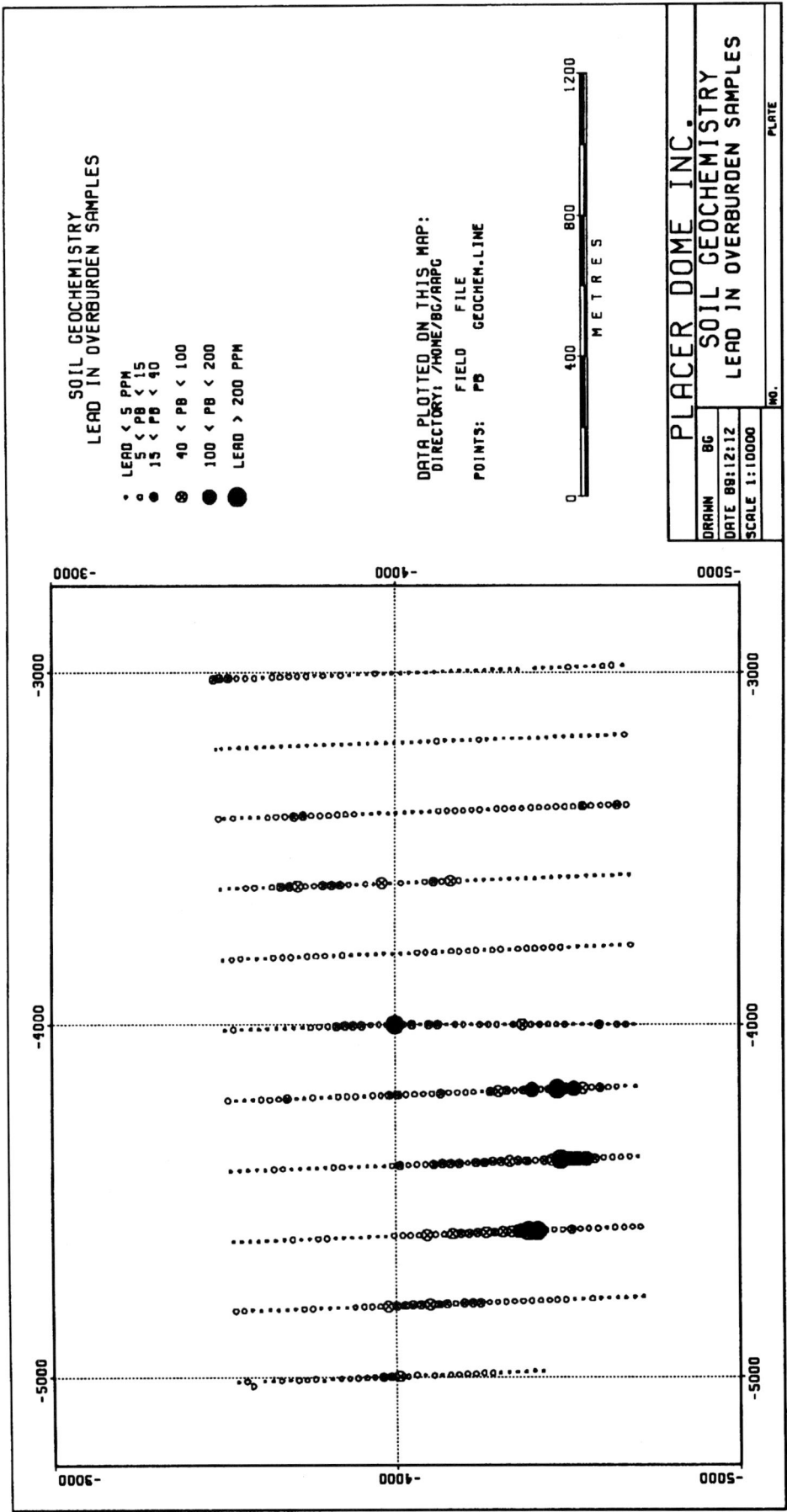

Figure 10. Maps showing samples along lines. Data from an exploration project in Indonesia. (A) Lead concentrations represented by symbol shape and size.

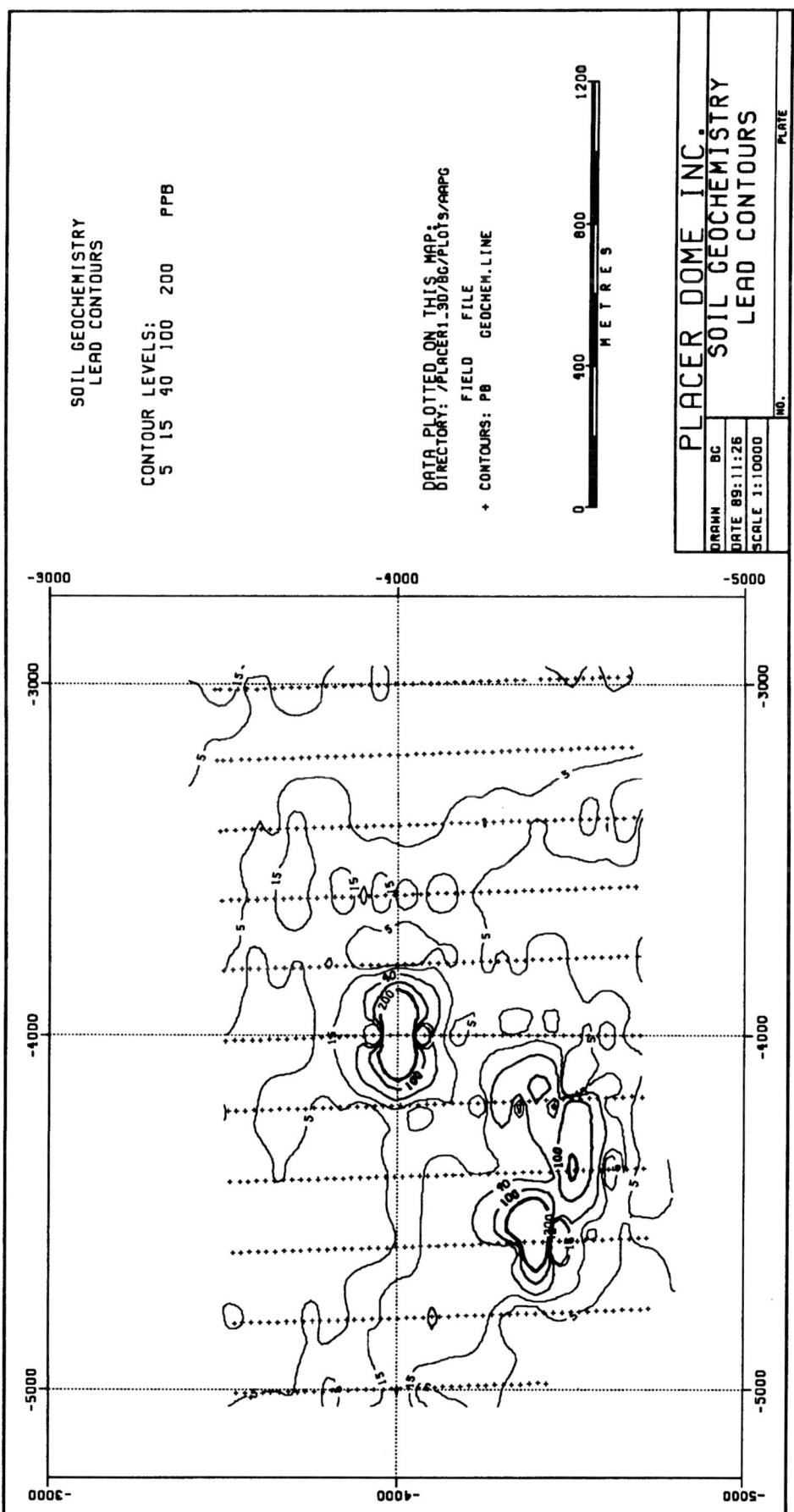

Figure 10. (B) Contour map of lead concentrations.

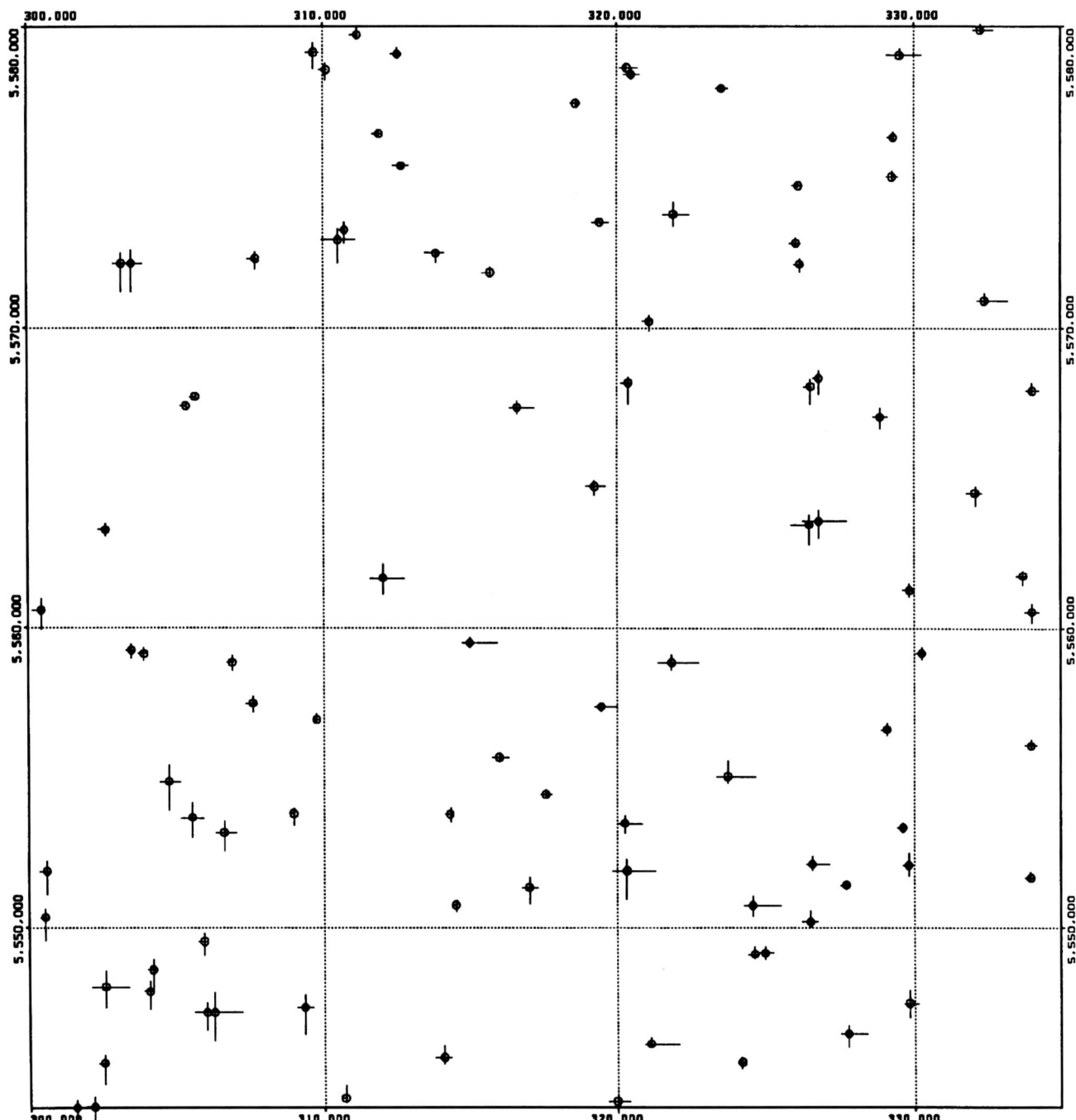

Figure 11. Multivariate symbols for geochemical data. A portion of the regional survey from Figure 8 is shown. The lengths of the arms of a cross are controlled by the amount of copper, lead, zinc, and mercury in each sample.

COMBINED DATA DISPLAYS

The combined display of satellite imagery and aeromagnetics in Figure 6 is a fairly sophisticated type of data presentation. The same principle of showing quite different types of data on the same map can also be applied to much simpler formats. The key to success is to make the different types of data visually distinct.

In the simplest case, showing locations of other features on a map of geochemical or geophysical data can be very useful. As a standard drafting application, topographic and cultural features might be added. In Figure 12, for example, elevation contours, roads, and drainage features are all on a map of geophysical profiles. This makes it much easier to relate the anomalies to specific ground locations.

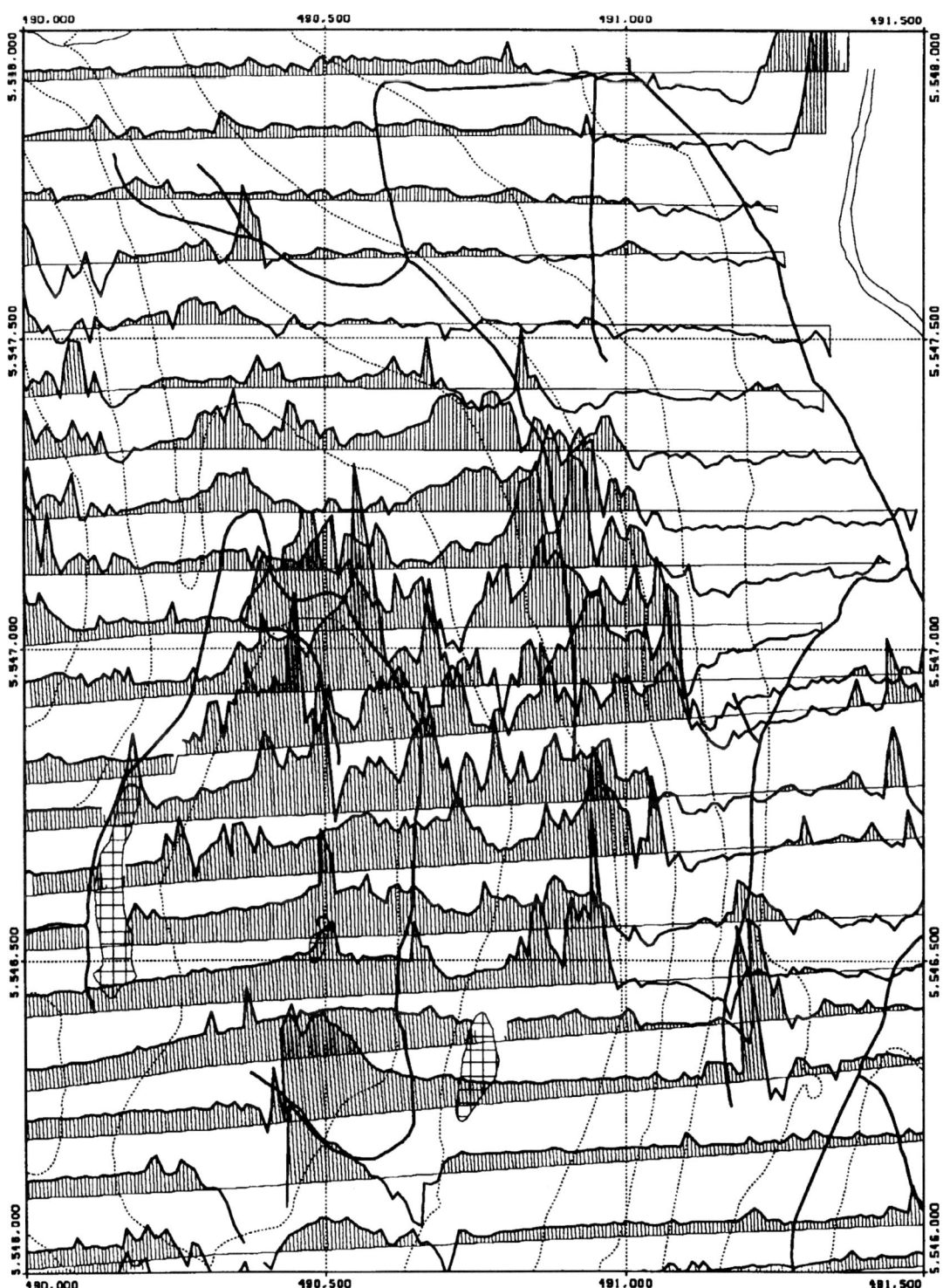

Figure 12. Composite map with analytical data and physical features. Data from an exploration project in southwestern British Columbia, with topographic contours (dotted lines), lakes (hatching), streams (double line), and roads (solid lines) shown over stacked profiles of magnetic readings.

It is also useful to show more than one type of exploration data on a map. Figure 13 is a map combining geochemical sampling with locations of known mineral deposits. The triangles are known mineral occurrences with gold present, while the scaled circles represent the amount of gold found in stream sediment samples. Some areas show both a concentration of anomalous samples and occurrences

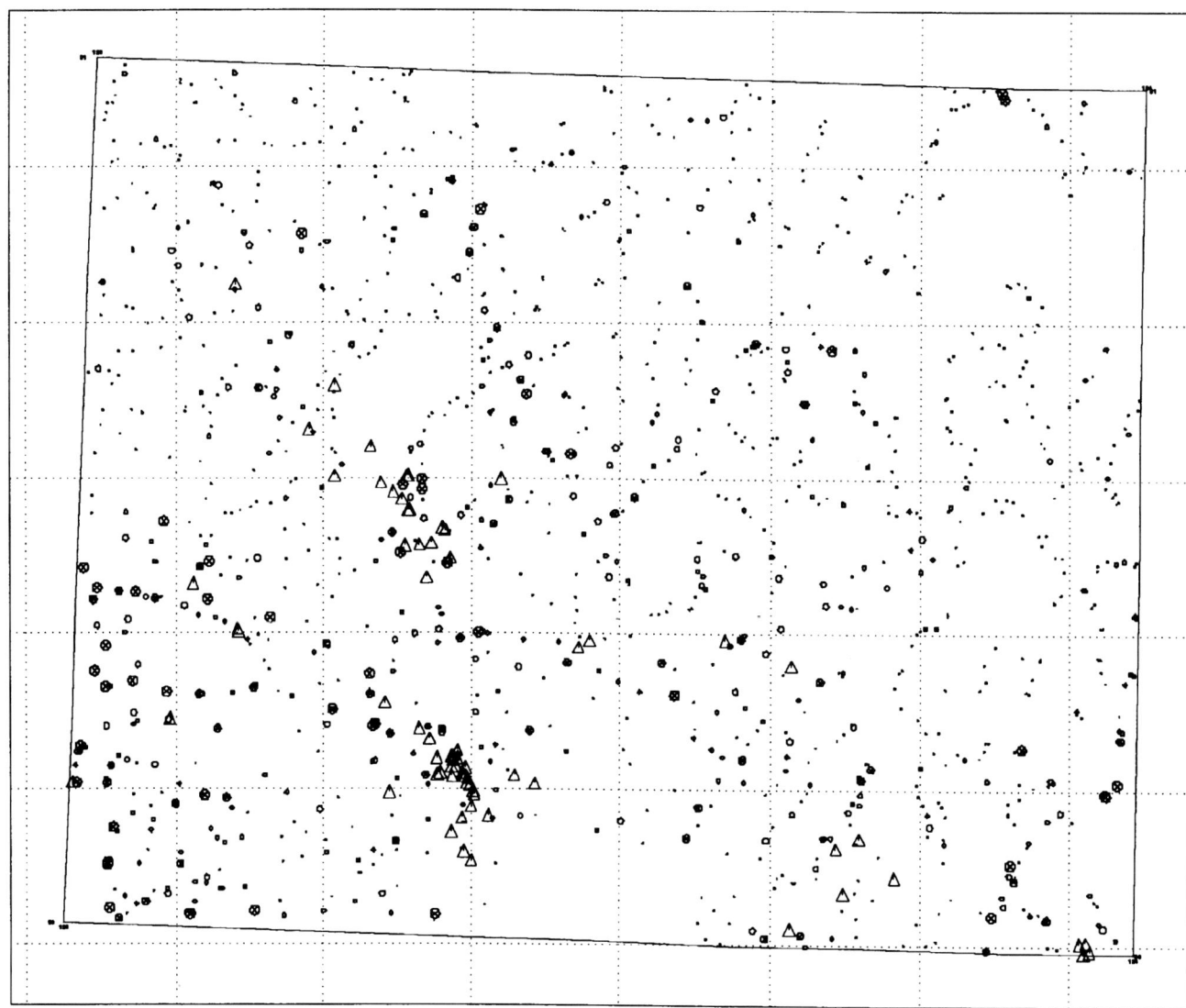

Figure 13. Composite map: geochemical data (circles) and mineral occurrences (triangles). Data for central Vancouver Island.

of gold, but others have one without the other.

While a combined display may improve the definition of anomalies, interpretation must incorporate knowledge of the data characteristics (method of collection, analytical limits, potential errors, etc.). In Figure 14, the geochemical data for gold in stream sediment samples are shown together with contours of aeromagnetic data. Note that the anomalous gold values tend to lie on the edges of magnetic anomalies. While this might appear to have geologic significance, the nature of the data is probably more important here. The magnetic structures are aligned with the mountain ranges (due to cores of intrusive igneous rocks). Geochemical sampling is concentrated in lower areas, with anomalies offset from their source (that is, the anomalous material may well have come from mountain peaks, but is collected in the valleys).

CONCLUSION

With present computer software and hardware, it is relatively easy to produce complex maps which summarize a great volume of raw data. Several types of data may be shown on the same map, provided the format minimizes overlap, and allows each separate type to be readily distinguished. This might be a combination of data forms (posted text, symbols, line drawings, contours), and display style (different line style, text size, colors, map or perspective view, etc.). One common application is to use a topographic model in conjunction with exploration data. This type of display is useful for putting the exploration targets into geographic perspective, as illustrated in Figure 15.

The possibilities are virtually unlimited. The key to success is to have a fast, flexible system (both hardware and software are important here), so that various

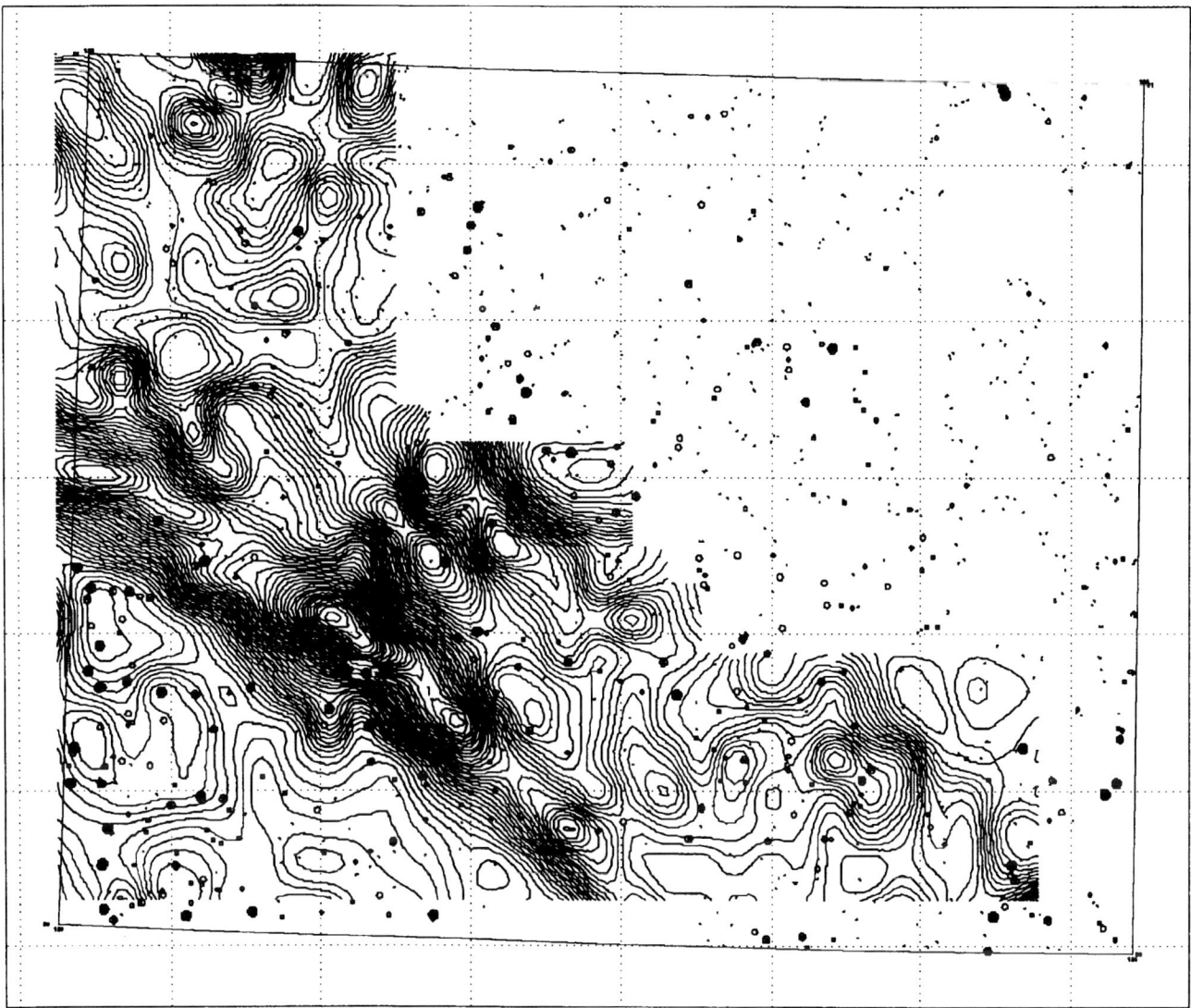

Figure 14. Composite map: geochemical and geophysical data for central Vancouver Island. Gold concentration is shown by symbol size and shade (larger, darker symbols indicate higher gold content), with contours of aeromagnetic data overplotted.

options can be tested quickly. An interactive approach is mandatory. Due to the complex nature of the maps produced, this process will always demand a considerable level of training and experience from the computer user.

Geographic information systems (GIS) are a rapidly developing technology which provide this capability. In addition to the extensive graphical display options, they include database management facilities needed for easy access to topographic models, cultural details, and other cartographic elements which can be added to analytical maps. Beyond the powerful graphics, the ability to study complex spatial relationships will provide a strong incentive to use GIS. For example, exploration favorability might be defined by distances from features such as faults and by the geochemical and geophysical observations.

An example application in mineral exploration is discussed by Bonham-Carter et al. (1988). The use of GIS systems in exploration is still largely a research area, but is likely to become much more common in the near future.

ACKNOWLEDGMENT

Michael Ehling developed many of the color display procedures used at Placer Dome, and provided the color illustrations used here. Other associates at PDI have contributed indirectly, as their suggestions have been incorporated into the company's mapping software.

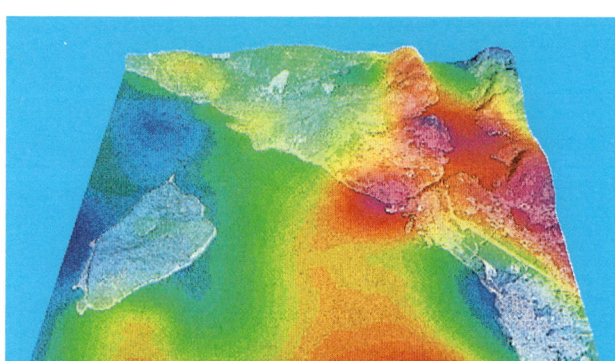

Figure 15. Magnetic data displayed as colors on a topographic surface. Data for the region around Powell River, British Columbia.

REFERENCES CITED

Agterberg, F. P., 1989, Computer programs for mineral exploration: Science, v. 245, p. 76–81.

Bonham-Carter, G. F., F. P. Agterberg, and D. F. Wright, 1988, Integration of geological datasets for gold exploration in Nova Scotia: Photogrammetric Engineering and Remote Sensing, v. 54, no. 11, pp. 1585–1592.

Chambers, J. M., W. S. Cleveland, B. Kleiner, and P. A. Tukey, 1983, Graphical methods for data analysis: Boston, Duxbury Press, 395 p.

Green, W. R., 1985, Computer-aided data analysis: New York, John Wiley and Sons, 268 p.

Green, W. R., 1991, Exploration with a computer: Oxford, Pergamon Press, 225 p.

Jones, T. A., D. E. Hamilton, and C. R. Johnson, 1986, Contouring geologic surfaces with the computer: New York, Van Nostrand Reinhold, 314 p.

Kowalik, W. S., and W. E. Glenn, 1987, Image processing of aeromagnetic data and integration with Landsat images for improved structural interpretation: Geophysics, v. 52, no. 7, p. 875–884.

Kruse, F. A., 1984, Munsell color analysis of Landsat color-ratio-composite images of limonitic areas in southwest New Mexico, in Proceedings of International Symposium on Remote Sensing of Environment, Third Thematic Conference, Remote Sensing for Exploration Geology: Ann Arbor, Environmental Research Institute of Michigan, p. 761–773.

Kruse, F. A., and G. L. Raines, 1984, A technique for enhancing digital color images by contrast stretching in Munsell color space, in Proceedings of International Symposium on Remote Sensing of Environment, Third Thematic Conference, Remote Sensing for Exploration Geology: Ann Arbor, Environmental Research Institute of Michigan, p. 755–760.

Mather, P. M., 1987, Computer processing of remotely sensed images: New York, John Wiley and Sons, 352 p.

Rock, N. M. S., 1988, Numerical Geology: A source guide, glossary, and selective bibliography to geological uses of computers and statistics, Lecture Notes in Earth Sciences 18: Berlin, Springer-Verlag, 427 p.

Sabins, F. F., 1987, Remote sensing: Principles and interpretation, 2nd ed.: New York, W.H. Freeman, 449 p.

Chapter 3

The Shape-Assist Technique: Incorporating Stream Channel Interpretations into Computer-Generated Surface Models: Clark County, Kansas

John F. Fierstien
J.M. Huber
Houston, Texas, U.S.A.

Arthur V. Brewster
J.M. Huber
Amarillo, Texas, U.S.A.

ABSTRACT

The shape-assist technique of computer mapping enables a geologist to sketch the form or shape of a particular geologic feature, then use computer modeling to align that shape to the existing data. Long, narrow Morrowan-aged (Early Pennsylvanian) stream channels in Clark County, Kansas provide an excellent data source to demonstrate this technique. This method provides a quick way to incorporate the geologist's interpretation and the results can be used to accurately model related surfaces above and below the channel.

INTRODUCTION

Long, narrow, or lenticular geologic features, such as stream channels, sand bars, etc., have always been difficult to contour satisfactorily with the computer. Dahlberg (1975) documented several comparisons between hand-drawn and machine-contoured maps which illustrate this point. His studies concluded that round or ovoid features were easily adaptable to machine contouring. The computer would detail the geometry of the rounded feature as well as, and in some cases, more accurately than hand contouring by a geologist. The geologist, in fact, would often over interpret the surface.

When it came to a lenticular sand deposit, however, the computer maps were unrealistic, and the hand-drawn maps better predicted the actual sand geometry. This result was due to the computer's tendency to draw round or ovoid features which did not represent the actual geometry of the sand deposit.

This being the case, a geologist wishing to use the computer to map lenticular features often faces a real dilemma. He must either edit the rounded features generated by the computer using an interactive editing program or he must hand contour his interpretation, digitize the hand-drawn contours, and create a grid from the digitized values. Both alternatives are time consuming.

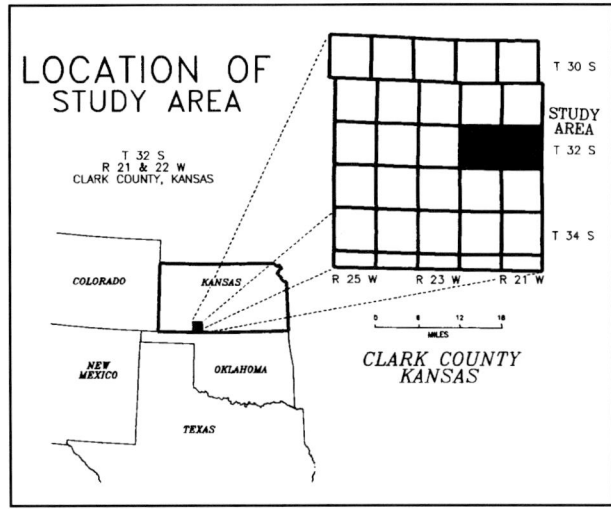

Figure 1. Location Map. The study was undertaken in a two township area in Clark County, western Kansas.

The shape-assist method described below has all the benefits of a computer-generated map, but encompasses the geologist's interpretation in crucial areas. It also has the additional advantage of being faster to generate than a traditional digitized map, and it honors all data.

GEOLOGIC BACKGROUND

To demonstrate the shape-assist method of geological modeling, we selected a two-township area in eastern Clark County, Kansas (Figure 1) where Morrowan-aged streams cut long, narrow channels into Mississippian limestones. Well logs indicate these channels often reach depths of more than 80 to 100 ft (25 to 30 m) and are filled with shale and fine-to-coarse-grained sandstones. These channel sands form stratigraphic oil and gas traps. Geologists exploring for these oil and gas fields must first map the channel system, and it is this channel system on which we plan to demonstrate the shape-assist method.

The Morrow Formation in this area is mainly composed of interfingering shales and silts which average 20 to 30 ft (7 to 9 m) thick and which thicken locally and abruptly into sandstone and shale-filled channels. These Morrow shales and silts rest unconformably on dense limestones of Mississippian age, which can be divided into three essentially conformable stratigraphic units: the St. Louis, Ste. Genevieve, and Chester formations. These Mississippian rocks were regionally tilted following deposition and erosionally bevelled to create an angular unconformity with the overlying Morrow. The Morrow drainage system locally scoured long narrow channels across these exposed, slightly dipping rocks, sometimes cutting completely through one unit and into another.

BUILDING THE CHANNEL SYSTEM

First Look

The most sensitive interval for delineating these Morrow channels is the total Morrow isopach. Our first step in incorporating a channel interpretation into the Morrow is to construct a simple map of this isopach interval using standard gridding techniques. It is important to look at an objective interpretation of the area without any implied geological bias, for this helps guide selection of the channel position. Once this map has been prepared, the channel outline is interpreted and sketched over the unbiased computer-constructed contours (Figure 2).

Creating the Form

The next step is to build a grid that encompasses the interpreted channel. In order to capture channel shape, the interpreted channel system is drawn as a series of contour lines outlining the edges and thalweg of the channel. These contours need not be extremely precise at this stage but can be sketched rather freely through the data. The idea is to create the form we wish the channel to take when the maps are complete. The channel edges and thalweg are then digitized. For simple geometries, the process of adding and digitizing interpretive contours may be done interactively, however, for complex projects we find that paper maps and a digitizing tablet work best.

In order to construct a grid of the channel form, values must be assigned to the edge and thalweg contours. Because we plan to superimpose this form onto the Morrow isopach, we used channel-edge values that were typical of Morrow thickness across the study area and a thalweg value typical of Morrow thickness within the channel. We thus assigned channel edges a contour value of 30 ft (19 m) and the thalweg a value of 80 ft (25 m). These are the only contours that need be transferred into the computer to create the channel form. These contours are then gridded using a least-squares algorithm. We call this resulting surface our form grid (Figure 3). Note: our digitized contour lines actually bifurcate as our interpreted channel bifurcates. Although this violates the rules of contouring, the computer grids these lines similarly to a line of data points and therefore produces acceptable results.

Since the channel system only encompasses the east half of the study area, grid values from our form grid in the western half of the study area were therefore missing. In order to extend the form grid over the entire area, we first constructed a grid of 30 ft (9 m) values over the entire area (the average value of the Morrow isopach outside the channel), then

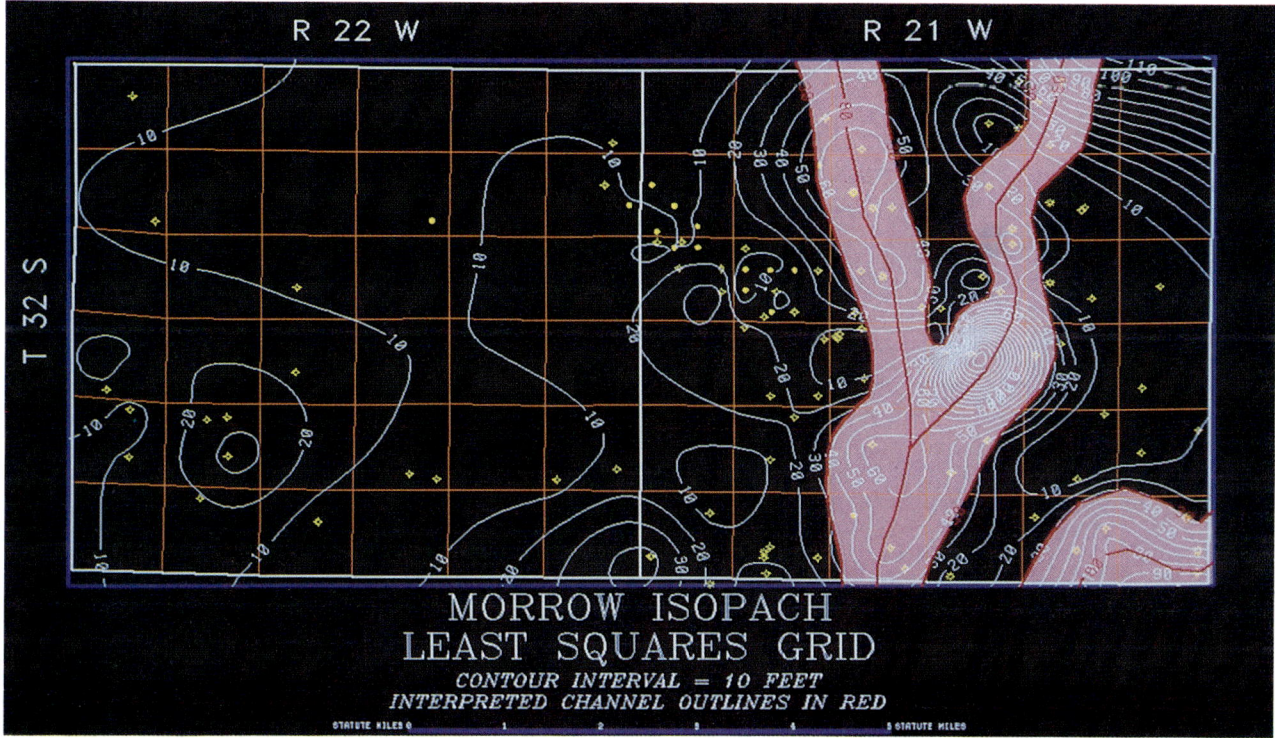

Figure 2. Morrow Formation Isopach Map. Morrow isopach was constructed using least squares gridding techniques. The interpreted channel contours were overlain, digitized and are shown in red.

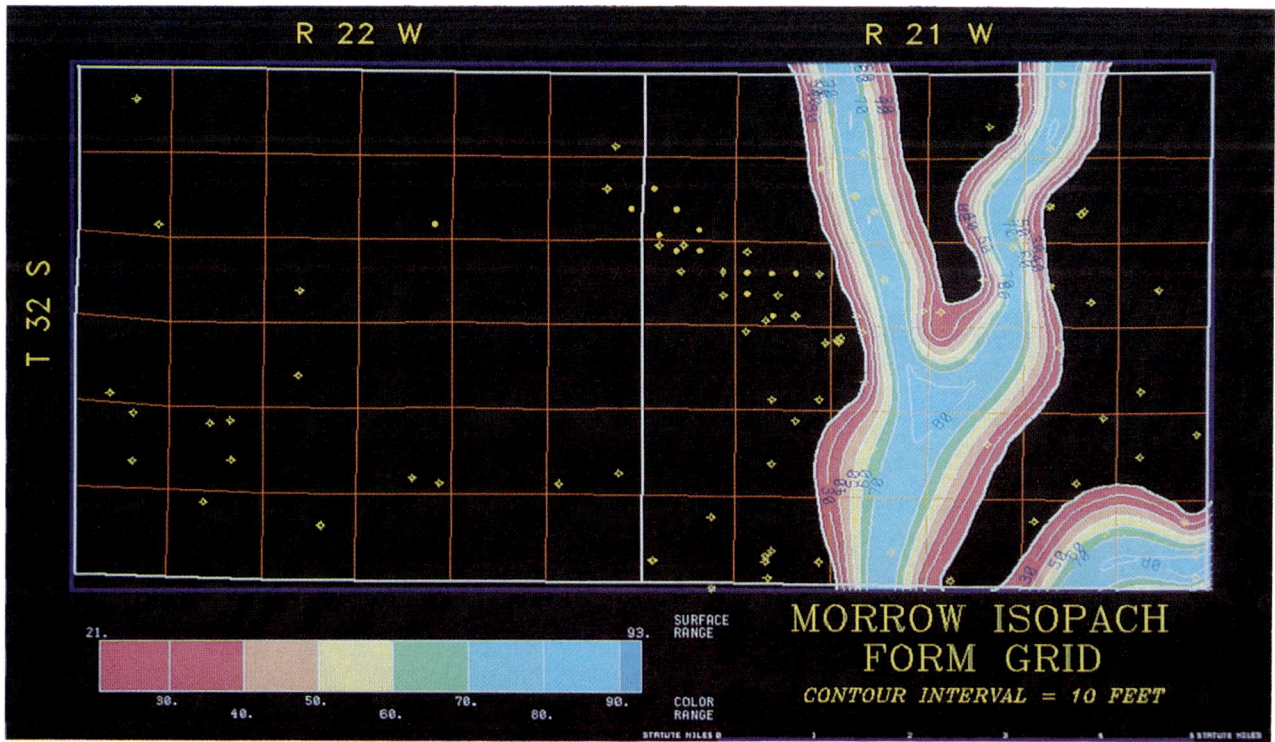

Figure 3. Form Grid Map. The form grid was constructed by gridding the digitized channel form contours. The edges of the grid were controlled to prevent unwanted effects in the western side of the study area.

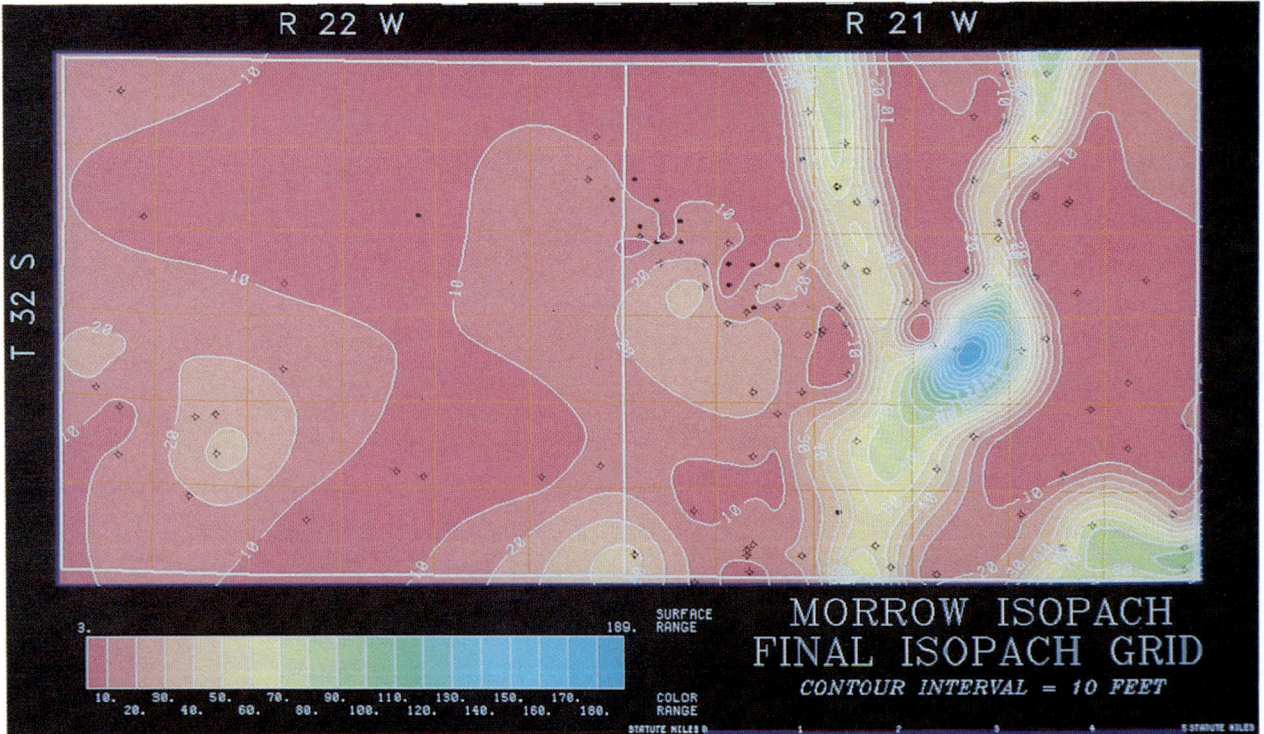

Figure 4. Final Morrow Formation Isopach Surface. This grid was constructed by first interpolating the form grid values onto the well data. The difference between the actual well data values and the form grid is gridded added to the form grid to construct the final isopach.

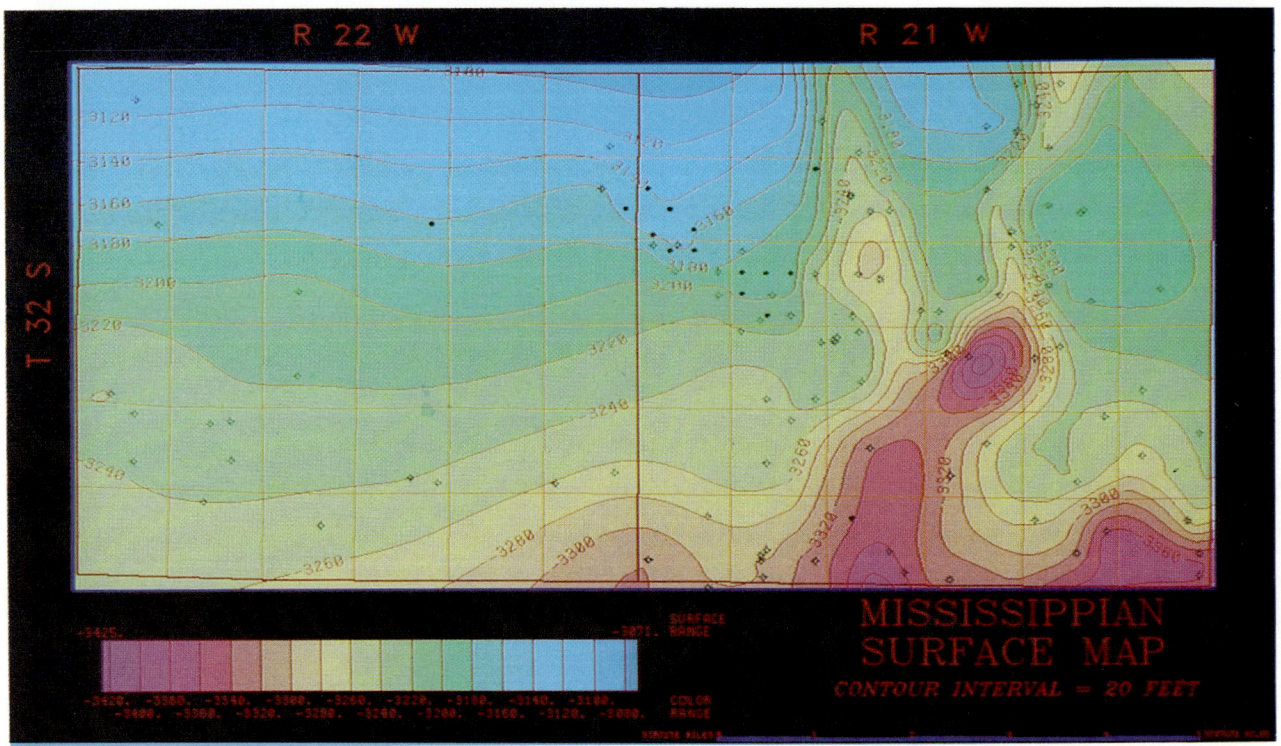

Figure 5. Mississippian Structure. The Mississippian surface is constructed by subtracting the final Morrow isopach grid from the Morrow structure grid.

patched in the initial form grid. This kept our form grid from extreme and uncontrolled values away from the channel.

An alternative approach for controlling the form grid away from the interpreted channel data is to add more data to the interpretation. This can be accomplished in many ways; probably the easiest is to add a few contours or data points which have values equal to the channel-edge contour values. These "away-from-the-channel" data are then combined with the interpreted channel to construct the final form grid. It should be noted, however, that any irregularities (shapes) that are built into this grid have the potential for being transferred to the final surface. This is the reason we chose to use a smooth-valued grid away from the channel interpretation.

Transferring Channel Shape to the Isopach Grid

The last step in the shape-assist method is to transfer the channel shape of the form grid to the Morrow isopach. This requires calculating the difference between the actual Morrow thickness and the form grid, gridding these values, then adding this residual grid back onto the form grid to obtain the final Morrow isopach grid.

Determining the difference between actual Morrow thickness data and the form grid is done by back-interpolating values from the form grid at all existing data point locations. These values are placed in the original data file as a new data field. The back-interpolated form values are then subtracted from the original Morrow thickness values to create new residual values. This residual data represents the discrepancy between the form grid and the original well data. The residual value is then gridded using standard least squares techniques and added to the form grid to create a final Morrow isopach grid (Figure 4). This Morrow isopach grid has all the advantages of a hand drawn map in that it represents our interpretation of the channel and honors all the data. In addition, this map was easily constructed, and could have easily accommodated much larger amounts of data, or a larger study area, or more complex channel systems. For this interpretation, only six contour lines needed to be digitized and our final map honors all of the log data as well as the geologist's interpretation.

BUILDING THE STRUCTURAL GRIDS

We believe the real advantage to computer mapping does not lie in the prediction of a single structure or surface, but rather in the ability to quickly and accurately interrelate surfaces, a tedious and nearly impossible task to do accurately by hand. The Morrow isopach was not the ultimate goal of our study, in fact it was only the beginning. We wanted to ensure that the remaining maps to be constructed incorporated our interpretation of the Morrow as well.

Mississippian Erosional Surface

Using the techniques described by Jones, et al. (1986), the top of Mississippian erosional surface (base of Morrow) was best constructed by building-down from the top of the Morrow surface, using the newly constructed Morrow isopach to shape the erosional surface. First, the top of Morrow structure is constructed from the existing well control using standard least squares gridding techniques. Because irregularities in this initial grid will influence all remaining grids, any modifications which need to be made to this computer-generated surface should be made at this time. Then the Morrow isopach is subtracted from the top Morrow structure, resulting in a grid which represents the erosional surface of the Mississippian cut by the Morrow stream channel (Figure 5).

DEFINING THE SUBCROP

The Chester, Ste. Genevieve, a local Ste. Genevieve marker, and St. Louis all dip slightly and subcrop as part of the Mississippian surface. The basal Morrow cuts into the Chester, Ste. Genevieve and the Ste. Genevieve marker in the study area. In order to map these three surfaces correctly it is important allow the top of the Mississippian (basal Morrow) to intersect the three Mississippian units.

Upper Mississippian Structure

The Ste. Genevieve had the most well penetrations in the area and resulted in the best surface grid of the four Mississippian surfaces. We then built down to the Ste. Genevieve marker and the St. Louis using the conformable surface technique described in Jones, et al. (1986). Once these surfaces were constructed, we built up to the Chester, again using the Ste. Genevieve marker as control. The result was four conformable structure grids representing the initial (pre-truncation) models of the Mississippian surface.

Truncation by Upper Mississippian Erosional Surface

Each of these four final structure grids was then truncated against the Mississippian surface by comparing the structural grids with the Mississippian surface grid, and outputting a final grid that retains the minimum values from both. This step created four new structural grids. The relationship of these surfaces is illustrated in Figure 6.

The Morrow channel, in places, downcuts entirely through the Ste. Genevieve Formation. Using our interpreted surface, we were quickly able to map the top of the Ste. Genevieve structure, show channel down-cutting into the Ste. Genevieve, and even identify an area of missing Ste. Genevieve within the channel (Figure 7). The blanked area of the Ste.

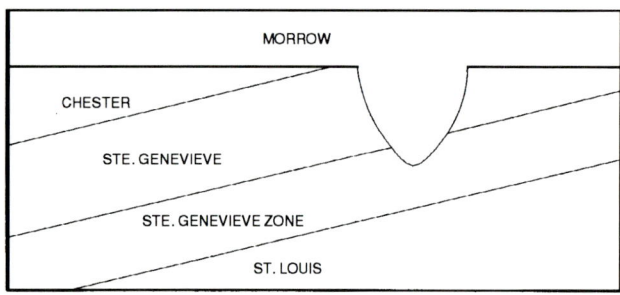

Figure 6. Generalized Cross Section. The relationship between the Morrow, Mississippian surface, Chester, Ste. Genevieve, Ste. Genevieve marker, and St. Louis structures are shown.

Genevieve structure was created by blanking all negative grid values from the intersecting Ste. Genevieve marker and the Mississippian surface. This represented an area where the Ste. Genevieve was completely eroded.

The subcrop patterns can be shown by posting lines of intersection between the Mississippian erosional surface and the pretruncated structural grids of each Mississippian unit. This is done by subtracting the Mississippian surface from the pretruncated surface and outputting the zero contour line of this resulting grid. This contour line represents the line of the subcrop. This line can be posted on the structure contour map of the respective Mississippian formation.

OTHER INTERPRETATIONS

Once this study was completed we decided to test the versatility of the shape-assist technique by adding a second channel which ran west to east across the study area. Although we found no well data to suggest the presence of this channel, we wished to evaluate this technique in areas of sparse (and in this case, non-existent) data. Figures 8 and 9 show this new interpretation and results.

Appendix B of this volume present data for sand thickness in stream channels of the Upper Red Fork Formation in Oklahoma. Several algorithms were applied to model these data including an algorithm designed to more effectively handle the zero edge associated with thickness data. Although the data are dense, none of these algorithms acceptably models the stream channels. This is because any interpretation of these data will force the channels to pass between zero thickness data, something

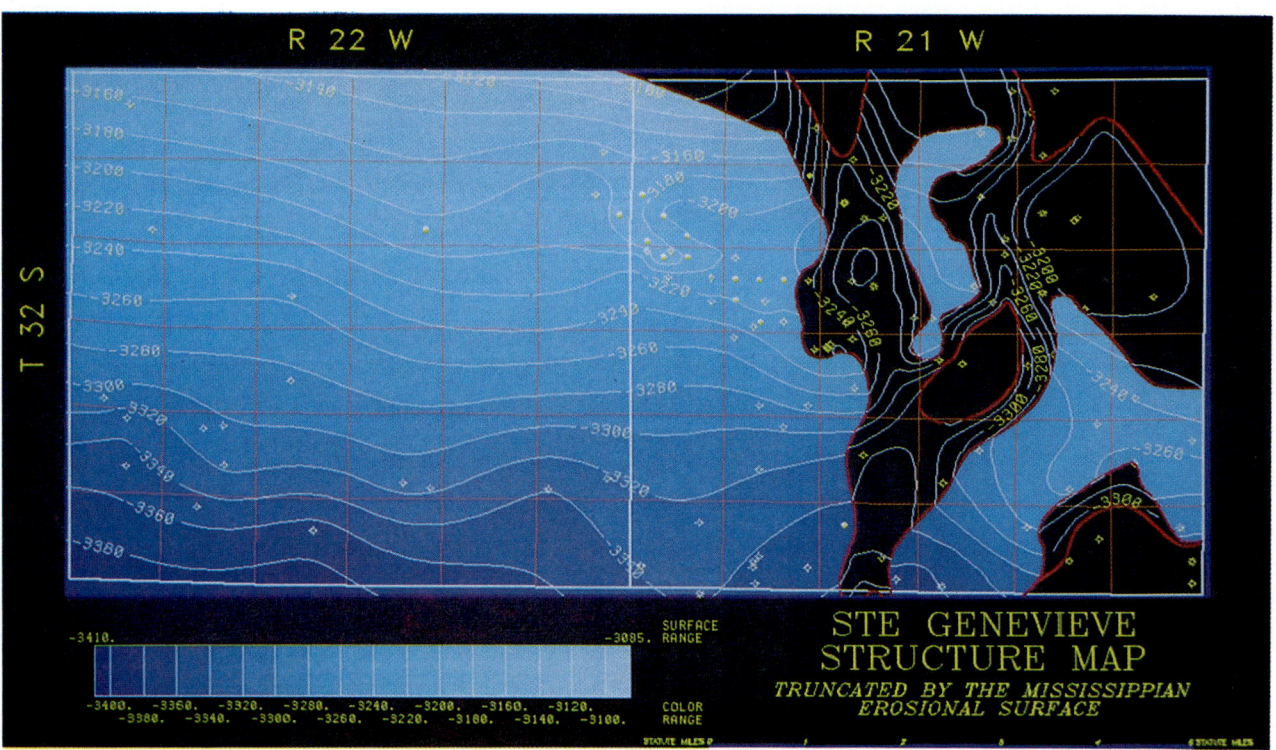

Figure 7. Ste. Genevieve Formation Structure. The Ste. Genevieve structure was constructed by calculating the minimum value between the Mississippian surface and the Ste. Genevieve structure. This surface was then blanked (or removed) where it extended below the Ste. Genevieve marker. This area is shown on the map as white areas without contours because the Morrow channel has completely eroded the Ste. Genevieve in these areas.

impossible for an algorithm to do without human intervention. Figure 10 shows thickness contours from the isopach gridding algorithm with channel interpretation lines overlain on top. Using this interpretation and the shape-assist technique we created the map of channel thickness shown in Figure 11.

In addition to stream channels, we have also used the shape-assist technique to successfully map bar sands and structural surfaces that have a shelf edge or a rapid change in gradient which was poorly controlled by the data.

SUMMARY

The shape-assist technique allows geologists to place an interpretive bias on any map with a minimal effort, taking advantage of the interactive ease and speed of the computer. The technique is also useful for modeling a variety of depositional environments, such as stream channels, bar features, and shelf edges that have poor data control. This technique can be used for any surface which requires some interpretive help to make a more geologically reasonable map. Furthermore, this technique allows the user to completely integrate the resulting map with other surfaces above and below the interpreted horizon.

ACKNOWLEDGMENTS

We would like to thank MASERA Corporation, Tulsa, OK for allowing us to use their Redfork stream channel data (Figures 10 and 11).

REFERENCES CITED

Dahlberg, E. C., 1975, Relative effectiveness of geologists and computers in mapping potential hydrocarbon exploration targets, Mathematical Geology, v. 7, p. 373.

Jones, T. A., D. E. Hamilton, and C. R. Johnson, 1986, Contouring geologic surfaces with the computer: New York, Van Nostrand Reinhold, 314 p.

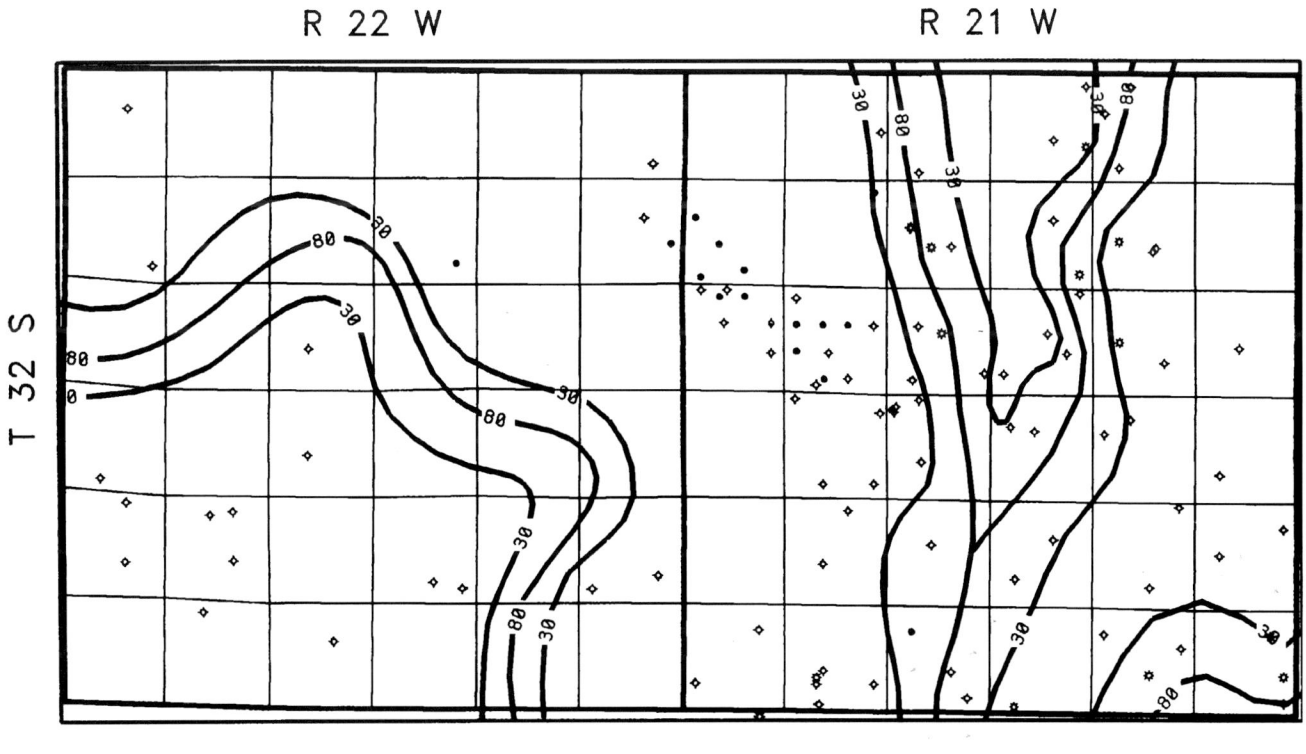

Figure 8. Interpretive Channel Map. A second channel was interpreted on the western side of the study area to see what results the shape-assist technique would produce.

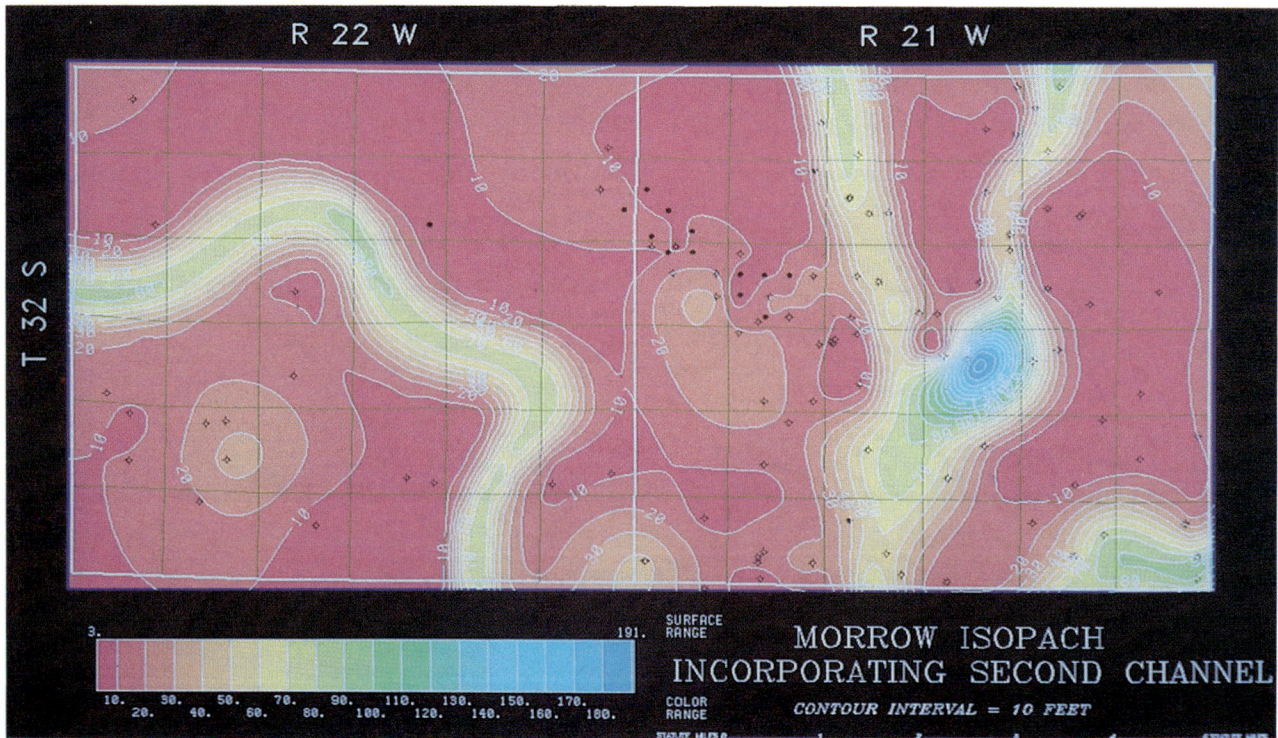

Figure 9. Final Morrow Isopach. The second interpretation of the Morrow isopach showing the new channel interpretation on the western side of the study area.

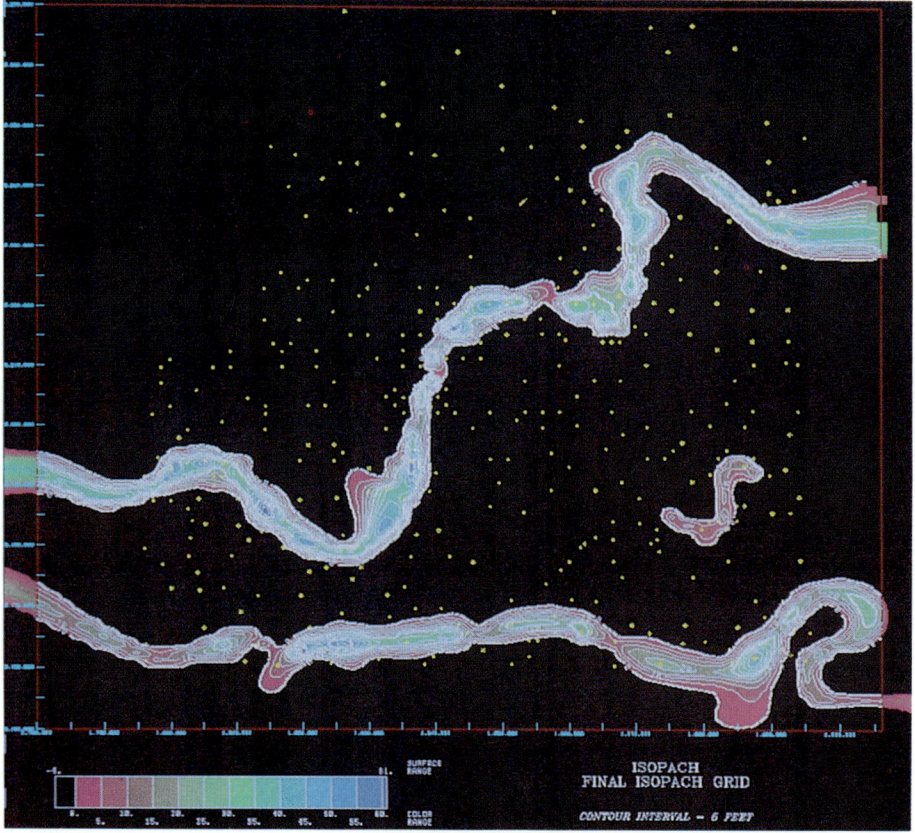

Figure 11. Stream Channel Sand Thickness. Contours represent the channel sand thickness built using the channel interpretation shown in Figure 10 and the shape-assist technique.

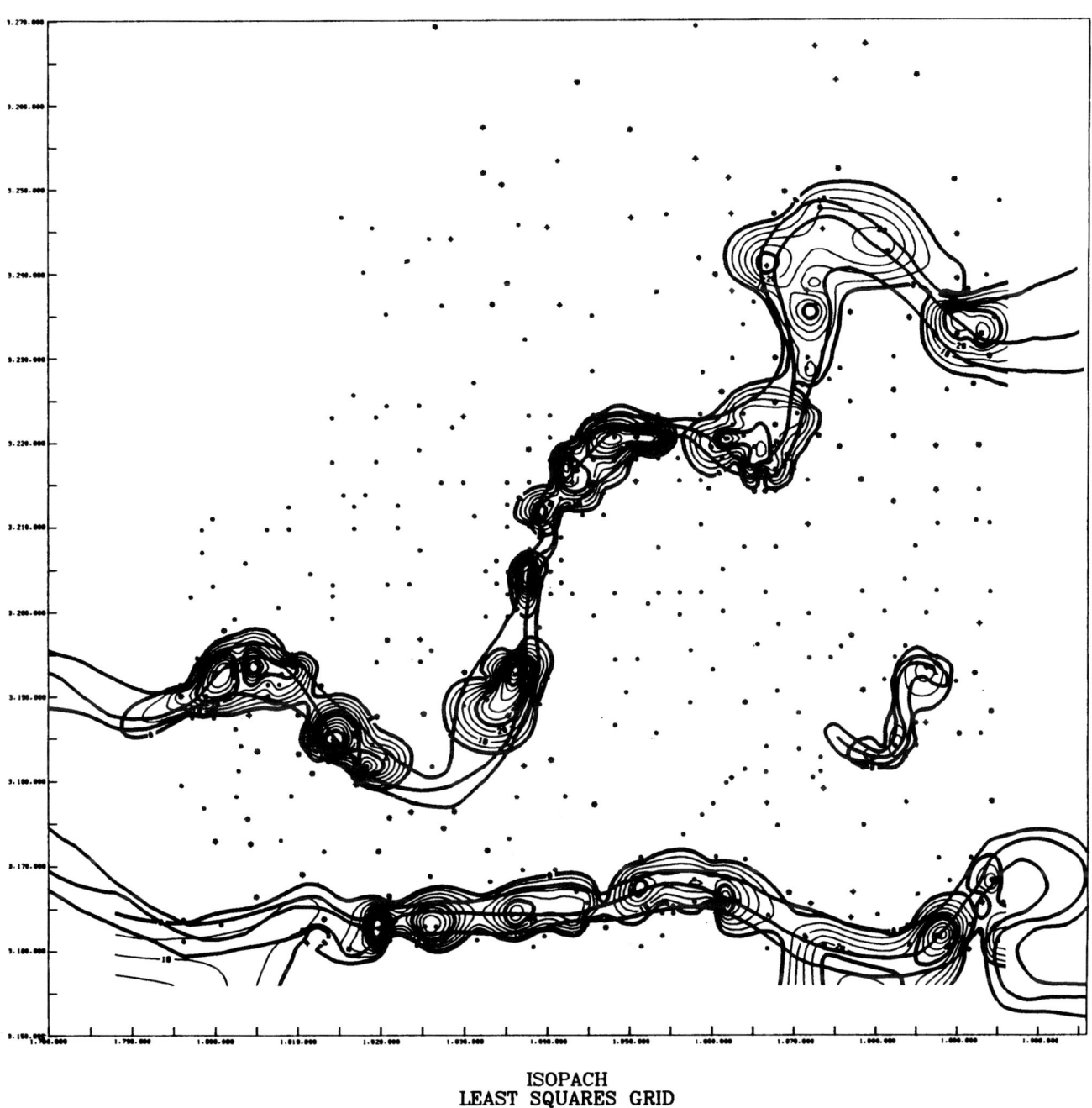

ISOPACH
LEAST SQUARES GRID

CONTOUR INTERVAL = 5 FEET
INTERPRETED CHANNEL OUTLINES OVERLIE CONTOURS

Figure 10. Sand Thickness and Channel Interpretation. Contours represent sand thickness of stream channels in the Upper Red Fork Formation gridded with an algorithm designed to handle the zero thickness edge. Lines represent the channel interpretation.

Chapter 4

Computer-Generated Surfaces and Grids of the Geometry, Rock Type, and Chemistry of a Bedded Mineral Deposit

Walter J. Bawiec
*U.S. Geological Survey,
Reston, Virginia, U.S.A.*

Wilbur D. Grundy
*U.S. Geological Survey,
Denver, Colorado, U.S.A.*

ABSTRACT

Of particular importance to the geologic community is the ability of computers to calculate surfaces, extrapolate the subsurface geometry of rock units, and exhibit the spatial distribution and intensity of related parameters such as rock chemistry. Methods and techniques that describe subsurface-bedded ore deposits are used to define lithologic units of interest, determine the geometries of these units by using computer-generated grids, document the spatial distribution and intensity of related attributes, and produce derivative products through the mathematical manipulation of these grids.

INTRODUCTION

The mining industry is a sector of the economy that has inherent uncertainty and risk. Some of this risk, such as the factors that control demand for specific commodities, is outside of the influence of the mining industry and unpredictable. However, by employing computers and reliable data bases, some factors can be controlled and quantified and therefore, examined in greater detail. For example, certain physical variables can be measured at the site of potentially economic mineral deposits. These physical parameters, which in the past were equivocal, can now be evaluated with more certainty.

The geology of subsurface-bedded ore deposits can be evaluated by using data gathered from a gridded drill-hole exploration program. The input data required for a geologic evaluation, the computer techniques used to generate the information needed for an economic evaluation, and the results presented here can easily be translated to other types of bedded ore deposits.

Initially, the geometry and extent of a bedded phosphate-enriched subsurface zone was delineated to determine the geological feasibility of developing this deposit. Studies were conducted to define distribution and grade of the phosphate ore, proportions and rock type associated with the ore, magnitude and spatial distribution of potential chemical contaminants within the ore, and amount of phosphate resource present. The objective of this study is to bring a large amount of disjointed but related information into one com-

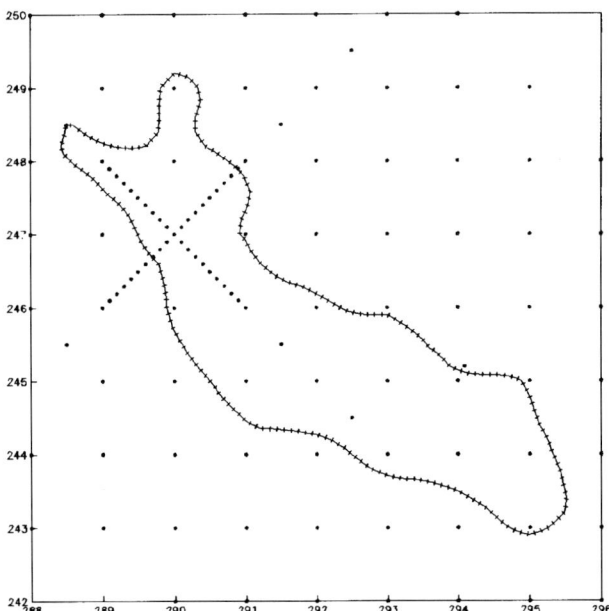

Figure 1. Area of proposed open pit mine and drill hole locations.

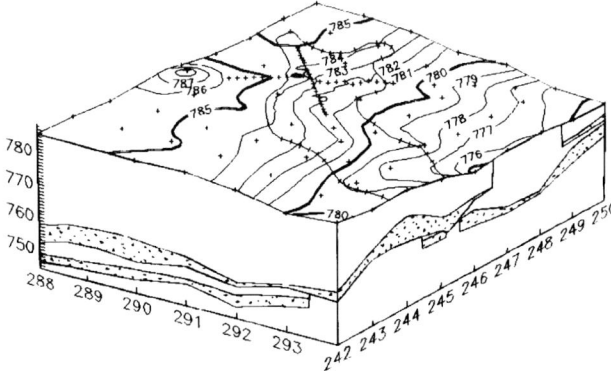

Figure 2. Upper and lower phosphate ore beds, proposed open pit, and drill hole locations. Contour interval is one meter.

plete, comprehensive, and accurate product. The data obtained from this deposit is proprietary; therefore, the location, grade, and tonnage is concealed. The actual numbers for these parameters are irrelevant to both the concept of the method being discussed here and the demonstration of the use of computer-generated surfaces. Also, characteristics other than geologic attributes must be evaluated to determine if this phosphate deposit can be economically and successfully developed.

DATA DESCRIPTION

The geometry, rock type, chemistry, and mineral resources of the phosphorite beds can be evaluated by using the data collected within a grid drilling program. The area of study is defined by an 8x8-kilometer (km) area shown in Figure 1. This area is bounded by coordinates that are 288 to 296 km east and 242 to 250 km north from an arbitrary point of origin. The foot-shaped area in the figure is an outline of thinly covered phosphorites where the stripping ratio (thickness of overburden divided by thickness of the ore bed) for an open-pit mine would be less than 2.25 to 1.

Drill holes shown in Figure 1 are percussion drill holes, which were sampled continuously down hole at a 0.5-meter (m) interval spacing. The drilling was done on a 1-km square grid and was supplemented by an X-shaped pattern of drill holes on a 141 m or wider spacing (northwest part of Figure 1). This close-spaced drilling, necessary for geostatistical analysis, is located in the richest and thickest part of the phosphorite deposit and is the most favorable site for exploitation by an open-pit mine. Because the X-shaped pattern may not be representative of the entire 64-km² area, all data in the area were used in variogram modeling.

There are two beds of enriched phosphorite (Figure 2). Each bed is defined by continuous drill core samples that are greater than 1-m thickness and have a grade of phosphorite (P_2O_5) greater than a cutoff determined by the developer. The samples from within each bed are classified by degree of cohesion into three rock types: hard compact, semifriable, and friable. Cohesion of the ore is an important factor in determining method of concentration into a shippable product.

Chemical information from analysis of core samples on oxides other than P_2O_5 were determined for use in selective mining, planning the mix of ore mined, and other engineering aspects of extraction and processing of the ore.

ENGINEERING PRODUCTS

A minimum set of engineering and geological parameters are needed to competently and confidently evaluate a mineral deposit of this type. These parameters include the characteristics described below. The goal of the chemical and statistical analyses of the drill hole data is to supply geological-engineering parameters in the form of computer-generated grids to be used for input to mine-modeling computer software. Information required for the analyses includes:

1. For each ore bed, computed grids of
   - ore thickness,
   - percent by weight for each chemical oxide of interest,
   - fraction of ore thickness contained in each of the three rock types (hard compact, semifriable, and friable),
   - included waste (thickness of material within the ore bed, which is less than P_2O_5 cutoff grade), and
   - elevation of lower contact of each of the beds.

2. A grid of interburden thickness (rock separating the lower from the upper ore beds).
3. A grid of surface elevation within the area of interest.

The following constraints were placed on the computer-generated grids:
1. Each grid cell is 100×100 m (1 hectare).
2. Origin of the grid is at 242 north, 288 east (Figure 1).
3. There are 81 grid cells north-south and 81 east-west.
4. All data points are honored.
5. All the variables are gridded by the same set of equations to insure that values at each grid cell will sum to 100%.
6. To be used in the gridding, a drill hole contains a minimum of 1 m of phosphorite and a minimum P_2O_5 cutoff grade.
7. Phosphorite assays of less than P_2O_5 cutoff grade may be included in obtaining a composite grade of the hole if not exceeding 0.5 m thick and are overlain and underlain by material above P_2O_5 cutoff grade such that the total interval will meet the minimum percent P_2O_5 grade cutoff.
8. Below cutoff grade rock not meeting criterion 7 above is classified as included waste.

The above grids are used in estimating the in situ P_2O_5 tonnage, grade, stripping ratios, and in predicting the chemical composition of possible ore production. The grids also are the basis for producing computerized graphic images useful in characterizing the spatial distribution of the data, for selective mining to control the mix of ore to the mill, and possibly in geological interpretation of the phosphorite deposit.

GRIDDING METHOD

The geostatistical interpolation technique known as kriging was used to grid the variables required to evaluate this ore deposit. The theory of kriging is described in full by Journel and Huijbregts (1978). Kriging is based on the premise that two samples taken close together tend to have more similar values than samples located farther apart. This phenomenon is indicative of spatial correlation, which indicates that closely spaced samples share information. Accounting for this shared information can produce a better product than if this information were ignored. To accomplish this, kriging produces estimates of the value at an unsampled location by weighting samples from adjacent locations. The weights are computed by using the spatial correlation. Nearby sets of data points are weighted more heavily than those of remote points. The weights are computed such that the resulting estimates are unbiased (the average of the estimation errors is 0) and have a minimized error variance (the squared errors are made as small as possible). The spatial correlation function is quantified by using a variogram, which is a plot of one-half the mean squared difference of paired sample measurements as a function of the distance between samples.

Because of the spatial correlation, kriging has the following advantages over other interpolation methods:
- Smoothing—Kriging smooths, or regresses, estimates based on the proportion of total sample variance accounted for by random noise. The noisier the data set, the less the individual samples represent their immediate vicinity and the more they are smoothed.
- Declustering—The kriging weight assigned to a sample is lowered to the degree that its information is duplicated by nearby, highly correlated samples. This helps mitigate the effect of oversampling high-grade areas.
- Precision—Given that the correct spatial correlation function is used, kriging will compute the more precise estimates possible from the available data. In practice, this is only approximated, as the spatial correlation function must be estimated from available data.
- Trend-handling—If data values show a tendency to increase or decrease in a systematic way across an area (trend or drift), this information can be incorporated into the kriging algorithm, further improving the quality of the estimates. If a trend is used in the gridding, the procedure is called universal kriging (as opposed to ordinary kriging) in which no trend is assumed to exist over the space.

VARIOGRAM-DRIFT MODELS

Grids of thickness of the phosphorites, interburden, overburden, included waste, and proportion by rock type are computed directly from the data. Grids computed indirectly represent percent oxides of silica (SiO_2), aluminum (Al_2O_3), iron (Fe_2O_3), magnesium (MgO), Calcium (CaO), and phosphorite (P_2O_5). The International Atomic Energy Agency (IAEA, 1985, p. 14–17) procedure used for the indirect computation of grids is a standard engineering practice that is employed if the thickness of the ore varies from place to place within the deposit. The IAEA method requires that at each data point or drill hole, the thickness of rock meeting the cutoff criteria is multiplied by the average grade of the composited oxides of interest. This forms a new variable called accumulation (Journel and Huijbregts, 1978) that is then gridded. Each computed grid node value for the accumulation is then divided by the corresponding grid node value for thickness, yielding the desired percentage value for the particular oxide.

A linear variogram was used for the area of the proposed open pit, as required to meet the demands of the open pit design and evaluation software. Thickness and grade were kriged over the entire 64-km^2 area of interest by using a spherical model and parameters estimated from the data.

To comply with the requirement that all grid values in the area of the open pit must be computed with the same set of equations and must also honor the data points, it was necessary to compute all these grid values by using a linear variogram passing through the origin with a slope of unity. Such a variogram has the following equation:

$$\gamma(h) = h, \quad (1)$$

where $\gamma(h)$ is the variogram value for points h kilometers apart. Because trend or drift is evident in most of the maps of the open pit area, it was found necessary to use universal kriging to compensate for this drift. For small distances, the drift is assumed to be of the form:

$$z = d + ex + fy, \quad (2)$$

where z is the variable under study, d, e, and f are coefficients, and x and y are eastings and northings. The kriging algorithm does not require values of the constants d, e, and f. The combination of the linear variogram with a linear drift allows creation of unbiased grids complying with the above constraints (David, 1977, p. 274). Unfortunately, the kriging variances (theoretical errors associated with the estimate) have no meaning when kriging in this fashion.

Grids of P_2O_5 thickness and accumulation for the entire study area were computed by using spherical model variograms having parameters estimated from the data set. The variogram was computed along the direction of the grain (N 45 W) of the phosphorite deposits. The variogram form selected is the spherical model, which has the following equation:

$$\begin{aligned}\gamma(h) &= c[1.5 \times h/a - 0.5(h/a)^3] & \text{for } h \leq a \\ \gamma(h) &= c & \text{for } h > a \quad (3) \\ \gamma(h) &= 0 & \text{for } h = 0,\end{aligned}$$

where

h is the distance of separation of two points,
a is the range of the variogram, and
c is the sill of the variogram.

The range is the distance beyond which samples are not spatially correlated with one another. The sill is a maximum value of the variogram and is a good approximation of the population variance. The range can also be thought of as the distance from the origin of the variogram to the sill. No evidence of a nugget effect (failure of the variogram to pass through the origin of the graph) was found for the thickness or accumulation for either bed.

The spherical variogram parameters assigned for the various bed properties are tabulated below:

Variable	a	c
Thickness of upper bed	1.60 km	8.0 m²
Accumulation of upper bed	1.55 km	4000% × m
Thickness of lower bed	2.40 km	1.8 m²
Accumulation of lower bed	2.40 km	750 % × m

These variograms are used with a complete linear drift model whose form is described above.

To compute a grid value for a cell, a search radius of 3 km was used to determine the data points that would be eligible for use in the computations. If more than 12 points were within 3 km of the center of the cell, only the nearest 12 points were used. If fewer than four samples were located within the search radius, the cell was not assigned a grid value. If either the upper or lower beds were not present, the grid cell value for the missing bed or beds was not computed. A grid of 4x4 equally spaced points was calculated for each cell, or block, to estimate the average value of the thickness or accumulation within that 100x100-m block.

GRAPHIC DISPLAY OF GRIDDED DATA

After computation of a grid for a variable of interest, the graphic representation of that data is displayed. As shown in Figures 3 to 11, these computed grids or surfaces are displayed as both contour maps and three-dimensional perspective views to enhance visualization and to give insight into the variable being examined. Contour maps, Figures 3 to 11, are oriented with north at the top of the page. Three-dimensional perspective and block diagrams, Figures 2 to 11, are viewed with a 20-degree inclination above the center of the figure and from the southeast, north is rotated 35 degrees in a clockwise direction. All grid manipulations (addition, subtraction, multiplication, and division) of the kriged surfaces and all figures presented in this paper were created by using Interactive Surface Modeling software (Dynamic Graphics, Inc., 1988).

Displayed computed surfaces show the geometries of the phosphorite ore beds (Figure 2), an example of the spatial distribution of a rock type (Figure 6) and chemical oxide (Figure 7), and phosphate distribution and grade (Figures 9 to 11). The proposed open pit location is displayed in these figures as a point of reference.

GEOMETRY OF THE PHOSPHORITE DEPOSIT

Computer programs were used to examine all analyses of phosphate samples and to select samples of interest. Computer-generated down-hole histograms were plotted to identify zones of interest. Within the most prospective area of the phosphate deposit (Figure 1), the P_2O_5 grade found within the 0.5-m interval core samples ranges from 0 percent to very enriched P_2O_5. For this study, phosphate ore beds are defined by using two criteria: grade at least equal to a minimum cutoff percent P_2O_5 and thickness of at least 1 meter. By using these criteria as each sample in each drill hole was examined for P_2O_5 grade, enriched

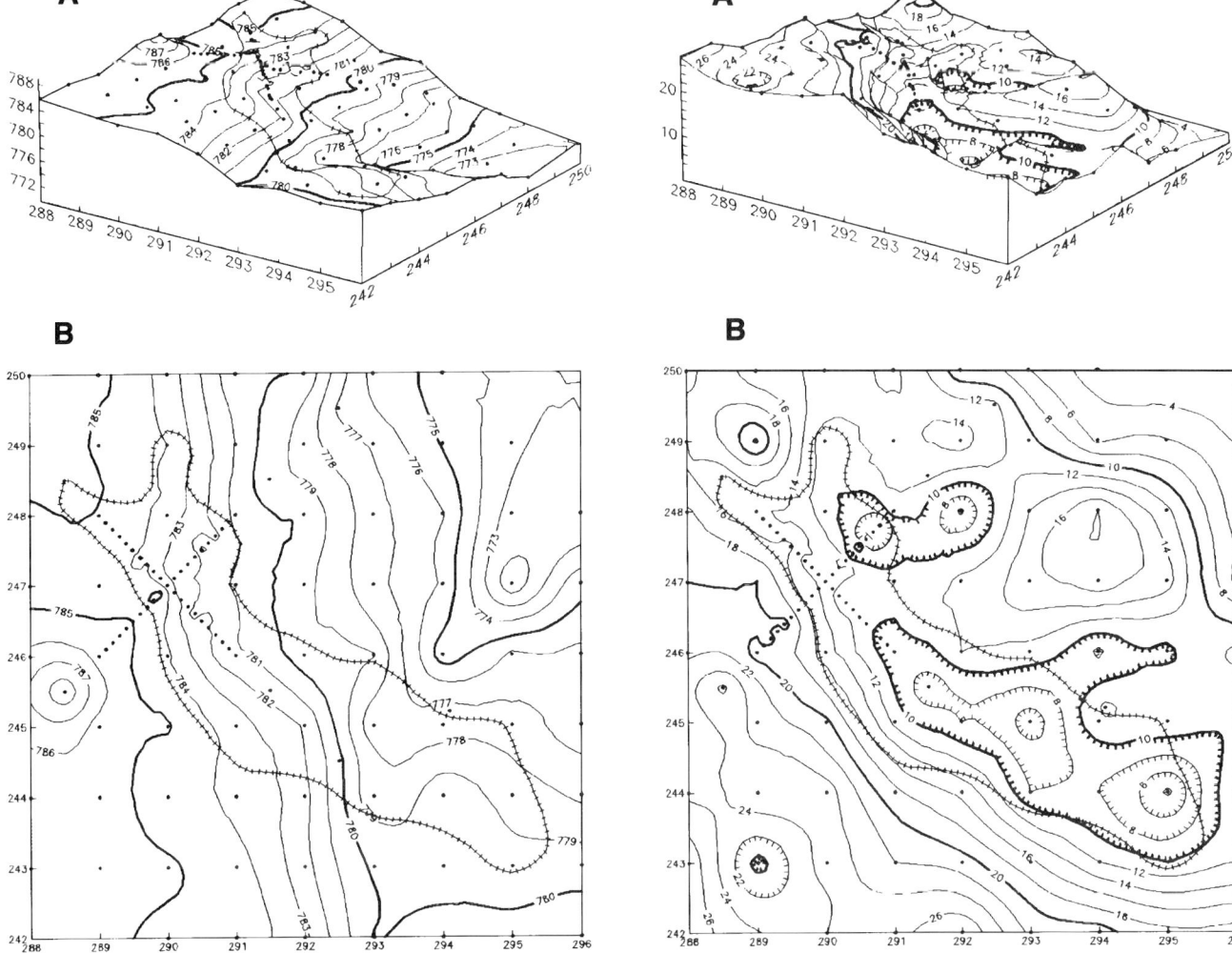

Figure 3. Computer-generated diagrams of surface topography, open pit and drill hole locations in, (A) three-dimensional and, (B) map views. Contour interval is one meter.

Figure 4. Computer-generated diagrams of thickness of overburden, open pit and drill hole locations in, (A) three-dimensional and, (B) map views. Contour interval is two meters.

zones became apparent within the subsurface. This down-hole zonation consists of a phosphate-poor overburden, a phosphate-enriched upper ore bed, a phosphate-poor interburden zone, and a phosphate-enriched lower ore bed (Figure 2). Both the upper and lower ore beds show some evidence of discontinuity; that is, they either fall below the cutoff grade or become thinner than 1 meter. Also, a northwest-trending grain to the structure of the deposit is apparent.

The surface topography (Figure 3) shows a gently sloping surface, which has a maximum of 788 m above sea level in the west, sloping to 772 m above sea level in the northeast corner of the study area. Drill hole locations are shown. This surface is computed by kriging the collar elevations of the drill holes. The northeast sloping surface is bisected in the eastern section by a northward flowing dry water course, indicated by the v-shaped pattern of the contour lines.

Overburden is defined as waste rock that extends from the topographic surface to the top of the upper ore bed. In places where the upper ore bed does not meet the thickness-grade cutoff, overburden thickness is measured from the topographic surface to the top of the lower ore bed. As can be seen in Figure 4, overburden thickens in a northeast to southwest direction and ranges from 4 to 26 m thick. A northwest-southeast belt of thin overburden overlies the area of the proposed pit.

The thickness of the upper ore bed, the primary mining target for this project, is presented as contour and perspective diagrams in Figure 5. The upper ore bed is absent from the northeast quadrant of the area and is thickest in a northwest-southeast direction (Figure 2). Thickness of the upper ore bed was gridded by using the variogram calculated on page 39 that has a linear drift. An isopach map of the thickness of the

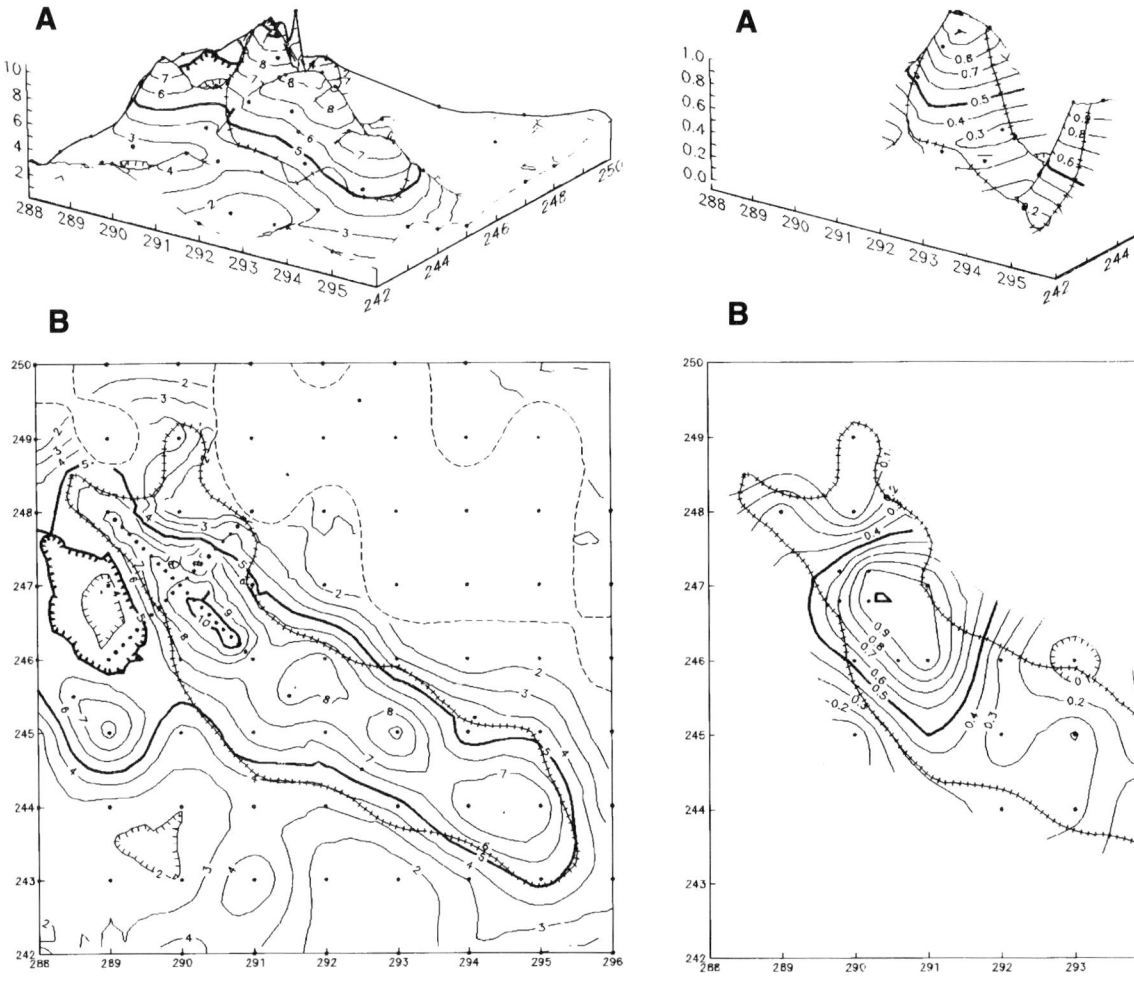

Figure 5. Computer-generated diagrams of the thickness of the upper ore bed, open pit, and drill hole locations in, (A) three-dimensional and, (B) map views. Contour interval is one meter.

Figure 6. Computer-generated diagrams of the distribution of upper bed hard compact lithology, open pit, and drill hole locations in, (A) three-dimensional and, (B) map views. Contour interval is 0.1 units.

upper ore bed shows a thickening area trending northwest-southeast and lying directly under the open pit outline.

ROCK TYPE CHARACTERIZATION

Between the top of the upper ore bed and bottom of the lower phosphate ore bed, 0.5-m core samples were grouped lithologically into the following categories: hard compact, semifriable, or friable. Characteristics of each of these rock types make them either desirable or detrimental to the mining, milling, or processing of the ore.

The ability to describe spatial distribution, both vertically and horizontally, of these rock types is important in examining the economic feasibility of the mineral prospect. This distribution is mapped by using the changing fractions or proportions of each rock type within each ore bed. These fractions are determined by dividing the thickness of the rock type of interest by the total thickness of the ore bed in the drill hole at that location. In this way the total thickness of the ore bed is 100%. Note that does not provide the absolute thickness in meters of that rock type, but rather a ratio, since the thickness of the ore bed varies from location to location.

As an example of rock type distribution within the upper ore bed, all samples are classified into one of three lithologic types described above. The lengths of these sample intervals in the ore bed are then cumulated for each rock type within each drill hole and divided by the total thickness of the ore bed within that drill hole. These percentages of rock types are then gridded and graphically displayed. The hard-compact lithology (Figure 6) rock type appears to be dominant in the central and eastern regions of the proposed open pit area.

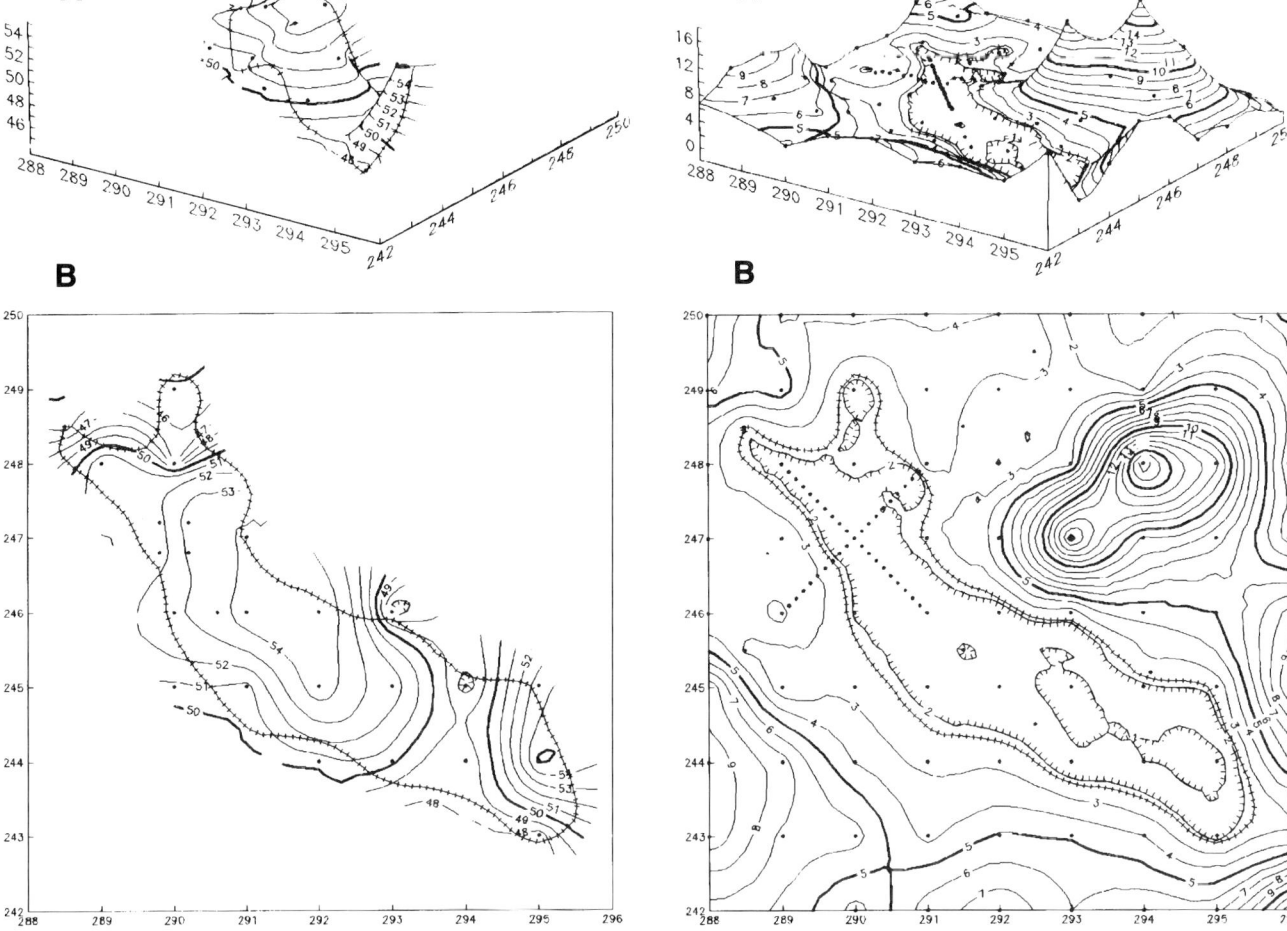

Figure 7. Computer-generated diagrams of the distribution of CaO in the upper ore bed, open pit, and drill hole locations in, (A) three-dimensional and, (B) map views.

Figure 8. Computer-generated diagrams of the ratio of total overburden, interburden, and included waste to the total thickness of rock meeting the required cutoff specifications in, (A) three-dimensional and, (B) map views. Drill hole and open pit locations are also shown.

CHEMISTRY OF THE PHOSPHORITE BEDS

The chemical composition of the phosphate resources present is important to the suitability of the phosphate to be mined, concentrated, processed, and ultimately used as fertilizer. The correlation of chemical oxides contained within the phosphates is best displayed by using scatter diagrams to examine relationships between and among the chemical oxides. In some cases there is a strong positive correlation with P_2O_5, in other cases an inverse or negative correlation is present, and in others no correlation exists at all.

Understanding positive, negative, and noncorrelations between chemical oxides within the deposit is important for interpreting present conditions and genesis of the ore material. Standard statistical methods and scatter diagrams are used to examine these relationships.

SPATIAL DISTRIBUTION OF MAJOR OXIDES

Analyzing scatter plots of chemical oxide constituents and phosphate grade is necessary to understand relevant correlations. Also of importance to the mining, milling, and processing of the ore is the spatial distribution and magnitude of these chemical oxide constituents. A contour map and perspective diagram were created for calcium oxide (CaO) for the upper ore bed (Figure 7). These chemical oxide diagrams show the spatial distribution of the oxides, the spatial interrelation of the oxides, and isolated less desirable oxides.

Within the upper ore bed, the distribution of CaO ranges from 45 to 55% CaO. The highest CaO concen-

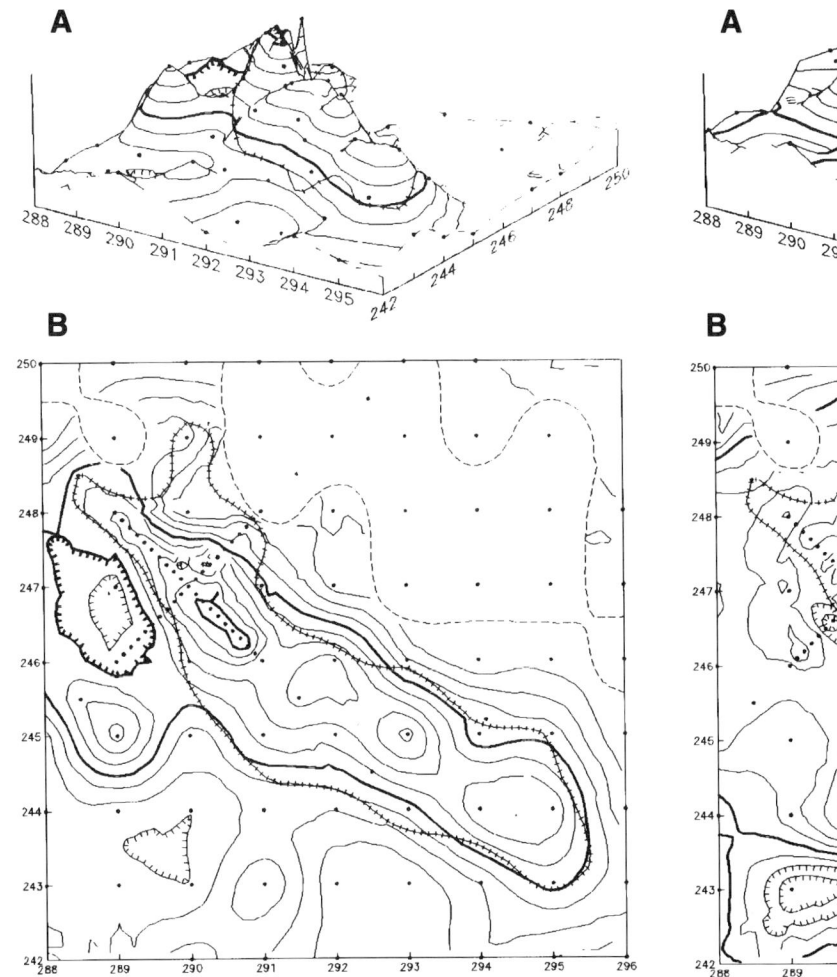

Figure 9. Computer-generated diagrams of kilotonnes of P_2O_5 ore per hectare in the upper ore bed, open pit, and drill hole locations in, (A) three-dimensional and, (B) map views.

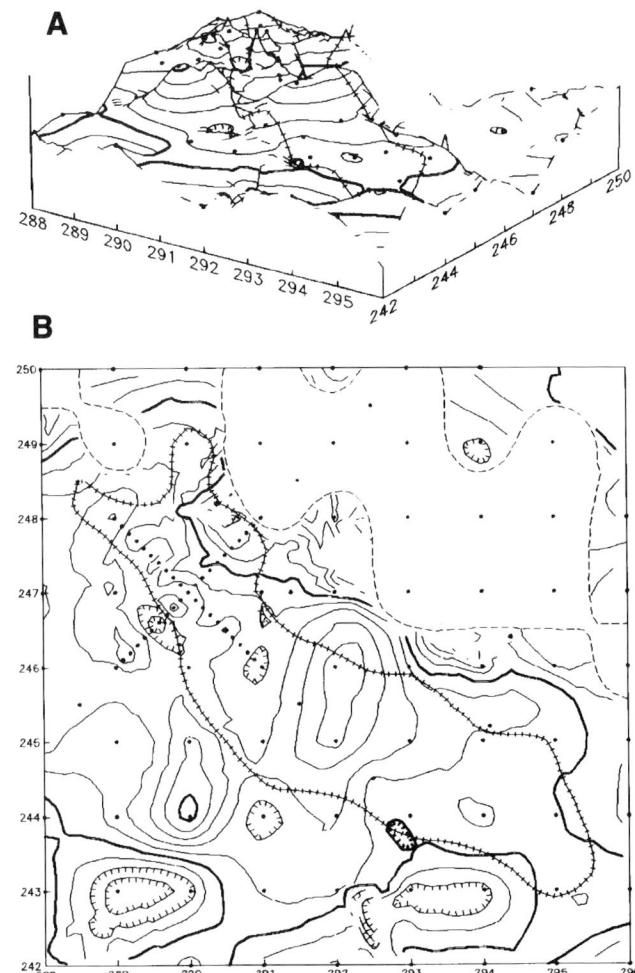

Figure 10. Computer-generated diagrams of percent P_2O_5 grade of ore per hectare in the upper ore bed, open pit, and drill hole locations in, (A) Three-dimensional and, (B) map views.

trations are localized into two areas and are separated by a CaO low in the southeast. This pattern mimics previously discussed rock-type patterns of the upper bed hard compact lithology (Figure 6).

IN SITU PHOSPHATE RESOURCES

In situ phosphate resources for the 64 km² area were estimated by using 146 percussion drill holes in which cuttings were assayed at 0.5-m depth increments for percent P_2O_5. This number of holes exceeds the proposed open pit area and allows an estimate of mineral resources both within the proposed pit area and beyond its limits. Contour maps and perspective plots were prepared for the 8x8-km area of interest for the upper and lower ore beds and the following variables:
1. thickness of phosphatic material above cutoff percent P_2O_5,
2. average grade of material exceeding cutoff percent P_2O_5,
3. kilotonnes of ore per hectare (10,000 m²) derived from item (1)) above, and
4. kilotonnes of P_2O_5 per hectare derived from items (1) and (2) above.

A map was prepared to show the stripping ratio of the area. This ratio was evaluated by summing the overburden, interburden, and included waste and dividing by the total thickness of rock meeting the required cutoff specifications (Figure 8). This was done through the manipulation of computer-generated grids. This strip-ratio map shows an elongate northwest-trending area in which the stripping ratios are less than 2.25 to 1. The highest stripping ratios approach 17 to 1 and trend northeastward from the shallow open pit area. The ratios increase to a value of 9 to 1 in the south and southwest parts of the map.

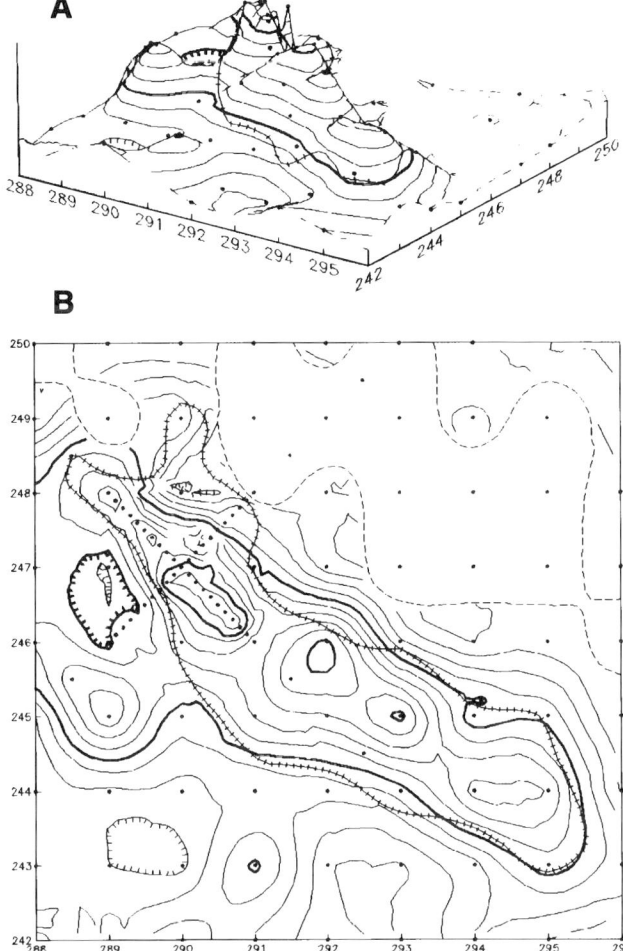

Figure 11. Computer-generated diagrams of kilotonnes of P_2O_5 per hectare in the upper ore bed, (A) Three-dimensional and, (B) map views.

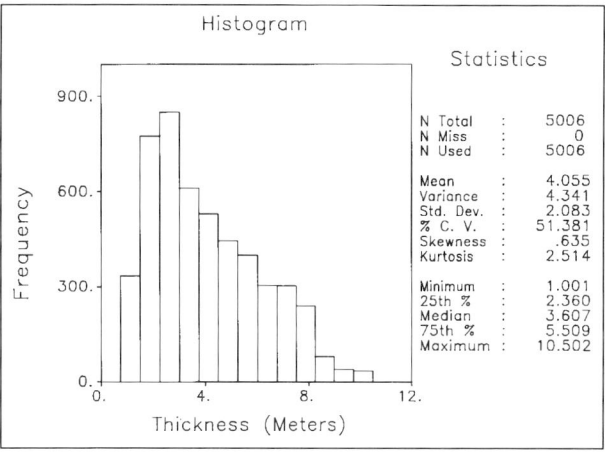

Figure 12. Histogram of ore bed thickness for each hectare in the upper ore bed.

Grids of tonnes of ore per hectare were computed for those qualifying cells having at least 1 m of P_2O_5 above cutoff grade. Figure 9 shows a contour and perspective plot of the kilotonnes of ore per hectare for the upper ore bed. The belt of highest values closely controls the shape of the proposed pit area. Grid values of the estimated thickness in meters were multiplied by 10,000 m² (the area of each grid cell) to obtain a volume grid of phosphatic rock in cubic meters. This volume grid was then multiplied by 2.0 (the assumed bulk density of the ore—2 tonnes per cubic meter), yielding the tonnage of ore for each grid cell. The metric tonnages were converted to kilotonnes (a more convenient number for posting on maps) by dividing the computed value of each grid cell by 1,000.

Where:
  each cell ≥ 1 m thickness
  each cell ≥ cutoff percent P_2O_5
  each cell = 100 m × 100 m = 10,000 m² = 1 hectare

assumed bulk density of ore = 2.0 tonnes per cubic meter

For each cell:

  cubic meters (m³) = thickness (m) × cell size (10,000 m²)
  tonnage of ore (tonnes) = cubic meters (m³)
    × bulk density (2.0 tonnes/m³).   (4)

The grids of percent P_2O_5 (Figure 10) for each grid cell were computed by dividing the kriged grid cell estimated value for accumulation by the corresponding kriged grid cell value for thickness in meters of the ore bed. Trends of P_2O_5 grade (Figure 10) do not show any strong parallelism with the proposed pit outline. The highest grade material trends north-northeast along the 292 east coordinate.

For each cell:

$$\text{percent } P_2O_5 = \frac{\text{kriged accumulation}}{\text{kriged thickness (m)}} \quad (5)$$

Kilotonnes of P_2O_5 per hectare were obtained by multiplying the tonnage of ore per hectare grid values by the corresponding grid values for P_2O_5 grade. The product thus obtained was divided by 1,000 (to make the final conversion into kilotonnes of P_2O_5 per hectare). Kilotonnes of P_2O_5 per hectare (Figure 11) show a strong correlation with the contours of kilotonnes of ore (Figure 9). They show the potential pit is positioned over the portion of the upper ore bed where the largest amount of P_2O_5 per unit area will be obtained.

For each cell:

  tonnage of P_2O_5 = kilotonnes of ore per hectare
    × percent P_2O_5 per hectare.   (6)

The previously discussed plots of kilotonnes of ore per hectare, P_2O_5 grade per hectare, and kilotonnes of contained P_2O_5 per hectare were generated from kriged grids of accumulation and ore thickness. By using available statistical packages, descriptive statistics can be produced for the 81x81 cell grid for each generated variable. Figure 12 displays the types of statistical information that can be useful. The upper ore bed contains rock meeting the cutoff thickness-grade specifications in 5,006 of the 6,561 100x100 m cells in the area of interest. The mean thickness of these 5,006 cells is 4.055 m, the standard deviation is 2.083 m and the median is 3.607 m.

Note that the method shown here is for estimating the total tonnage of phosphorite that meets the cutoff criteria of thickness and grade. Determinations of the portion of this resource that is profitably extractable is the next step in determining the economic feasibility of developing a mining operation; this step is beyond the scope of this paper.

REFERENCES CITED

David, M., 1977, Geostatistical ore reserve estimation, Developments in Geomathematics 2, Amsterdam, Elsevier Scientific Publishing, 364 p.

Dynamic Graphics, Inc., 1988, Interactive surface modeling, Release 6.93: Berkeley, Dynamic Graphics Inc.

International Atomic Energy Agency, 1985, Methods for the estimation of uranium ore reserves, an instruction manual: Technical Report Series no. 255, 92 p.

Journel, A. J., and C. J. Huijbregts, 1978, Mining geostatistics: London, Academic Press, 600 p.

Chapter 5

Computer Mapping of Pinnacle Reefs, Evaporites, and Carbonates: Northern Trend, Michigan Basin

David E. Hamilton
Landmark/Zycor Inc.
Austin, Texas, U.S.A.

Skye K. Henize
Landmark/Zycor Inc.
Austin, Texas, U.S.A.

ABSTRACT

Data from 590 wells in Antrim and Kalkaska counties of Michigan were used to model pinnacle reefs in the northern trend of the Michigan basin. The data consisted of tops for 6 units: Niagaran Formation, Gray and Brown; Salina Formation, A-1 Evaporite and Carbonate, and A-2 Evaporite and Carbonate.

Building grids and contour maps of the top of Niagaran Brown (Pinnacle) surface using standard algorithms produced a surface that projected below and above the interreef surface. By using a combination of standard algorithms, filters, and surface and data operations the Pinnacles and interreef (nonreef Niagaran Brown) surface were acceptably gridded. The Salina evaporites and carbonates are usually modeled using either direct gridding or addition of a thickness grid to an adjacent surface. However, these approaches did not produce surfaces that reflected the geologist's interpretation. Simple modifications to the thickness-gridding approach allowed more accurate representation of the geology. The evaporites were gridded assuming they were deposited parallel to a paleo-water surface. The carbonates were gridded assuming they draped the surface existing at the time of deposition.

INTRODUCTION

The goal of this project was to generate regional grids of pinnacle reefs and their associated surfaces in the northern trend of the Michigan basin. The project was undertaken to demonstrate methods for modeling surfaces that can not be acceptably modeled using standard algorithms. Understanding how to quickly generate grids and maps of these and similar problem surfaces is important to the oil and minerals industries and to personnel developing new algorithms to solve these problems.

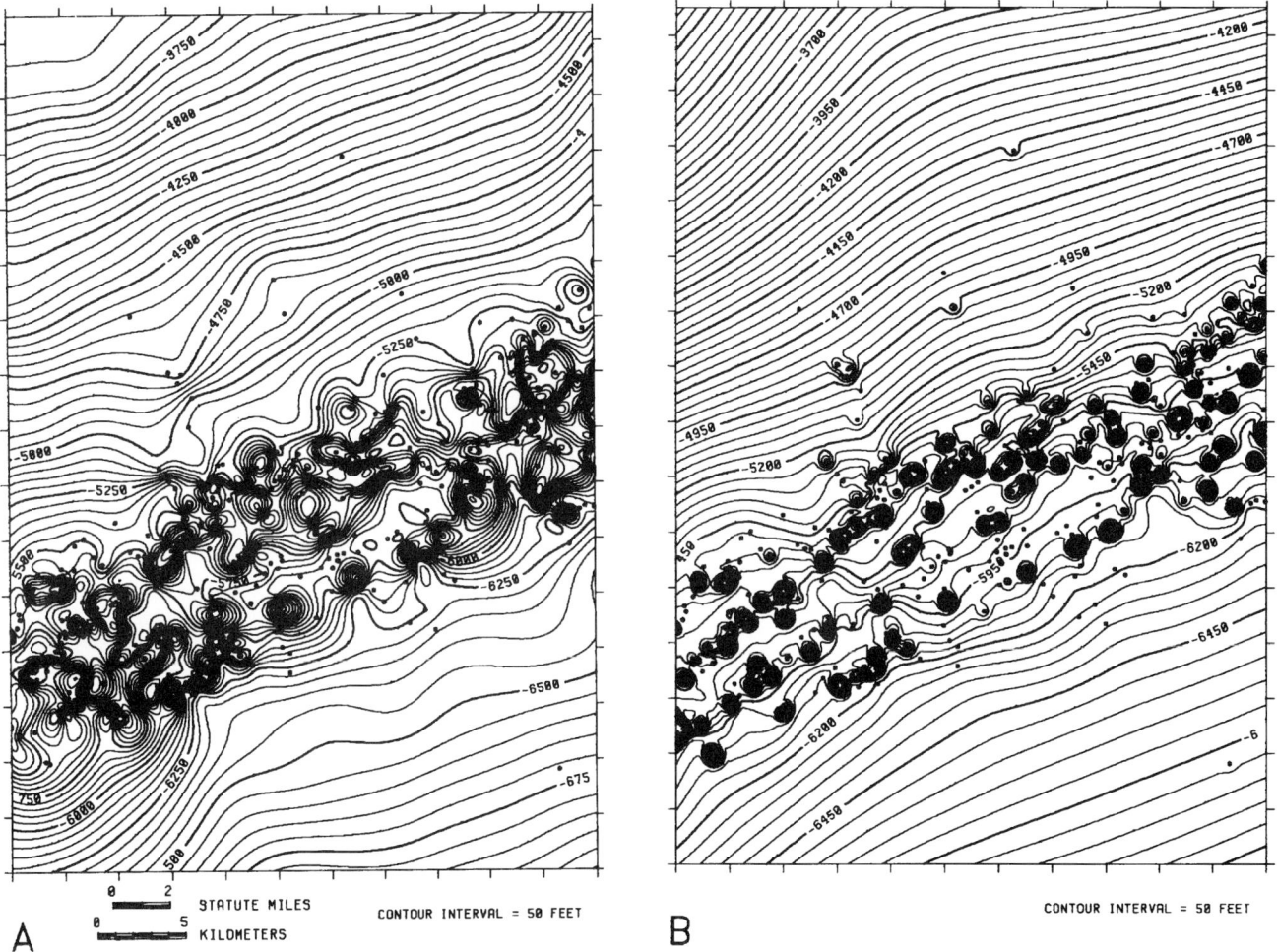

Figure 1. Contours generated from top of Niagaran Brown (pinnacle) grids. Dots represent data for this surface. (A) Grid built using a least squares algorithm with a biharmonic filter. It is difficult to tell which closures are fictitious projections below the surface upon which the pinnacles grew and which are pinnacles. (B) Grid built using the multi-step approach described in this paper.

Modeling pinnacle reefs using standard algorithms produces poor, if not unacceptable, results (Figure 1A). For example, in Figure 2A the right most pinnacle is too broad at its base and two pinnacles in the center of the section have been merged together. The reason for these poor results is because the form of the top-of-pinnacle surface deviates significantly from the form the algorithms were designed to model. Recognizing this difference between reality and design, and using tools available in most computer-mapping systems, allowed more acceptable models to be created (Figure 1B and 2B).

Similarly, using standard methods to model the evaporites and carbonates between and above the reefs also produced poor results. For example, in Figure 2A and B, gridded surfaces for the evaporite and carbonate immediately above the pinnacles do not match the interpreted form. Making subtle changes to the standard methods based on geologic understanding and interpretation significantly improved the grids (Figure 2C).

Much of this paper discusses how algorithms respond to data and the philosophy behind the technique used. Although we are applying the same fundamental mapping techniques used for hand mapping, subtle variations and constraints related to the computer must be understood to be effective in their implementation. In essence, we are trying to present our "mind set" for computer mapping through its application to this problem.

GEOLOGY AND DATA

Geologic Setting

Reefs in the northern trend of the Michigan basin extend for 170 mi (270 km) in a gentle arch from Lake Michigan eastward to Lake Huron (Figure 3). They occur in a band varying in width from 10 to 20 mi (16 to 36 km). Reef heights vary from 100 ft (30 m) for

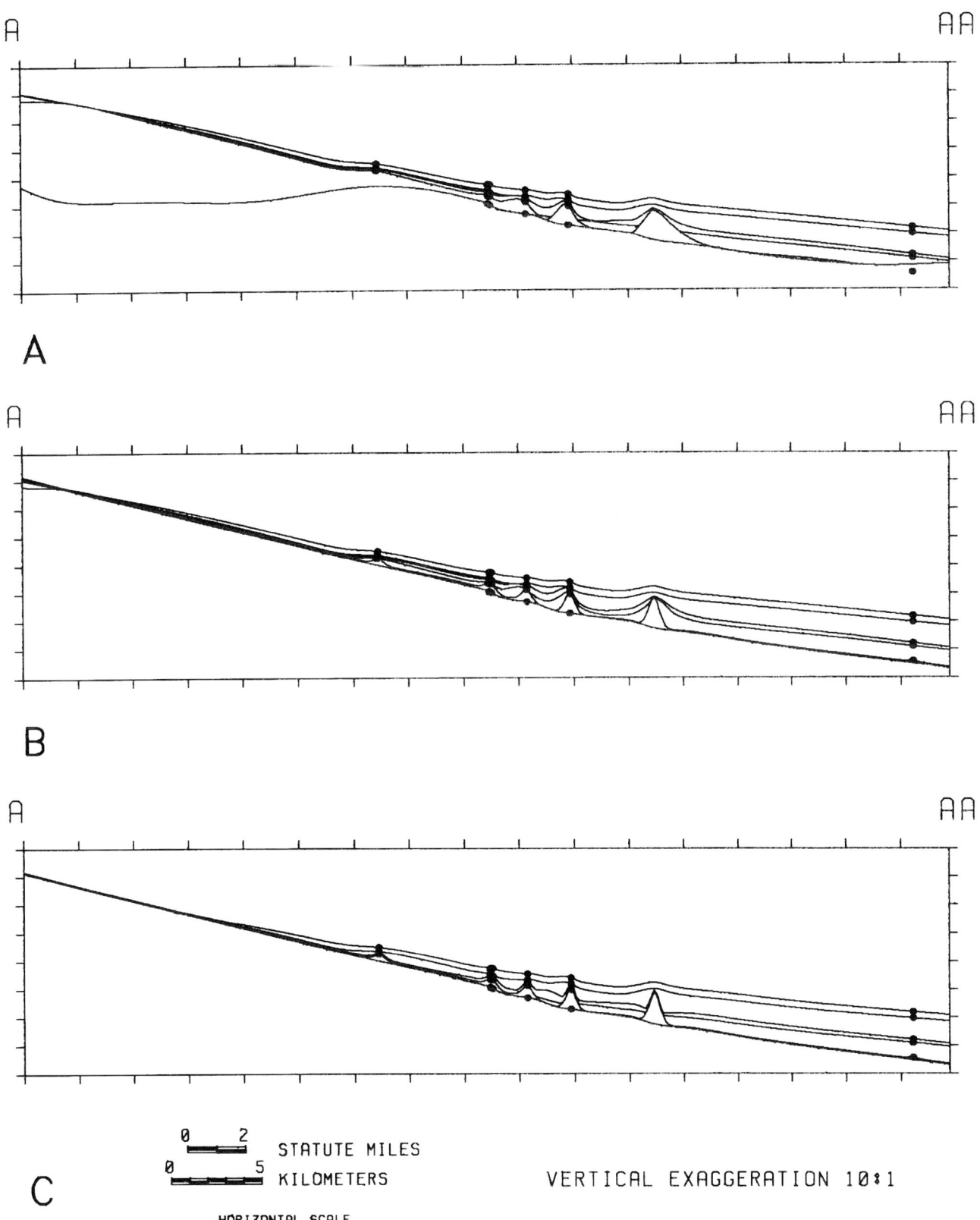

Figure 2. Cross sections showing (bottom-to-top) Niagaran Gray, Niagaran Brown, A-1 Evaporite, A-1 Carbonate, A-2 Evaporite, and A-2 Carbonate. Dots represent top picks. (A) All grids built using standard least-squares gridding. (B) The Niagaran Gray and Niagaran Brown built using shape-assist procedures and the A-1 and A-2 surfaces built using standard conformable techniques. (C) The same Niagaran Gray and Niagaran Brown surfaces as in Figure 2B with the A-1 and A-2 surfaces built using the shape-assist technique. See Figure 3 for location of this section.

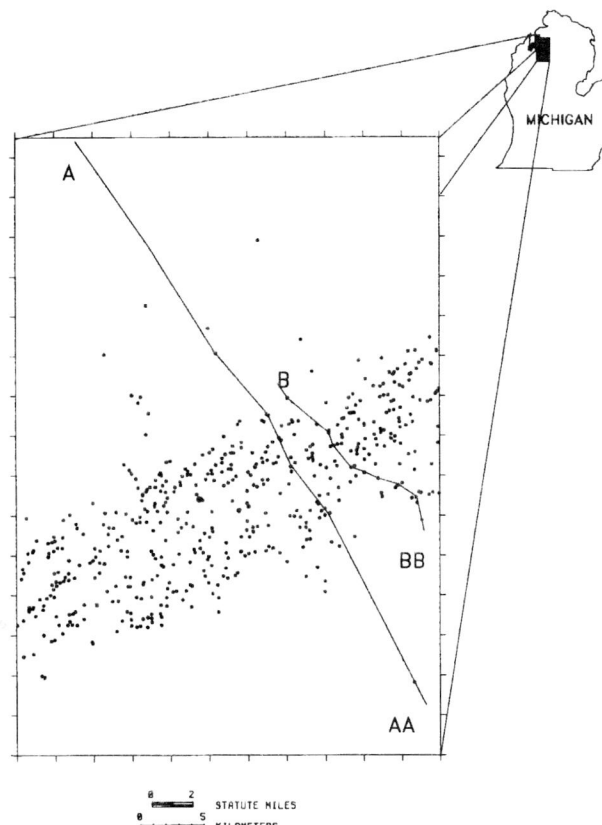

Figure 3. Base map showing location of study area and distribution of wells used in the study. Cross section lines A-AA and B-BB will be used throughout this paper.

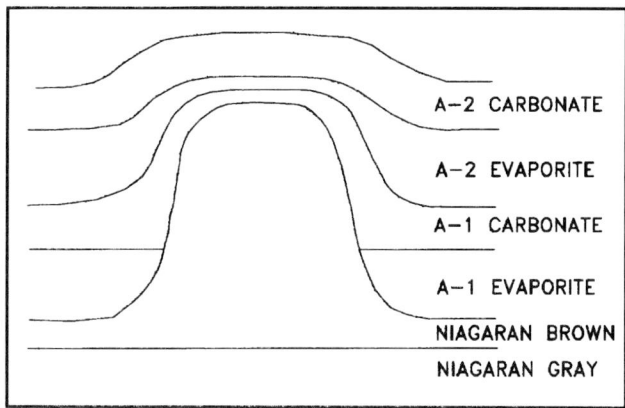

Figure 4. A generalized section showing relationships between rock units in the study area.

aborted reefs to about 800 ft (245 m), with heights increasing toward the basin center. Reef widths vary from 500 to 4000 ft (150 to 1220 m) and have roughly circular forms. Dips on reef flanks range from 30 to 60 degrees. In the area modeled, reefs occur in linear groupings oriented 15 to 20 degrees off trend, creating an en echelon pattern (Figure 1B).

Mesolella et al. (1974) summarized the general characteristics of the reef and interreef material. Typical reefs consist of crinoids and carbonate sands and muds near the base, branching finger and tabulate corals, algal zones, stromatolitic carbonate, and near the top a stromatoporoid zone. Between the reefs is Niagaran nonreef material consisting of dense, argillaceous, micritic carbonate. The four units immediately above the Niagaran are the A-1 Evaporite and Carbonate and the A-2 Evaporite and Carbonate (Figure 4).

Depositional Model

Middle Silurian reefs and associated units of the Michigan basin formed in a shallow sea positioned about 10 degrees south of the equator. The pinnacles grew upon the Niagaran Gray, a dark-gray carbonate mud (dolomite layer) which formed a gently undulating surface at the time of pinnacle growth. Several schools of thought have developed concerning the growth of the Niagaran pinnacles and deposition of the Salina evaporites and carbonates. Mesolella et al. (1974) summarized these into three models. The first of their models is used as our basis for generating grids and maps of these units.

The model assumes that the pinnacle reefs grew to their full height of several hundred feet during Niagaran time. Increased salinity, caused by evaporation in conjunction with restricted flow, terminated reef growth and initiated deposition of the A-1 Evaporite. A major unconformity is thought to separate the pinnacles from the evaporites and carbonates above (Gill, 1977). During evaporite deposition, major portions of the reefs were exposed and eroded to create a talus apron around the reef bases. Transgression and flooding created less saline environments and initiated deposition of the A-1 Carbonate. This process was repeated for A-2 Evaporite and Carbonate deposition.

The evaporite and carbonate unit geometries for this model are described by Gill (1975), Briggs and Briggs (1974), and Huh et al. (1977). Those studies have found the A-1 Evaporites to be in off-reef areas (Gill, 1977) and to wedge out against pinnacles and the basin edge. A-1 Evaporite thicknesses reach 475 ft (145 m) near the center of the basin and thin to zero at the basin edge and against the pinnacles. The A-1 Carbonate unconformably overlies the A-1 Evaporite. The carbonate may be up to 160 ft (48 m) thick along the edge of the carbonate platform, thins over the reefs, and gradually thins toward the basin center. Depositional environments of the overlying A-2 Evaporite and Carbonate are similar to those of the A-1 units. The A-2 units are always found capping the reefs.

Data

Data from Antrim and Kalkaska counties of northern Michigan were provided by Petroleum Informa-

tion, Incorporated. The data consisted of API number, latitude, longitude, total depth (TD), and tops for six surfaces:
- Salina A-2 Carbonate
- Salina A-2 Evaporite
- Salina A-1 Carbonate
- Salina A-1 Evaporite
- Niagaran Brown–Pinnacle reefs
- Niagaran Gray–base upon which reefs grew

Latitude and longitude were converted to state-plane coordinates (feet) for northern Michigan. Depths (feet) were converted to elevations (subsea values are negative). If a unit was missing, the top pick for that unit was given a *missing* value. Standard quality control procedures were applied to the data. Fewer than four percent of the data points were flagged as obvious errors; those were removed from the file. No attempt was made to clean the data other than to inform P.I. of the few problems that did occur.

MODELING THE NIAGARAN GRAY SURFACE

The Niagaran Gray is a gently undulating surface upon which the pinnacle reefs grew and which dipped gently into the basin at the time of reef growth. Since reef growth, some tilting has occurred, resulting in a present day-dip of 10 to 15 degrees to the southeast. Most of the data for this surface result from preferential drilling of the pinnacles and therefore lie in a northeast-southwest trending line through the center of the study area (Figure 3). Only a few outliers exist and few if any of these lie in the northwest or southeast areas.

Gridding Parameters and Problems with Surface Extrapolation

Modeling procedures generally varied from surface to surface. However, most of the gridding parameters were the same for the entire project. For the Niagaran Gray and the other surfaces the following parameters remained the same:
- X-Y limits
- Grid increment was 500 ft (152 m) (produced a 274 × 376 grid)
- Reach was large; the algorithm could look as far as needed to find data
- Eight sectors were used
- A maximum of four nearest data points could be collected in each sector
- Biharmonic filter was used

A parameter that often varied between surfaces in this study was number of refinements. A refinement is division of the grid increment by 2 and then interpolation of values for the new grid nodes from values in neighboring old grid nodes. After each refinement the filter is used to smooth and tie to data. Using two refinements means an initial grid increment of 2000 ft is used to create a grid with a final increment of 500 ft.

By starting with a coarse grid increment, regional trends of the data are built into the model. As the grid is refined the local form begins to dominate where data exists. Two or three refinements generally were used, depending upon which gave best results (i.e., larger and smoother closures, fewer or no surface features that are unsupported by data, smooth surface form across large open areas).

A least-squares algorithm with biharmonic filter was used to build a grid of the Niagaran Gray. In the area of data, the algorithm worked well; however, as the grid extrapolated into nondata areas the grid became erratic, reversing dip, flattening, and sometimes steepening (Figures 5 and 6A). This was expected since sectors were used to collect data and only four points were allowed per sector. Different data were collected from each sector for differing grid nodes. Thus, different data were used to calculate the least squares fit surface, causing the surface to have significantly different orientations. Projection of these variously oriented surfaces far from data resulted in highly erratic values for the calculated grid nodes.

Constructing the Grid

Extrapolation is the primary problem to be overcome when modeling the Niagaran Gray, so the first and simplest approach is to limit grid-node calculation to nodes near data (Figure 6A). Some gridding parameters can restrict extrapolation to a specified distance from the data. If moderate to large extrapolations are desired, then one of the following techniques could be used.

Using an Algorithm that Projects Slope

Building a grid of this surface that extrapolates properly requires an algorithm that projects the trend of the surface beyond data limits as far as contours are desired. Some algorithms use surrounding data points to calculate strike and dip of the surface at each data location. These strikes and dips are then used to project the surface away from the data. Algorithms of this type will often produce acceptable extrapolation for tilted surfaces (Figures 5 and 6B). However, subtle dip changes near the data edge will sometimes cause these algorithms to project those edge dips rather than project the general trend of the data.

Using a Trend and Residual Method

The trend and residual method is commonly used to build grids of tilted surfaces. We used this method for the Niagaran Gray and it involved five steps:
(1) Build a first-order trend grid through the original Niagaran Gray data, extrapolating to the edge of the map area or as far as extrapolations are desired.
(2) Interpolate values from the trend surface at each data location and put them in a new field (column of the data file) called INTERPOLATED.
(3) Subtract values in the INTERPOLATED field from the original values for the Niagaran Gray surface,

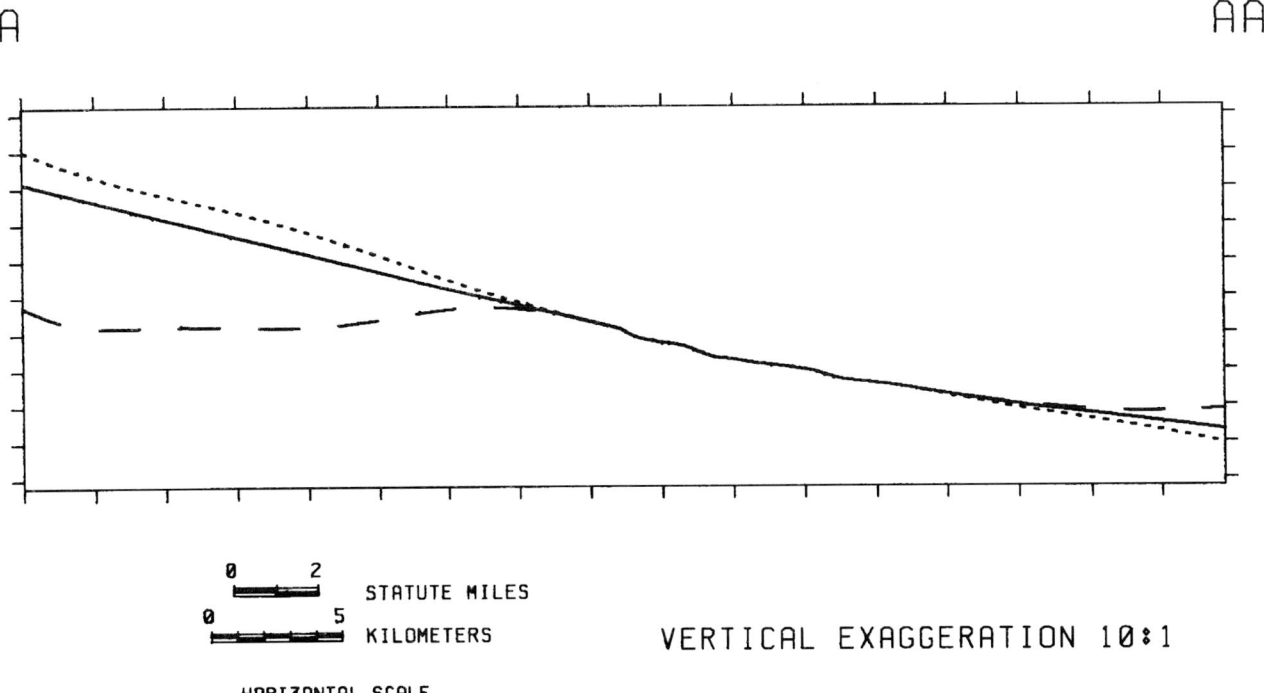

Figure 5. Cross section A-AA showing the Niagaran Gray surface. Dashes represent a grid built using a least-squares algorithm and a biharmonic filter. Dots represent a grid built using an algorithm that projects slope and a biharmonic filter. A solid line represents a grid built using a trend and residual method. See Figure 3 for location of this section.

creating a new field called RESIDUAL.
(4) Build a grid of the RESIDUAL field's values using a standard gridding algorithm. The residual data represented a horizontal surface undulating about a value of zero, a form which is handled well by a least-squares algorithm with biharmonic filter.
(5) Add the residual grid to the trend grid to create a grid of the Niagaran Gray surface (Figure 5).

Adding Interpretive Data

The final surface appeared acceptable and projected the trend of the data to the north. Unfortunately, looking at the difference between the Niagaran Gray surface values to the north (no Niagaran Gray data existed there) and the Niagaran Brown data, we found the Niagaran Brown to be several hundred feet higher. Since average thickness of the non-pinnacle Niagaran Brown was 20-30 ft (6–9 m), this would imply that each data point to the north represented a pinnacle. Only small patch reefs a few tens of feet high existed in this area. The dip of the Niagaran Gray surface must therefore steepen to the north but the grid did not because no data forced the steepening.

Since the projected slope algorithm steepens the Niagaran Gray to the north (Figure 5) we considered using that method, however, the projected surface was several hundred feet too high. To solve this problem, three interpretive (dummy) data points were added to the north with values 20 ft (6 m) lower than Niagaran Brown values in that area. These interpretive data, along with the original Niagaran Gray data, produced a grid (trend and residual method) that properly steepened to the north (Figure 6B).

MODELING THE NIAGARAN BROWN (PINNACLE) SURFACE

Problems Caused by Surface Form and Data

Simply modeling the pinnacle reef data using a least-squares algorithm with a biharmonic filter did not produce acceptable results (Figure 1A). This and other commonly used algorithms are designed to handle surfaces that are horizontal and smoothly varying (gently undulating), but the pinnacle surface does not fit this. For example, if data exist for two adjacent pinnacles and does not exist for the interreef surface, the algorithm will smoothly connect the two pinnacles (Figure 7). This tendency to connect data also broadens pinnacles at their bases by connecting the tops of pinnacles to distant points on the interreef surface. The more a surface deviates from horizontal and gently undulating, the less effective these algorithms are in modeling it.

Another feature of many algorithms is the ability to project above highs or below lows. In most geologic

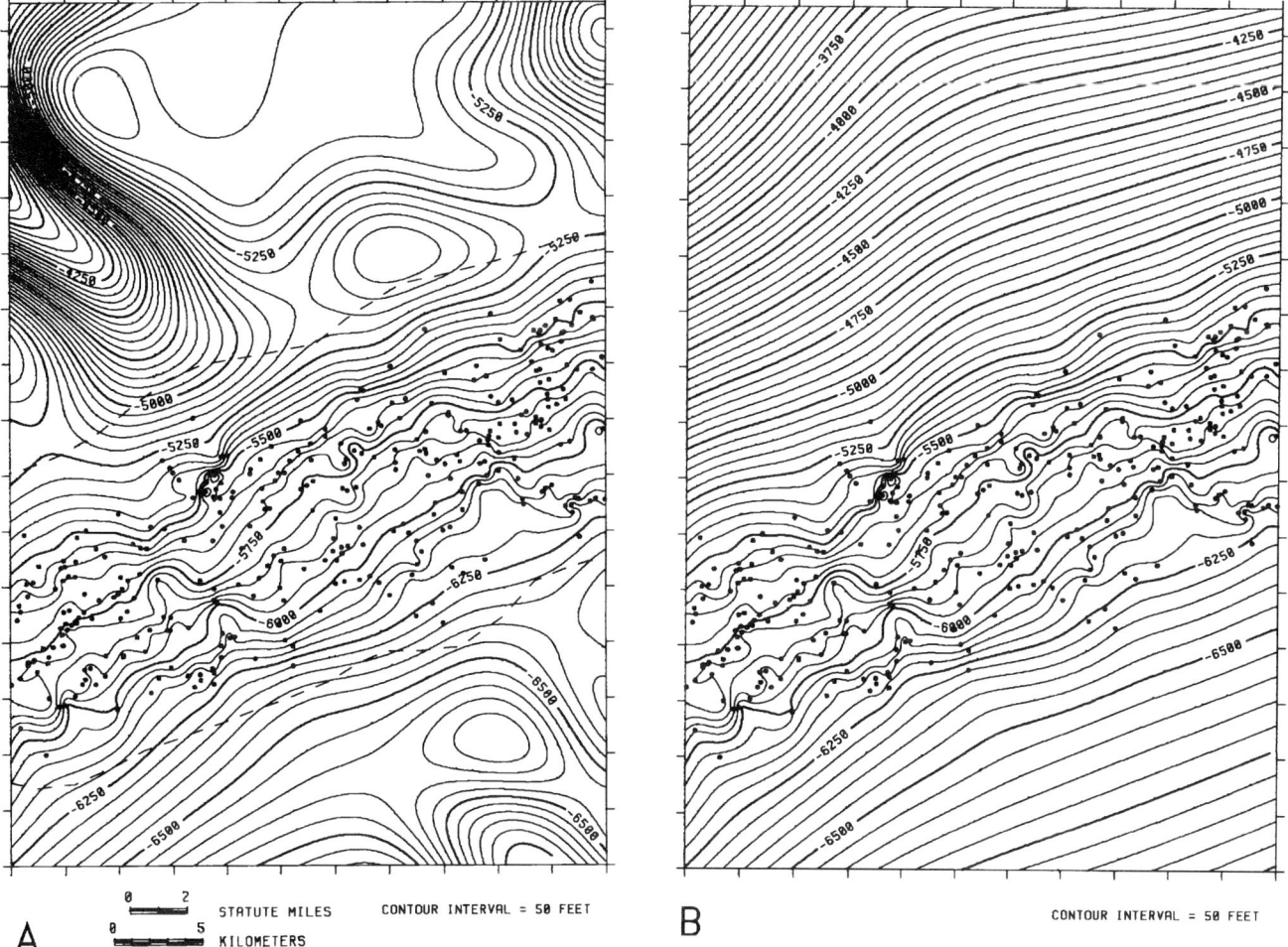

Figure 6. Maps showing contours of the Niagaran Gray surface. Dots represent data for this surface. (A) Grid built using a least-squares algorithm with a biharmonic filter and allowed to project to the edges of the study area. The dashed line defines the area where contours are reasonable and beyond which contours from the least-squares algorithm should not be displayed. (B) Grid built using a trend and residual method and interpretive data on the north edge of the map. The three interpretive data points are evenly distributed just north of the map edge.

settings, a well is unlikely to have penetrated the exact top or base of a structure and the algorithms should allow projection if the data support it. This overprojection, however, causes steep slopes created by points on the top and side of a pinnacle to project below the interreef surface (Figure 7). The result is a series of holes in the interreef surface beside many pinnacles.

Constructing the Grid

Suppose we consider the pinnacle surface to be composed of two forms: pinnacles and interreef. The interreef surface could be modeled easily by an algorithm if data for the pinnacles were removed. This is because the interreef surface has characteristics similar to the surface upon which the pinnacles grew: a gently undulating surface dipping basinward. An algorithm would require few points to model that simple surface.

We therefore used a three-step modeling procedure. First, separate the pinnacle data from the interreef surface data. Second, build an interreef surface grid using the separated interreef data, allowing it to project under the pinnacles. Third, superimpose the pinnacles on that surface.

Separating Pinnacle and Interreef Surface Data

Splitting the data so that there are two files, one for the pinnacles and one for the interreef surface, can be done in several ways. If the data base contains codes which define the character of the surface at each data point (i.e., present, not penetrated, lost core, reef, interreef, etc.), then the codes could be used to separate the data. If these codes are not available, the data could be separated by manually editing the file, a tedious and time-consuming task. More desirable is an automated approach that quickly separates data.

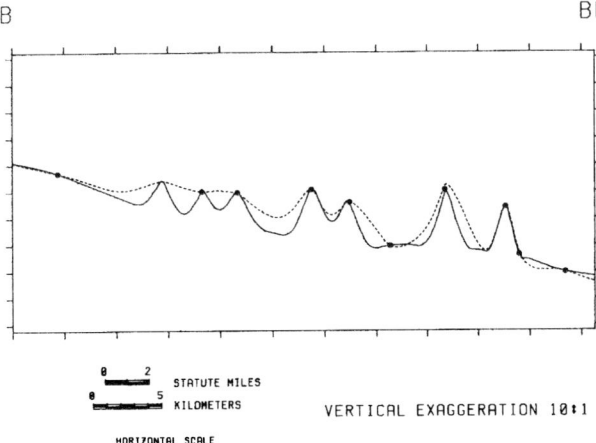

Figure 7. Cross section B-BB showing top of Niagaran Brown (Pinnacle) grids built using a least squares algorithm with a biharmonic filter (dashed line) and using the multi-step approach described in this paper (solid line). The least squares approach tends to connect pinnacles, broaden their bases, and project above highs and lows. Dots represent data for this surface. See Figure 3 for location of this section.

This can be done by building a grid that passes through the middle of the data (i.e., below the pinnacle tops and above the interreef surfaces). Those data points above the grid are then put in one file and those below are put in another.

To build a grid that passes through the middle of the data, we considered building a polynomial trend grid of second or third order. This would work well if the data were consistent and predictable; however, this technique does not guarantee that the tops of pinnacles stay above the trend and the interreef data below it. We also considered using a moving weighted-average algorithm and a coarse grid increment. This method averages several neighboring data points together, pushing the surface through their middle. The advantage of this approach is that the surface follows the data, regardless of undulations of the surface.

If the pinnacle surface were the only surface being modeled, the weighted-average approach would be acceptable. However, the Niagaran Gray surface has already been built and provides the best means for generating a grid through the middle of the Niagaran Brown data. The thickness of Niagaran Brown interreef is 20–30 ft (6–9 m). Therefore, 100 ft (30 m) was added to the Niagaran Gray grid, shifting it well above the Niagaran Brown interreef data, and creating the middle of data grid (Figure 8).

Once the grid through the middle of data is built, we must separate data lying above and below it. The following steps can be used to do this:
(1) Interpolate Z-values from the grid at each data location to create a new field (INTERPOLATED).
(2) Subtract the INTERPOLATED field from the original PINNACLE field, creating a new field (DIFFERENCE).
(3) Set to missing, data values in the DIFFERENCE field that are negative, creating a new field (ABOVE).
(4) Set to missing, data values in the DIFFERENCE field that are greater than or equal to zero, creating a new field (BELOW).
(5) Compare the PINNACLE field (original data) with the ABOVE field and replace values of the ABOVE field with PINNACLE field values. Leave missing those values in the ABOVE field that are missing.
(6) Repeat the operation performed in step 5 for the BELOW field.

The two fields, ABOVE and BELOW, now contain tops for the pinnacles and interreef surfaces, respectively.

Building the Interreef Surface Grid

Data in the BELOW field could be used to construct a grid for the interreef surface. This grid would project underneath the pinnacles and reasonably represent the interreef surface.

However, the interreef grid would have problems similar to the Niagaran Gray grid in areas of extrapolation. That is, because the surface is tilted, extrapolations would fold the surface back on itself. Since the interreef surface was essentially conformable to the Niagaran Gray, the Niagaran Gray grid was used to assist in its construction by applying the conformable isopach method described by Jones et al. (1986). Five steps are used to do this:
(1) Interpolate values from the Niagaran Gray grid and replace missing values in the Niagaran Gray's original data field with these.
(2) Subtract the modified Niagaran Gray field from the BELOW field to create a new field (THICKNESS).
(3) Build a grid of the THICKNESS field allowing it to project to negative values.
(4) Clip (convert values beyond a threshold to that threshold) the thickness grid to a minimum value of zero.
(5) Add the clipped thickness grid to the Niagaran Gray grid, creating the interreef surface grid (Figure 8).

The resulting grid not only extrapolates better but more closely follows the undulations of the Niagaran Gray in the area of the pinnacles.

Growing the Pinnacles

Growing pinnacles on the interreef surface grid could be done in several ways; we used a filtering algorithm to pull the interreef surface grid up to the top-of-pinnacle data. The interreef surface grid, and all data for the Pinnacle surface, were used to filter the grid while honoring data. This causes grid nodes near top-of-pinnacle data to be pulled up to those data, nodes near base-of-pinnacle data to be held in their present position, and nodes distant from any data to be smoothed slightly or left unchanged.

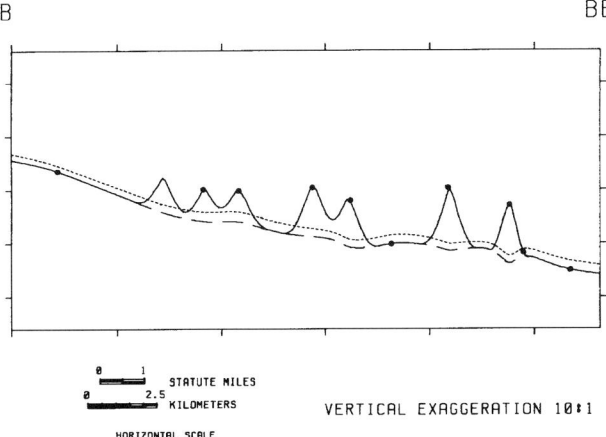

Figure 8. Cross section B-BB showing the grids used in the multi-step method for building the Niagaran Brown (Pinnacle) grid. Dotted line represents the middle-of-data grid (50 ft [15 m] above Niagaran Gray). Dashed line represents the interreef surface grid. Solid line represents the top of Niagaran Brown grid. Dots represent data for this surface. See Figure 3 for location of this section.

Selecting which filter to use depends upon the pinnacle shape desired. The program used had two filters available: biharmonic and laplacian. Biharmonic filtering was selected because it tended to give more rounded tops to the pinnacles. Laplacian filtering gave a more pointed volcano-like top and did not match our concept of a pinnacle reef.

The biharmonic filter could be executed several times with the same data and grid. The effects of the filter extend further with more executions, making pinnacle sideslopes gentler and pinnacle bases broader. Deciding when to stop filtering was a judgment based upon our interpretation of pinnacle form. We found that most of the pinnacles fit our interpretation after five passes of the filter (Figures 1B, 7, and 8).

This filtering method produced small pinnacles which were little larger than the area encompassed by the data; it is a conservative approach but adequate to define pinnacle positions and general form. If a more optimistic shape is desirable (eg. for regional estimation of hydrocarbons in place) an alternative approach to growing pinnacles involves five steps:

(1) Build a pinnacle-top grid using the ABOVE field in the data file and a method similar to that used to build the interreef surface grid.
(2) Build a grid whose node values represent distance from that node to the closest data point. Do this by using the same data used to build the pinnacle-top grid.
(3) Blank (set to missing) grid nodes in the pinnacle-top grid that are greater than a specified distance from data points in the ABOVE field. Use the distance-to-closest point grid to do this. This distance should be specified as the radius (or slightly smaller) of the average pinnacle.
(4) Combine the interreef surface grid with the blanked pinnacle-top grid, replacing all missing values in the pinnacle-top grid with values from the interreef grid. The resulting grid will have sharp "cliffs" at the edges of the pinnacles.
(5) Apply a filtering algorithm to the combined grid and to all Niagaran Brown data. This will smooth rough edges on the pinnacles, properly honor the side of pinnacle data, and hold the surface to the top- and base-pinnacle data. Again, the amount of filtering will depend upon the interpreted shape of the pinnacles.

This alternative method can be made fully automatic and thus allow testing of various reef radii, number of filter passes, and positioning of the middle-of-data grid. The problem with the method is that the reef-top surface always reaches a specified distance from wells that penetrate the reef top and is often strangely shaped due to several penetrations in the same pinnacle. The filter will remove some of this effect but the final result will almost always be an overly wide pinnacle.

If it is desirable to incorporate an interpreted shape for the top of each pinnacle, then the outline of each pinnacle top can be digitized as a polygon. The original pinnacle-top grid is then set to missing values outside the pinnacle boundary polygons. From that point, all steps described for the alternative method are the same.

MODELING THE EVAPORITE AND CARBONATE SURFACES

Problems with Standard Gridding Methods

Direct gridding of each surface above the pinnacles did not create acceptable results (Figure 2A). Data density varied between surfaces, allowing considerable extrapolation, so grids from one surface often projected through another. Although these were clipped when they encountered the next surface, in some cases resulting maps had a nongeologic look. For example, when gentle folding raised or lowered a pinnacle and no evaporite picks were found nearby (no wells drilled between the pinnacles), the evaporite surface would tend to project through to the next evaporite pick and did not demonstrate the same folding seen in the pinnacle surface.

The evaporates and carbonates above the pinnacle surface can be thought of as being deposited during the same general period of geologic time. Because of this they might be treated conformably for computer modeling and the conformable-isopach technique used to build their grids (Jones et al., 1986). To do this, the surface that would build the best grid is selected as control and a grid is built directly from its data. Adjacent surfaces are built by calculating thickness between them and the control surface, gridding thickness, clipping thickness to a minimum of zero, and adding or subtracting the thickness grid to or from the control surface grid. These new grids are then treated as control grids. Adjacent surfaces are built in the same manner until all surfaces in the sequence are built (Figure 2B).

The grids built using this technique, although better than those built by independent gridding, were still unacceptable. For example, when a carbonate that draped over deeper units was used as control to build an adjacent evaporite surface, the evaporite took on the draping characteristics of the carbonate. Conversely, when an evaporite was used as control to build a draping carbonate, the carbonate tended to terminate abruptly against lower surfaces rather than drape over them.

Constructing the Grids

Evaporites

The geologic interpretation for the evaporites is that they filled areas between or above pinnacles in a manner that wedged out against the pinnacles or thinned over them. The evaporites thinned to zero at the basin edge and thickened significantly basinward. It is likely that their thickness was significantly less than paleo-water thickness toward the basin center but approached that thickness near pinnacles and the basin edge (Nurmi and Friedman, 1977). This implies that the top of evaporite was roughly parallel to paleo-water level in the vicinity of the pinnacles.

The shape-assist technique (Fierstien and Brewster, 1992) uses a control surface and thickness to transfer shape from the control surface to the surface being built (regardless of geologic conformability), so it would be desirable to use paleo-water level as the control surface for building the evaporite surface. Unfortunately, paleo-water level is not available and an acceptable substitute must be found with essentially the same shape as water level (planar, although a tilt or other broad perturbation would not be a problem). In addition, the control surface must have experienced deformational history similar to the evaporite. Two surfaces presently available in grid form are the Pinnacle and Niagaran Gray. The Pinnacle surface is not flat and is related to paleo-water level only in that many pinnacles rose to near a paleo-water level. If no other surface existed, then gridding the crests of pinnacle data as control might be reasonable to try. The Niagaran Gray surface, the shelf upon which the pinnacles grew, dipped gently into the basin. Thickness of water above this surface gradually thickened basinward. Any deformation of the evaporite would also have been experienced by the Niagaran Gray, so the Niagaran Gray is the logical choice to assist in defining the evaporite shape.

The following four steps were used to build the evaporite surfaces:
(1) Interpolate values from the Niagaran Gray grid, replacing missing values in the Niagaran Gray's original data field with these.
(2) Subtract the modified Niagaran Gray field from the field for the A-1 Evaporite, creating a new field (THICKNESS).
(3) Build a grid of the THICKNESS field, allowing it to project to negative values.
(4) Add the thickness grid to the Niagaran Gray grid to create the A-1 Evaporite grid. Because the thickness grid contains negative values, the A-1 Evaporite grid could cross grids below it; this was corrected in the final modeling step (see Incorporating Baselap below).

The same process is used to build the A-2 Evaporite grid. Either the Nigaran Gray or the newly created, unbaselapped A-1 Evaporite could be used as control for building the next higher evaporite. We used the Niagaran Gray as control.

The A-1 Evaporite pinches out against the basin edge to the north. No data exist for this unit in the north, so its shape is controlled by the Niagaran Gray grid and the evaporite's thickness at data locations near the basin edge. The basin-edge thickness will be projected to the north, causing the resulting evaporite grid to parallel the Niagaran Gray when it should pinch out. To prevent this, evaporite values were added to the three interpretative data locations previously specified. The interpretive values were below the interpretive Niagaran Gray values at those locations. The distance below was determined by visually continuing the trend of the evaporite from the desired line of pinchout to a position below the Niagaran Gray. The resulting grid contained an acceptable pinchout (Figure 9A). A similar technique was used for the A-2 Evaporite.

Carbonates

The A-1 Carbonate unit unconformably overlies the A-1 Evaporite and the pinnacles. Discussions with geologists who have worked the area indicated that several hundred feet of relief exist on the top of the carbonate. High points on top of reefs and lows in the interreef areas are common in the data. This amount of relief, even with significant thinning of the carbonate over the reefs, supports the interpretation that significant compaction has occurred since carbonate deposition (Jodry, 1969). Techniques for incorporating compaction into grids are discussed by Hamilton and Riehl (1992). Treating the carbonates as if they were sheets of material draped over the basin floor allows approximation of their compacted form while avoiding the complex process of incorporating estimated compaction. Although the A-2 Carbonate has significantly lower relief, it is interpreted to be from a similar depositional environment and we use the same modeling techniques.

Applying the same philosophy as for the evaporites, we must find a surface that has similar shape and deformational history as the carbonates. Four surface grids are available to choose from: Niagaran Gray, Pinnacle, A-1 Evaporite, and A-2 Evaporite. None of these has a shape that mimics that of the draping carbonates. The surface that most mimics the carbonate

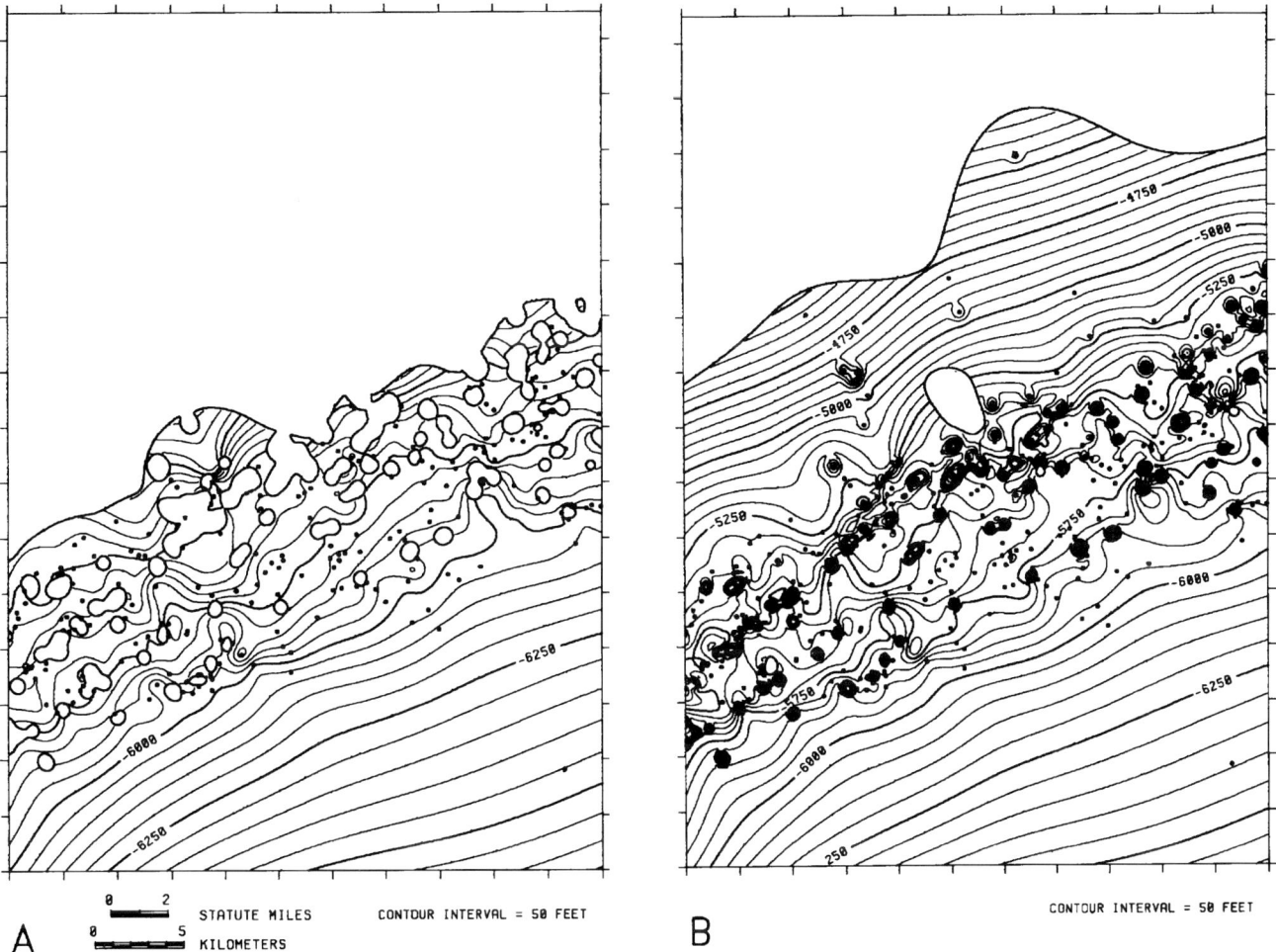

Figure 9. Maps showing contours and subcrop lines. Dots represent data for these surfaces. (A) The A-1 Evaporite. (B) The A-1 Carbonate.

would be the surface upon which it was deposited, i.e., a combination of two surfaces: the top of pinnacles, and top of evaporite deposited between the pinnacles. Therefore, our control surface for building the carbonate grids must be a combination of these two surfaces.

The following five steps were used to build the carbonate surfaces.
(1) Compare the Pinnacle grid and the A-1 Evaporite grid, creating a new grid that is the maximum of the two.
(2) Interpolate values from the combined Pinnacle-Evaporite grid and place them in a new field (BASE CARBONATE).
(3) Subtract the BASE CARBONATE field from the A-1 Carbonate field, creating a new field (THICKNESS).
(4) Build a grid of the THICKNESS field, allowing it to project to negative values.
(5) Add the thickness grid to the combined Pinnacle-Evaporite grid built in step 1, creating the draping A-1 Carbonate grid. Because the thickness grid contains negatives, the A-1 Carbonate grid could cross grids below it; this will be corrected in the final modeling step (see Incorporating Baselap below and Figure 9B).

The same steps are repeated for the A-2 Carbonate using the maximum of the Pinnacle, A-2 Evaporite, and A-1 Carbonate grids as the control grid. The resulting grids were acceptable representations of the geologic interpretation (Figure 2C).

A sharp angle often occurs at the intersection of the Evaporite and Pinnacle (i.e., where the slope of the combined grid makes a sudden bend). This sharp angle is transferred to the carbonate surface through the addition of smooth thickness to the angular control grid. Because of the 10:1 vertical exaggeration in the cross sections (Figure 2C), the problem is exaggerated but was considered minor and acceptable for the goals of the project. However, the problem may be reduced by applying a filter to the control grid before using it. During filtering, the interpolated data of step 2 (BASE CARBONATE field) should be used to force the surface to honor control points.

Incorporating Baselap

After constructing all grids for this project, they needed to be compared to ensure they did not cross. These crossings were expected since each grid was constructed independently. Because this project had no major unconformities, each surface could be thought of as lapping onto the surface below it. The techniques for incorporating baselap into a suite of grids for generation of either cross section or map display are described by Jones et al. (1986).

For cross section display, the baselapping process started at the bottom and compared the next higher grid to the grid below it to create a new higher grid that was the maximum of the two. Figure 10 is a cross section showing grids before and after these operations were performed.

For map display, the baselapping process again started at the bottom, but this time values of the higher surface were set to missing where they were below the lower surface. To create subcrop lines for the map display, an additional grid was constructed. This grid is commonly referred to as plus-minus grid since it contains positives on one side of the subcrop line and negatives on the other. It was created by subtracting the lower surface grid from the unbaselapped higher grid. The zero contour from this difference grid was then drawn on the contour map for the higher surface (Figure 9). The plus-minus grid could also be used for isochore display (Figure 11) and volumetrics. The one consideration is that only zero and positive contours are displayed, and only positive volumes reported.

QUALITY CONTROL

Evaluating Error

The final grids should honor the data used to construct them. Some mapping programs have procedures which automatically report statistics about error between data and a surface. If the procedure is not available, it can be simulated by: (1) Interpolating on the grid, creating a new field, (2) Subtracting that field from the original field for that surface, creating a difference field, and (3) Generating statistics for the difference field. Errors for smoothly varying surfaces with evenly distributed data and an appropriate grid increment will typically be less than 0.01% percent of the contour interval. Data that are highly variable both in value and areal distribution will have similar accuracy for most of the data but may have a few points that are in error by as much as 5% of the contour interval (1/5 to 1/2 a percent of the total data range). When more than one data point lies in the same grid cell, most programs average them in some way. If they are on opposite sides of the grid cell, they are usually honored. However, when they are close relative to cell size, they will be averaged and the error at either point will be about half the difference in value of the two points.

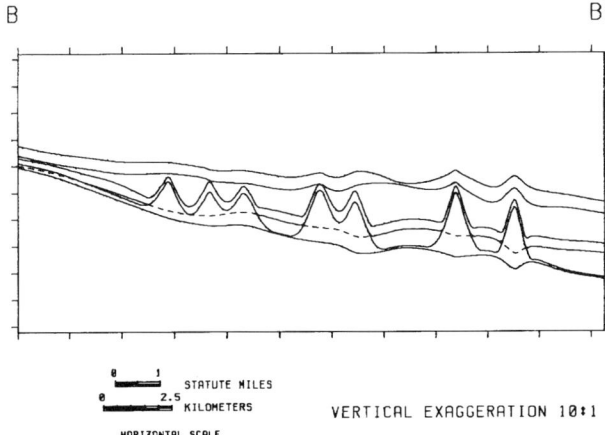

Figure 10. Cross section B-BB showing (bottom-to-top) Niagaran Gray, Niagaran Brown, A-1 Evaporite, A-1 Carbonate, A-2 Evaporite, and A-2 Carbonate. The surfaces before baselap operations are dashed and the surfaces after baselap operations are solid. See Figure 3 for location of this section.

For all surfaces except the Pinnacles the largest error was less than 2 ft (0.6 m), with standard deviations near 0.2 ft (0.6 cm). The Pinnacle surface had errors at two points of 30 ft (9 m) and –40 ft (12 m). These two points were very close together on the side of a pinnacle. Other points on the Pinnacle surface had small error ranges similar to the other surfaces. All errors were considered acceptable for the goal of the project and well within the likely error of data picks and positions.

Adjusting for Limits Data

The same error detection procedure can be used to determine if a surface projects above or below picks for higher or lower surfaces. When this was done, the Pinnacle and A-1 Carbonate surfaces were found to violate picks above them by several tens of feet in 6 wells. Subcrop maps near the violated surfaces show that grids for the violated surface data pinch out against the Pinnacle or A-1 Carbonate grids even though data exist for the violated surface in the pinchout area. This is because the surface beneath had projected higher than it should have. These violations happened when there was no pick in the violating surface but there was a pick for the higher surface. Some of these were because of high undulations in the lower surface which caused projection of trends away from data. Others were because the lower surface was built connecting two picks while a pick from the higher surface existed between them that was lower in elevation. Since the gridding algorithms we used did not use data for surfaces above and below to restrict where the grid went, a multi-step process we called the limits procedure was used to do this (Jones et al., 1986).

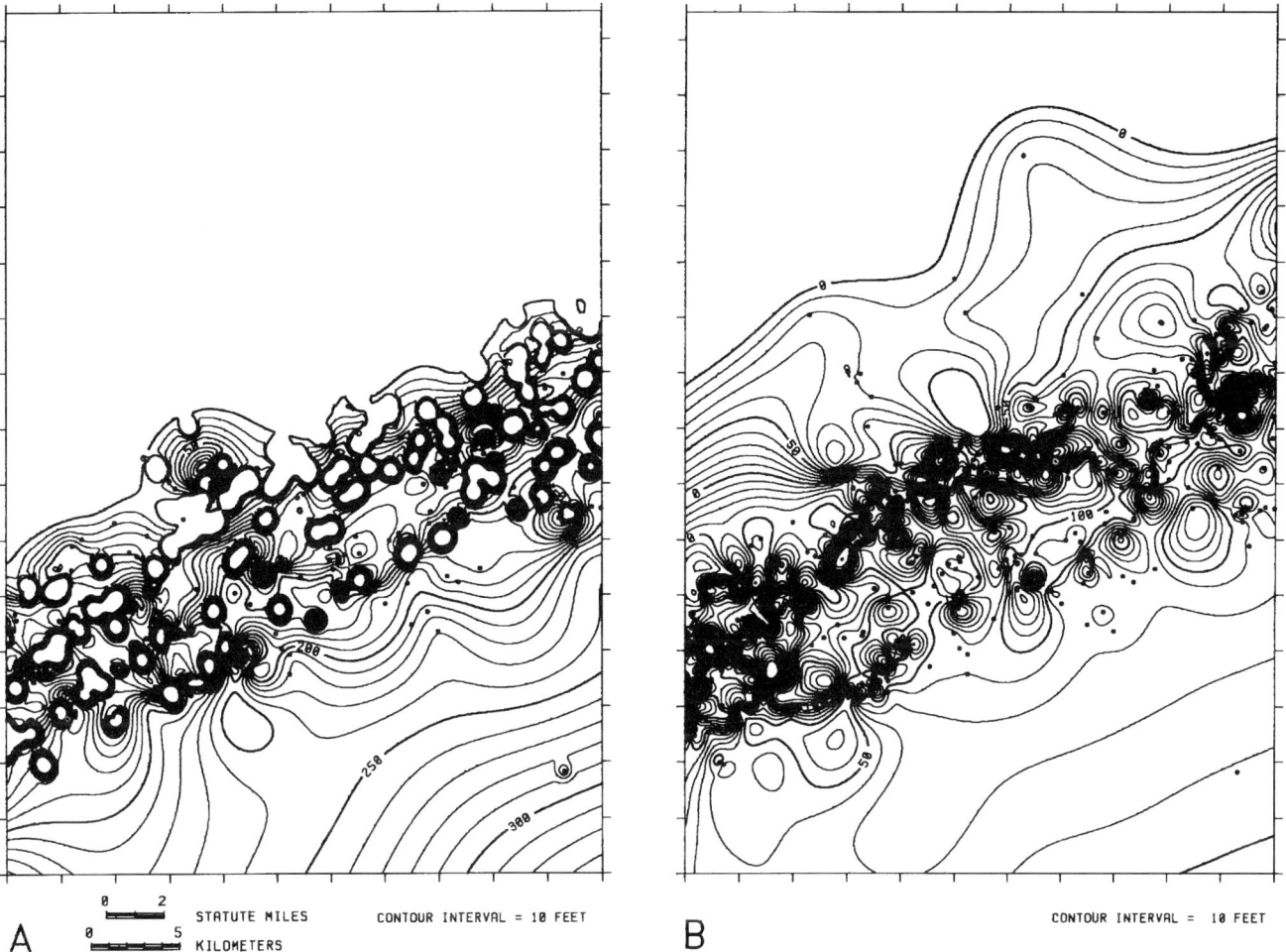

Figure 11. Map showing contours of thickness. Dots represent data for these units. (A) The A-1 Evaporite. (B) The A-1 Carbonate.

This limits procedure was applied after the initial grid was built. That grid was tested against the lowest pick of all picks above that surface in every well. When a violation was found, 1 foot (0.3 m) was subtracted from the violated pick's value and that new value placed in the field of the surface being gridded. The grid was then rebuilt using the original data and these new values. Tests on the new grids showed that they now honored all data for picks above and below them.

SUMMARY

We have generated regional models of the pinnacle reefs and their associated surfaces. To do this we employed a variety of mapping tools that allowed significant improvements to be made in the results with a minimum of effort. Those tools included:

(1) Remove a trend to simplify surface form so standard gridding algorithms are more effective.
(2) Split data into two subsets based on whether values lie above or below a grid.
(3) Use filters and data to modify a grid by rounding edges, expanding feature size, lifting or pushing a surface to data, and correcting small data errors.
(4) Use the shape of one surface to assist in constructing another. This technique traditionally has been applied to conformable surfaces, but it need not be limited to adjacent surfaces and can be used for surfaces whose Z-values are measured in different units. For example, thickness might be used to assist in defining the shape of a porosity grid although this was not done in this study.
(5) Use data above and below the surface being modeled to act as limits above and below which the grid being modeled cannot project.
(6) Use interpretive data (dummy points) to control the shape of surfaces in areas where no data are available.

Have we created the best possible model? Regardless of the project, the answer is almost always NO. The real question is: Has the modeling approach achieved the goal of the project? To repeat what was stated in the introduction, our goal was to create a set

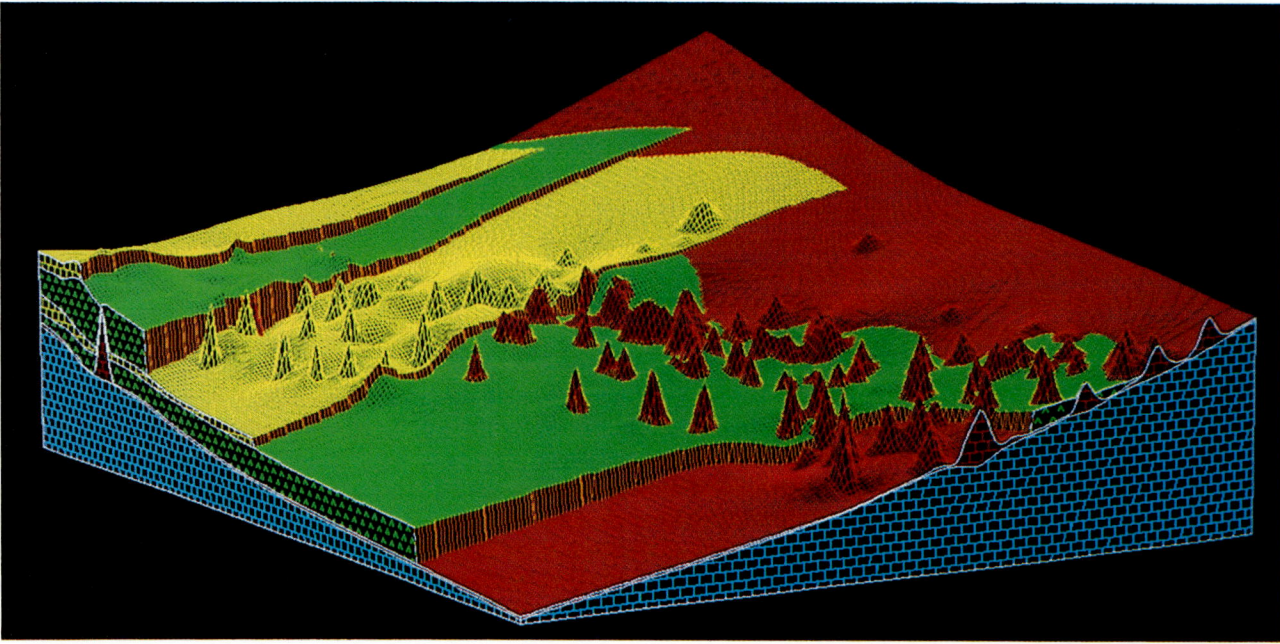

Figure 12. Three-dimensional block diagram of the modeled units with the upper units peeled back to show lower units.

of surface models that provide a regional look at pinnacle reefs and their associated surfaces. The model had to be acceptable and informative to geologists, but without working the problem to such an extent that the effort could not be justified by the results. We feel we have accomplished this goal (Figure 12). We did not use a grid increment fine enough to honor every data point to a precision of +/−0.01 ft (0.3 cm), although this could easily have been done. We did use methods and gridding parameters that preserved the relevant information content of each data point and effectively blended the geologic interpretation with those data.

ACKNOWLEDGMENTS

We would like to thank Petroleum Information Incorporated, Denver, CO., for allowing us to use their Northern Trend, Michigan basin data. We would also like to thank Landmark/Zycor, Austin, Texas, for providing the programs and computers used to do this study and for giving us the time to complete the study.

REFERENCES CITED

Briggs, L. I., and D. Z. Briggs, 1974, Niagara Salina relationships in the Michigan basin, in R.V. Kesling, ed., Silurian reef-evaporite relationships: Michigan Basin Geological Society Field Conference, p. 1–23.

Fierstien, J. F., and A. V. Brewster, 1992, The shape-assist technique: Incorporating stream channel interpretations into computer-generated surface models: Clark County, Kansas, (this volume).

Gill, D., 1975, Cyclic deposition of Silurian carbonates and evaporites in Michigan Basin, discussion: AAPG Bulletin, v. 59, p. 535–538.

Gill, D., 1977, The Belle River Mills gas field: productive Niagaran reefs encased by sabkha deposits: Michigan basin, Michigan Basin Geological Society Special Papers 2, 187 p.

Hamilton, D. E., and W. G. Riehl, 1992, Techniques for modeling compaction applied to the pinnacle reef/carbonate/evaporate environment, (this volume).

Huh, J. M., L. I. Briggs, and D. Gill, 1977, Depositional environments of pinnacle reefs, Niagara and Salina Groups northern shelf, Michigan basin, in J. H. Fisher, ed., Reefs and evaporites—concepts and depositional models, AAPG Studies in Geology 5: Tulsa, AAPG, p. 1–21.

Jodry, R. L., 1969, Growth and dolomitization of Silurian reefs, St. Clair County, Michigan: AAPG Bulletin v. 53, p. 957-981.

Jones, T. A., D. E. Hamilton, and C. R. Johnson, 1986, Contouring geologic surfaces with the computer. New York, Van Nostrand Reinhold, 314 p.

Mesolella, K. J., K. D. Robinson, L. M. McCormick, and A. R. Ormiston, 1974, Cyclic deposition of Silurian carbonates and evaporites in Michigan basin: AAPG Bulletin, v. 58, p. 34–62.

Nurmi, R. D., and G. M. Friedman, 1977, Sedimentology and depositional environments of basin-center evaporites, Lower Salina Group (Upper Silurian), Michigan basin, in J. H. Fisher, ed., Reefs and evaporites—concepts and depositional models: AAPG Studies in Geology 5: Tulsa, AAPG, p. 23–52.

Chapter 6

Techniques for Modeling Compaction Applied to the Pinnacle Reef/Carbonate/Evaporite Environment

David E. Hamilton
Landmark/Zycor Inc.
Austin, Texas, U.S.A.

William G. Riehl
Landmark/Zycor Inc.
Austin, Texas, U.S.A.

ABSTRACT

Evaporites and carbonates deposited above pinnacle reefs in the northern trend of the Michigan basin demonstrate surface geometries commonly associated with compaction. This is due primarily to postdepositional sediment loading and dewatering of the carbonates. Geometries seen in densely drilled areas include: (1) dips away from reef flanks into thicker, more compacted areas, (2) structural highs over reefs where thinner portions of a unit have compacted less than thicker portions, and (3) mimicking of the surface upon which a compacted unit was deposited by the structural top of that compacted unit.

Computer-generated surface models in areas of sparse data do not demonstrate these geometries when standard modeling techniques are used, but compaction features can be incorporated into computer-generated surface models. The technique for doing this is a series of steps which apply an estimate of percent compaction to a unit. The steps involve restoring the compacted unit tops to their precompacted positions, building the surface models, and then compacting those models to create the present-day structures. The procedure also allows for propagating the compaction to intervals above the compacted unit (i.e. distributing the compaction through time).

The compaction procedure described here was applied to carbonates above and adjacent to pinnacle reefs in the Michigan basin. The compaction effects were also incorporated into surface models of evaporites immediately above the compacted units. The procedure improved surface models in dense-data areas and produced reasonable compacted surface forms in sparse-data areas. The geometries in areas of sparse data were similar to those in dense-data areas.

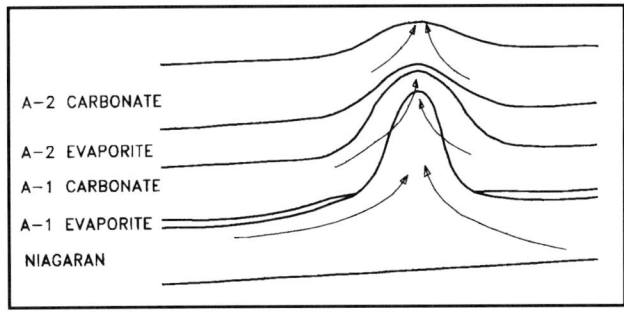

Figure 1. Idealized cross section showing compacted geometries (off-reef dips, and lows between and highs above reefs) of surfaces above the reefs. Arrows show theorized flow of fluids during compaction (After Jodry, 1969).

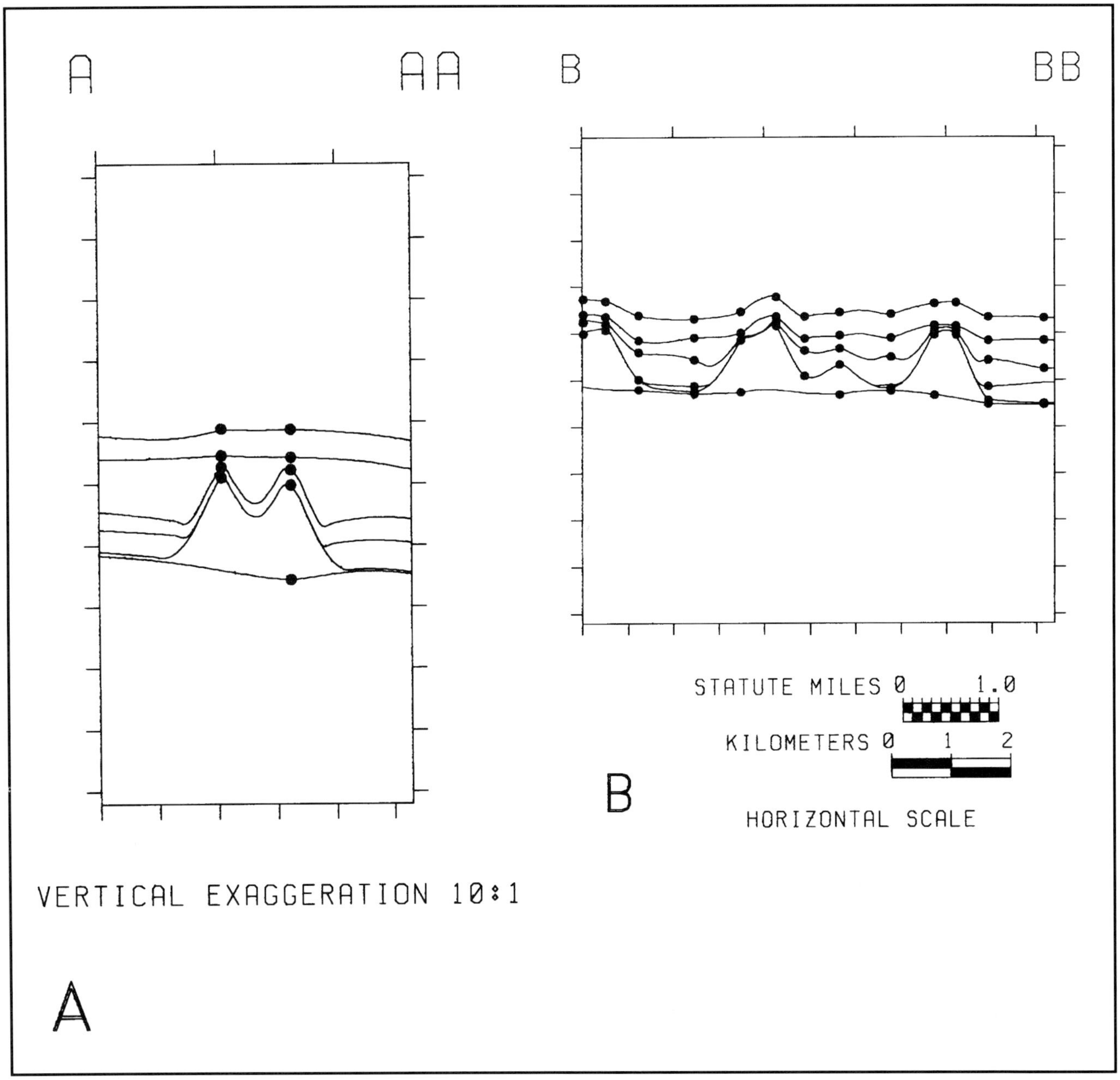

Figure 2. Cross sections showing surface models constructed without incorporating compaction. (A) In areas of sparse data, direct gridding (without compaction) of tops for the A-1 Carbonate and A-2 Evaporite and Carbonate produces grids that do not show off-reef dips, and lows between and highs above reefs. (B) In areas of dense data, the grids show these compaction related geometries.

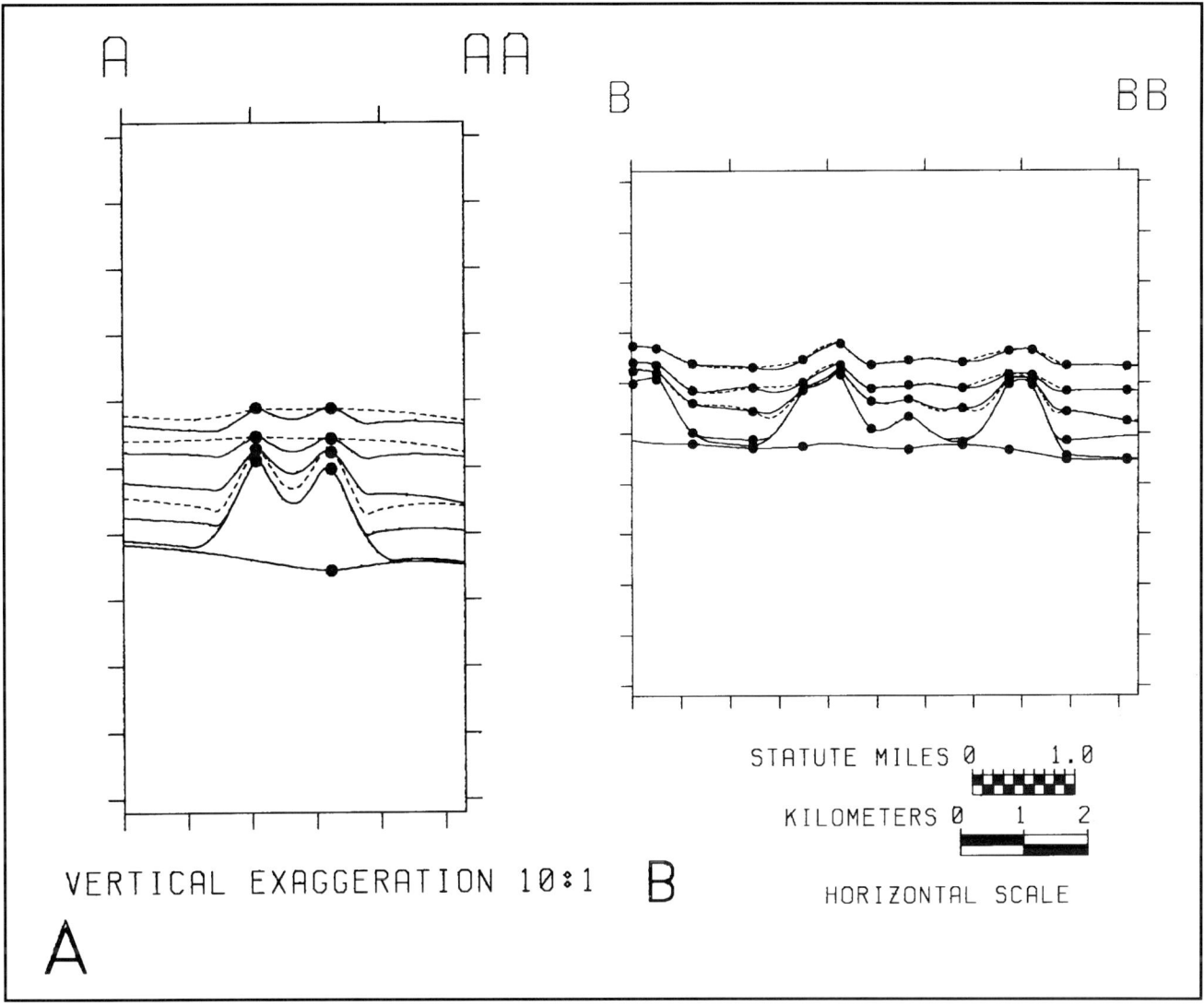

Figure 3. Cross sections showing surface models. Solid lines represent grids built using the compaction procedure and dashed lines represent grids built without compaction. (A) In areas of sparse data, tops of units above the reefs show off-reef dips and lows between and highs above reefs. (B) In areas of dense data, grids built with compaction are similar to grids built without compaction. See Figure 4 for the location of these sections.

INTRODUCTION

Carbonates above pinnacle reefs in the Michigan basin have experienced compaction (Jodry, 1969). Because of this, structural surfaces for the carbonates and units above them are expected to show compaction features such as drape over pinnacles and sag between them as seen in the hand-drawn cross section of Figure 1. Computer-generated surfaces for evaporites and carbonates above pinnacle reefs in the northern trend, Michigan basin, show these compaction features in dense-data areas (Figure 2 section B-BB). However, these surface forms are not seen for those same surfaces in sparse-data areas (Figure 2 section A-AA), although data are honored in both cross sections.

All areas of the northern trend where units have experienced compaction should have compaction geometries similar to those just described. However, in areas of sparse drilling, computer-generated surface models smoothly connect top picks, creating structure grids that show little or no compaction. This discrepancy has commonly been ignored, resolved by hand contouring, digitizing, and gridding the compacted thickness, or corrected by interactively editing thickness or structure grids. Where significant, compaction should not be ignored. Unfortunately, these methods for resolving the problem are time consuming and difficult to apply uniformly.

The compaction procedure described in this paper produces better surface models for the structural tops of these compacted units. Figure 3 shows surface models generated with this technique. The sparse-data area (Figure 3 section A-AA) demonstrates the

Table 1. Distribution of Compaction.

Unit and Total Compaction		Interreef 35%	A-1 Carbonate 60%	A-2 Carbonate 60%
Percent experienced by each unit	A-2 Carb.		20%	100%
	A-2 Evap.		60%	
	A-1 Carb.	20%	100%	
	A-1 Evap.	60%		
	Interreef	100%		

features previously missing from the model: off-flank dips, highs over pinnacles, and forms which mimic the underlying surface.

The technique presented here follows four steps: (1) estimate the percent of compaction, (2) restore the tops to precompacted positions, (3) build the surface models, and (4) compact those models to create the present-day structures. Before the method can be applied, a grid must be built for the base of the compacted unit.

GEOLOGIC BACKGROUND AND DATA

The geologic setting for pinnacle reefs and their associated surfaces in the northern trend, Michigan basin, is described by Drost and Shaver (1985) and Mesolella et al. (1974), among others, and is summarized by Hamilton and Henize (1992). Geologic units of importance to this paper are shown in Figure 1.

Compaction in this area occurs in the carbonates between and above the reefs. The reefs are considered to be rigid bodies that compact only slightly. As the carbonates compact, their tops and the tops of units above them sag between, and bulge above, the reefs (Gill, 1977).

In a study of Silurian reefs in St. Clair and Macomb counties, eastern Michigan, Jodry (1969) suggests that the evaporites were deposited in shallow seas as nearly flat-lying beds. Today these beds dip away from the flanks of the reefs, so the evaporites or units below them must have compacted. The evaporites were considered to be essentially incompressible and to have experienced little dewatering; compaction was therefore accounted for by dewatering the underlying carbonates. Jodry suggests that compaction of the carbonates beneath the evaporites can account for all of the reef flank dips and relief on those evaporites (Figure 1).

The model that Jodry proposes stresses a multi-stage growth of the pinnacles; this was not used by Hamilton and Henize (1992) to model the pinnacles. However, Jodry's compaction concepts can be used to guide portions of the modeling process. The A-1 Evaporites acted as a seal above the Niagaran interreef carbonate to prevent water loss and as a weight to squeeze water from the carbonate into the reefs and into the sea above. Thirty-five percent compaction of the Niagaran interreef carbonate is considered adequate to account for all dips and relief on the A-1 Evaporite. Compaction began near the end of interreef carbonate deposition, continued through A-1 Evaporite deposition, and probably extended into A-1 Carbonate deposition.

Similar processes occurred in the A-1 and A-2 carbonates. The top of the A-2 Evaporite, assumed to have been originally horizontal, now dips away from the pinnacles. To restore dips to horizontal, Jodry applied a series of compaction factors and concluded that the A-1 Carbonate was compacted 60% during lithification. Compaction began near the end of A-1 Carbonate deposition, continued throughout A-2 Evaporite deposition, and probably extended into A-2 Carbonate deposition. The A-2 Carbonate was deposited between and above the pinnacles. By applying the same 60% compaction factor, Jodry was able to bring dips of the A-2 Carbonate back to horizontal.

Table 1 summarizes the compaction model presented by Jodry and displays estimates of the percent of total compaction that was experienced by each unit. For example, the A-1 Carbonate compacted 60 percent and its top is assumed to reflect all of that compaction. The A-2 Evaporite top is assumed to reflect 60 percent of the A-1 Carbonate compaction and the A-2 Carbonate top to reflect only 20 percent of that. The carbonate compactions derived by Jodry are for eastern Michigan and their effects on the units above are only guesses; however, these estimates provide a good basis to begin modeling. Rapid experimentation with the computer can allow these numbers to be fine tuned.

The data used by Hamilton and Henize (1992) was also used in this study. Those data included 590 wells covering Antrim and Kalkaska counties in northern Michigan and was provided by Petroleum Information Inc. The data consisted of API well number, latitude, longitude, total depth, and tops for the six Niagaran and Salina surfaces. Figure 4 shows the data

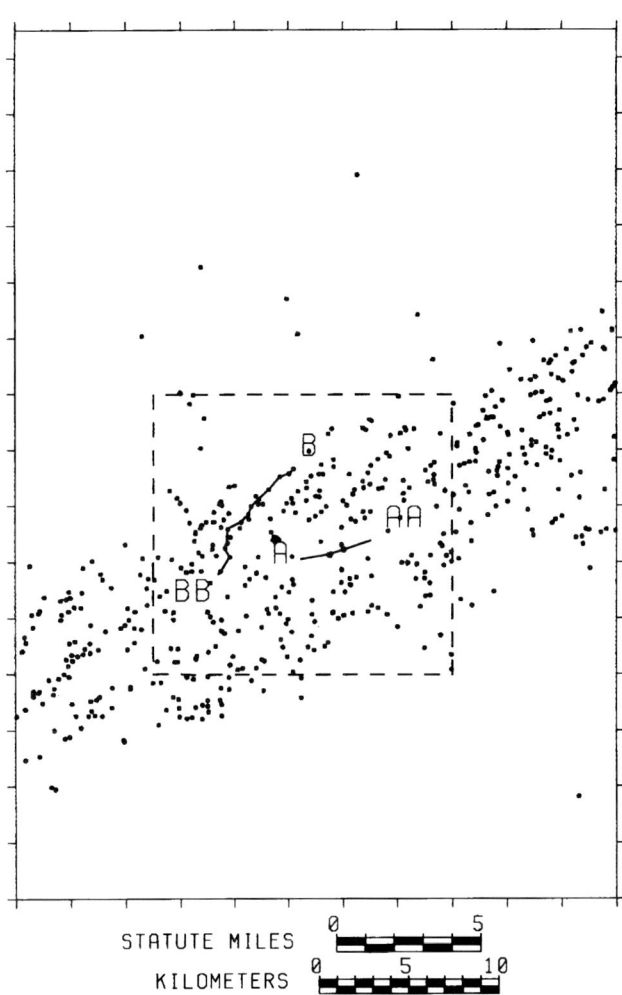

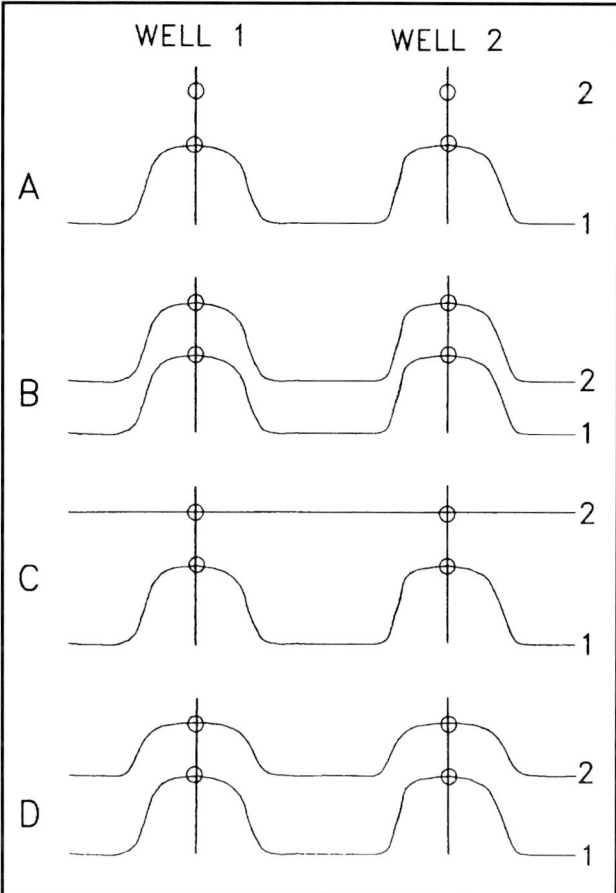

Figure 4. Base map showing data distribution and cross-section lines used in other figures. The dashed box represents the area contoured in other figures.

Figure 5. Sections showing geologic data and three interpretations for those data. (A) The geology and available data. Surface 1 is an irregular structure for which a grid exists. Surface 2 lies above surface 1 and is defined by two top picks. (B) Draping or conformable interpretation: calculate thickness between surfaces 1 and 2 and add that thickness to surface 1. (C) Depression-filling with no compaction: directly grid the tops for surface 2. (D) Depression-filling with compaction cannot be modeled effectively by direct gridding of either thickness or top picks.

distribution, lines where cross sections are cut, and an outline of the area where contour maps are drawn.

ESTIMATING COMPACTION AND THE COMPACTION PROCEDURE

To incorporate compaction into surface models, an estimate of the amount of compaction must be made. Theoretical compaction calculations (Schlater and Christie, 1980) utilize measured porosity/depth relationships, packing model studies, backstripping techniques, overall basin subsidence, heat flow calculations, and plate stretching models to estimate corrected positions of original tops and original thicknesses. We generally take a more empirical approach by estimating compaction and then using those estimates in the modeling process. If the estimate is valid, results in sparse-data areas will match results in dense-data areas. A second test of the empirical approach is that used by Jodry (1969) which checks dips of surfaces after removal of compaction. These dips should approximate those at the time of deposition.

A simple example will best explain the concepts used in our compaction technique and where they are most appropriately applied. Figure 5A shows a cross section through surfaces 1 and 2. Surface 1 is a highly irregular surface which is well defined everywhere (perhaps a seismic reflector); assume a grid of it has been built and is displayed on the section. Material was deposited on surface 1 and surface 2 forms the structural top of this deposited material. Surface 2 is defined by picks in two wells on this line of section.

Three interpretations show the range of geometries that are possible for this data. The first interpretation

(Figure 5B) assumes that a blanket of material of relatively uniform thickness was draped over surface 1. This could be modeled using procedures for draped or conformable surfaces (Hamilton and Henize, 1992) which would add a grid of the unit's thickness to surface 1. The second interpretation (Figure 5C) assumes that material completely filled depressions and covered highs so that the surface defining the top of the material had little or no topographic expression (that is, no compaction has occurred since deposition). This could be modeled by directly gridding surface 2's top picks. The third interpretation (Figure 5D) assumes again that material filled lows and culminated in a flat to low relief upper surface. Since deposition, however, the material has been compacted by 60%. There is no obvious method to directly model tops or thicknesses and produce a grid that matches this interpretation.

The method used in this paper to build grids of a compacted surface is the same as that used to produce a hand-drawn cross section through that compacted surface. Figure 6 shows the steps used to create a hand-drawn section where 60% compaction is assumed to have occurred.

1) Draw the base of the unit and post the picks for the top of the unit (Figure 6A).
2) Move the tops of the compacted unit to their pre-compacted positions. To determine how far to move the tops, the vertical thickness between the top and the underlying surface is multiplied by the ratio of lost to present thickness [%compaction/(100 – %compaction)]. For this example, each present-day vertical thickness would be multiplied by 1.5 (60%/40%) to get lost thickness; the top pick is shifted upward by that amount (Figure 6B).
3) Draw a profile connecting the precompaction top of unit picks and incorporate any interpretation about the form of the surface as it existed at the time of deposition (Figure 6B).
4) Measure vertical thickness between the base of the unit and the precompaction top of the unit at evenly distributed points along the section. Multiply that thickness by the percent of compaction to give the lost thickness. This lost thickness is measured down from the precompacted top of unit and a tick placed on the section (Figure 6C).
5) Draw a smooth profile connecting the ticks made in step 4. This line defines the compacted surface as it would look based on 60% compaction (Figure 6C).

This technique is based on three assumptions: (1) compaction has taken place, (2) the thickness of the unit being compacted varies across the model area, and (3) the percent compaction is uniform across the area. The assumption of varying thickness is critical; if there is no variation, the result will be essentially the same as direct gridding (Figure 5C).

A word of caution: the restored (precompaction) surface for the top of unit should not be built by modeling restored thickness and adding that to the base of unit (draping technique). Compacting a precompaction-structure surface built in this manner would

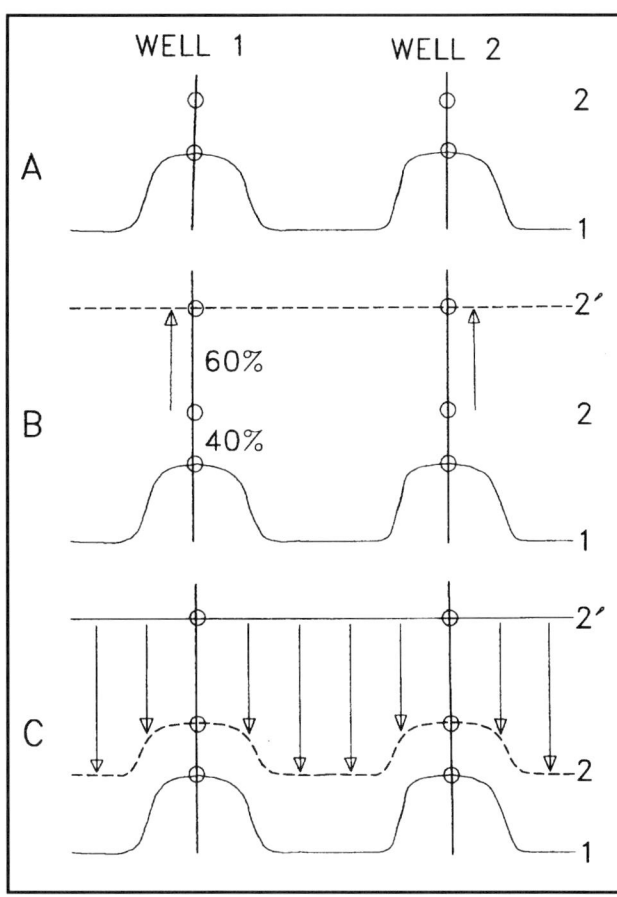

Figure 6. Steps used to incorporate compaction into a hand-drawn cross section. (A) Starting point for incorporating compaction, a profile exists for base of the compacted unit and picks for its top. (B) Shift top of compacted-unit picks up by the amount of lost section (current thickness × % compaction / (100 – % compaction)). Draw a profile through the shifted tops. (C) Measure vertical thickness between base and uncompacted top of unit at equal increments along section. Multiply thickness by % compaction / 100. Measure this distance down from uncompacted top and put a tick mark. Then draw a profile that smoothly connects the tick marks.

produce results very similar to that using the draped surface approach (Figure 5B).

Computer steps for incorporating compaction into structural surfaces above one or more compacted units are described in Appendix 1 of this chapter. These procedures will handle any number of compacted units and prorate that compaction to any number of overlying units. Significant simplification of the original procedures used by the authors has been incorporated into these procedures to reduce the number of processing steps. Those modifications related to preprocessing the base of second and later compacted units. No noticeable loss in model quality was seen due to these modifications, so the extra steps were not considered worthwhile.

Compacting the Carbonates and Overlying Surfaces

Hamilton and Henize (1992) describe procedures for building grids of pinnacle reefs, evaporites, and carbonates using the same data used in this paper. Details from their procedures are not described to avoid repetition here. The pinnacle reef grid built by Hamilton and Henize was used here. Grids for the evaporites and carbonates were reconstructed for this paper using the compaction procedures in Appendix 1 of this chapter.

Building a compacted grid for the top of the interreef carbonate would have been difficult since interreef carbonate picks were mixed with the Niagaran Brown pinnacle picks. Hamilton and Henize (1992) modeled the two units as one. It would take many steps to separate the two units, creating individual grids for each. Since the interreef carbonate is very thin in the modeling area and experienced only 35% compaction, we felt the improvement in surface form would be minimal relative to the effort required. We chose not to incorporate compaction into the interreef carbonate.

Modeling the compaction associated with the A-1 and A-2 carbonates requires that a grid exist for the base of each of those units. Since the tops of the A-1 Carbonate and the A-2 Evaporite must be built before a base of the A-2 Carbonate will exist, A-1 Carbonate compaction must be addressed first. The base of A-1 Carbonate is composed of two surfaces: the Top of A-1 Evaporite and the top of Pinnacles where they projected above the Evaporite. Therefore, the A-1 EVAPORITE-BLAP grid, which represents both the top of Interreef Carbonate and top of Pinnacles, is used as the base of the compacted unit.

The A-1 Carbonate

The compaction procedure described above is applied to the A-1 Carbonate unit. Tops for the A-1 Carbonate and the A-2 Evaporite are involved in this procedure. The A-1 Carbonate is assumed to be compacted by 60%. After the top picks for the A-1 Carbonate and the A-2 Evaporite were restored to their precompaction positions, the method used by Hamilton and Henize (1992) to build the A-2 Evaporite grid was used to build grids of both surfaces. Those PRECOMPACTION TOP grids were then compacted with the method described above. The resulting grids could potentially project though the Pinnacles and had to be baselapped onto the A-1 EVAPORITE-BLAP grid by outputting the maximum of the two grids, creating the A-2 EVAPORITE-BLAP and A-1 CARBONATE-BLAP grids. It should be noted that the draping technique used by Hamilton and Henize (1992) to build the A-1 and A-2 carbonate surfaces was not used here because it is a thickness-addition method and would not have created the desired compaction effects.

The A-2 Carbonate

The base of the A-2 Carbonate, although primarily the top of the A-2 Evaporite, could potentially be the top of Pinnacles which project above the Evaporite. Therefore, the A-2 EVAPORITE-BLAP grid, which represents both the top of A-2 Evaporite and top of pinnacles, is used to represent the base of the compacted unit.

The same compaction procedure described above was applied to the A-2 Carbonate unit. Since there were no surfaces to be modeled above the A-2 Carbonate, only its tops were used in the modeling procedure. The A-2 Carbonate was assumed to be reduced by 60%. After the top picks for the A-2 Carbonate were restored to their precompaction positions, the method used by Hamilton and Henize (1992) to build the A-2 Evaporite grid was used to build this grid. That PRECOMPACTION TOP grid was then compacted with the method described above. To prevent the compacted top-of-unit grid from projecting through either the pinnacles or the A-2 Evaporite, it was compared to the A-2 EVAPORITE-BLAP grid and the maximum node values of the two grids (baselap) were output to create the A-2 CARBONATE-BLAP grid.

Review of Compaction Results

The resulting grids are shown in cross-section as solid lines in Figure 3. The dashed lines are the surface models built without using the compaction method. As discussed above, in dense-data areas (Figure 3, section B–BB) there is some change in surface form, although the model is essentially the same as before. In sparse-data areas (Figure 3, section A–AA) significant improvement is seen. The surfaces show flank dips away from reefs and the position and amount of compaction is associated with the position and relative thickness of the carbonates. The A-2 Evaporite and Carbonate surfaces form obvious highs over the reefs, in essence *feeling* the reef, just as they do in dense-data areas and in hand-drawn sections. The top of A-1 Carbonate is thicker when 60% compaction is applied than when the draping technique is used. This is because the draping method will project data-defined thins above reefs into the nondata areas beside the reefs. We found that about 85% compaction would produce an A-1 Carbonate surface grid similar to that built using the draping method.

Figure 7 shows sections through grids beneath the A-1 Carbonate, A-1 Carbonate top compacted 60%, and the A-1 Carbonate top restored to its precompaction position for 60% compaction. As Jodry implied, when surfaces are decompacted their dips should be similar to dips at the time the unit was being deposited. Section B–BB seems to indicate that the picks have been decompacted too much, since nearly all restored picks between reefs are higher than picks at the reefs. We found, after looking at about 30 sections, that restored-surface geometries varied significantly from area to area. Some areas required small compactions (30–40%) and others nearly 80% compaction to bring compacted top picks into alignment with uncompacted (above reef) picks. No effort was made to map these variations.

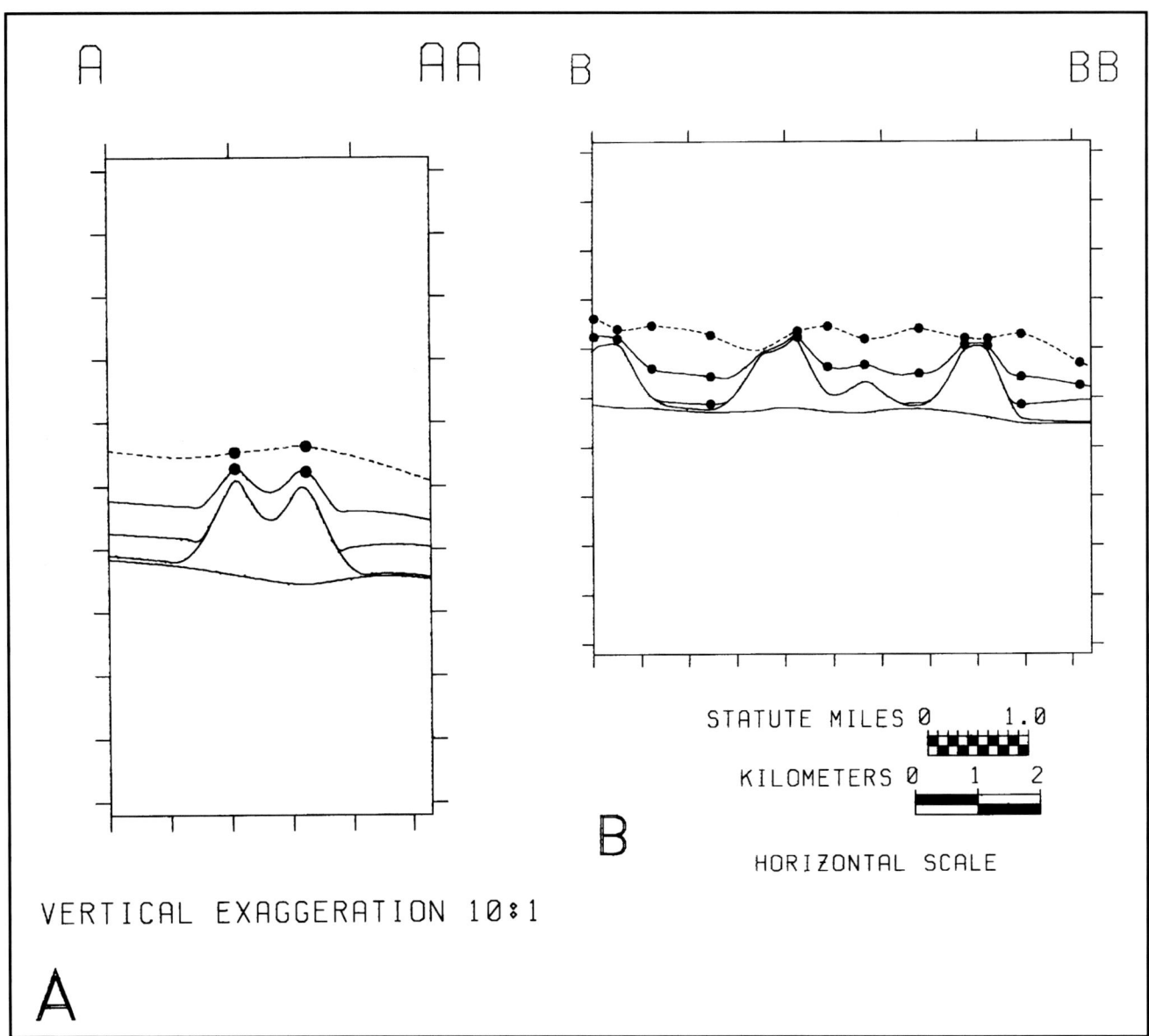

Figure 7. Cross sections showing from bottom to top: Niagaran Gray, Niagaran Brown, A-1 Evaporite, A-1 Carbonate compacted 60%, and the A-1 Carbonate restored to its precompaction position for a 60% compaction (dashed line). See Figure 4 for the location of these sections.

Grids were built using 40% and 80% compaction. As expected for low data-density areas (Figure 8A and C), significant changes are seen between 40 and 80% compaction and between them and the 60% compaction of Figure 3. High data density areas (Figure 8B and C) show few changes with varying compaction, although sags develop and disappear between data. Perhaps these sags can be used to indicate reasonable compaction amounts. Adjusting compaction involved changing two numbers, rerunning the procedures, and generating displays, and required about an hour for each compaction.

Figures 9A and 9C show contour maps of structure and thickness for the A-2 Evaporite when the underlying carbonate was compacted 60% and 60% of that compaction was reflected in the evaporite. Figures 9B and D show contour maps of structure and thickness for the A-2 Evaporite using no compaction. Significantly more form is seen in structure contours of grids built using compaction. These reflect the sag of the A-2 Evaporite between the pinnacles as the A-1 Carbonate compacted. Thickness maps for the A-2 Evaporite were actually simpler for the compacted model than for the uncompacted model, as expected. The draping method built an A-1 Carbonate grid more complex than the A-1 Carbonate built with the compaction method. The uncompacted A-2 Evaporite grid was built independently of the pinnacles and A-1 Carbonate and thus was quite simple in form. The compacted A-2 Evaporite had the complexity of both

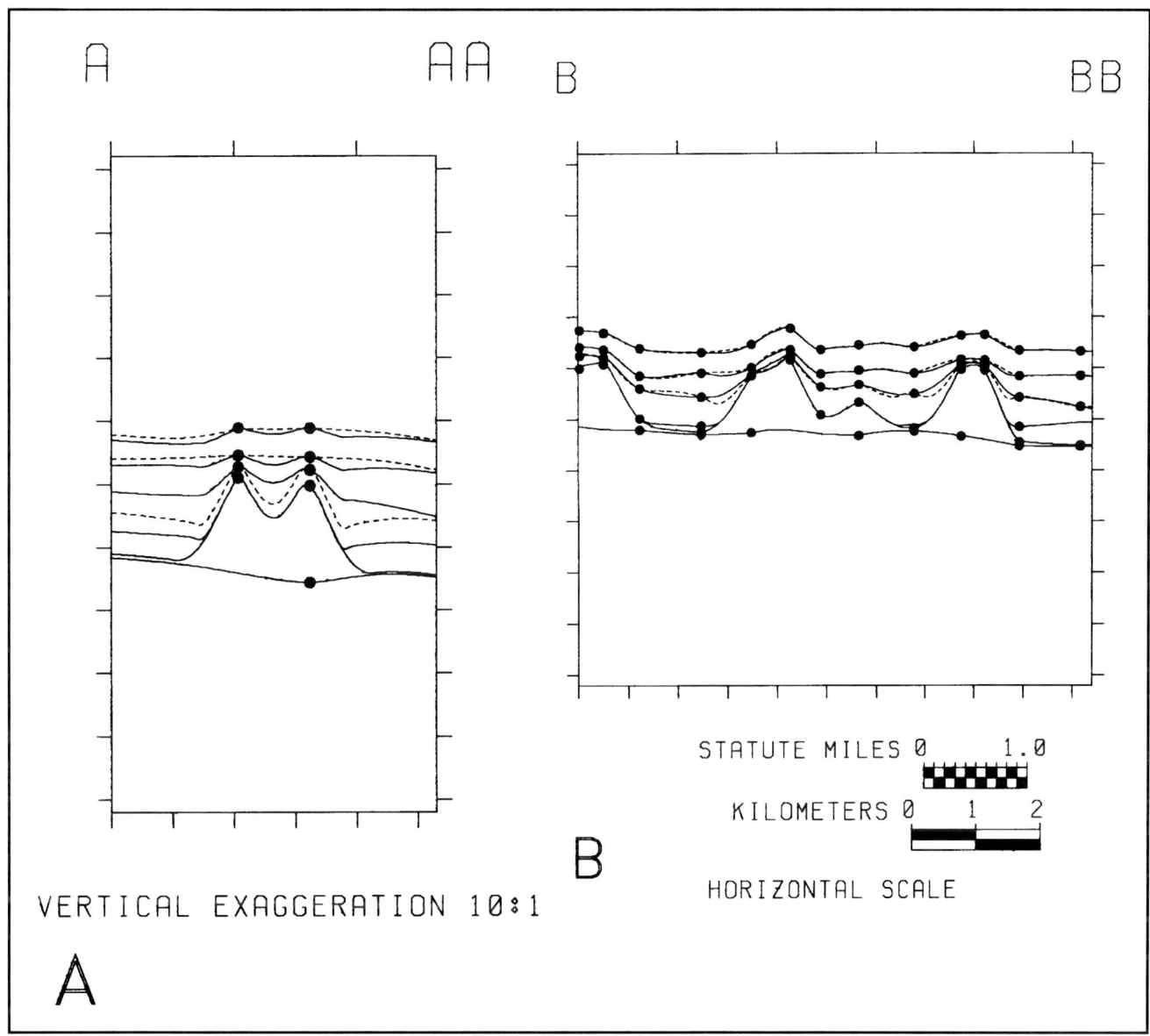

Figure 8. Cross sections through grids built by varying the amount of compaction experienced by the carbonates. The percent of the A-1 Carbonate compaction that reached the A-2 Evaporite and A-2 Carbonate was 60% and 20% respectively in each case. (A & B) 40% compaction applied to the A-1 and A-2 Carbonates. (C & D) 80% compaction applied to the A-1 and A-2 Carbonates. See Figure 4 for the location of these sections.

the pinnacles and compacted A-1 Carbonate built into it using the compaction procedure. Subtracting the uncompacted A-1 Carbonate (complex) from the uncompacted A-2 Evaporite (simple) resulted in a complex thickness grid, while subtracting the compacted A-1 Carbonate (complex) from the compacted A-2 Evaporite (complex), created a simple thickness grid.

CONCLUDING REMARKS

The technique in this paper was developed for mapping carbonates and evaporites above and adjacent to pinnacle reefs in the northern trend, Michigan basin. Compaction occurred in the carbonates, with the units immediately above also being affected by the compaction. Incorporating compaction significantly improved maps of these structural surfaces. Any irregularities existing in the surface below the compacted unit were reflected in that unit's top and in units above it.

The compaction procedure provides two modeling advantages. First, restoring picks to their precompaction position usually simplifies the surface by removing irregularities caused by differential compaction. Since surface modeling programs generally produce better models of gently undulating surfaces

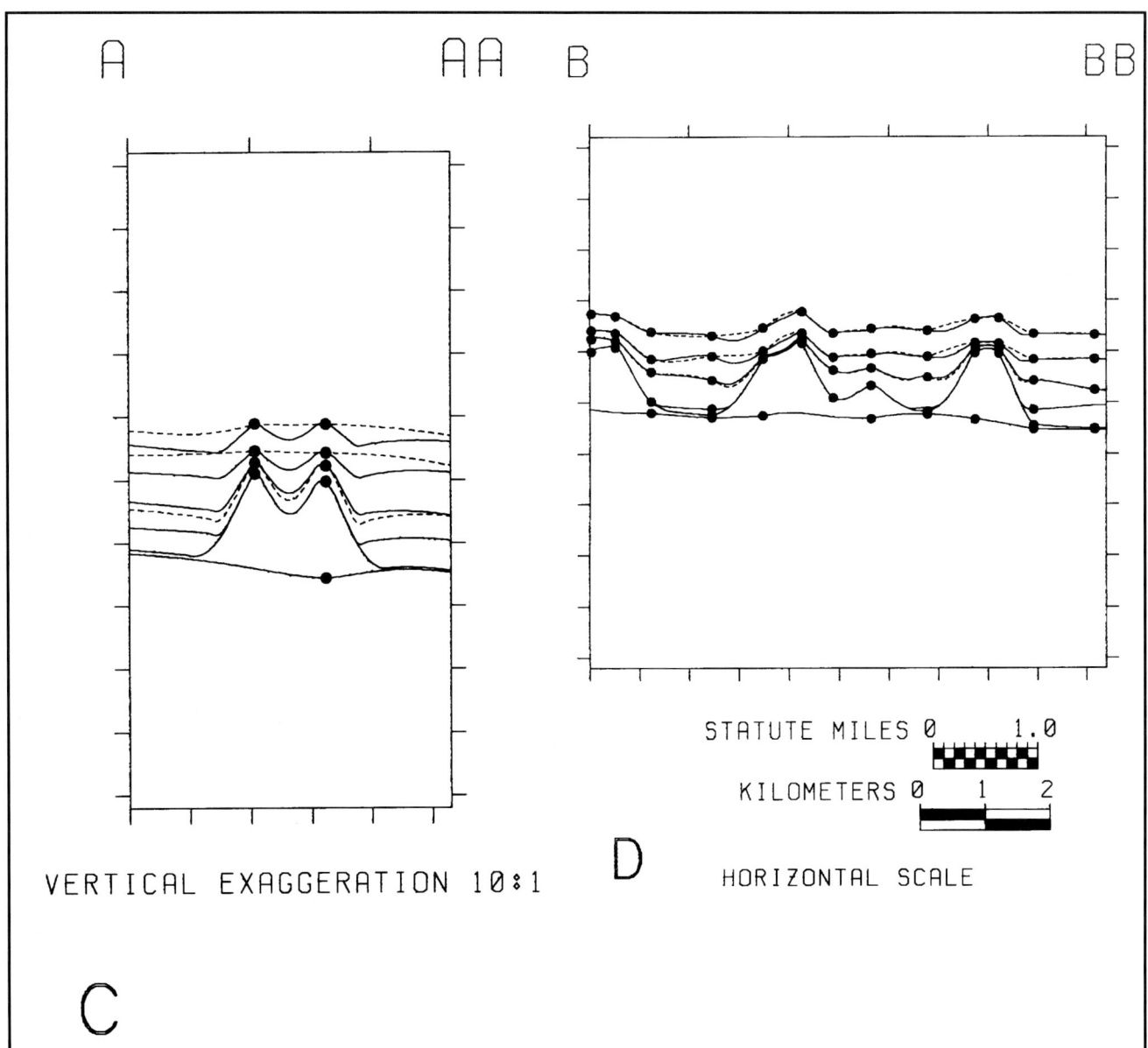

Figure 8. Continued.

than of highly irregular surfaces, they have a better chance of producing an acceptable precompacted structure than compacted structure. Second, the interpretation of compaction and its amount is really a type of data, but is commonly ignored. The compaction procedure provides a means for using this data during grid construction.

Any or all units in a model may be compacted. Those compactions may be distributed (prorated) through time to any number of units above the compacted units. In this study, the A-1 Carbonate experienced 60% compaction. 60% of the vertical drop in elevation due to compaction of the A-1 Carbonate was distributed to (experienced by) the A-2 Evaporite above it, and 20% of the vertical drop distributed to the A-2 Carbonate above that. The A-2 Carbonate also experienced a 60% loss in thickness due to compaction.

Three assumptions are made when using this technique. The first is that compaction has taken place. The second is that the compacted unit's thickness varies over the area being modeled. If thickness is uniform across the area or gradually thickens in one direction, then compacted surface geometries will not exist in the rocks nor will they appear in the structure grids. The third assumption is that rate of compaction is uniformly distributed over the area. This assumption may be relaxed. If compaction varies across the area, then a grid of compaction could be built and appropriate changes made to the compaction proce-

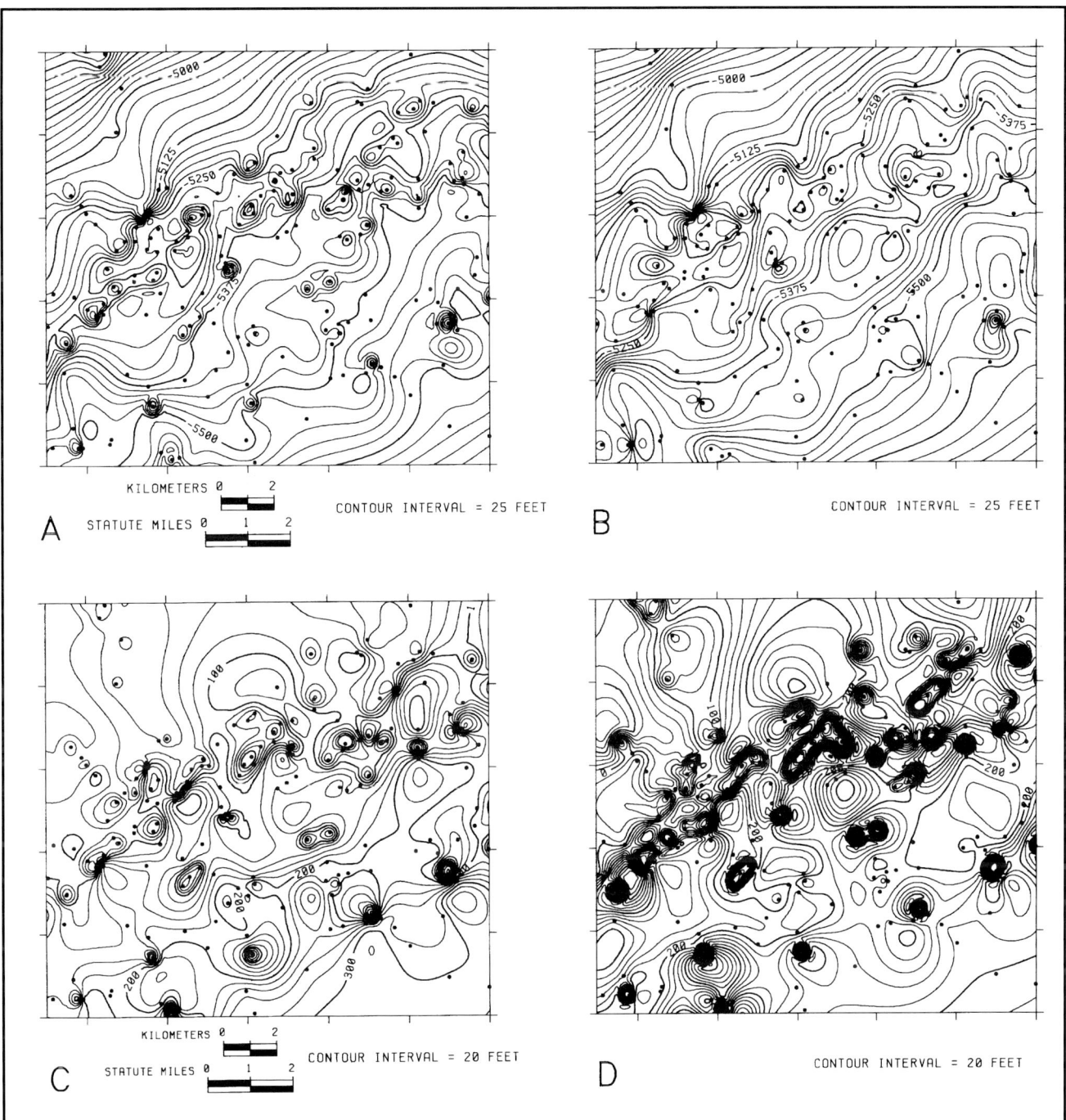

Figure 9. Contour maps of the A-2 Evaporite top and thickness. Compaction was 60% in the A-1 Carbonate with 60% of that passed to the A-2 Evaporite. (A) Compaction reflected in top of A-2 Evaporite. (B) No compaction affecting the top of A-2 Evaporite. (C) Compaction reflected in thickness of A-2 Evaporite. (D) No compaction affecting the thickness of A-2 Evaporite.

dure.

Many steps currently are required to incorporate compaction into a compacted unit's structural top and into affected surfaces above that unit. These steps fit smoothly into the procedures commonly used to incorporate baselap, truncation, and conformity into stratigraphic frameworks (Jones et al., 1986). Whether the effort required to implement these steps can be justified depends upon the size of the project, importance of surface form, and use to which the resulting surface models will be put. Because the same steps are used each time, a program could be written that automatically implements those steps. The input to that program would be the amount of compaction, proration of compaction to other units, base-of-unit grid, top picks, and parameters that control final grid construction. If automated, this technique would be simple to run and have the potential for being a standard method by which a compacted unit is modeled.

We feel that this technique is appropriate for application to any compacted unit. The technique should be applied when a compacted unit overlies incompressible features such as igneous bodies, sand channels, salt domes, high-relief unconformities, and other features of significant size. Where thickness variations of the compacted unit are large, application of this technique will improve the resulting structure grids and their use in volumetrics, studying paleo-structures, and in evaluating hydrocarbon migration routes. A final use for this method is the rapid testing of several compaction factors to determine which is more reasonable.

APPENDIX 1. COMPACTION PROCEDURE—DETAILED STEPS

Three procedures are used to incorporate compaction into structural surfaces above one or more compacted units: (1) Build a grid for the top of compacted unit, (2) Calculate and grid lost thickness (due to the compacted unit) for all units above the compacted unit, and (3) Build grids for tops of overlying units up to and including the base of the next compacted unit. Each of these procedures is implemented sequentially, once for each compacted unit in the stratigraphic framework. Implementation starts with the lowest compacted unit and works up through the last compacted unit. Because the procedures are repeated several times, grids from a previous iteration will be used by a later iteration. However, to be general, even the first iteration of the procedure checks to see if those grids exist. This asking for nonexistent grids at first seems bewildering until the entire set of procedures have been read. Figure 6 should help clarify these steps.

Definitions

CUMULATIVE LOST THICKNESS: A grid for a surface which is used to cumulate the vertical movement downward that surface has experienced due to thickness lost by compacted units below that surface. Each compacted unit may have an affect on surfaces above it, usually a percentage of the compaction it has experienced. The downward movement a higher surface experiences is the percent of the compacted unit's compaction that occurred after that higher surface was deposited, multiplied by the total lost thickness experienced by the compacted unit.

RESTORED TOP: The top pick field for any surface above one or several compacted units (including the top of a compacted unit) whose values have been shifted to their position before compaction (relative position at time of deposition).

PRECOMPACTION TOP: A grid for any surface above one or several compacted units built using tops that have had the compaction effects of those units removed (RESTORED TOPs).

Procedures

(1) Grid the Compacted Unit.
 (A) Interpolate values from the base of the compacted unit grid (must have been previously built) and put the results in a new BASE OF COMPACTION field.
 (B) Subtract the BASE OF COMPACTION field from the present-day top of compacted unit field, creating a new THICKNESS field.
 (C) Convert the THICKNESS field values to thickness lost due to compaction using the following equation:

$$\text{Lost thickness} = \frac{\text{Present Thickness} \times \% \text{ compaction}}{(100\% - \% \text{ compaction})} \quad (1)$$

Do this by multiplying the THICKNESS field by [% compaction/(100% − % compaction)] and placing the results back into the THICKNESS field.
If this compacted unit has a CUMULATIVE LOST THICKNESS grid, then:
– Convert the THICKNESS field values to thickness lost due to compaction as described above.
– Interpolate values from the CUMULATIVE LOST THICKNESS grid, putting the results into a temporary WORK field.
– Add the WORK field to the THICKNESS field, putting the results back into the THICKNESS field.
 (D) Restore the present-day top picks for the compacted unit to their precompacted position. Do this by adding the THICKNESS field (lost thickness) to the present-day top of compacted unit field (Figure 6B), creating a new RESTORED TOP field.
 (E) Build a grid of PRECOMPACTION TOP for the compacted unit using the RESTORED TOP field.
Note: do not add a thickness grid to the unit base to build the top of compacted unit grid.
 (F) Build a grid of PRECOMPACTION THICKNESS for the compacted unit by subtracting the base of the compacted unit grid from the PRECOMPACTION TOP grid for that unit.
If this compacted unit has a CUMULATIVE LOST THICKNESS grid, then:
– Subtract the base of the compacted unit grid from the PRECOMPACTION TOP grid for that unit, creating a temporary WORK grid similar to the step described above.
– Subtract the CUMULATIVE LOST THICKNESS grid from the WORK grid, creating a SECOND WORK grid.
– Clip the SECOND WORK grid to a minimum value of zero, creating a PRECOMPACTION THICKNESS grid.

(G) Build a grid of THICKNESS LOST due to compaction by multiplying the PRECOMPACTION THICKNESS grid by the percent compaction as a ratio (similar to Figure 6C).
If a CUMULATIVE LOST THICKNESS grid exists for this unit (Step 2B) then add it to this thickness lost grid after doing the above manipulation.
(H) Build a grid of PRESENT-DAY STRUCTURE for the compacted unit by subtracting the THICKNESS LOST grid from the PRECOMPACTION TOP grid (Figure 6C).
Note: this PRESENT-DAY STRUCTURE grid may cross surfaces below or above it since baselap and truncation procedures have not yet been applied to it.

(2) Calculate and build grids of *Lost Thickness* for units affected by Step 1 compaction. The following steps are repeated for each unit above the compacted unit that has been affected by the compaction.
(A) Build a grid of CUMULATIVE LOST THICKNESS by multiplying the LOST THICKNESS grid (for the underlying compacted unit) by the percent (as a ratio) of the compaction that was experienced by this unit.
(B) If a CUMULATIVE LOST THICKNESS grid already exists for this surface (effects from a deeper compacted unit), then add this CUMULATIVE LOST THICKNESS grid to that one; otherwise, keep this new CUMULATIVE LOST THICKNESS grid for later use with this surface.

(3) Grid all affected surfaces above the compacted unit up to and including the base of the next higher compacted unit. The following steps are repeated for each unit that has been affected by the compaction.
(A) Interpolate values from the CUMULATIVE LOST THICKNESS grid, putting the results in a temporary WORK field.
(B) Add the WORK field to the structural top field for this surface, moving its picks to their precompacted position.
(C) Build a PRECOMPACTION TOP grid for this unit using the restored top picks.
Note: Thickness gridding between this top and the base of the compacted unit would probably not produce final results significantly different from uncompacted thickness gridding.
(D) Build the present-day surface grid (unbaselapped) by subtracting the CUMULATIVE LOST THICKNESS grid built in Step-2(B) above from the PRECOMPACTION TOP grid for this surface.
If another compacted unit exists above the surfaces just constructed then go to Step-1 and repeat the process. However, before doing that you will need to construct and baselap any, as yet unmodeled, surfaces between the surface just constructed and the top of that next-higher compacted unit.
These procedures will handle any number of compacted units and prorate that compaction to any number of overlying units.

ACKNOWLEDGMENTS

We wish to thank Petroleum Information Incorporated, Denver, CO. for allowing us to use their northern trend, Michigan basin data. We would also like to thank Landmark/Zycor, Austin, TX. for providing the programs and facilities to do this study.

REFERENCES CITED

Drost, J. B., and R. H. Shaver, 1985, Comparative stratigraphic framework for silurian reefs—Michigan basin to surrounding platforms: Michigan Basin Geological Society Special Paper 4, p. 73–94.

Gill, D., 1977, The Belle River Mills gas field: productive Niagaran reefs encased by sabkha deposits: Michigan Basin Geological Society Special Paper 2, 187 p.

Jodry, R. L., 1969, Growth and dolomitization of Silurian reefs, St. Clair County, Michigan: AAPG Bulletin, v. 53., p. 957–981.

Jones, T. A., D. E. Hamilton, and C. R. Johnson, 1986, Contouring geologic surfaces with the computer: New York, Van Nostrand Reinhold, 314 p.

Mesolella, K. J., K. D. Robinson, L. M. McCormick, and A. R. Ormiston, 1974, Cyclic deposition of Silurian carbonates and evaporites in Michigan basin: AAPG Bulletin, v. 58, p. 34–62.

Schlater, J. G., and P. A. F. Christie, 1980, Continental stretching: An explanation of the post-mid Cretaceous subsidence of the central North Sea basin: Journal of Geophysical Research, v. 85, no. 37, p. 3711–3739.

Chapter 7

Modeling Anisotropy in Computer Mapping of Geologic Data

Gregory Kushnir
GeoGraphix, Inc.
Denver, Colorado, U.S.A.

Jeffrey M. Yarus
PTC Marathon Oil Company
Littleton, Colorado, U.S.A.

ABSTRACT

Geologic terrains commonly contain trends arising from structural and depositional causes. Geological mapping algorithms often suppress or distort such trends (anisotropy) because interpolation schemes generally assume spatially uniform (isotropic) influences of nearby data points in all directions. In such cases, the resulting map can look unrealistic and therefore be unusable unless edited. Proper assessment of the data prior to mapping can determine the direction and strength of anisotropy. Incorporation of this information in the gridding algorithm permits a more faithful representation of the underlying geological reality. Improvement in the areas of sparse control arises because the structural analysis coupled with other geological insights puts important constraints on the estimation procedures. In addition, drainage area configurations (Voronoi polygons) more closely conform to the geologic model, when anisotropic properties are incorporated in their construction.

Two methods for incorporating directional trends into surface models are presented, a regional approach where the entire dataset is adjusted before gridding, and a local approach where directional bias is built into the weighting function of each data point. Because the regional approach is built into the gridding algorithm, it is more efficient than some techniques commonly used, which require the user to perform a series of steps. The local approach allows different directions and magnitudes of bias to be applied to different areas of a map more easily. Also, the surface at boundaries between areas of varying trend is smooth, overcoming a major problem with methods currently used.

INTRODUCTION

Geologic properties (structure, porosity, etc.) often exhibit a trend or a preferred direction of orientation across a given study area. Within the context of computer mapping, this phenomenon of preferred direction may be defined as anisotropy. Computer-generated maps of irregular and sparse data distributions that do not account for anisotropy may contain geologically unrealistic features. Some of these features or artifacts include unsupported noses (*pinocchio effects*), converging contours (*funneling, bottlenecking, hourglassing, eyeglassing*), local anomalies in areas of sparse control, and improper projection of closures away from a data cluster (Yarus and Lewis, 1989). In some cases the resulting map does not even reflect realistic geological concepts, let alone the anisotropic properties of the study area. Many of these problems result from the underlying assumption of isotropy in all of the commonly used computer gridding algorithms.

Geologists wishing to introduce their personal interpretation of the depositional, diagenetic, or structural history of an area must be able to control anisotropy. This allows their professional judgments to be reflected in the final computer-generated map. The resulting map will be the best possible representation, encapsulating the experience of the geologist along with the strength of the gridding and contouring algorithms. This approach is similar to that used by geophysicists for modeling and processing seismic data.

ANISOTROPY

Assumptions

When geologic variables are anisotropic, the lateral (X/Y) relationships between measured Z-values are not equal in all directions. This lateral (X/Y) anisotropy is assumed to describe the behavior of the geologic variable being analyzed (in some instances a geologic variable may have a component of strong vertical anisotropy, however vertical anisotropy is not discussed in this article). Anisotropy implies that the relationships between data points are stronger in certain directions and weaker in others. This is important because control points are used to produce estimates at intermediate locations, like grid nodes.

Normally gridding algorithms assume that the degree of influence a control point exerts diminishes uniformly in all directions. Thus, a closed line of equal influence (LEI) around a given control point would be a circle (Figure 1). In the presence of anisotropy however, the shape of the LEI would be an ellipse (Figure 2) (David, 1977). In general, any point on such a line is affected equally by the data point around which the LEI is drawn.

Assuming an ellipse for the LEI around a data point, two parameters can be introduced to accom-

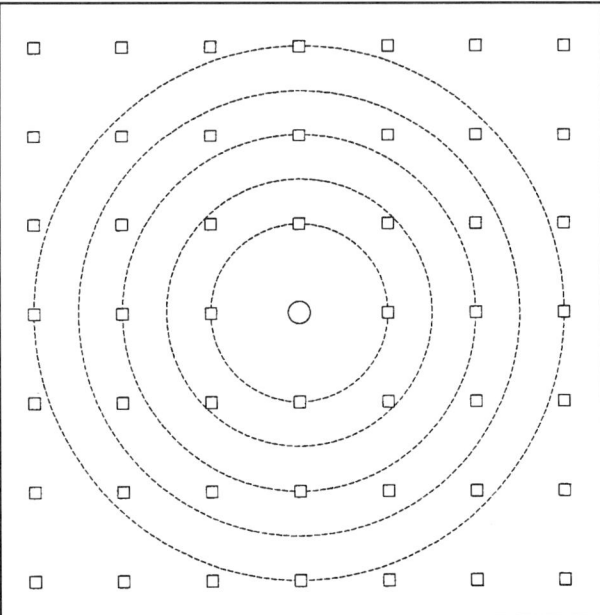

Figure 1. Lines of equal influence in the areas with isotropic properties.

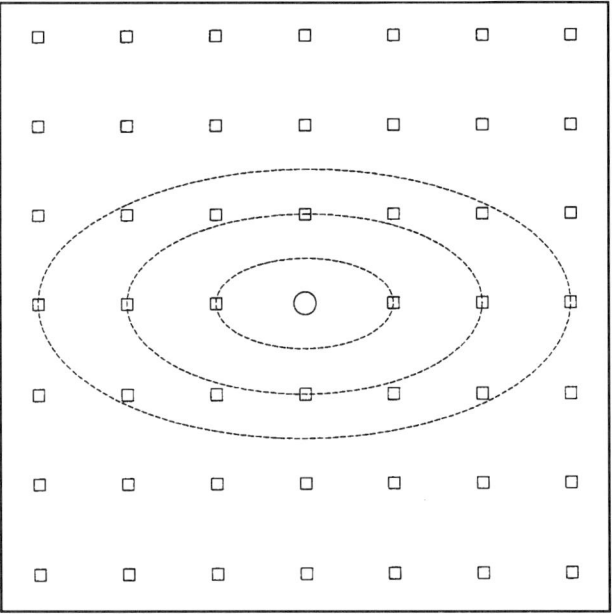

Figure 2. Lines of equal influence in the areas with anisotropic properties.

plish anisotropic mapping: 1. The direction of anisotropy — the compass value of the preferred direction, which is the orientation of the long axis of the ellipse (LEI), and 2. The magnitude of anisotropy — the ratio between the long and short axes of the ellipse.

Determining Anisotropic Direction and Magnitude

A priori

In many cases the anisotropic parameters are already known. Previously published geologic studies may contain information such as the direction, length, and width of anticlines, bars, channels and other geological features. Alternatively, such parameters may be known from experience in the area.

Empirical

There are two empirical methods for determining the anisotropic parameters. The first method is qualitative and requires an analysis of previously generated models or maps. The orientation and length-to-width ratio of contour closure are analyzed to estimate the direction and magnitude of the anisotropy.

The second method is quantitative and requires calculation and analysis of a semivariogram. Although it is not the intent of this paper to discuss all the details of variography and geostatistics, a review of certain basic procedures and definitions will benefit some readers. For a more detailed discussion see Matheron (1963), Davis (1986), and Hohn (1988).

As applied here, the semivariogram is a curve representing the degree of continuity, or anisotropy, of the mapping variable (Matheron, 1963). Anisotropy is measured as a function of the variance between pairs of points oriented in a particular direction, at progressively increasing distances or *lag* intervals. A lag is a predetermined range of separation distances between the pairs (Hohn, 1988). Although normally variography is used to calculate the weights used in geostatistical contouring (Kriging), the objective here is solely to determine the direction and magnitude of greatest anisotropy of the data. This is done by constructing a series of semivariograms characterizing the changes in variance of the mapping variable along certain directions (Figure 3). Since the semivariogram is directionally sensitive, the relationship will not be the same in all directions when the data are anisotropic. The variance for each lag is calculated by the equation:

$$\gamma(h) = \frac{\Sigma [Z(x) - Z(x+h)]^2}{2n} \quad (1)$$

where: (h) represents the semivariance, Z is the value of a map variable at a specific location x, h is the distance that characterizes the lag interval, and n is the number of sampled differences (Hohn, 1988).

The shape of the semivariogram is not unexpected. Pairs of points that are very close together (lag interval with small h) consistently generate differences near 0, while pairs of points with greater separation (large h) generate differences that are more variable, thus increasing the value of the semivariance. Where the data have no trend [data are stationary, (Olea,

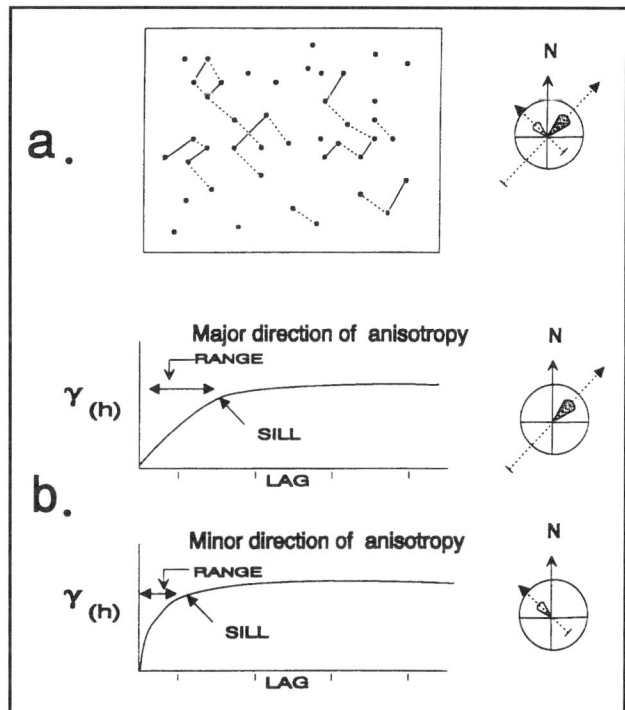

Figure 3(A). Set of hypothetical control data points. Pairs of points falling into one lag interval (separation distances) are shown for both the major (solid lines) and minor (dashed lines) directions of anisotropy. All pairs selected have an azimuth that falls within a specified range shown on the compass. (B) Two hypothetical semivariograms, one in the direction of maximum anisotropy and another in the direction of minimum anisotropy. The magnitude is the ratio between the two ranges.

1984)], there is a critical lag interval beyond which no appreciable change in the semivariance occurs (semivariance variance). This critical distance is called the *range,* and the principal direction of anisotropy is identified by the semivariogram with the largest range (Hohn, 1988). In a regional sense, this direction will often be parallel to the grain of the map. The semivariogram with the shortest range will be at some angle to this, mainly perpendicular to the above, and the ratio between the two range values will be a measure of the magnitude. The point at which the variogram begins to flatten is called the sill (Figure 3).

Types of Anisotropy

Anisotropy can occur on a number of different scales and in both simple and complex forms. By simple anisotropy the authors refer only to those instances where anisotropic features do not overlap. Complex anisotropy is characterized by overlapping or overprinting where more than one primary direction of anisotropy is present in the same feature or surface. Examples of simple anisotropy include struc-

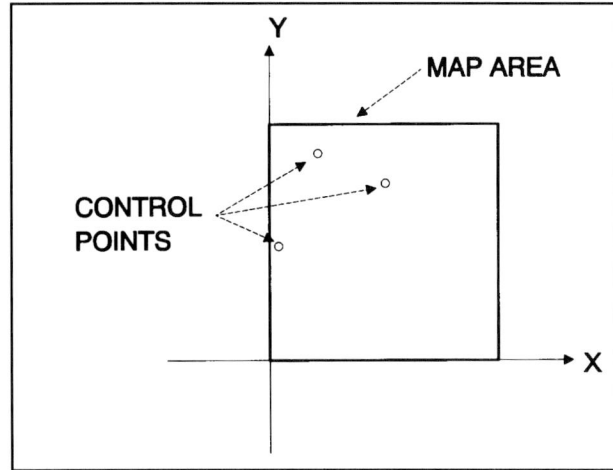

Figure 4. Original Coordinate System: axes X and Y.

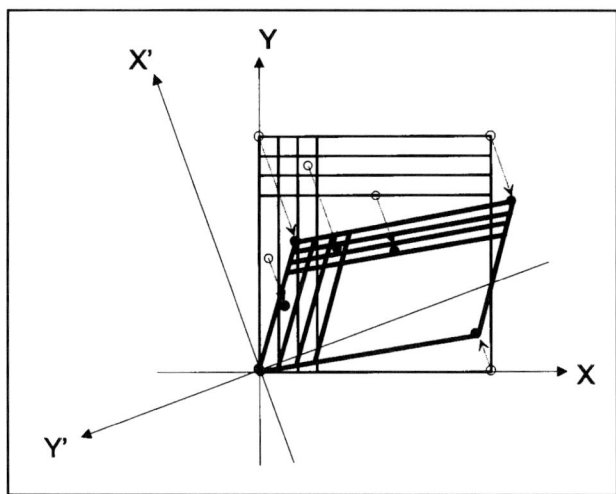

Figure 6. Compress New Coordinate System along axis X' with ratio 2 (magnitude of anisotropy = 2).

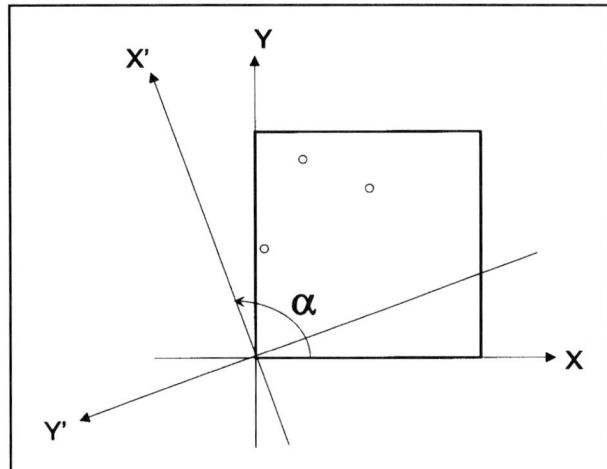

Figure 5. Rotate Original Coordinate System (axes X and Y) with the given angle α into the New Coordinate System: axes X' and Y'.

tural surfaces like regional dip and depositional surfaces like marine bars or dunes. Examples of complex anisotropy could include any structural surface which has been severely modified by multiple tectonic events through time.

The authors have identified six general forms of anisotropy, ranging from simple to complex. These are: (a) Regional anisotropy, in one direction with one magnitude. (b) Localized anisotropy, in multiple directions with varying magnitudes (without overprinting). (c) Combination of regional and localized anisotropy without overprinting. (d) Combination of regional and localized anisotropy with overprinting. (e) Localized anisotropy with overprinting. (f) Multiple regional anisotropy with overprinting.

This article addresses only the simple cases of anisotropy (a) through (c). Anisotropy with overprinting requires knowledge of the correct historical order and the degree of influence exerted by each anisotropic episode. Procedures for evaluating and modeling anisotropy with overprinting are being investigated.

COMPUTER MAPPING USING ANISOTROPY DIRECTION AND MAGNITUDE

Regional Anisotropic Data Mapping (One Direction With One Magnitude)

Introduction

As discussed above, most surface modeling algorithms assume an isotropic influence from surrounding data. In the presence of trends or anisotropy, resulting maps may be unrealistic. Regional anisotropy can be incorporated into surface models without developing new gridding algorithms using a procedure called Regional Anisotropic Data Mapping (RADM). The concept behind this procedure is to create a temporary coordinate system (space) in which the data are isotropic (Shurubor et al., 1976; Aronov et al., 1982; Kushnir, 1984; Yarus and Kushnir, 1990). To do this, the data points that align with the given direction of anisotropy are moved closer together relative to the control points that are perpendicular to this direction. For the purpose of this discussion, the temporary coordinate system will be referred to as compressed space, and the actual or real coordinate system will be referred to as original space.

Figures 4-6 illustrate the steps necessary to create the appropriate compressed space. The first step is to rotate the original coordinate system (Figure 4) by the given angle (direction of anisotropy measured from

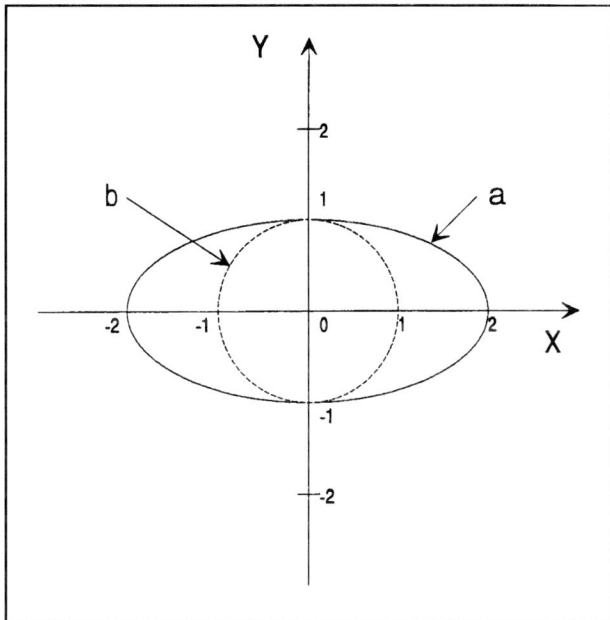

Figure 7. Affine transformation: (a) initial ellipse representing line of equal influence; (b) resultant circle representing line of equal influence, after compression of initial ellipse along axis X (direction of anisotropy 0 degrees) with a ratio of 2 (magnitude of the anisotropy = 2).

the positive X-axis) as shown in Figure 5. This aligns the X-axis, now labeled X′, with the preferred direction of anisotropy. The second step is to compress the space by reducing or squeezing the scale of the X′-axis relative to the Y′-axis as in Figure 6. The amount of relative reduction (ratio) is determined by the magnitude of the anisotropy. As previously mentioned, this has the effect of relocating all the control data points to the new compressed coordinate system (space). The following equations calculate the compressed X′ and Y′ coordinates for each control data point:

$$x' = (x \times \cos \alpha + y \times \sin \alpha) / \text{ratio};$$
$$y' = -x \times \sin \alpha + y \times \cos \alpha \qquad (2)$$

As a result of this compression [known also as an Affine Transformation (David, 1977)], the elliptical representation of the LEI for a given data point is transformed into a circle (Figure 7). This essentially removes the one direction, one magnitude anisotropy from the data and allows standard gridding algorithms to perform optimally.

It is important to emphasize that once the data points are converted to compressed space, the coordinate system changes (Figures 4–6). There remains, however, a mathematical relationship between the two systems. Thus, any point in one space has a known location in the other. In this way, grid node positions in original space are easily transferred to their compressed space (isotropic) locations. Grid node estimates are calculated in compressed space. This step is critical, because as previously mentioned, gridding algorithms perform optimally when the data are isotropic. Once the interpolation is completed, grid node values are transferred back to their appropriate known positions in the original coordinate system for contouring, visual display, and any further analysis. The resulting grid will exhibit the user specified regional anisotropy.

This approach is similar to the biased gridding technique described by Jones et al. (1986). The RADM approach is a simpler application of bias gridding. Because of this, it is more efficient and can be used as a preprocessing technique in any gridding algorithm.

Geologic Example: Structural Study

To demonstrate the Regional Anisotropic Data Mapping procedure, data from the Powder River basin Minnelusa Formation B-Sand was used. This formation is largely comprised of longitudinal, siliciclastic dunes. The dune field physiography is superimposed on a regional monocline.

Figure 8 presents a computer-generated map on the top of the B-Sand unit using the standard gridding assumption of isotropy. Experienced geologists working this area believe the map lacks the strong expression of both the regional monocline and orientation of dune features parallel to strike. This is corroborated by field workers working from outcrop data who have reported strong orientations in the Pennsylvanian dunes (Hunter, 1981). If it true that the surface is comprised of a series of longitudinal dunes roughly parallel to regional strike, the grain of the surface in Figure 8 needs to be accentuated in that direction.

To perform RADM, it is necessary to determine the anisotropic parameters for the structural datum. Anisotropy of the structural surface for the entire study area is considered first. Variography, as described above, was used to estimate the azimuth and magnitude of the anisotropy (Figures 9A and 9B).

The presence of drift (tilt or regional dip of the surface) is immediately apparent in Figure 9B as indicated by the monotonically increasing semivariogram with no sill (Olea, 1982; Olea, 1984). To determine the precise azimuth and magnitude, the drift must be removed and the semivariance of the residuals calculated (Yarus et al., 1990). Residual values were calculated by subtracting a first-order trend surface from the original data. The final semivariograms of the residual variance for the structural surface are shown in Figures 10A and 10B. In this case, the major range appears to be oriented N20W, parallel to the direction of structural strike (Figure 8). The value of the major range is 0.042, while the minor range has a smaller value of 0.011 (both numbers in relative units). The ratio of these two numbers is 0.042/0.011 or a magnitude in the strike direction is close to 4. All other combinations of semivariogram pairs produced range values greater than 0.011 or less than 0.042, thus generating smaller magnitudes.

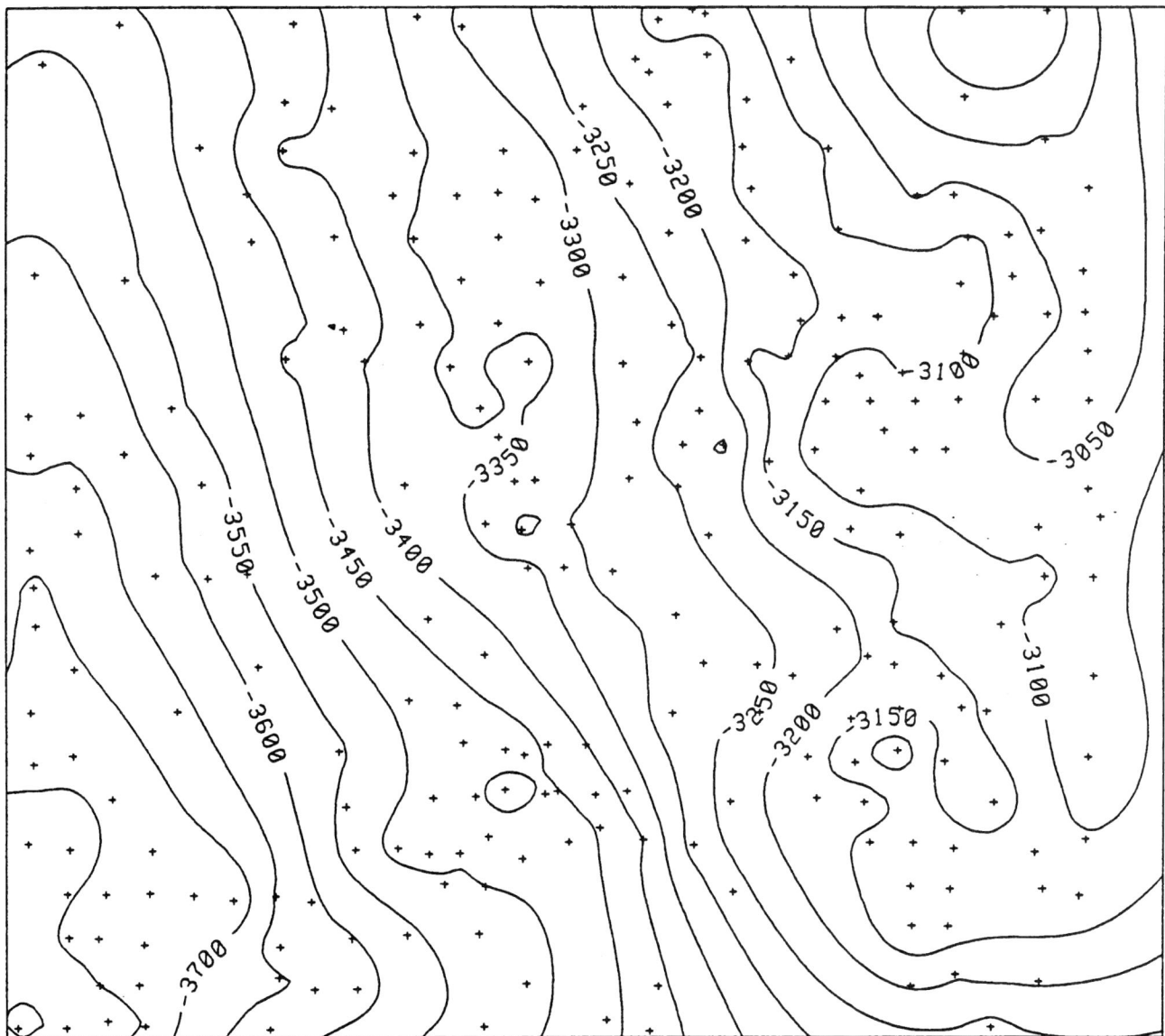

Figure 8. Computer-generated map on the top of Minnelusa Formation B-Sand without anisotropic information.

Figure 11 represents a computer-generated map on the above structural data with anisotropic information: direction N20W, magnitude 4. This figure was produced by first calculating the values of the grid nodes in compressed space using the RADM method, and then transferring the values back to their appropriate locations in the original coordinate system where the grid nodes and data are contoured. Thus the data are gridded in compressed space, but contoured and displayed in the original space. Inspection of Figure 11 reveals significantly enhanced structural grain over that of Figure 8, and provides a more faithful representation to the underlying geologic model. The improvement is the result of including only two additional pieces of information in the gridding algorithm: anisotropy azimuth and magnitude.

The Effect of Regional Anisotropic Data Mapping on Delaunay Triangles and Voronoi Polygons

The ability to improve gridding and contouring demonstrated in Figure 11 is the result of changing the spatial relationship between control points using the compression of the original space. The most important reason to use this methodology is to enable gridding algorithms to include one form of geologic information (anisotropy) in its interpolation procedures. This results in not only having more control over the appearance of the map, but also in the underlying mathematical structure of the two-dimensional model. This can be demonstrated by examining the effect RADM has on the calculation of Delaunay triangles (Cline and Renka, 1984; Davis, 1986) and Voronoi

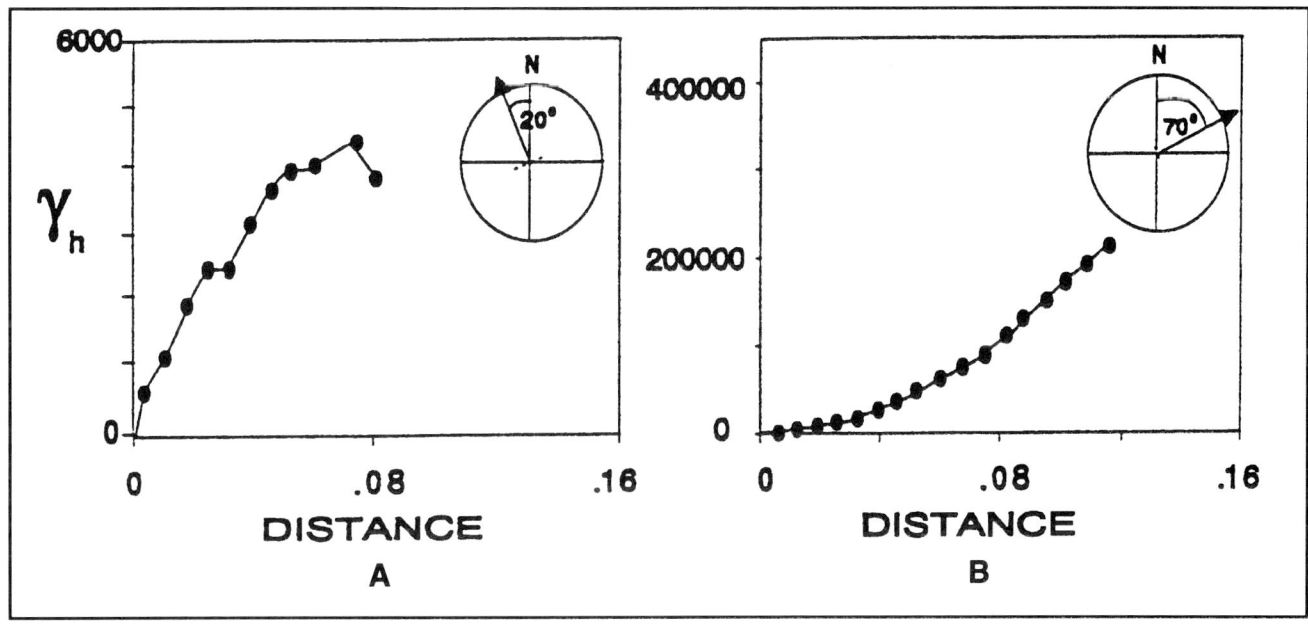

Figure 9. Semivariogram pair for structural datum over the entire study area. Semivariogram 9A is parallel to strike and 9B is parallel to dip. Note the presence of drift in the dip direction.

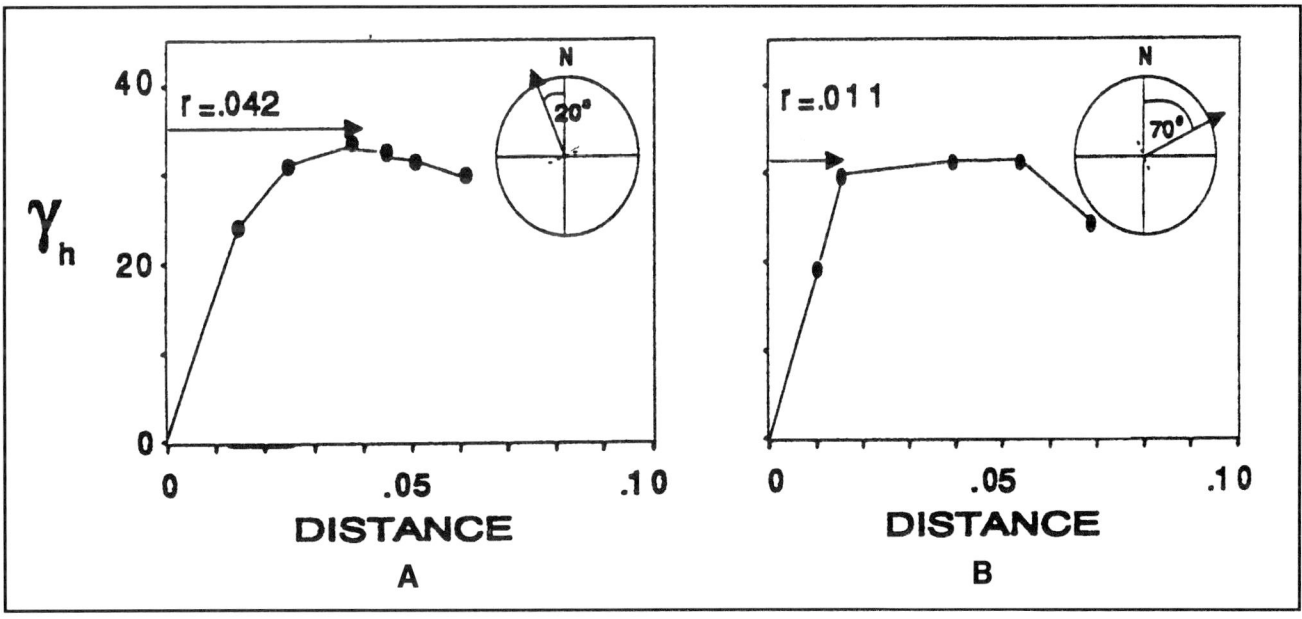

Figure 10. Semivariogram pair calculated from residual structure data. (A) Direction of maximum anisotropy is parallel to strike, N20W. (B) The minimum direction of anisotropy is parallel to dip, N70E.

polygons (Davis, 1986; Arakelyan, 1986; Watson and Philip, 1987). The main concept of Delaunay triangulation is to create as much as possible an equiangular set of triangles. Voronoi polygons are used in neighborhood based interpolation and for estimating drainage areas.

Performing calculations in compressed space offers certain advantages. As previously mentioned, most algorithms will perform more optimally when the data are isotropic. This includes triangulation and neighborhood based interpolations. Triangulation is particularly interesting from the standpoint of how the Delaunay triangle network is constructed. Figure 12 demonstrates the Delaunay set of triangles in the orig-

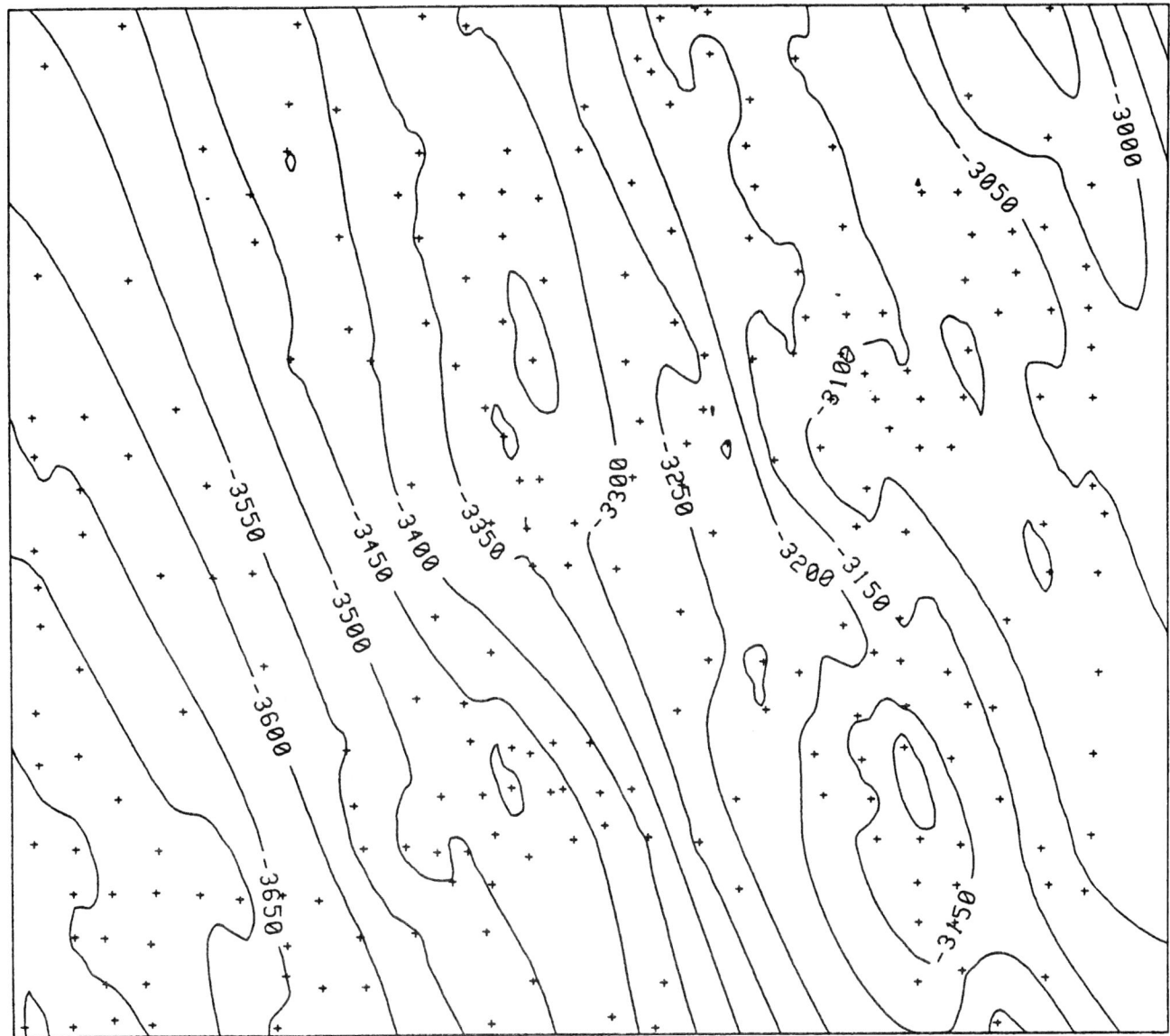

Figure 11. Computer-generated map on the same data as map on Figure 8 but with anisotropic information: direction N20W, magnitude 4.

inal space for the actual distribution of the data points. The construction of these triangles is a mathematical procedure that uses only information about the geographic distribution of the data points. No geologic information is utilized in the procedure (Cline and Renka, 1984; Davis, 1986). If Delaunay triangulation is performed in compressed space, a new set of triangles is selected because the geographic position of each data point relative to another is changed. For example, Figure 13 shows the result of Delaunay triangulation performed in compressed space using the anisotropic parameters discussed earlier in the structural study (azimuth N20W, magnitude 4), but displayed in original space. The triangles therefore appear elongated in the direction of maximum anisotropy.

The critical improvement of this technology is that the resulting configuration of triangles now carries with it information about the geologic properties of the surface and not just the distribution of its data points. As a result, a final map may look very different.

Voronoi polygons are a form of information that can be displayed on a map, but not contoured. They are derived from Delaunay triangles. These polygons define the shape of the area of influence exerted by each control point. Within a polygon, the control point exerts maximum influence over the interpolated values of grid nodes. Like grid node estimates, the Voronoi polygons are dependent upon the spatial distribution of the control points. In fact, the only information necessary to determine the size, shape, and position of the polygons is the geographic distribution of the control points. The mapping variable (structure, isopach, porosity, perme-

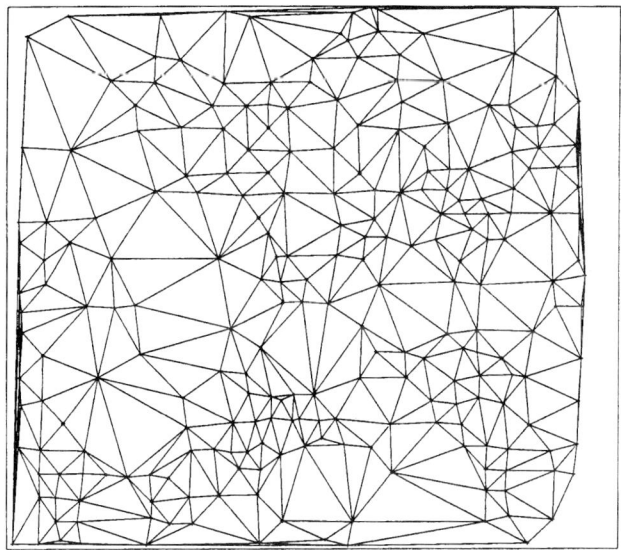

Figure 12. Delaunay triangles for non-compressed data.

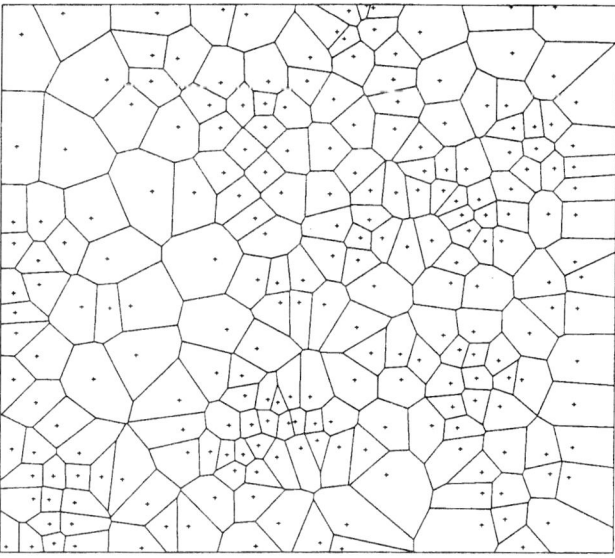

Figure 14. Voronoi polygons (tessellations) for non-compressed data.

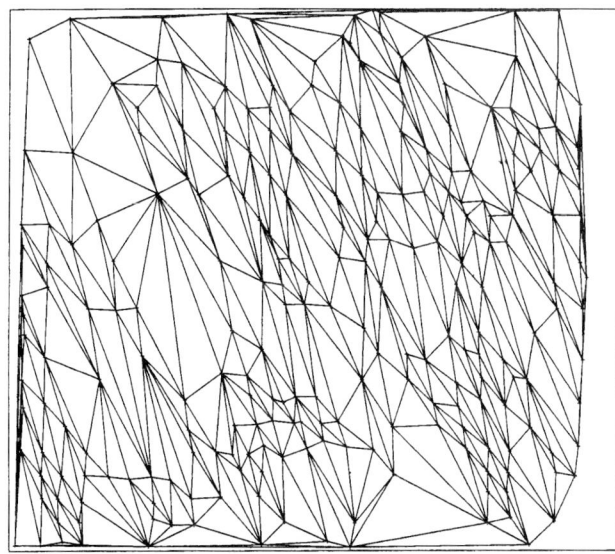

Figure 13. Delaunay triangles for compressed data with anisotropic information: direction N20W, magnitude 4.

ability, etc.) has no influence at all. As a consequence, the polygons tend to be useful only when the mapping variable is isotropic, and the density of control points is high and uniformly distributed throughout the area of investigation. In terms of drainage area calculations, this would imply that a given mapping variable—permeability for example—is isotropic. This may be an unreasonable assumption, thus negating the utility of the polygons. Similarly, the effect on neighborhood based interpolation can also be unreasonable, although the reasons for this may be less obvious to some readers. A detailed discussion of neighborhood based interpolation can be found in Arakelyan (1986) and Watson and Philip (1987), which describes the dependency of spatial arrangement of data points and the selection of Voronoi polygons.

However, because the spatial distribution of the control points is critical to the construction of the polygons, a change in the spatial arrangement will produce a different set of polygons. As discussed earlier, RADM changes the spatial arrangement in order to model anisotropy of the data. Consequently, when the mapping variable is known to be anisotropic, the RADM can be used to produce a set of Voronoi polygons that more realistically reflect the underlying anisotropy of a given variable. Generally stated, the polygons can be reshaped to reflect the presence of anisotropy when it exists. To effect this change, the polygons are created in compressed space, and displayed in original space.

For example, Figure 14 displays the set of polygons calculated in original space for the data used earlier in the structural study above. Each polygon in this figure circumscribes one control point, defining its area of influence. If the polygons in Figure 14 were to be used in drainage area calculations, there would be an underlying assumption that any pertaining geologic data related to flow directions are isotropic (no preferred flow direction). If the data were anisotropic, having a preferred direction of flow (fractures, porosity trends, etc.), the polygon solution displayed in Figure 14 would not accurately depict the underlying geologic model.

Figure 15 shows the Voronoi solution using the RADM technique. The polygons are created in compressed space using the anisotropic parameters previously determined (azimuth N20W, magnitude 4), and displayed in original space. Compared to Figure 14, the Voronoi solution presented in Figure 15 distinctly shows

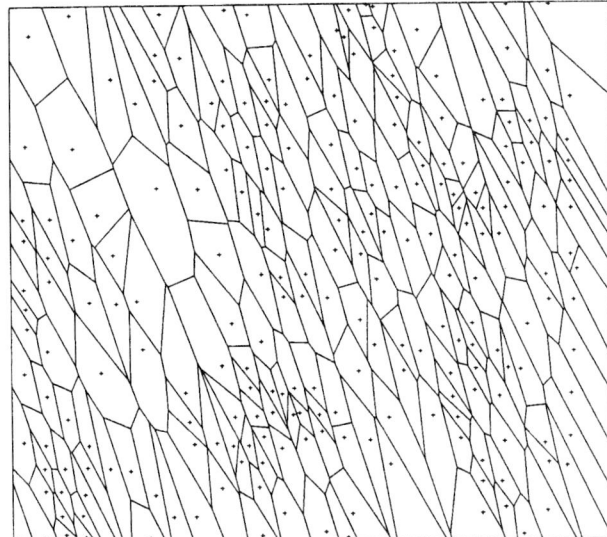

Figure 15. Voronoi polygons (tessellations) for compressed data with anisotropic information: direction N20W, magnitude 4.

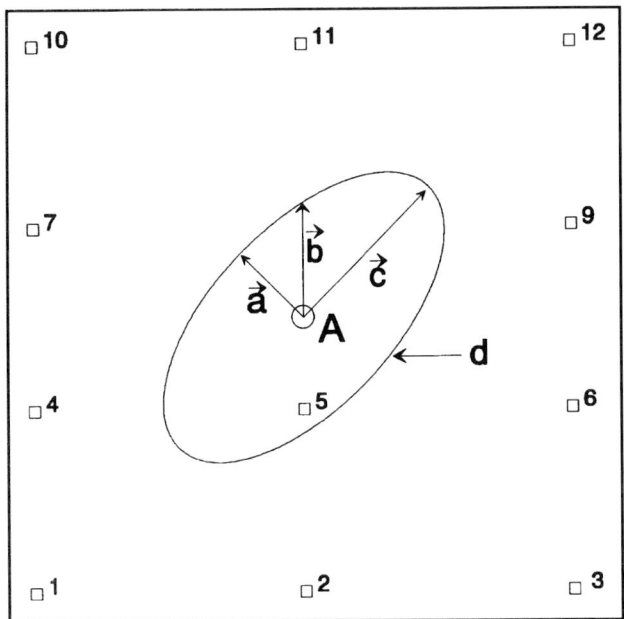

Figure 16.
O — Data point A.
▫ — Grid node.
$|\vec{a}|$ — Weight in direction perpendicular to direction of anisotropy.
$|\vec{b}|$ — Weight in intermediate direction.
$|\vec{c}|$ — Weight in direction of anisotropy.
d — Ellipse representing line of equal influence.

a prevailing NW–SE direction of orientation of Voronoi polygons. Since the polygons reflect more precisely the underlying anisotropic model, drainage area estimates can be made more accurately.

Additional applications of Voronoi polygons are varied and go beyond the scope of this paper. The following are a few of the applications regarding reservoir heterogeneity the authors have considered. Some readers may find these interesting. For example, the polygons can be color-coded to reflect the distribution of reservoir quality. Each color could reflect a range of values for a particular variable such as permeability, porosity, resistivity, cumulative production, etc. Further, this type of map may be useful in selecting offset drilling locations and in planning the direction of horizontally drilled wells.

Localized Anisotropic Data Mapping (One Or More Local Directions)

Introduction

RADM is most applicable in simple geologic settings where there is only one dominant direction of anisotropy. In many situations, however, the geologic setting may be more complex, composed of numerous dominant features each with varying directions and magnitudes of anisotropy. In the example above from the Minnelusa, the dominant structural component was the regional monocline that obscures much of the detail of the dune physiography (Figure 11). A better way to examine the detailed character of the dunes, without the influence of the regional structure, would be by isopaching the objective Minnelusa sand. This situation may require certain features on the map to have different orientations than others. That is, not all dunes will have the identical alignment.

Localized Anisotropic Data Mapping (LADM) is based on the same principle as RADM, that in the presence of anisotropy, the relationship between data points will be stronger in some directions than in others. However, in this case the LEI is not assumed to be identical for every data point as it was previously. Instead, the LEI for each data point is individually determined thereby uniquely defining the influence, or weight, that it will have on appropriate grid nodes. Consequently, the coordinate system cannot be uniformly changed as it is in RADM, because the anisotropic parameters (azimuth and magnitude) can vary from point to point.

An example best demonstrates how LADM works. Figure 16 shows data point A surrounded by grid nodes 1 through 12. The ellipse **d** represents the line of equal influence. The influence on a grid node from a given data point A will not only be dependent upon the distance to this grid node, but also its position relative to the orientation of a circumscribing ellipse **d** around data point A on Figure 16 (Kushnir, 1984; Jones et al., 1986). For example, the Z-values at grid nodes 3 and 10 will be dependent upon the distance from data point A, and the weight determined by the length of vector **a**. The Z-values at grid nodes 1 and

12 will be dependent upon the distance from data point A, and the weight determined by the length of vector **c**. Similarly, the Z-values at grid nodes 2 and 11 will be dependent upon the distance from data point A, and the weight determined by the length of vector **b**. In general, for a given data point, a set of control points are selected, weighted, and mathematically accumulated according to a chosen algorithm. Thus LADM modifies the influence of a given data point on the appropriate grid node. The final interpolated value for a grid node depends on the algorithm that is used. Therefore, no specific changes can be described in this paper, but the example should help to guide the alteration of these algorithms.

As a practical matter, the simplest way to implement LADM is by defining groups of data points that have the same anisotropic properties. However, when changes are subtle, anisotropic properties can be assigned to each individual data point. This would allow a surface to exhibit gradual changes from one area to another. Handling data in this way is difficult in that it requires more upfront modeling, but should present no particular problem to the computational efforts of a computer.

In the past, other procedures were used that attempted to produce results similar to RADM technique (Jones et al., 1986). This was done by defining polygons around each local area, which would be gridded independently from other areas using an RADM-like method. The independently gridded areas were then combined into one surface model that had anisotropic properties varying from area to area. Unfortunately, the boundaries between local areas were difficult to connect and generated sharp transitions, sometimes appearing more like faults than smooth surfaces. Moreover, when a polygon does not have enough points for qualitative gridding, the resultant map may not be correct in this local area.

Using the LADM approach to model local anisotropy creates a final surface model that has smooth transitions from one local area to the next. These smooth transitions result from the fact that grid nodes on or near the boundaries of two local areas will use data from both to generate the appropriate, representative weights. The result is a logical blending of anisotropies at their boundary.

Geological Example: Gross Isopach Study

Figure 17 represents the initial computer-generated isopach map (gross isopach map between the top of the Minnelusa B-Sand and the top of the Minnelusa B-Dolomite formations) of the study area prior to including anisotropic information. It suggests, along with other geological information, that there are a number of individual geologic features (dunes). Four dune-type features can be crudely identified, one in each quadrant of the map. However, not all the dunes may have the same orientation. To determine the orientation of each feature, the map was divided into four equal quadrants, and each quadrant subjected to variography (Figure 17).

Results show two prominent directions of anisotropy: N70E and N60W. Figures 18(A,B) and 19(A,B) show the variogram pairs for the NE and SE quadrants respectively, which demonstrate these predominant orientations. Variography for the SW quadrant indicates that the data are isotropic, while results for the NW quadrant are similar to that in the SE quadrant. In places where the data are sparse, as in the northwestern quadrant of the study area, the user may wish to model the semivariogram using additional information, or information from a similar area where there is more control. For the purposes of demonstrating the versatility of the LADM procedure, these identified local trends and magnitude are sufficient.

To build a map using the LADM procedure, the information from each quadrant is used as a first approximation of the orientation and magnitude of each specific geological feature. The exact locations of the features are subject to some interpretation, but are grossly defined and shown in Figure 20: a large polygon trending northwest–southeast and connected by a narrow neck, and a small one oriented east-north-east–west-northwest. The large polygon defines two en echelon dunes oriented in a northwest–southeast direction. Data inside the large polygon were assigned anisotropic parameters consisting of an azimuth of N60W and a magnitude of 2. The small polygon defines a third feature, which has east, north-easterly orientation. Data inside this polygon were assigned anisotropic parameters consisting of an azimuth of N70E and a magnitude of 3. Thus the data points in each polygon have their own orientation and magnitude. Data outside the polygons were treated as isotropic (LEI was a circle). The resulting map (Figure 21) accommodates knowledge about the geology through the appropriate anisotropic parameters.

The resulting map (Figure 21) is an improvement over that of Figure 20. However, the strength of the magnitude can be further adjusted, depending on the effect desired. In this respect the LADM technique is very flexible. For example, if the geologic model calls for stronger anisotropy than demonstrated in Figure 21, the magnitude can be increased to produce the desired effect. Figure 22 shows the result of increasing the magnitude to 4 in each of the polygons.

Local and Regional Anisotropic Data Mapping

Finally, if the model suggests that regional anisotropy is present outside the defined polygons, this information can be introduced as well. Figure 23 shows the result of introducing regional anisotropy to the area outside the polygons in addition to the local anisotropic features described above. Data outside the polygons were assigned the following anisotropic parameters: azimuth = N20E, magnitude = 4.

SUMMARY AND CONCLUSIONS

Conventional gridding algorithms do not perform well on geological data with underlying trends, par-

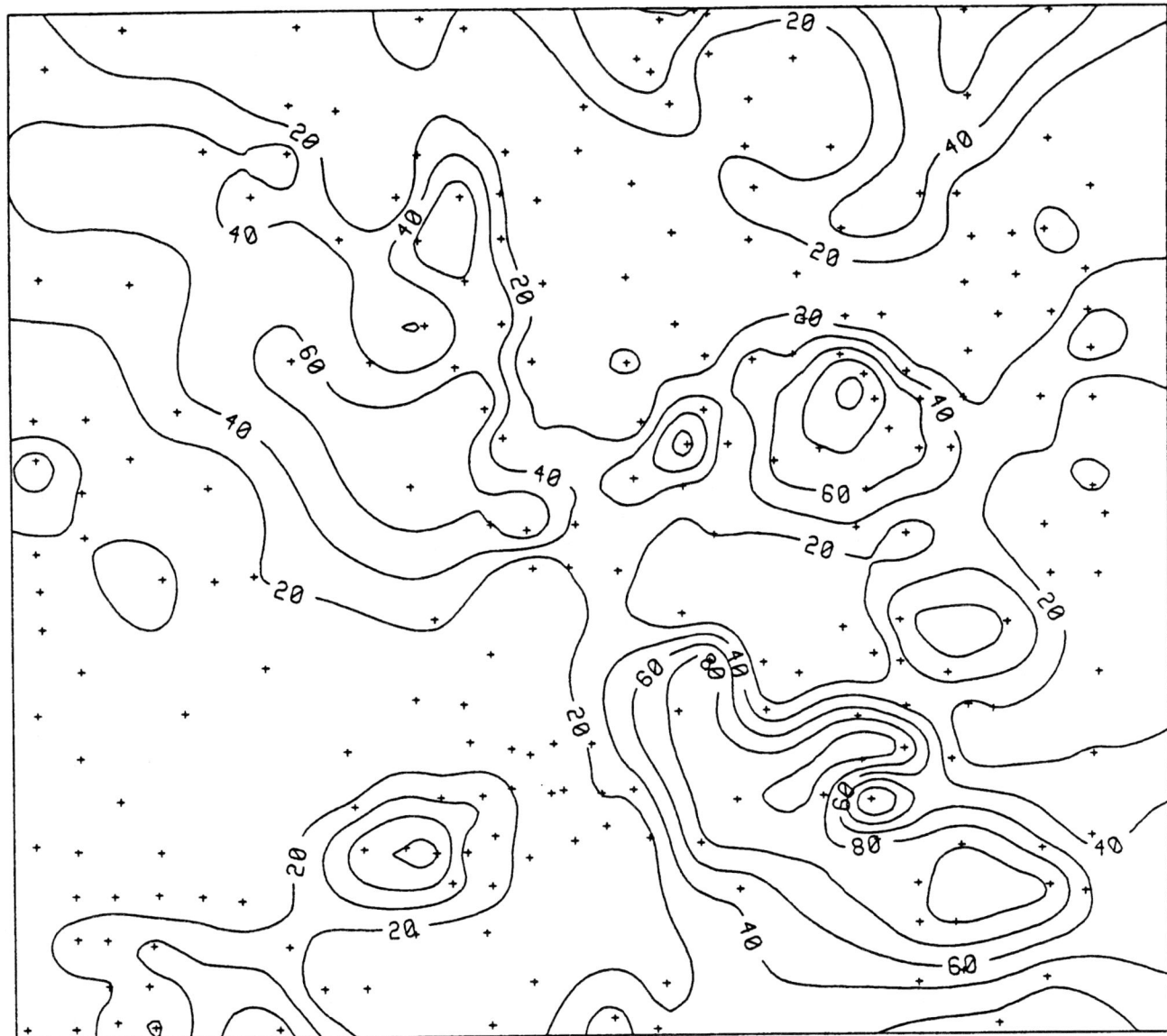

Figure 17. Gross isopach map between formations B-Sand and B-Dolomite without anisotropic information.

ticularly when the data array is less than optimal. One critical reason for this is the underlying assumption of isotropy (no trends), which most gridding algorithms make about the mapping variable. Because of this, resulting maps often do not faithfully resemble the geologic model, and can appear unrealistic. Anisotropic Data Mapping (ADM) is a technique which incorporates common geological information, direction and magnitude, which describes the anisotropic properties of the mapping variable. The procedure is versatile, capable of handling both regional anisotropy (RADM), local anisotropy in multiple directions (LADM), or both. The introduction of these two properties into the interpolation procedure results in more realistic maps which honor the underlying geologic model. Further, ADM changes the spatial relationship of the data points. This results in not only having more control over the appearance of the map, but also in the underlying mathematical structure of the two-dimensional model. For example, RADM incorporates information about the anisotropy of mapping variable in the process of creation of Delaunay triangles and Voronoi polygons. Voronoi polygons can be used in drainage area calculations, spotting development and infill drilling locations, horizontal drilling, and areas requiring reservoir characterization. ADM does not yet handle overprinting, or areas where trends overlap.

ACKNOWLEDGMENTS

The authors would like to thank Plains Petroleum Company for providing data from the Powder River

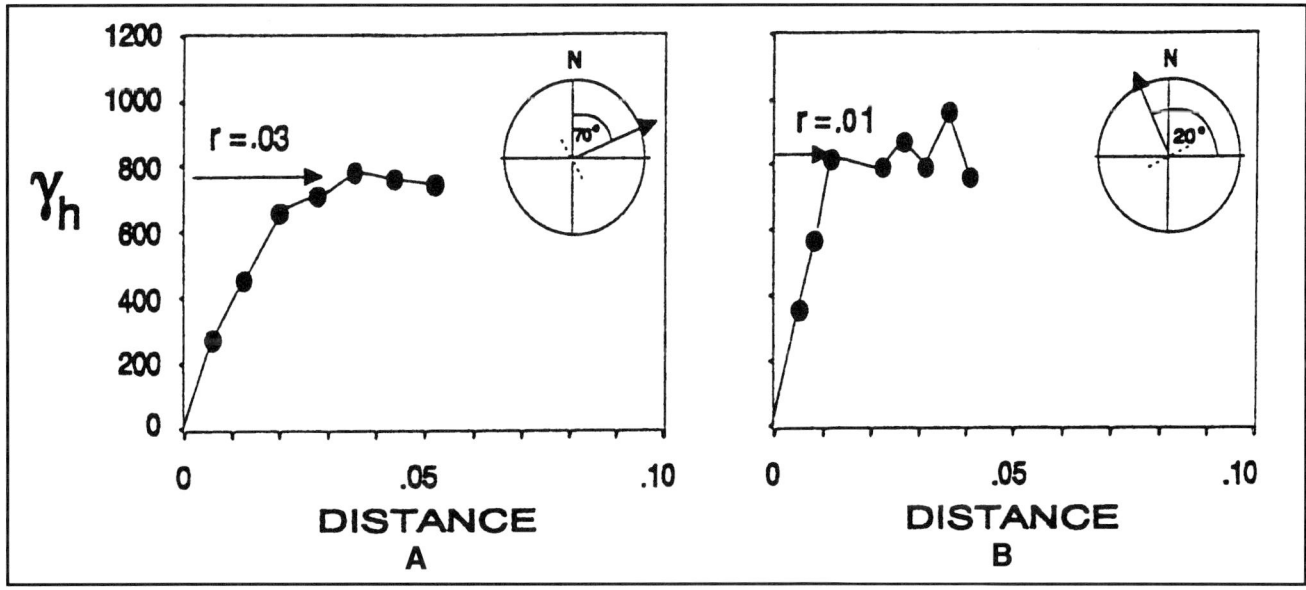

Figure 18. Semivariogram pair showing direction of maximum anisotropy, N70E (A) and the corresponding perpendicular N20W (B) in the NE quadrant.

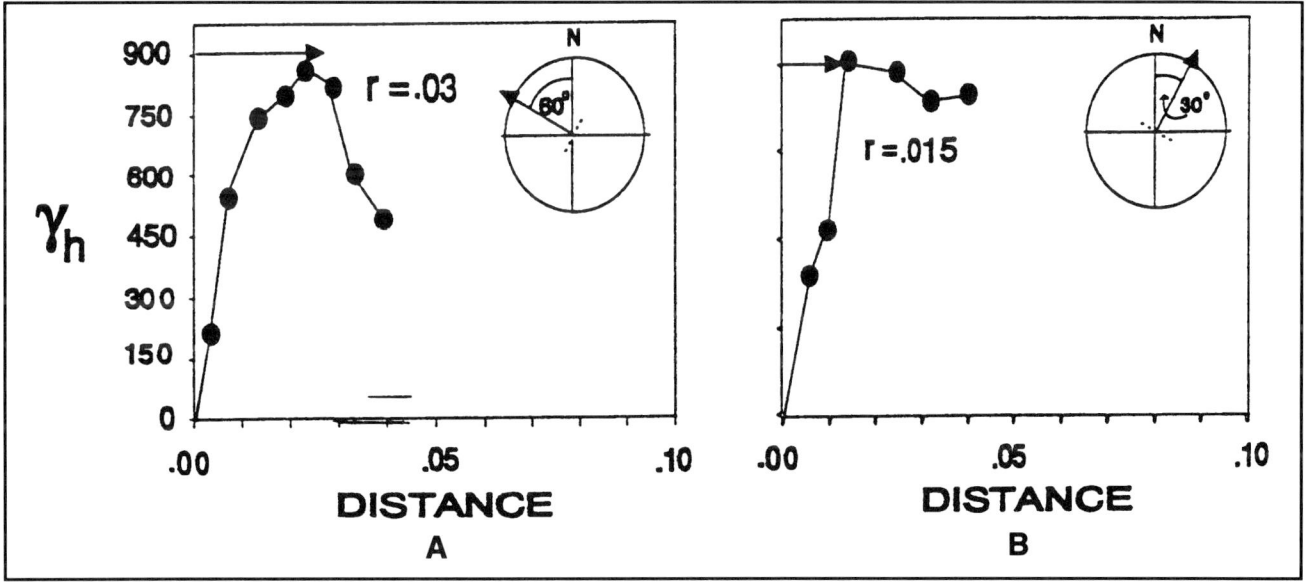

Figure 19. Semivariogram pair showing direction of maximum anisotropy, N60W (A) and the corresponding perpendicular N30E (B) in the SE quadrant.

basin. In addition, we would like to thank David E. Hamilton and Thomas A. Jones for their comments and suggestions during the review of this paper.

Acknowledgment from Gregory Kushnir

I would like to thank GeoGraphix, Inc., especially President David L. Armitage and Software Development Manager Kevin F. Cartin, for the support and opportunity to pursue the research and development of the subject of this article. I would like to thank Dr. Vladimir I. Aronov (GEOINFORMSYSTEM, the USSR) for knowledge and guidance in computer mapping, and Dr. Vladimir A. Arakelyan (VNIGNI, the USSR) for initial conversations about the substance of this article. Finally, I would like to thank Dr. Jeffrey M. Yarus for numerous discussions and support.

Acknowledgment from Jeffrey M. Yarus

I wish to thank Marathon Oil Company for allowing my participation in this paper. In particular, I

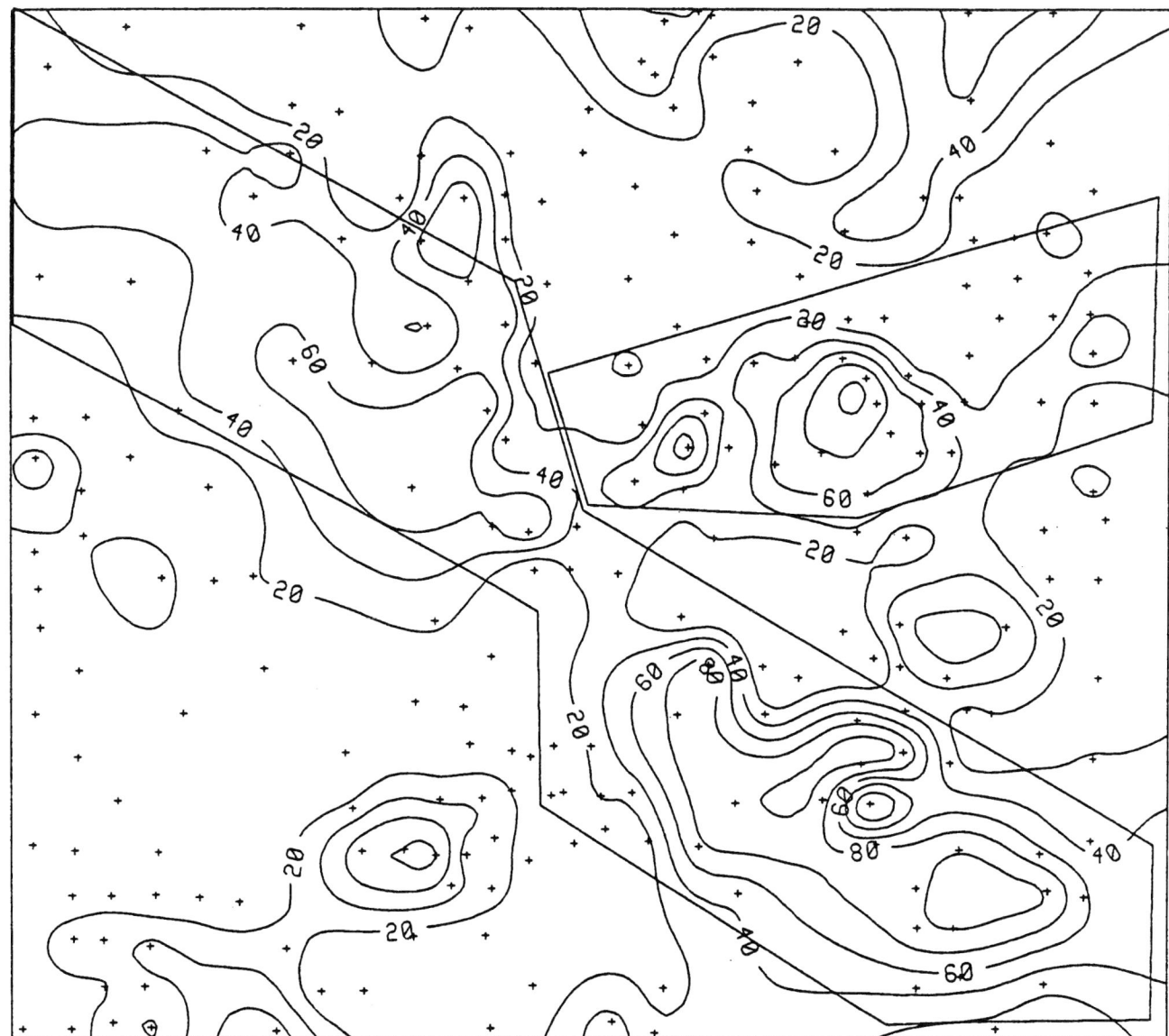

Figure 20. Gross isopach isotropic map with polygons.

would like to acknowledge Dr. Timothy Coburn and Dr. Charles Vestal for their comments, suggestions, and patience. In addition, I am indebted to Dr. John Davis and Dr. Robert Ehrlich, whose insights and teaching skills enabled me to make a small contribution to the field of mathematical geology. Finally, I would like to thank GeoGraphix, Inc. for supporting and implementing these ideas, and allowing me the privilege of working with Dr. Gregory Kushnir.

REFERENCES CITED

Aronov, V. I., M. M. Ellansky, V. A. Arakelyan, T. N. Kolchitskaya, and G. F. Kushnir, 1982, Automatization of oil and gas reserve estimation: Moscow, VIEMS, 66 p.

Arakelyan, V. A., 1986, New method of interpolation of geological properties for construction of contour maps with the computer, GIDROPROEKT: Moscow, p. 114 - 117.

Cline, A. K., and R. L. Renka, 1984, A storage-efficient method for construction of a thiessen triangulation: Rocky Mountain Journal of Mathematics, v. 14, no. 1, p.119 - 139.

David, M., 1977, Geostatistical ore reserve estimation: New York, Elsevier Scientific Publishing Company, 364 p.

Davis, J. C., 1986, Statistics and data analysis in geology: New York, John Wiley and Sons, 646 p.

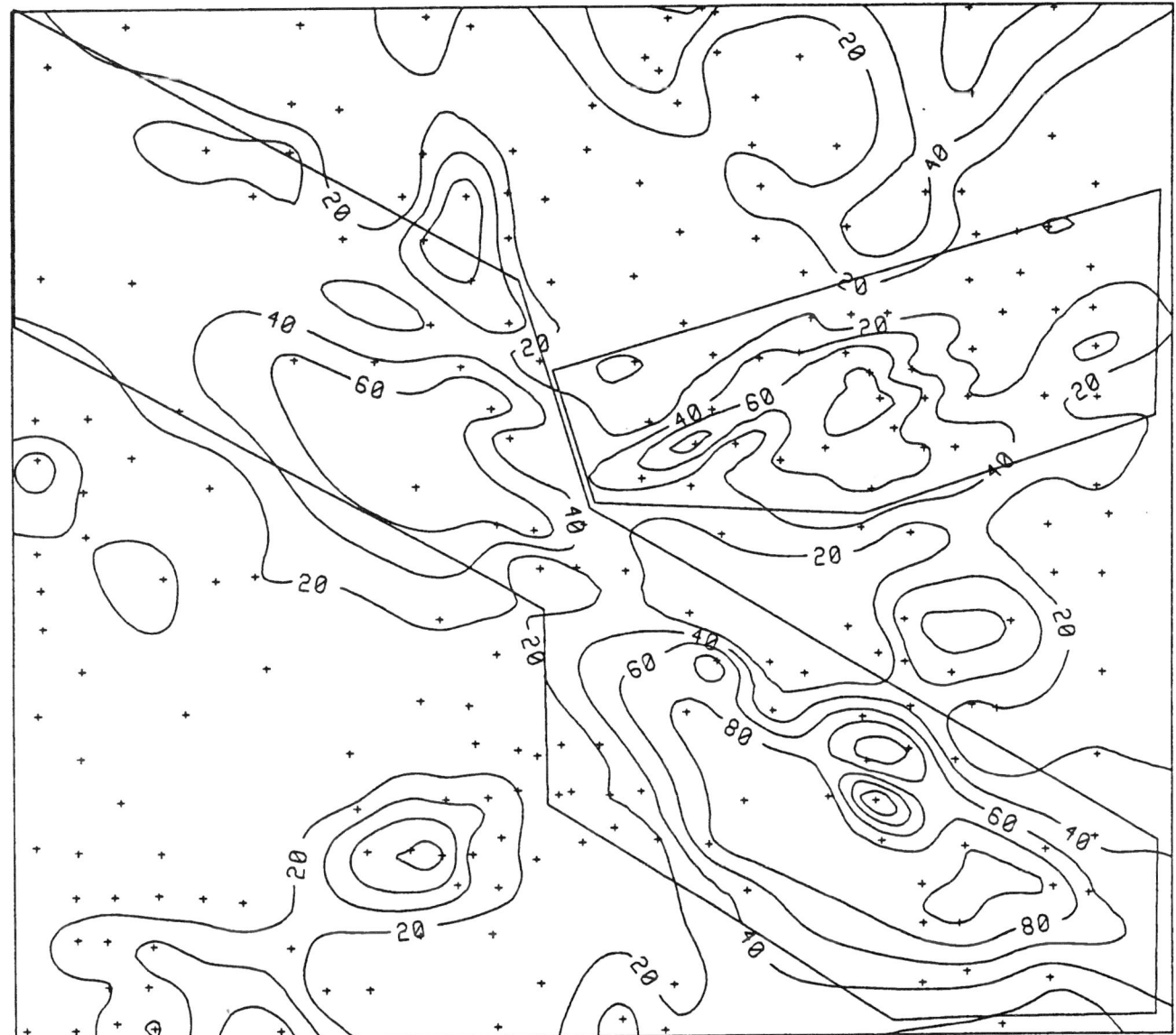

Figure 21. Computer-generated map with anisotropic information: Large polygon anisotropy = N60W, magnitude 2; Small polygon local anisotropy = N70E, magnitude 3.

Hohn, M. E., 1988, Geostatistics and petroleum geology: New York, Van Nostrand Reinhold, 264 p.

Hunter, R. E., 1981, Stratification styles in eolian sandstones: some Pennsylvanian to Jurassic examples from the Western Interior USA, SEPM Special Publication 31, pp.315-329.

Jones, T. A., D. E. Hamilton, and C. R. Johnson, 1986, Contouring geological surfaces with the computer: New York, Van Nostrand Reinhold, 314 p.

Kushnir, G., 1984, Geometrization of oil and gas fields on development stage using mathematical methods and the computer: PhD dissertation, Moscow, VNIGNI, 176 p.

Matheron, G., 1963, Principles of geostatistics: Economic Geology, v. 58, p. 1246 - 1266.

Olea, R., 1982, Optimization of the High Plains Aquifer Observation Network, Kansas: Groundwater Series 7, Kansas Geological Survey, p. 36 - 73.

Olea, R., 1984, Systematic sampling of spacial functions: Series on Spatial Analysis No. 7, Kansas Geological Survey, p. 4 - 5.

Shurubor, Y. V., N. N. Markov, M. L. Rusanova, Z. M. Fakhrudinova, and M. K. Lenskykh, 1976, The experience of using of anisotropy geological properties for structural mapping with the computer, Journal Oil and Gas Geology and Geophysics 8: Moscow, VNIIOENG, p. 48 - 49

Watson, D. F., and G. M. Philip, 1987, Neighborhood-based interpolation: Geobyte, v. 2, no. 2, p. 12 - 16.

Yarus, J. M., and G. Kushnir, 1990, Anisotropy and

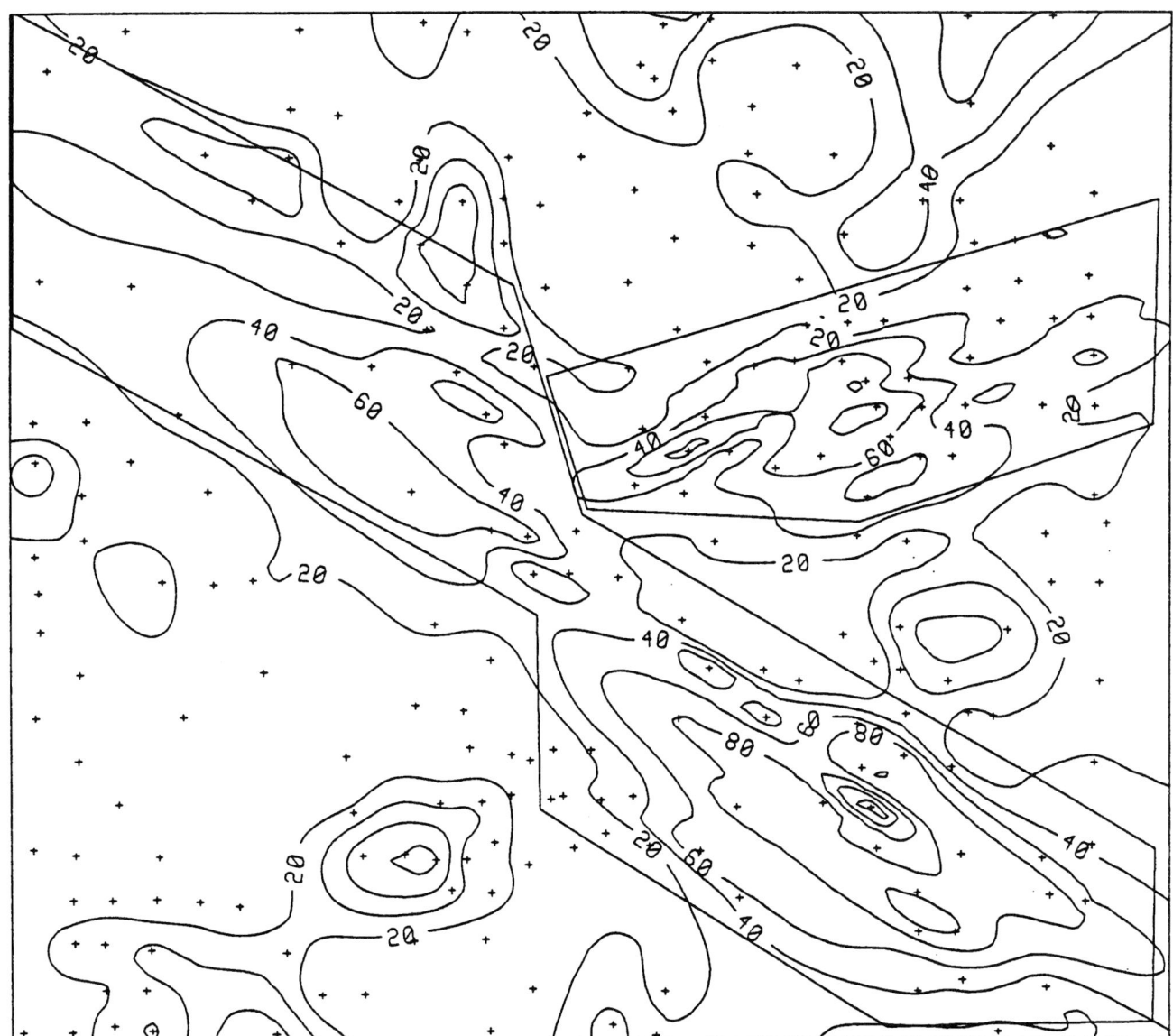

Figure 22. Computer-generated map with anisotropic information: Large polygon anisotropy = N60W, magnitude 4; Small polygon local anisotropy = N70E, magnitude 4.

computer mapping — reading between the lines (abs.), AAPG Bulletin, v. 74, no. 8, p. 1339.

Yarus, J. M., and J. P. Lewis, 1989, Machine contouring, pitfalls and solutions: Guide for the explorationist (abs.): AAPG Bulletin, v. 73, no. 3, p. 429.

Yarus, J. M. et al., 1990, More information from computer contoured maps: regional component mapping and kriging— reading between the lines (abs.): AAPG Bulletin, v. 74, no. 5, p. 794 - 795.

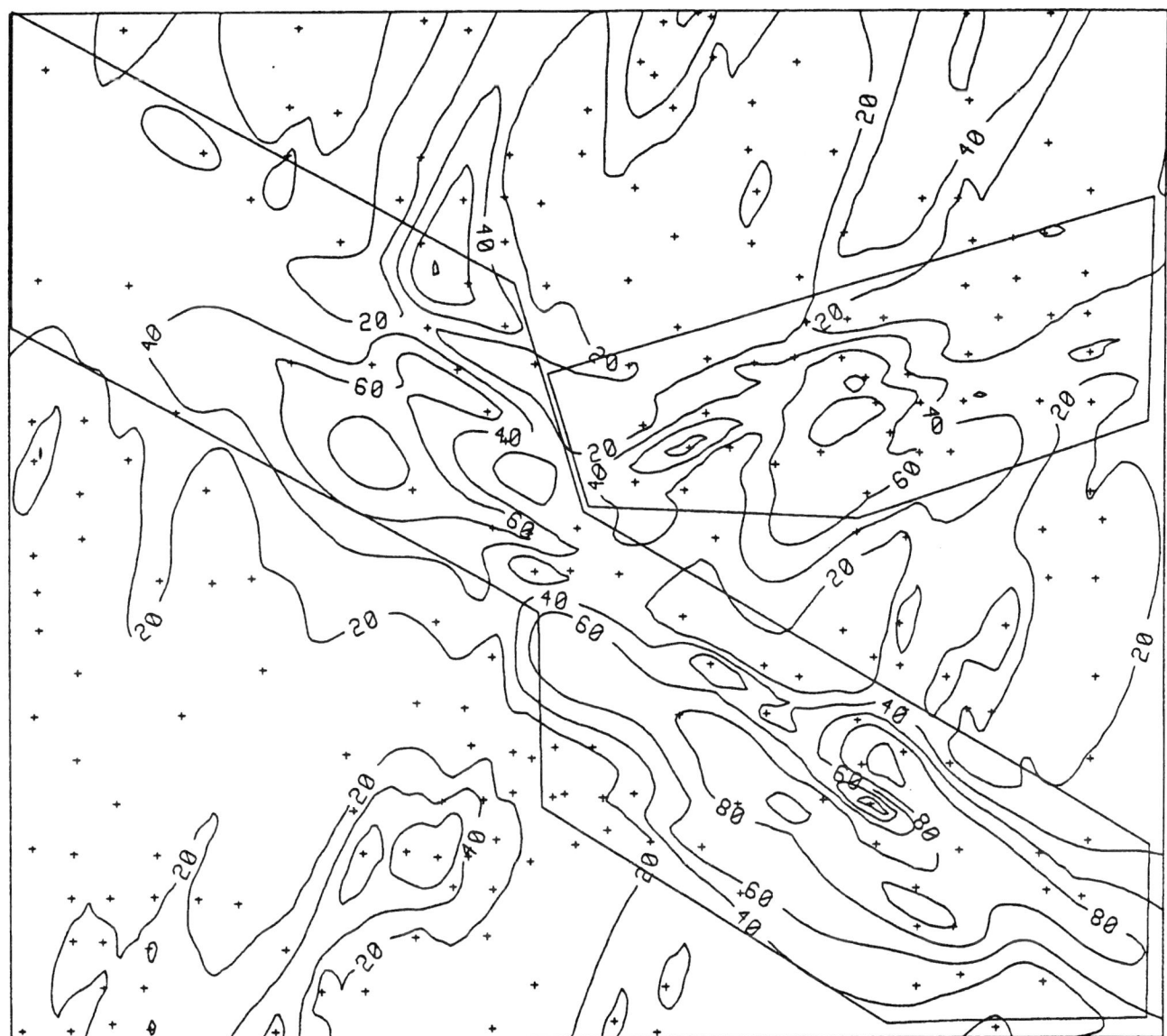

Figure 23. Computer-generated map with regional and local anisotropic information: Regional anisotropy = N20E, magnitude 4; Large polygon local anisotropy = N60W, magnitude 4; Small polygon local anisotropy = N70E, magnitude 4.

Chapter 8

Surface Modeling for Sedimentary Basin Simulation

John C. Tipper
Geologisches Institut der Albert-Ludwigs-Universität
Freiburg im Breisgau
Germany

ABSTRACT

Successful surface modeling requires, first of all, that an appropriate spatial framework be selected. The more diverse the surface modeling application, the more important this selection is. Of the four most obvious spatial frameworks—digital arrays, rectangular grids, parametric grids, and triangular meshes—triangular meshes seem to offer the most promise for many applications that involve both surface-fitting and the numerical solution of sets of equations. The use of triangular meshes is illustrated here in the context of a three-dimensional simulation of sediment build-up in a sedimentary basin. Triangle-based interpolation is first used to specify the initial conditions for the simulation, and then a triangle-based finite-difference technique gives the new sediment surface for successive steps in time.

INTRODUCTION

Geological applications of surface modeling can conveniently be characterized as either surface-fitting or equation-solving. Many of the contoured maps and diagrams that are used routinely in sedimentary basin analysis are good examples of surface-fitting, e.g. isopach and structure contour maps. These rarely have to conform to any underlying mathematical model, but are just interpretations of what the particular geological surfaces would look like if they were realized physically and exposed at the surface, constrained somewhat by the data on which they are based. In contrast, the surfaces that are modeled in equation-solving applications have known underlying forms. Thus when numerical techniques are used to study, for example, flow in a ground-water basin, the calculated piezometric surface for each aquifer must everywhere satisfy the differential equations that govern the flow. Surface-fitting and equation-solving are quite different contexts for surface modeling, and it is hardly surprising that the techniques used in each context are usually quite distinct.

The problem with using different types of techniques in the two contexts is that the results may not be strictly compatible. This can be a problem in some circumstances. In the ground-water example, for instance, calculation of the piezometric surfaces may be carried out using a rectangular finite-difference mesh, whereas the isopach maps of the aquifers concerned (which would usually be constructed beforehand) may be contoured by Kriging. The results would then be compatible only at the nodal points of the mesh. This problem of compatibility may not matter greatly in situations where different sets of results are only to be compared visually, or where the precision of any of the data is rather limited. In other situations, however, where surface-fitting and equation-solving applications are closely linked, it is essential that their results be absolutely compatible. In this paper I explore one such situation, sedimentary basin simulation, and show how an appropriate spatial framework can give this compatibility. Finally, I look at some of the different spatial frameworks that can be used for surface modeling, and point out various advantages and disadvantages that they have when used in this type of situation.

SIMULATING SEDIMENTARY BASINS

Computer simulation of sedimentary basins is a field that has developed rapidly in the last two decades. Harbaugh and Bonham-Carter (1970) documented many of its hesitant beginnings: the simulations described by Helland-Hansen et al. (1988), Jervey (1988) and Tetzlaff and Harbaugh (1989) are good examples of what is routinely done today. The reason for this rapid development is that the value of computer simulation is now widely appreciated. Only with simulation can systems as complex as sedimentary basins be studied effectively and in detail, systems that (1) are fully four-dimensional (the three spatial dimensions, and time), (2) have space- and time-dependent parameters, and (3) involve processes which may be continuous or discontinuous, which may operate at a wide range of (generally time-dependent) rates, and which are, almost without exception, delightfully non-linear. Perhaps it is not surprising that sedimentary basin simulation can provide almost the complete spectrum of surface modeling situations.

For those aspects of basin simulation for which there is sufficient mathematical theory, formal numerical modeling can be carried out. The standard techniques (finite-difference, finite-element, boundary-element) are then used to solve the appropriate sets of equations. For other aspects of basin simulation, however, formal numerical modeling is not possible: there simply is no adequate underlying theory. Thus, whereas detailed thermomechanical modeling of basement subsidence can be used to predict how basement configurations evolve in time (Beaumont et al., 1982), in studying the way that a sediment pile is built up by clastic depositional processes over long time-spans, the most promising simulation approaches still seem to be those based on the sorts of rather ad hoc models that Harbaugh and Bonham-Carter (1970) pioneered.

It might be thought that the formal numerical model must always be the goal in basin simulation and any less sophisticated model can be acceptable only as a temporary expedient. This is by no means so. The greater precision of the formal model is generally more apparent than real, because it is rarely the modeling method itself that limits the precision of the results. In most cases, the limiting factor is the almost inevitable imprecision with which the parameters of the model (whatever the model) and its initial and boundary conditions must be specified. The keys to successful basin simulation seem to be the same, irrespective of the degree of sophistication in the model that is used: (1) to be able to specify with sufficient precision the initial and boundary conditions of the simulation, (2) to be able to reproduce faithfully and in sufficient detail any required basin geometry, and (3) to do this within a framework that allows the parameters of the simulation to be varied both in space and in time.

A BASIN SIMULATION MODEL

The simulation model introduced here is concerned with how sediment builds up in a basin over time. Its object is not, however, to focus on the detailed patterns of that sedimentation, but rather to enable the build-up of the sediment to be seen from a stratigraphic viewpoint (cf. Culling, 1960, p. 336). The time-scale of the simulation model is thus that of the growing sediment pile (of the order of millions of years), not that of the individual depositional and erosional events. This model contrasts markedly in its intent with the SEDSIM models developed successfully by Tetzlaff and Harbaugh (1989) for simulating clastic sedimentation. Those models are for simulation on a much more restricted scale than that used here: horizontal distances of kilometers rather than tens or hundreds of kilometers, and time spans of tens of thousands rather than millions of years.

The theoretical basis for the simulation model (Tipper, 1991a; 1992) is the stochastic nature of sedimentation systems. In any local area of a sedimentary basin, events of three different types (deposition, non-deposition, and erosion) recur in time in a complex and relatively unpredictable sequence. This sequence varies in detail within that area, and the sedimentation system that is responsible can best be characterized just by attributing probability distributions to the relative frequency and duration of each event type. These probability distributions change across the basin, from environments that are predominantly depositional to those dominated by erosion, and from environments that are active (either depositional or erosional) to those in which non-deposition is the norm.

Continued sediment build-up at any point in a basin demands that the environment there be predominantly depositional, although short-term accumulation can certainly occur in environments that are predominantly erosional. A useful measure of how likely build-up is to occur is the time that any deposited horizon can be expected to survive before being removed by erosion: this follows directly from the probability distributions that describe the sedimentation system at that point. The reciprocal of this survival time is a variable which can be mapped across the basin. High in areas that are highly erosive and tending to approach zero in sediment sinks, it measures what can be described as the degradational energy level of the environment. It is termed here the *degradation potential*.

With the degradation potential defined everywhere over a sedimentary basin, the processes of deposition, erosion, and sediment transport in the basin can now be seen as aspects of potential-driven flow, analogous to the way that fluid moves through a porous medium. Thus the transport of sediment laterally across the basin floor can be taken to be controlled by the conductance of the transport path and by the difference in degradation potential at the path ends. Deposition and erosion correspond to sediment going into and out of storage. The equilibrium condition, in

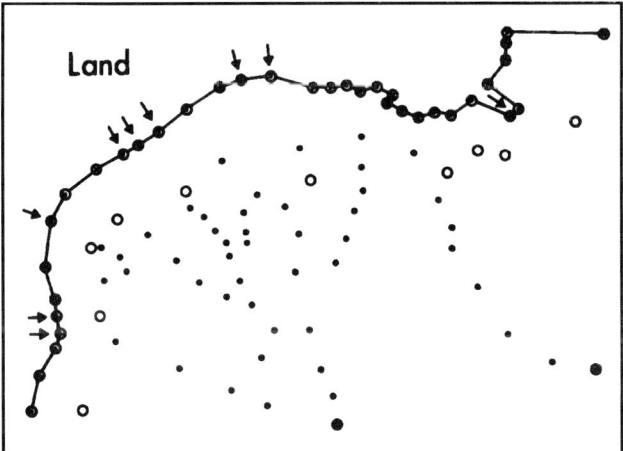

Figure 1. Data locations for a simulation of clastic sedimentation in a basin fed from its margins at the arrowed points. Solid circles indicate points at which initial basin floor elevations are specified. Open circles indicate points at which initial potentials are specified. Coincident solid and open circles indicate points at which both elevations and potentials are specified. Source fluxes are also specified at the arrowed points.

which nowhere is there erosion or deposition and all supplied sediment is by-passed through the basin, is what is conventionally recognized as the attainment of base level.

This potential-driven flow formulation (see Appendix 1, this chapter, for details) is implemented here by using a finite-difference approximation over an irregular triangular mesh (Macneal, 1953; Tyson and Weber, 1964; Heinrich, 1987), the equations that describe the time-evolution of the degradation potential surface and of the basin floor topography being solved at successive steps in time. The spatial framework for this simulation model, the triangular mesh, is the key to the model's success, for not only does it (1) allow the finite-difference technique to be used to solve the potential flow equations, and (2) permit space- and time-variation of conductance and storage parameters, but it also (3) readily enables standard surface-fitting techniques to be used to specify initial and boundary conditions for basin geometries that are as complex as necessary. It was just these factors that were earlier identified as critical to successful basin simulation.

THE TRIANGULAR MESH

The triangular mesh is ultimately based on the points at which the raw data for the basin simulation are initially supplied. This data set will generally have to be augmented, however, before the actual simulation can begin, either to make the data set complete, to vary its density, or to satisfy one particular requirement of the finite-difference technique, that no triangle in the mesh be obtuse-angled.

Completing the Dataset

The raw data are of three types: (1) initial elevations of points on the basin floor, (2) initial values of the degradation potential, and (3) source fluxes, if necessary, to allow for autochthonous sedimentation and for sediment supply into the basin from its margins. A simple data set (Figure 1) illustrates how incomplete and how erratically distributed these data will usually be. What must be done is to create a single triangular mesh, encompassing all the data points, that will allow each type of data to be defined everywhere on the basin floor.

To do this, the Delaunay triangulation is first constructed for the set of points at which initial elevations are supplied, and an interpolation over this triangulation used to obtain the elevations for all the other points (Figure 2a), i.e. at any point for which there is just a value of the potential. This exercise is then repeated for the initial potentials (Figure 2b), obtaining interpolated values of the potential for every point at which only the elevation is supplied. In each case the Delaunay triangulation is made to be conditional on the known basin boundary (Figure 2c), i.e. no line of the triangulation is permitted to cut the boundary (which may be of any shape). Both interpolation exercises will usually involve just conventional surface-fitting, and there are many appropriate triangle-based algorithms (McCullagh, 1981). The one used here is a modification of an algorithm developed by Klucewicz (1977, 1978; see also Barnhill and Gregory, 1975), and is suitable for any arbitrarily located data. There may, of course, be circumstances in which additional constraints must be put on the form of either the initial elevation surface or the initial potential surface. Elevation discontinuities, for instance, may be known to exist (McCullagh, 1980), and if this is the case other interpolation strategies will have to be used. It is also possible that either the elevation data or the potential data will be incomplete at the margins of the basin. Extrapolation from data within the basin will then be required, and this must be carried out with considerable care. It is not really possible to give general advice on how this extrapolation should be done, as each case must be treated on its merits.

The choice of the Delaunay triangulation (Miles, 1970; Lawson, 1977) as the basis for the triangular mesh is important for two reasons. The first is that the Delaunay triangulation is locally equiangular (Sibson, 1978), and so using it tends to minimize the overall distortion produced in any interpolation (McCullagh, 1981; Watson and Philip, 1984a). The second reason is that the Delaunay triangulation, as it is the geometric dual, or complement, of the Voronoi diagram, defines the natural neighborhood relationships of any set of data points in the plane (Figure 3). Any interpolating function that uses the Delaunay triangulation should thus be expected to give a more natural expression of the underlying data than would the same function if it were to use some other triangulation. There is, in fact, a 'Natural Neighbor Interpolant' (Sibson, 1981), which

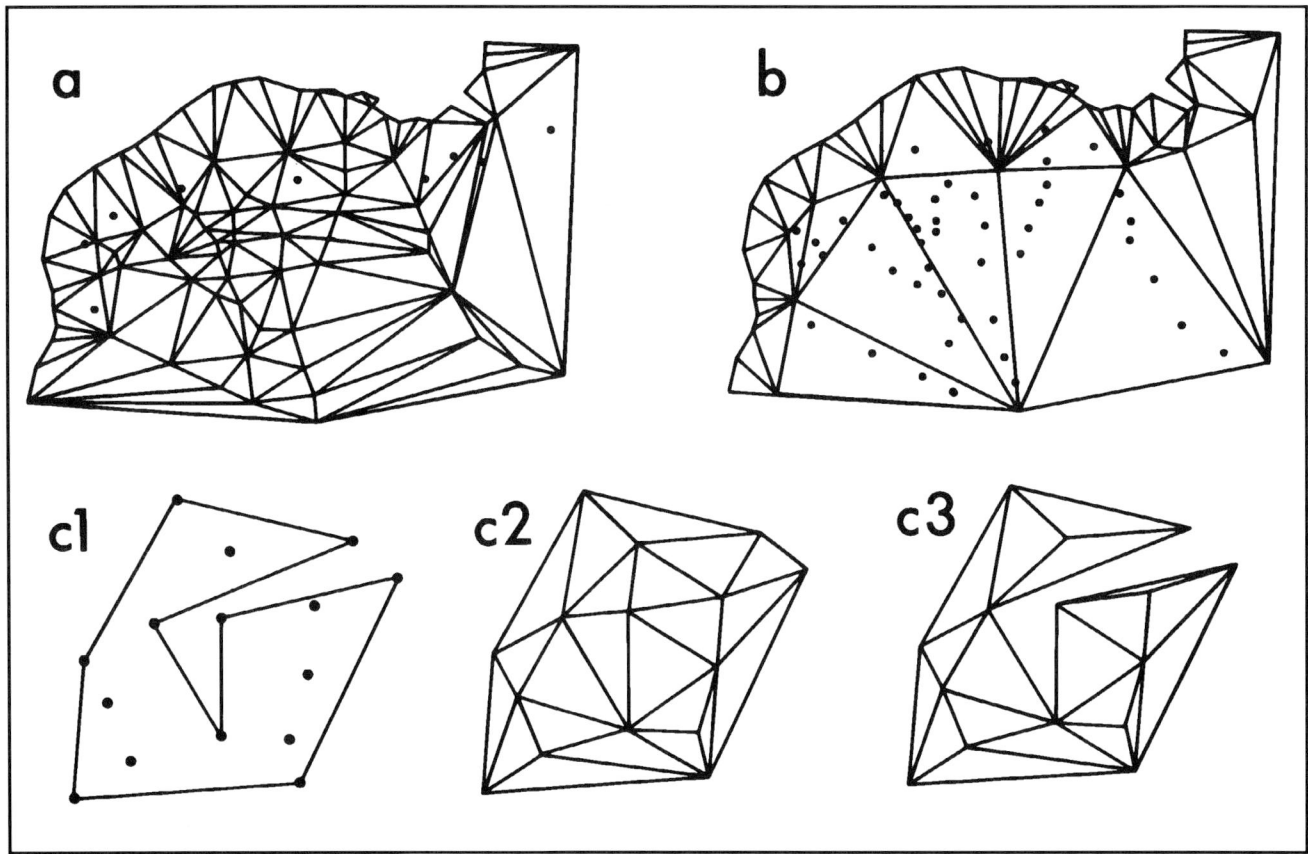

Figure 2. Initial Delaunay triangulations for the data from Figure 1. (a) Initial Delaunay triangulation of points at which elevations were specified. Solid circles mark points at which only initial potentials were specified. (b) Initial Delaunay triangulation of points at which potentials were specified. Solid circles mark points at which only initial elevations were specified. (c) Boundary conditioning of a Delaunay triangulation: (c1) A set of 15 points in a closed domain with a re-entrant boundary; (c2) Delaunay triangulation of the 15 points with the boundary ignored; (c3) Delaunay triangulation of the 15 points, made conditional on the boundary, i.e. with no line in the triangulation cutting the boundary. Note the differences between (c2) and (c3).

gives smooth interpolation in the interior of any Delaunay triangulation in a way that uses all the available data in an optimal way. Its one drawback is that it does produce problems at the margins of the triangulation, and the Klucewicz interpolation scheme is preferred for this present work.

Varying the Data Density

Once both the initial elevations and the initial potentials are obtained at every data point in the basin, a single triangular mesh can be constructed. This mesh (Figure 4a) is the Delaunay triangulation of the complete set of data points, again made to be conditional on the known basin boundary. The corresponding Voronoi diagram (Figure 4b) shows how the basin is effectively divided in this way into a mosaic of non-overlapping tiles. The spatial resolution of this mosaic, i.e., the detail that it can reproduce, is controlled by the density of the data points on which it is based.

The number of data points directly determines the number of tiles in the mosaic; there is one tile for each data point. Thus changing the total number of data points is one way to change the mosaic's spatial resolution. Ultimately, however, it is the configuration of the data points rather than their total number that is the more important consideration, because it is the configuration that determines the size of the tiles. Areas in which the data points are tightly clustered have smaller tiles than those in which the data points are widely spaced. The spatial resolution of the mosaic is a local property, one that depends on the local density of the data points.

Varying the local data point density to alter the mosaic's spatial resolution may be necessary for either of two reasons. The first is to reproduce more faithfully particular aspects of basin floor topography or basin margin geometry. The second is to control the precision of the finite-difference solution. Increasing the data point density in any given area allows more topographic and geometric detail, and provides high-

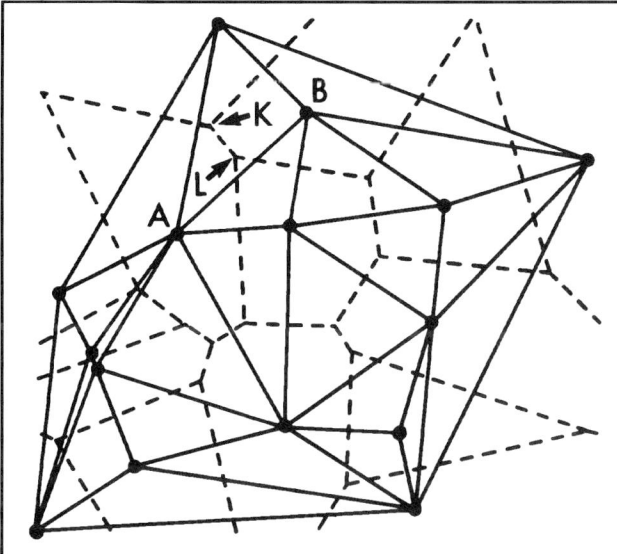

Figure 3. Delaunay triangulation (solid lines) and Voronoi diagram (broken lines) for a set of 15 data points (solid circles). Each convex polygon (tile) of the Voronoi diagram bounds the region of the plane within which all points are closer to the corresponding data point than to any other data point. Each tile defines the natural neighborhood of its data point, so the Delaunay triangulation (formed by joining all pairs of data points which have adjacent tiles) defines the natural neighborhood relationships of the whole data point set. (For significance of points A, B, K, L, see caption of Figure 5).

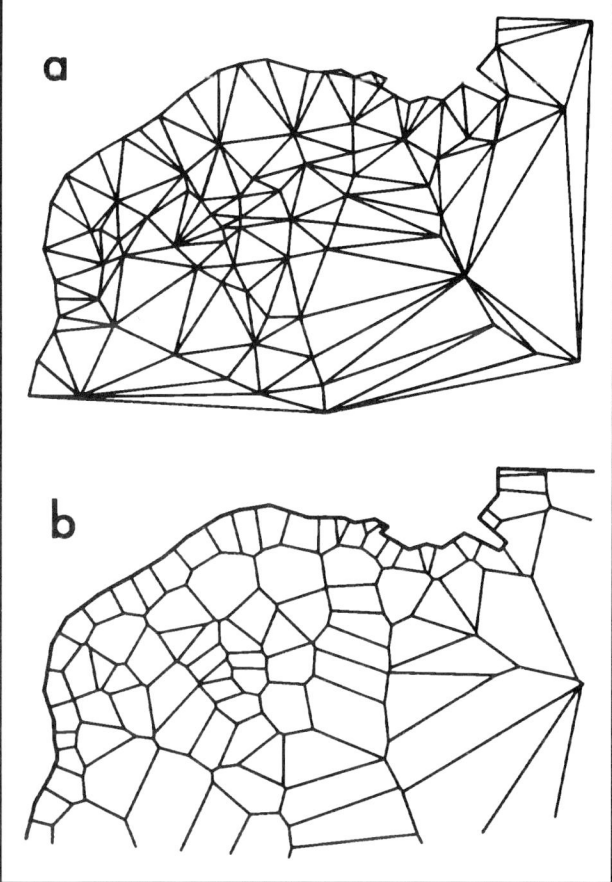

Figure 4. (a) Delaunay triangulation for the complete data set shown in Figure 1. (b) Corresponding Voronoi diagram, with basin margin (coastline) shown in heavier line.

er precision in the solution: decreasing the data point density decreases the computational load. A balance of these factors must be struck, with everywhere just the minimum data point density being used to give acceptable results. (Note that, whenever new data points are added, the corresponding elevation and potential values are obtained by interpolation using the original elevation and potential meshes.)

Removing Obtuse-Angled Triangles

In order for the numerical results of the simulation to have any meaning, it is essential that the potential field solutions obtained by the finite-difference technique be physically realizable, i.e., that real sediment movements could take place in a basin and result in those calculated changes in the potential field. To see how this constrains the form of the triangular mesh, consider the two ways in which the finite-difference technique treats sediment transport as occurring (Figure 5). Firstly it treats it as occurring only between adjacent tiles across their shared side. Secondly it treats it as occurring only along paths that join the data points of these adjacent tiles, i.e. along the lines of the triangular mesh. For these treatments to be consistent, every line in the triangular mesh must cut the side in the mosaic that is common to the tiles whose data points the particular mesh line joins. This implies that no triangle in the mesh may be obtuse-angled.

To ensure that this requirement is met, the triangular mesh that was previously obtained (for instance, Figure 4a) may need to be modified by having new points added to it until all the triangles are acute-angled. For computational efficiency, as few points as possible should be added: the Delaunay basis must also be maintained. Tipper (1989) described a domain triangulation algorithm which offers one way of carrying this out, and it appears to give generally satisfactory results. There is no guarantee of optimality in this algorithm, however, either in terms of minimizing the number of new points, or of maintaining the prior balance of point densities across the whole triangulation. The most common problem tends to occur in areas of sharply varying data point density (Figure 6), where an excess of small triangles may sometimes be generated. To avoid this it is sensible to ensure that the data point density prior to removing the obtuse-angled triangles always varies smoothly across the

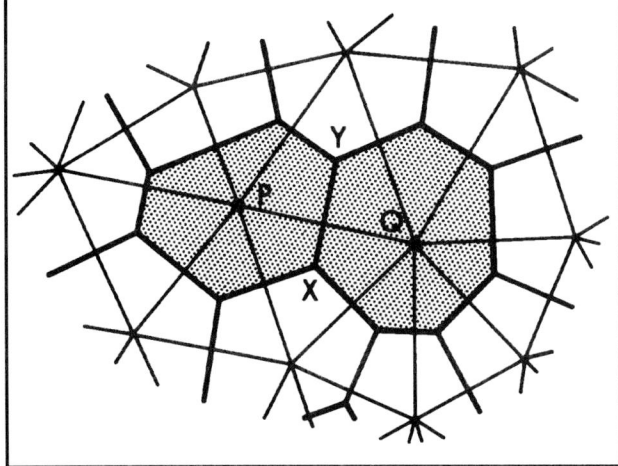

Figure 5. Sediment transport across a Voronoi mosaic: Delaunay triangulation shown in lighter lines. The transport path (PQ) between the two shaded tiles is the line in the Delaunay triangulation that joins their data points (solid circles). The transport is across XY, the side shared by the tiles. All the triangles are acute-angled, so PQ and XY cut internally. Compare with Figure 3, in which not all the Delaunay triangles are acute-angled. There the transport path AB does not cut the shared tile side KL unless KL is projected.

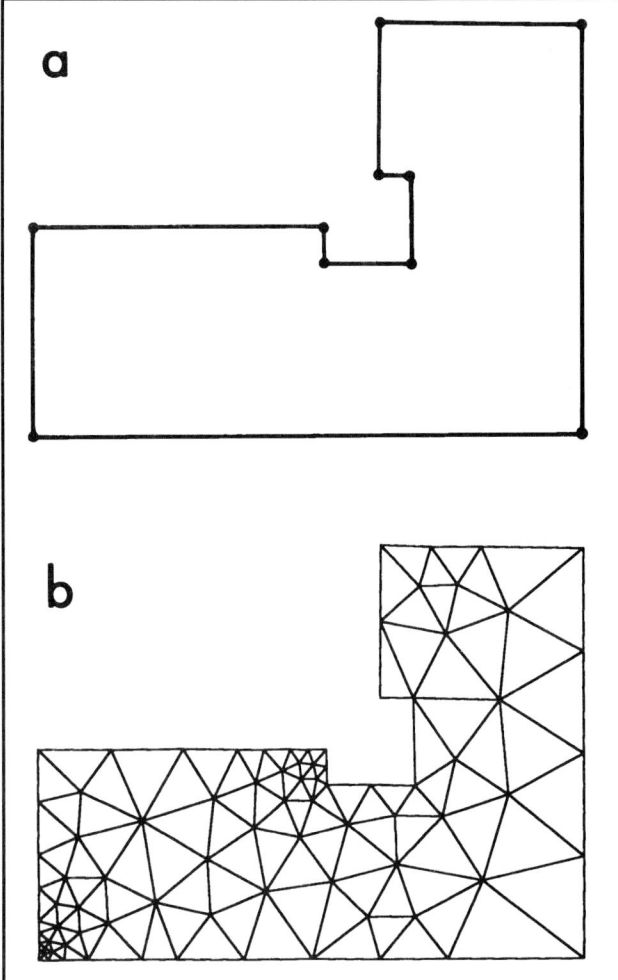

Figure 6. A problem in domain triangulation (Tipper, 1989). (a) Original domains, with data points marked by solid circles. (b) Final Delaunay triangulation, with points added to make every triangle acute-angled. Note that extreme local variability in the original data point density may result in unpredictable final point densities.

basin. This corresponds to the gradually graded triangular meshes commonly required for finite-element analysis (Bykat, 1976).

After all obtuse-angled triangles have been removed, it may be thought necessary to add further points manually, for whatever reason. This will mean, of course, that the mesh must then be checked again for obtuse-angled triangles. Even more points may have to be added, and this process may be repeated several times until satisfactory results are achieved. Once the final mesh has been obtained (Figure 7), the initial elevations and initial potentials are calculated for every one of the newly added data points, and at last the simulation itself can begin.

THE SIMULATION IN OPERATION

The simulation operates in discrete time, and at each time-step the following calculations are made: (1) the storage and conductance parameters are determined everywhere in the basin; (2) the equation for the new potential field is then solved for each data point (Appendix 1, this chapter); (3) the new basin floor elevations are then obtained (Appendix 1, this chapter); (4) adjustments are made to allow for compaction, basement subsidence, and isostatic balancing. Results for various runs of the simulation are given by Tipper (1991a, 1992).

The storage parameters are determined individually for each tile of the mosaic, as functions of the existing potentials and elevations of that tile and of its immediate neighbors. The conductance parameters are likewise determined individually for each transport path (each line in the triangular mesh), as functions of the existing potentials and elevations of the tiles at the path ends. The form of these functions is flexible, and they can be varied as required (for instance, to reproduce different transportation and deposition mechanisms). What is really important, however, is not the form of these functions, but the fact that their calculated values have local meaning only. Thus each tile has its own storage parameter, and each transport path has its own conductance parameter. This local basis for the simulation parameters, along with the fact that they can conveniently be recalculated at each time-step, is exactly one of the factors in successful basin simulation identified earlier. It

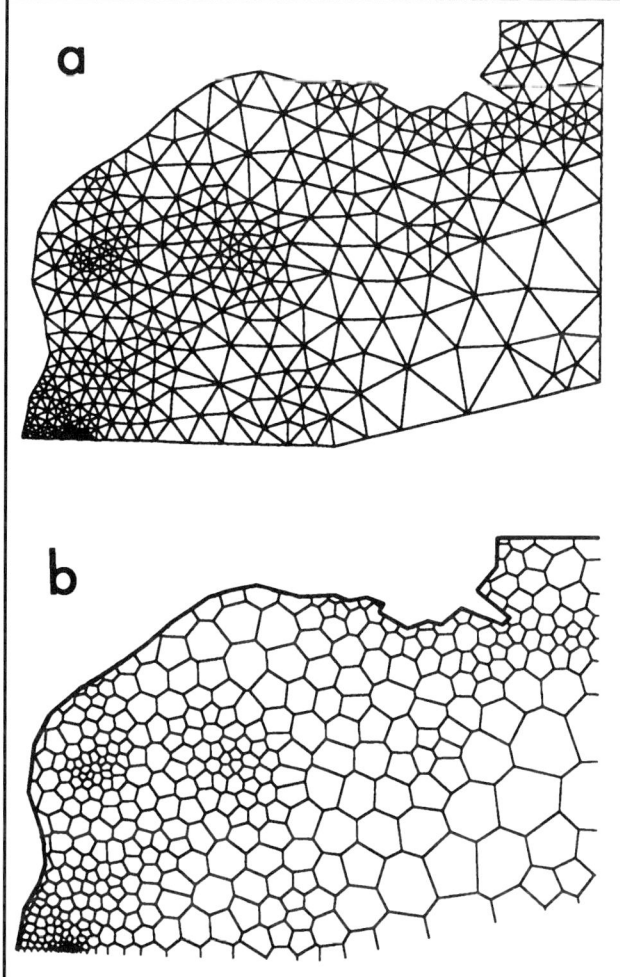

Figure 7. (a) Final Delaunay triangulation. Every triangle is acute-angled. (b) Corresponding Voronoi diagram, with basin margin (coastline) shown in heavier line.

is achieved naturally when the triangular mesh described here is adopted as the spatial framework for the simulation.

DISCUSSION: SPATIAL FRAMEWORKS FOR SURFACE MODELING

In the sedimentary basin simulation outlined here, both surface-fitting and equation-solving operations are essential; the one to help specify the initial conditions for the simulation, the other to obtain the new potential field and basin floor topography at each time-step. By using a triangular mesh as the spatial framework throughout, each type of operation can be carried out efficiently and give absolutely compatible results. Yet it would, of course, be perfectly possible to carry out this work using other spatial frameworks. All that is necessary is that the chosen spatial framework be sufficiently flexible to let all the required operations, both surface-fitting and equation-solving, be implemented while using it.

There are a large number of papers describing the different types of spatial framework that can be used for surface modeling: Burrough (1986) gives a useful introduction to much of it, and some recent applications are described in Raper (1989). Various classification schemes for these frameworks have also been developed. A common one, splitting the frameworks into those that are essentially volume representations and those that essentially represent volume-bounding surfaces, is illustrated by Fried and Leonard (1990). In contrast, four types of spatial framework are recognized here. In addition to triangular meshes, these are (in order of increasing complexity and increasing generality): (1) digital arrays, (2) rectangular grids, and (3) parametric grids (Figure 8). Digital arrays (Figure 8a, 8b) are dense regular arrays, such as are commonly used for complex terrain analysis (Digital Elevation Models). For surface-fitting they are highly efficient, as any surface is represented exactly by the data on which it is based, and by those data alone. One type of digital array (Figure 8a) represents surfaces as the implied boundary between 3-D cells (voxels) of different nature (Pflug, 1988): another (Figure 8b) represents a surface by the set of individual z-values for the 2-D cells (pixels) over which it stretches (Pfaltz, 1975). Note that each element of a digital array (either a voxel or a pixel) is the smallest element into which either the volume of interest or the area of interest is being partitioned. Within each voxel the nature of the material (the attribute of the voxel) is constant, and within each pixel the z-value (height) of the surface is constant. Digital arrays are computationally very attractive, as many of the operations that are involved in manipulating them are individually very simple and can be implemented in hardware (e.g., digital arrays are the natural spatial framework to use with any fast raster display or image analysis system). For any serious numerical modeling, however, in which standard techniques are to be used to solve large systems of equations, digital arrays are cumbersome and quite unsuitable. The only exception is when they are treated just as an extreme case of the second type of spatial framework, the rectangular grid.

Rectangular grids provide a very logical spatial framework for surface modeling. Besides the basic uniform grid (Figure 8c), there are also the non-uniform grid (Figure 8d), and the nested square grid (Figure 8e), these latter grids enabling irregular geometries and data point configurations to be better represented. A multiplicity of different techniques use rectangular grids, both for surface-fitting and for the numerical solution of equations, and these techniques are usually simple to understand, efficient to implement, and readily available in most computing environments. For all these reasons, rectangular grids are often taken as the first-choice spatial framework for most surface modeling applications.

Rectangular grids do, however, have several major disadvantages, all of which emphasize that they are

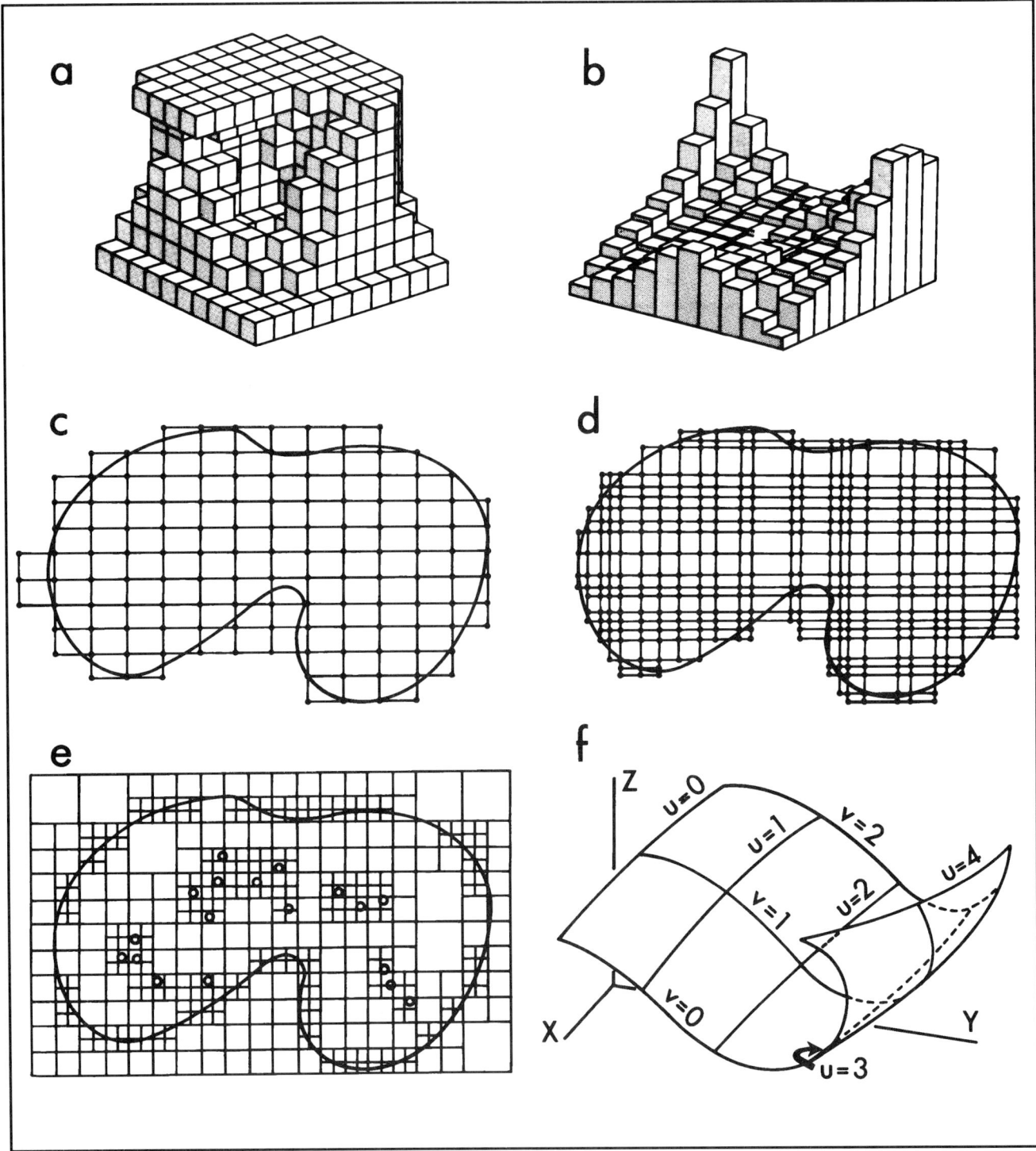

Figure 8. Other spatial frameworks for surface modeling. (a) Perspective view of object represented by 3-D digital array. Total array is of 10 ×10 ×8 voxels, each a cube of unit side. (b) Perspective view of surface represented by 2-D digital array. Total array is of 10 ×10 pixels, each a square of unit side. Each pixel is tagged with a continuously scaled z-value. The z-value is constant across each pixel, so the data point for the pixel is implicitly in the pixel center. (c) Uniform rectangular grid over an arbitrary planar domain. 134 grid nodes (solid circles). (d) The same domain, now represented by a non-uniform rectangular grid. 422 grid nodes. (e) The same domain, now represented by a grid of nested squares. Some internal data points have been added (open circles), and the square size in those regions has been reduced. Note also the reduction in square size to give more detail on the boundary. The degree of nesting, and so the detail that the grid can represent, is chosen to suit the particular application. (f) Surface represented by a parametric grid. The parameters u and v are valid only on the surface.

relatively inflexible. Thus incorporating even a single extra data point into an existing grid will at the very least (Figure 8e) require that four other nodes be inserted to maintain the grid's structure: at worst it may require that the complete grid be reconfigured (Figure 8c, 8d). Representing complex geometries is also difficult with rectangular grids, the usual solution being to greatly reduce the grid cell size (Figure 8d, 8e). Finally, any rectangular grid imposes a directionality on a surface (the directions of the two principal axes of the grid) and this directionality (most obvious where the grid cells are very elongate (Figure 8d), especially at the grid's margins) can again only be removed by greatly reducing the grid cell size. The penalty for the relative inflexibility of the rectangular grid is that denser grids must be used, i.e., the grids must have more nodes. Even though both surface-fitting and equation-solving operations tend to be computationally rather undemanding for rectangular grids (thus offsetting somewhat this node increase), on balance rectangular grids may not be at all the most efficient spatial framework to use in many situations, especially for complex data sets.

Parametric grids (Figure 8f) can be considered simply as twisted rectangular grids: a parametric grid, $(x(u,v), y(u,v), z(u,v))$, is a uniform rectangular grid in its own coordinate system, (u,v). The parametric grid is a far more general spatial framework for surface modeling than is the rectangular grid, and a far more flexible one. For surface-fitting especially, there are many existing techniques: Tipper (1979) provides a comprehensive summary of these. The major advantages of parametric grids over rectangular grids are that they do not necessarily impose a rigid directionality on the surface, readily allow new points to be inserted, and are far better able to represent complex geometries and intricate boundaries, using far fewer points. Parametric grids can, in fact, readily allow surfaces to be modeled that have multiple z-values, a very considerable advantage in many geological applications and one which they share with 3-D digital arrays (Figure 8a).

Using parametric grids rather than rectangular grids almost inevitably involves more computation, for almost any surface modeling application. The reason for this is simply that the parametric grid is inherently a more complex spatial framework than is the rectangular grid, and this is most obvious in two contexts. The first is when surfaces are modeled that do have multiple z-values. Interrogation techniques (for instance, for contouring, sectioning, and dissection) then become very much more complicated than for single z-valued surfaces, and many special cases often need to be handled with care. The second context is when parametric grids are used in solving any systems of equations that are not at all naturally formulated in the parametric coordinates. Continually having to transform from one coordinate system to another introduces a computational load that may sometimes be too great to be justified. Parametric grids are probably best reserved for those applications (usually surface-fitting applications) in which their generality makes them the clear first choice.

As was seen earlier in the context of the sedimentary basin simulation, triangular meshes can be a very effective spatial framework for some surface modeling applications. For applications that require both surface-fitting and the numerical solution of equations, their only serious competition is from rectangular grids. Their main advantage over rectangular grids is their greater flexibility: they can represent complex geometries with far fewer data points, and only local reconfiguration is necessary when extra data points are added. For surface-fitting, a variety of techniques that use triangular meshes are available. Although most of these tend to be computationally more expensive per triangle than techniques which use rectangular grids are per grid cell, considerably fewer triangles than grid cells are generally necessary to represent a surface, and so surface-fitting over a triangular mesh is at worst not much less efficient than over a rectangular grid. McCullagh (1981) has provided a useful comparison of the two approaches, in the context of contouring.

For solving equations numerically, the triangular mesh is the natural spatial framework for finite-element schemes (Hughes, 1987), just as the rectangular grid is the natural spatial framework for finite-difference ones (see de Marsily, 1986, for examples in the context of ground-water modeling). As finite-element schemes are time-implicit, however, and thus liable to involve rather lengthy matrix inversion at each time-step (Thacker, 1978a, 1978b), this unfortunately implies that triangular meshes will generally be computationally less efficient than rectangular grids unless (1) the number of triangular elements can be kept very much smaller than the number of cells in the rectangular grid, or (2) a triangle-based finite-difference scheme can be used instead of a finite-element one. Fortunately, several such triangle-based finite-difference schemes now exist (Heinrich, 1987): in addition to the one used in the basin simulation described here, the scheme developed by Thacker (1977, 1978b) is also worthy of note. (Care must, of course, always be taken to maintain stability in any time-explicit finite-difference scheme, by keeping the length of the time-step and the size of each triangle both sufficiently small. Appropriate checks are built into the actual basin simulation program.)

On balance, triangular meshes seem to offer great promise for many surface modeling applications, certainly for far more than those for which they are currently used. Three reasons can be offered for this relative neglect: (1) the entrenched position of rectangular grid-based methods, (2) the generally greater computational load that triangular meshes are thought (often quite correctly) to involve, and (3) the computational problem of actually constructing triangular meshes, especially for very large numbers of points (N>1000). This final reason deserves particular comment, for the triangulation problem is quite substantial, especially if the triangulation is to be non-arbitrary (Watson and

Philip, 1984b). For the Delaunay triangulation, for instance, at least N log N time is required, and few algorithms achieve this. Many algorithms, in fact, require N² time, which renders them quite useless if N is very large. A recently developed Delaunay triangulation algorithm (Tipper, 1990), which is close to N log N in overall performance, does, however, offer some significant advantages in that it can run in linear time (N) for some common data configurations. A development of it (Tipper, 1991b) also has the facility to create the Delaunay triangulation for multiply-connected domains with non-convex boundaries, again in a highly efficient way. Algorithms such as this hold the key to using the Delaunay triangulation as the spatial framework for surface modeling for very large data sets, and triangular meshes may now, for many applications which involve both surface-fitting and equation-solving, be looked on as the spatial framework of choice.

REFERENCES CITED

Barnhill, R. E., and J. A. Gregory, 1975, Compatible smooth interpolation in triangles: Journal of Approximation Theory, v. 15, p. 214-225.

Beaumont, C., C. E. Keen, and R. Boutilier, 1982, On the evolution of rifted continental margins: Comparison of models and observations for the Nova Scotian margin: Royal Astronomical Society Geophysical Journal, v. 70, p. 667-715.

Burrough, P. A., 1986, Principles of geographic information systems: Oxford, Clarendon Press, 250 p.

Bykat, A., 1976, Automatic generation of triangular grid: I—Subdivision of a general polygon into convex subregions. II—Triangulation of convex polygons: International Journal for Numerical Methods in Engineering, v. 10, p. 1329-1342.

Culling, W. E. H., 1960, Analytical theory of erosion: Journal of Geology, v. 68, p. 336-344.

de Marsily, G., 1986, Quantitative Hydrogeology: Orlando, Academic Press, 440 p.

Fried, C. C., and J. E. Leonard, 1990, Petroleum 3-D models come in many flavors: Geobyte, v. 5, p. 27-30.

Harbaugh. J. W., and G. Bonham-Carter, 1970, Computer simulation in geology: New York, John Wiley and Sons, 575 p.

Heinrich, B., 1987, Finite difference methods on irregular networks: Basle, Birkhäuser Verlag, 206 p.

Helland-Hansen, W., C. G. St. C. Kendall, I. Lerche, and K. Nakayama, 1988, A simulation of continental basin margin sedimentation in response to crustal movements, eustatic sea level change, and sediment accumulation rates: Mathematical Geology, v. 20, p. 777-802.

Hughes, T. J. R., 1987, The finite element method: Englewood Cliffs, Prentice-Hall, 803 p.

Jervey, M. T., 1988, Quantitative geological modeling of siliciclastic rock sequences and their seismic expression, *in* C. K. Wilgus, B. S. Hastings, C. G. St. C. Kendall, H. W. Posamentier, C. A. Ross, and J. C. Van Wagoner, eds., Sea-level changes: and integrated approach: SEPM Special Publication 42, p. 47-69.

Klucewicz, I. M., 1977, A piecewise C1 interpolant to arbitrarily spaced data: M.S. thesis, Salt Lake City, Utah, University of Utah, 45 p.

Klucewicz, I. M., 1978, A piecewise C1 interpolant to arbitrarily spaced data: Computer Graphics and Image Processing, v. 8, p. 92-112.

Lawson, C. L., 1977, Software for C1 surface interpolation, *in* J. R. Rice, ed., Mathematical Software III: New York, Academic Press, p. 161-194.

Macneal, R. H., 1953, An asymmetrical finite difference network: Quarterly of Applied Mathematics, v. 11, p. 295-310.

McCullagh, M. J., 1980, Triangular systems in surface representation, *in* R. T. Aangenbrug, ed., AUTO-CARTO IV, Proceedings of the international symposium on cartography and computing: Applications in health and environment: Falls Church, VA, American Congress on Surveying and Mapping, p. 146-153.

McCullagh, M. J., 1981, Creation of smooth contours over irregularly distributed data using local surface patches: Geographical Analysis, v. 13, p. 51-63.

Miles, R. E., 1970, On the homogeneous planar Poisson point process: Mathematical Biosciences, v. 6, p. 85-127.

Pfaltz, J. L., 1975, Representation of geographic surfaces within a computer, *in* J. C. Davis and M. J. McCullagh, eds., Display and analysis of spatial data: London, John Wiley and Sons, p. 210-230.

Pflug, R., 1988, Solid modeling of geological objects with 3-D rasters: Geologisches Jahrbuch, v. A104, p. 213-219.

Raper, J. F. (ed.), 1989, Three-dimensional applications in geographical information systems: London, Taylor and Francis, 189 p.

Sibson, R., 1978, Locally equiangular triangulations: The Computer Journal, v. 21, p. 243-245.

Sibson, R., 1981, A brief description of natural neighbor interpolation, *in* V. Barnett, ed., Interpreting multivariate data: New York, John Wiley and Sons, p. 21-36.

Tetzlaff, D. M., and J. W. Harbaugh, 1989, Simulating clastic sedimentation: New York, Van Nostrand Reinhold, 202 p.

Thacker, W. C., 1977, Irregular grid finite-difference techniques: Simulations of oscillations in shallow circular basins: Journal of Physical Oceanography, v. 7, p. 284-292 and 932-934.

Thacker, W. C., 1978a, Comparison of finite-element and finite-difference schemes. Part I: One-dimensional gravity wave motion: Journal of Physical Oceanography, v. 8, p. 676-679.

Thacker, W. C., 1978b, Comparison of finite-element and finite-difference schemes. Part II: Two-dimensional gravity wave motion: Journal of Physical Oceanography, v. 8, p. 680-689.

Tipper, J. C., 1979, Surface modeling techniques, Kansas Geological Survey Series on Spatial Analysis 4: Lawrence, Kansas Geological Survey, 108 p.

Tipper, J. C., 1989, Fast triangulation of planar

domains, *in* W. L. Hogarth and B. J. Noye, eds., Proceedings of the Conference on Computational Techniques and Applications (CTAC-89): Washington, D.C., Hemisphere, p. 193-200.

Tipper, J. C., 1990, A straightforward iterative algorithm for the planar Voronoi diagram: Information Processing Letters, v. 34, p. 155-160.

Tipper, J. C., 1991a, Modeling the fill of sedimentary basins: Exploration Geophysics, v. 22, p. 397-400.

Tipper, J. C., 1991b, FORTRAN programs to construct the planar Voronoi diagram: Computers & Geosciences, v. 17, p. 597-632.

Tipper, J. C., 1992, Landforms developing and basins filling: Three-dimensional simulation of erosion, sediment transport, and deposition, *in* R. Pflug and J. W. Harbaugh, eds., Computer graphics in geology: Berlin, Springer-Verlag, p. 155-170.

Tyson, H. N., and E. M. Weber, 1964, Ground-water management for the nation's future—computer simulation of ground-water basins: American Society of Civil Engineers Proceedings, Journal of the Hydraulics Division, v. 90 (HY4), p. 59-77.

Watson, D. F., and G. M. Philip, 1984a, Triangle based interpolation: Mathematical Geology, v. 16, p. 779-795.

Watson, D. F., and G. M. Philip, 1984b, Systematic triangulations: Computer Vision, Graphics, and Image Processing, v. 26, p. 217-223.

APPENDIX 1: THE BASIN SIMULATION MODEL

Consider a tile, B, in a Voronoi mosaic: assume that all the triangles of the corresponding Delaunay triangulation are acute-angled. Let this tile have n_B neighbors. Then for the (j+1)th time-step, the potential at B, h_B^{j+1} is calculated by:

$$h_B^{j+1} = h_B^j \left[1 - \frac{\Delta t}{A_B S_B^j} \sum_{i=1}^{n_B} Y_{i;B}^j \right] + \frac{\Delta t}{A_B S_B^j} \sum_{i=1}^{n_B} h_i^j Y_{i;B}^j + \frac{\Delta t}{S_B^j} \cdot Q_B^j \qquad (1)$$

where h_B^j is the potential at B after the jth time-step,

h_i^j is the potential at the ith neighbor of B after the jth time-step,

Δt is the length of the time-step,

A_B is the area of tile B,

S_B^j is the storage coefficient for tile B after the jth time-step,

$Y_{i;B}^j$ is the conductance of the transport path between B and its ith neighbor after the jth time-step,

Q_B^j is the source/sink sediment flux into B, if any.

The new basin floor elevation of tile B, z_B^{j+1}, is then calculated by:

$$z_B^{j+1} = z_B^j + S_B^j \left(h_B^{j+1} - h_B^j \right) \qquad (2)$$

where z_B^{j+1} is the basin floor elevation of tile B after the jth time-step.

Chapter 9

Testing Hydrocarbon Saturation Models For Use in Original Oil-in-Place Estimation: South Dome of Oregon Basin Field, Park County, Wyoming

Michael J. Heymans
Marathon Oil Company
Littleton, Colorado, U.S.A.

Douglas A. Steed
Marathon Oil Company
Houston, Texas, U.S.A.

David E. Hamilton
Landmark/Zycor Inc.
Austin, Texas, U.S.A.

Barbara A. Pavlov
Landmark/Zycor Inc.
Austin, Texas, U.S.A.

ABSTRACT

A computer mapping technique for improving the accuracy of original oil-in-place estimates for the Tensleep and Phosphoria formations in South Dome of Oregon basin field, Park County, Wyoming, is presented. Grids representing top and base structure, average porosity, net-to-gross ratio, and integrated average oil saturation from capillary pressure curves, are combined to create a grid of hydrocarbon pore thickness for each reservoir.

The average-oil-saturation model is created by correlating capillary pressure curves to mapped porosity. This model incorporates more information about the three-dimensional distribution of reservoir fluids than was used in methods applied previously. Because of this, it provides a more accurate representation of the fluids' areal distributions. Volumes resulting from use of this technique are compared with methods that use alternative average oil saturation modeling approaches: (1) a constant oil saturation, (2) a grid derived from average-oil-saturation at each well, (3) an average oil saturation grid from a single capillary pressure curve, and (4) a dry-oil production contact used in place of the free-energy surface. Maps show the steps in the process and allow comparison of the effects of the five oil saturation modeling methods.

INTRODUCTION

As a new discovery is developed, one of the simplest methods for estimating original oil-in-place (OOIP) is to multiply a mapped area, by a net reservoir thickness, by an average porosity, and by an average oil saturation. Although this can be done quickly, it does not account for the spatial variation of oil saturation present in all hydrocarbon reservoirs. Being able to map variations in oil saturation is important for estimating areas containing recoverable oil within a reservoir, analyzing well pattern performance, and for matching and predicting performance history through computer numerical simulations. These analyses are used in reservoir management decisions for maximizing economic oil recovery.

As the reservoir limits are delineated and sufficient well control becomes available, more detailed knowledge of volumes can be obtained by mapping based on porosity logs and planimetering mapped areas. The advent of computer mapping has eliminated laborious work with the planimeter and allows calculation and updating of original oil-in-place estimates. With the computer surface models (grids) can be built for reservoir attributes (structure, thickness, porosity, net-to-gross ratio, and oil saturation) and those attributes combined to create a hydrocarbon-pore-thickness (HPT) model. Variations in any of these attributes can significantly impact the HPT model and resulting hydrocarbon volume estimates. Improvement in these estimates can be realized if correlations between porosity and saturations as a function of height are used to provide an improved model of oil-saturation distribution.

It is well known that hydrocarbon saturation decreases with greater depth within oil and gas reservoirs. One means of measuring this non-linear relationship is through core tests of capillary pressure on representative reservoir rock. Although vertical resolution is improved over using a correlation between only porosity and saturation, a single average curve of saturation as a function of height is short of being the most realistic representation.

Most reservoirs are heterogeneous with regard to porosity, permeability, and fluid-saturation distributions. A single curve provides a change of saturation vertically but no variation in the horizontal plane. An improvement can be made by using curves of capillary pressure vs. saturation that are representative of the changes in rock properties. This increased resolution provides more detailed variation in hydrocarbon saturation both vertically and laterally. With improved resolution comes increased complexity of OOIP calculations, which are best done with the aid of a computer.

GEOLOGIC SETTING

The South Dome of Oregon basin field is located about 12 miles southeast of Cody, Wyoming, on the western edge of the Big Horn basin (Figure 1). South Dome is a slightly asymmetrical north–south-trending anticline with a western flank dip of 37 degrees and an eastern flank dip of 7 degrees. The reservoirs of interest are the Pennsylvanian-age Tensleep and Permian-age Phosphoria formations (Figure 2). The most severe structural deformation of these Paleozoic formations occurred during the Laramide orogeny about 165 million years after Phosphoria deposition.

The Tensleep was deposited as a wind-blown sand during withdrawal of Pennsylvanian seas from the Wyoming shelf area. The Tensleep consists of sandstone beds (individually up to 35 ft [10 m] thick) alternating with beds of dolomite, sandy dolomite, and dolomitic sandstone (up to 12 ft [4 m] thick). The sandstone beds form the reservoir, whereas the dolomitic units are nonreservoir. The Tensleep in the Big Horn basin is over 400 ft (120 m) thick in the south and thins around the basin rim. At the end of Tensleep time (Late Pennsylvanian to Early Permian), epeirogenic uplift and erosion in northwestern Wyoming and southern Montana resulted in stream incisement into the Tensleep in the South Dome area. The erosional surface at the top of the Tensleep near South Dome is capped by non-porous and impermeable green to red shale mixed with chert breccia.

The Phosphoria consists of mostly limestone and dolomite sequences with beds of shale, sandstone, and anhydrite. It has a thickness of 200 ft (60 m) at the north end of the basin and more than 350 ft (106 m) in the southwest. The most prolific reservoir facies is a fractured dolomite ranging in thickness from 10 to 75 ft (3 to 23 m), with an average of about 20 to 30 ft (6 to 9 m), at the top of the Phosphoria. Within the Phosphoria, erosional unconformities lie between the Meade Peak lower red bed and Grandeur and between the lower Ervay and upper Franson. A relatively flat erosional unconformity is also at the top of the Phosphoria. The Dinwoody Formation, consisting of sandy and silty limestone, dolomite, and green mudstone, provides the reservoir seal for Phosphoria reservoirs.

OIL DISTRIBUTION

Hydrocarbon Migration and Local Distribution

Oil that fills Big Horn basin reservoirs is believed to have been sourced from the Phosphoria in the Cordilleran geosyncline to the west. Stone (1967) suggested that oil migrated in Early Jurassic time into Phosphoria stratigraphic traps. Later, during the Laramide deformation, some of the oil moved from these stratigraphic traps into newly formed structural traps. It is also believed by the authors that the oil in the Tensleep originated in the Phosphoria and migrated into the Tensleep along local faults (Campbell, 1962; Curtis et al., 1958; Hunt, 1953; Partridge, 1958; Stewart et al., 1955; Thompson, 1954; Walton, 1947).

In South Dome, geochemical analyses of crude oil samples from the Tensleep and Phosphoria reservoirs support the hypothesis that oil in the Tensleep and Phosphoria were both derived from the same marine

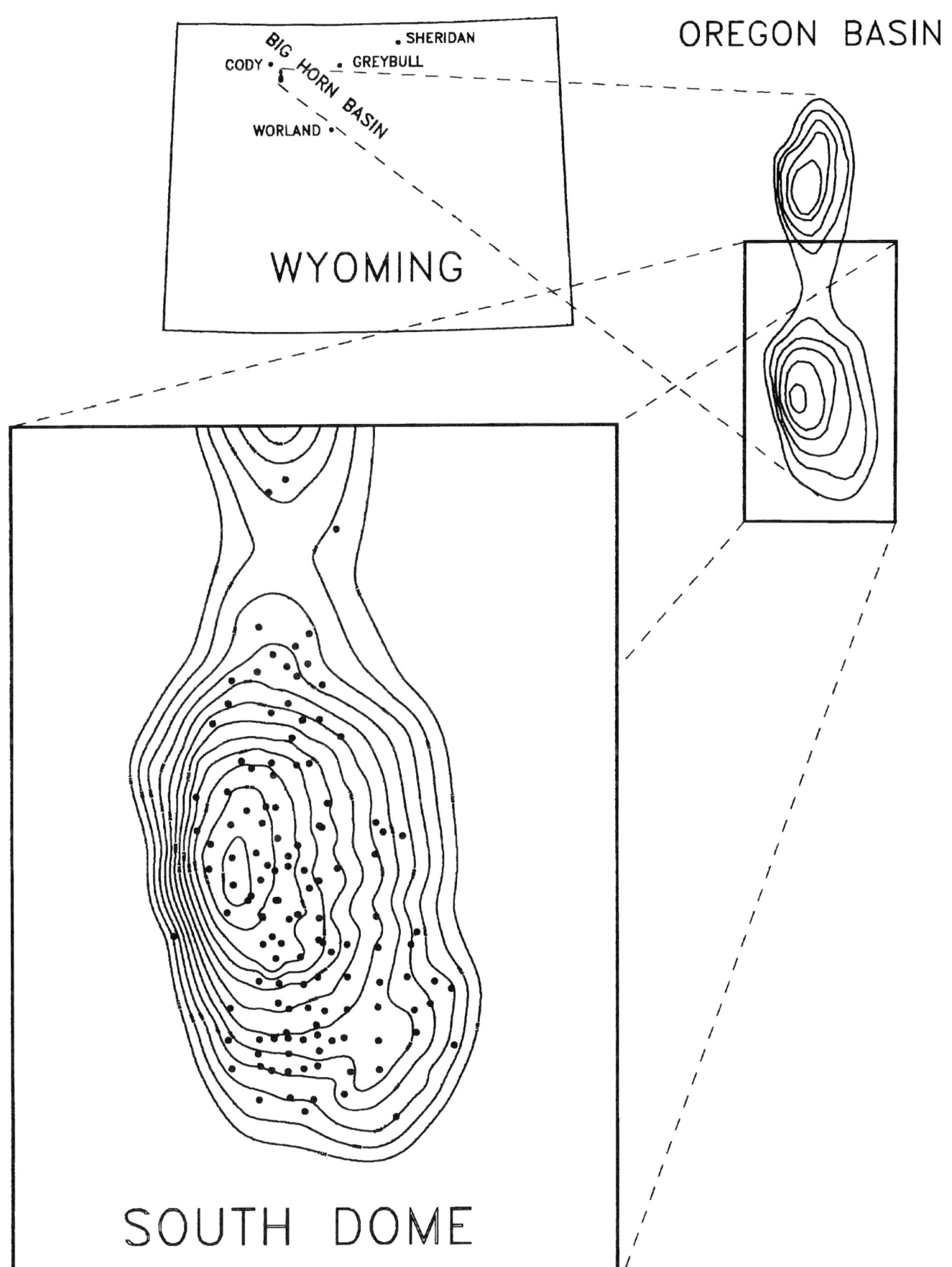

Figure 1. Location of the study area.

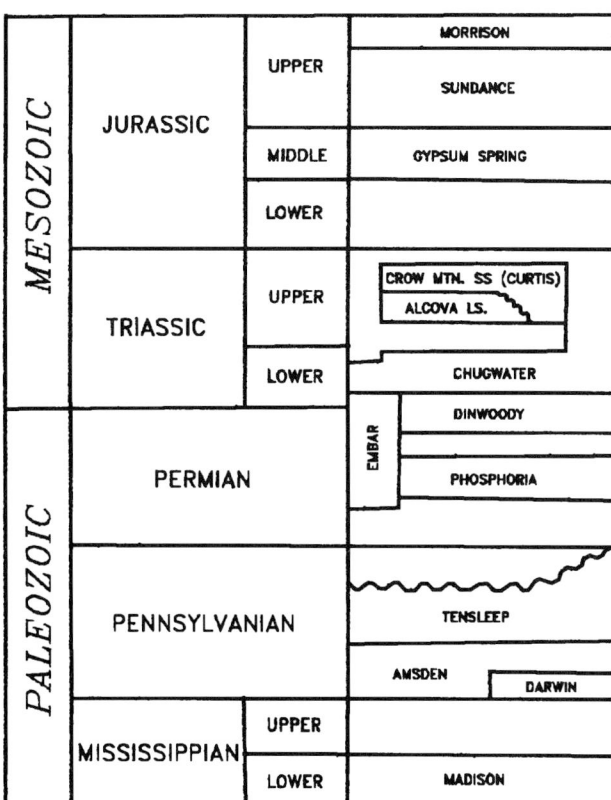

Figure 2. Relevant portion of the stratigraphic section for the Big Horn basin.

source—probably Phosphoria (Heymans et al., 1990). East–west-trending tear faults within the field area are believed by the authors to have been conduits for oil movement from the overlying Phosphoria into the Tensleep. Pressure history performance between both formations suggests they are presently in communication. These characteristics suggest a common free energy surface for the Tensleep and Phosphoria reservoirs at South Dome (Figure 3).

Distribution of Oil Saturation

A free energy surface is defined to be lowest point in the reservoir at which oil can be found in static equilibrium in a well bore at discovery. This oil-water interface, at +1110 ft (338 m) elevation, is below the initial dry-oil production contacts in the field, +1450 ft (442 m) elevation for Phosphoria and +1200 ft (366 m) for Tensleep (Figure 3). The region between the dry-oil production contact and the free energy surface is a transition zone where water, relative to oil, production increases with depth to 100% at the free-energy surface.

Transition-zone oil may contribute little economic oil production, but it is an important consideration for OOIP calculations. Although capillary forces prevent transition-zone oil from being very mobile, the oil still contributes to reservoir energy in the form of solution gas as reservoir pressure decreases. Therefore, in order to obtain an acceptable agreement between material balance and volumetric calculations of original oil in place, it is necessary to consider all oil from the free energy surface to the reservoir seal. This change in oil saturation as a function of height can be obtained from several sources, such as well logs, pressure retained cores, and capillary pressure core tests.

Determining the elevation of a free energy surface for older reservoirs can be difficult. For this study, a combination of field observations and core tests was used. It is well known that capillary pressure core test measurements on representative reservoir rock samples can be converted to saturation distributions as a function of height in a reservoir. However, determining the top of a transition zone from a capillary pressure curve is interpretive.

For this study a laboratory core test technique developed by Swanson (1981) was used to estimate the top of the transition zone and to extrapolate to determine the free energy surface. By replotting conventional air-mercury capillary pressure core test data on Swanson's published nomograph, it is possible to determine the saturation at which the non-wetting hydrocarbon phase is continuous. This effective saturation, the top of an oil-water transition, can be related to the oil saturation distribution as a function of height in a reservoir. Therefore, the elevation of a dry-oil production contact can be used to determine the free energy surface by converting the laboratory core test measurements (for the presence of reservoir fluids) to saturation distribution as a function of height. A representative capillary pressure curve of the Tensleep sand was used to estimate a conservative elevation for the free-energy surface because that formation had a thinner transition zone than the Phosphoria.

Capillary Pressure Curves Tied to Porosity

The capillary pressure curve described above was converted to oil saturation distribution as a function of height above the free-energy surface. However, a single curve of hydrocarbon saturation distribution as a function of height cannot adequately represent an entire reservoir. As the distribution of porosity and permeability changes in a reservoir because of pore type and facies differences, oil and water saturations will also change laterally and vertically. Therefore, a suite of saturation curves derived from variations in rock properties were used to approximate the changes of fluid saturation for this reservoir at discovery.

Calculating saturation curves involves a number of steps, the goal being to relate oil saturation to mappable attributes. Oil saturation distributions at discovery are controlled by several parameters, such as density difference between water and oil, thickness of the hydrocarbon column, pore entry size and distribution, and rock/fluid interaction. It can be shown that pore entry size and distribution are related to porosity and permeability. Porosity is commonly mapped for a

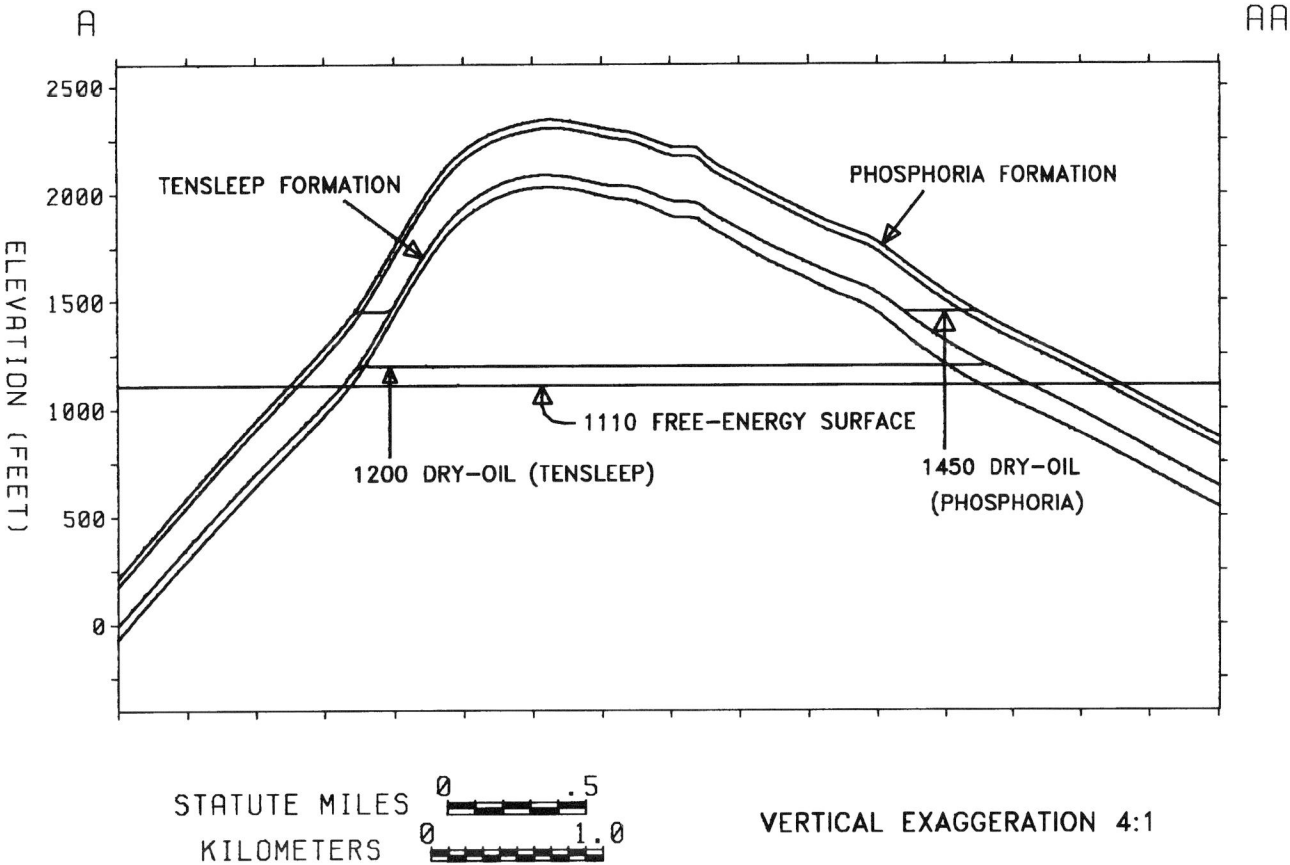

Figure 3. West-to-east cross section showing the top and base structures for the Phosphoria and Tensleep formations, free-energy surface, and dry-oil production surfaces for both reservoirs. See Figure 11 (Appendix 1 this chapter) for the position of the cross section.

reservoir and permeability is most easily related to capillary-pressure-curve-derived saturations; therefore, we must establish a relationship between porosity and permeability.

It can be shown that as porosity decreases, absolute permeability also decreases. A least-squares fit line drawn through data points on a plot of the logarithm of absolute permeability versus porosity is a common tool for correlating these two parameters. The absolute permeability value on the least-squares fit line for a given porosity represents the geometric mean of permeability for that porosity.

Swanson (1981) demonstrated that absolute permeability can be estimated from air-mercury capillary pressure core test measurements. Using this relationship and permeabilities derived from a porosity-permeability relationship, capillary pressure curves for geometric average permeability, each for a different porosity, were derived. These air-mercury capillary pressure curves were converted to reservoir conditions. Because each oil-saturation curve is associated with a porosity value, we now have a means to model oil saturation throughout the reservoir as a function of porosity and height above the free-energy surface.

WATER SATURATION (S_W) MODELS

As described above, oil saturation curves as a function of height versus porosity were derived from capillary pressure core test measurements converted to reservoir conditions and then to oil saturation. The algorithms used in this study require water saturation data to build the fluid saturation grids. Therefore, the oil saturation curves were converted to water saturation curves with height above the free energy surface on the ordinate and water saturation on the abscissa (Figure 4). These curves were digitized as contours of porosity.

Water saturation was modeled using four methods: (1) a constant water saturation, (2) a grid built using capillary pressure curves converted to curves of water saturation vs. height above free energy level for varying porosity, (3) a grid built using a single water saturation curve, also built from capillary pressure data, and (4) a grid built using average water saturation calculated at each well. These water-saturation grids were converted to oil saturation grids by subtracting each grid node value from one.

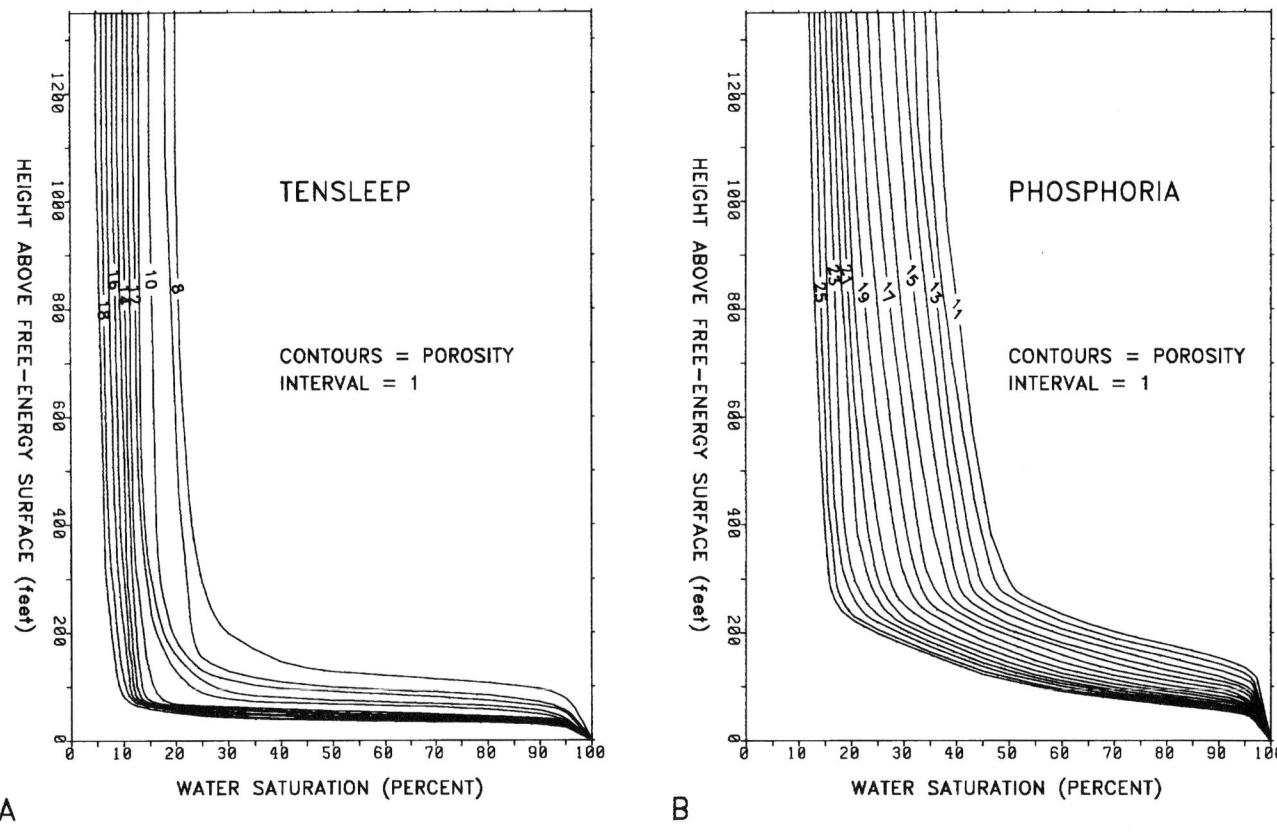

Figure 4. Plots relating porosity (curves) to water saturation (X-axis) and height above oil-water contact (Y-axis). (A) Tensleep reservoir. (B) Phosphoria reservoir.

Constant S_W

A constant value of 22.8% water saturation and a dry-oil water contact elevation of +1450 ft (442 m) were used for the Phosphoria Formation. Values of 11.2% and +1230 ft (375 m) were used for the Tensleep Formation.

S_W from Multiple Water Saturation Curves

A porosity grid, top and base structure grids, constant value for fluid contact, and the digitized water saturation curves as a function of reservoir height for varying porosity were used to build the water saturation grid for each reservoir. The resulting water saturation grids vary smoothly over the reservoir until the reservoir dips into the transition zone near the fluid contact. Once in this zone, water saturations quickly increase to 100% where the rock unit passes below the free-energy surface (Figure 5).

Reviewing the steps used by the average-water-saturation gridding algorithm will clarify how this model is constructed. The first step is to build a grid of the digitized water-saturation curves (Figure 4). That grid is then transformed into a grid whose Z-values are water saturation, and whose X-coordinate is porosity. The Y-coordinate remains height above the oil-water contact (Figure 6A). This new grid may be thought of as a look-up table from which water saturation is obtained for a given porosity and height above the fluid contact. If a location in the reservoir had 12% porosity and was 200 ft above the fluid contact, then interpolation at those coordinates on the look-up grid shown in Figure 6A will yield an average water saturation of 17%.

However, the reservoir at a location is not a point but has thickness and, therefore, a range of heights above the fluid contact (Figure 6A, vertical line). Instead of picking one point on the look-up grid, two points that define a line are picked. One end of the line has coordinates defined by porosity at the location in the reservoir and height of base of reservoir above the fluid contact. The other end of the line has "coordinates" defined by the same porosity value and the height of top of reservoir above the fluid contact at that location (Figure 6B, line designated as TOP and BASE).

A profile of water saturation generated along this line (Figure 6C) shows how saturation varies from base to top of reservoir at that point. The average height of this line above zero (integration of the area under the curve) will be the average water saturation at that point in the reservoir. Thus, the final step is to integrate, for every grid node, an average water saturation value using the water saturation look-up grid (Figure 5).

The resulting water-saturation grids have a sharp

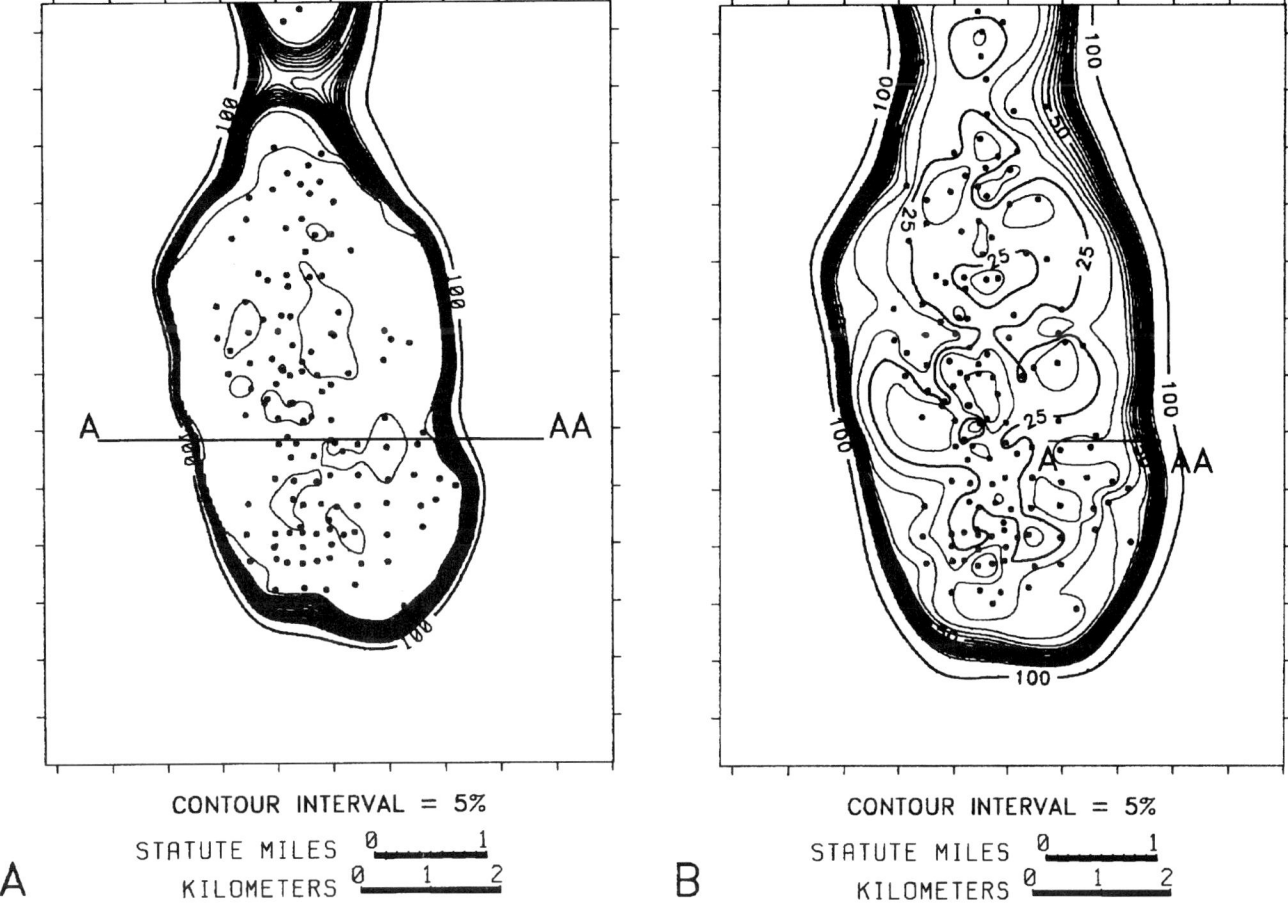

Figure 5. Contour maps of water saturation built using multiple curves of water saturation as a function of porosity versus height above free-energy surface. Saturation gradients flatten near the base of the transition zone reflecting the form of the water saturation curves used (see Figure 4). (A) Tensleep Formation. (B) Phosphoria Formation.

gradient defining the transition zone because height above the fluid contact is used to calculate S_w. They also show considerable lateral variation because average porosity was used to calculate S_w and lateral variations in porosity are reflected in the S_w grid.

S_W from a Single Water Saturation Curve

This grid is built using the same input and procedures as for S_w from multiple saturation curves above. However, instead of many water saturation curves, only one is digitized, that being the curve associated with average reservoir porosity. This curve then applies to the entire formation throughout the reservoir. The resulting S_w grid for the Tensleep Formation is shown in Figure 7. This S_w model is smooth because its variation is controlled only by height above the fluid contact.

S_W from Wells

Values of average water saturation were interpolated from the average water saturation grid built from capillary pressure curves (described above). These curves simulate the values that would have been calculated by hand at each well using a free-energy surface of +1110 ft (338 m). These values were gridded using a least-squares gridding algorithm and biharmonic filter (Figure 8). No attempt was made to incorporate a transition zone into the average water-saturation grid.

When wells occur in the area of rapidly increasing water saturation (transition zone), their water saturation values are significantly higher than those of wells above the transition zone. The dotted lines on Figure 8 define the outer and inner boundaries of the transition zone. Water saturation values and gradients in the transition zone are unique to the zone and should not be projected above or below it. The gridding process carries the influence of values in the transition zone well beyond its boundaries (Figure 8, top center and bottom center). Similarly, where no wells penetrate the zone, wells above the transition zone project their values into and below the zone (Figure 8, east and west sides).

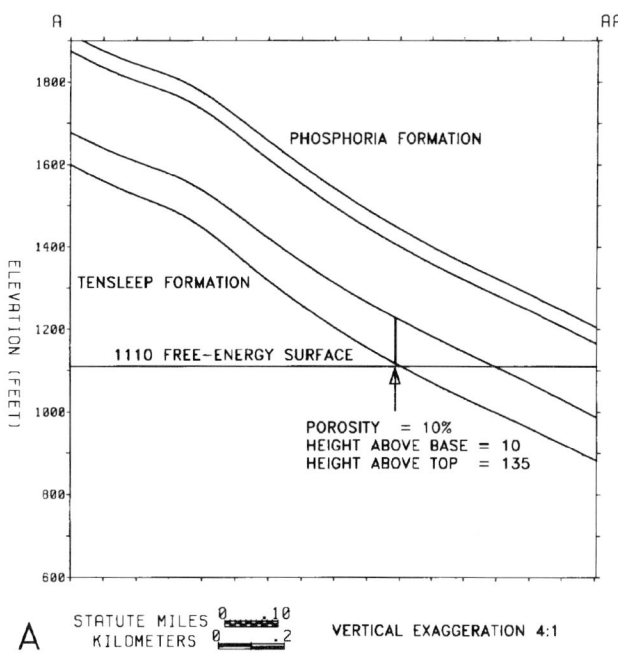
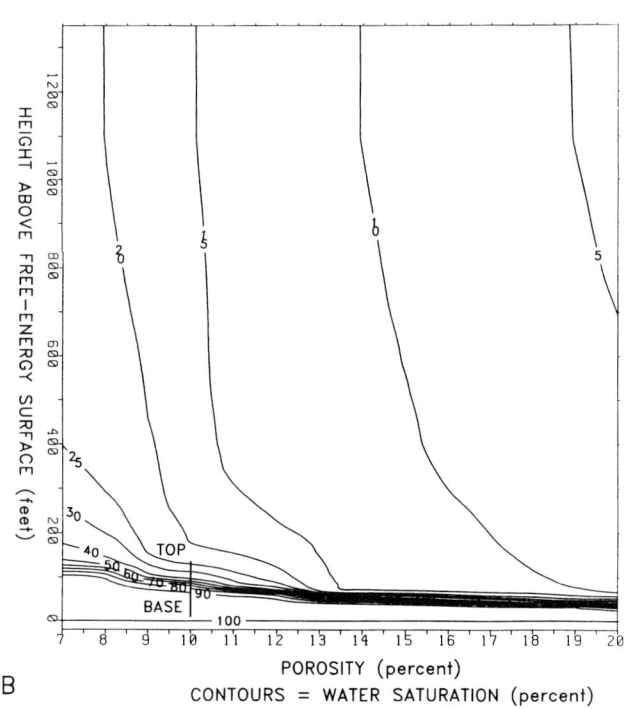
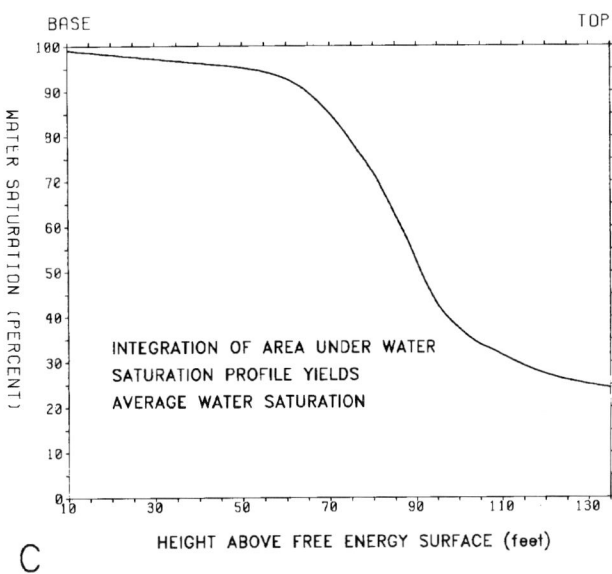

Figure 6. Steps in the water-saturation gridding procedure. (A) West-to-east cross section near the eastern edge of the Tensleep reservoir (see Figure 11, Appendix 1, this chapter, for position of the cross section) showing structure top, base, and free-energy surface. The vertical bar indicates the point at which the vertical range in water saturation will be averaged. (B) Grid of water saturation in terms of porosity (X-axis) and height above fluid contact (Y-axis). The line (BASE, TOP) represents the range in water saturation at a point in the reservoir. (C) A profile (BASE, TOP) through the saturation grid at a constant porosity from height of base of reservoir to height of top of reservoir. Integration under this curve gives average water saturation for the location in the reservoir.

BUILDING HYDROCARBON-PORE-THICKNESS GRIDS

Our goal when preparing input for volume estimation is to build a grid that represents Hydrocarbon Pore Thickness (HPT). To do this we calculate HPT using the equation:

$$HPT = GP \times N{:}G \times POR \times S_o \quad (1)$$

where:

HPT = Hydrocarbon Pore Thickness: a grid representing thickness of hydrocarbons. Integration of this grid over the area of interest produces Hydrocarbon Pore Volume.

GP = Gross Pay: a grid representing the total gross rock (reservoir) thickness that is above the fluid contact.

N:G = Net-to-Gross Ratio: a grid or constant representing the ratio of porous (net) rock thickness to gross rock thickness.

POR = Average Porosity: a grid or constant representing the average porosity of the porous rock.

S_o = Oil Saturation: a grid or constant representing the average oil saturation in the porous rock above the fluid contact. This is usually created from a grid of average water saturation ($S_o = 1 - S_w$).

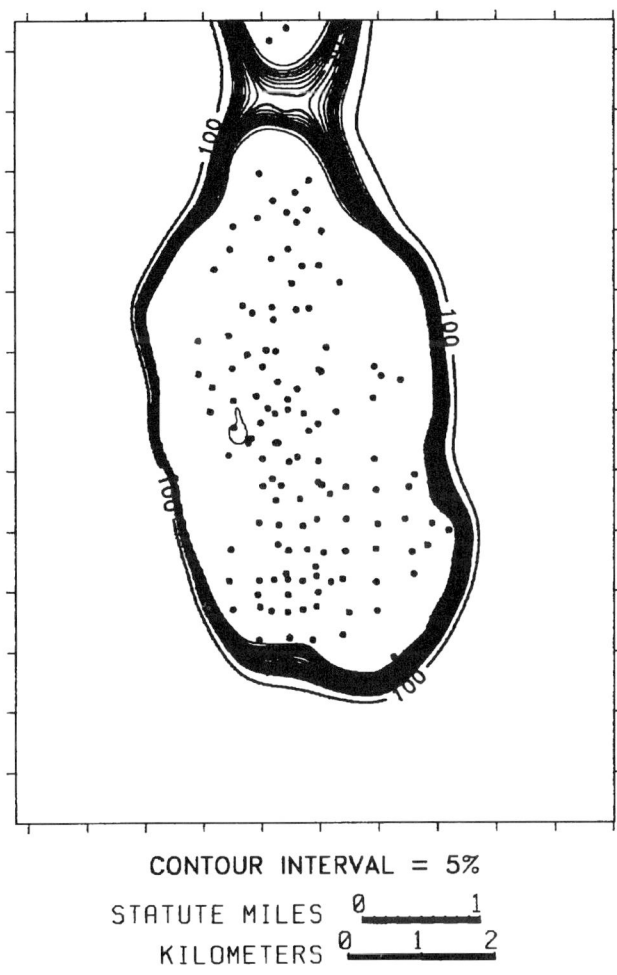

Figure 7. Contour map of water saturation built using a single water saturation curve (at 14% porosity) for the Tensleep Formation.

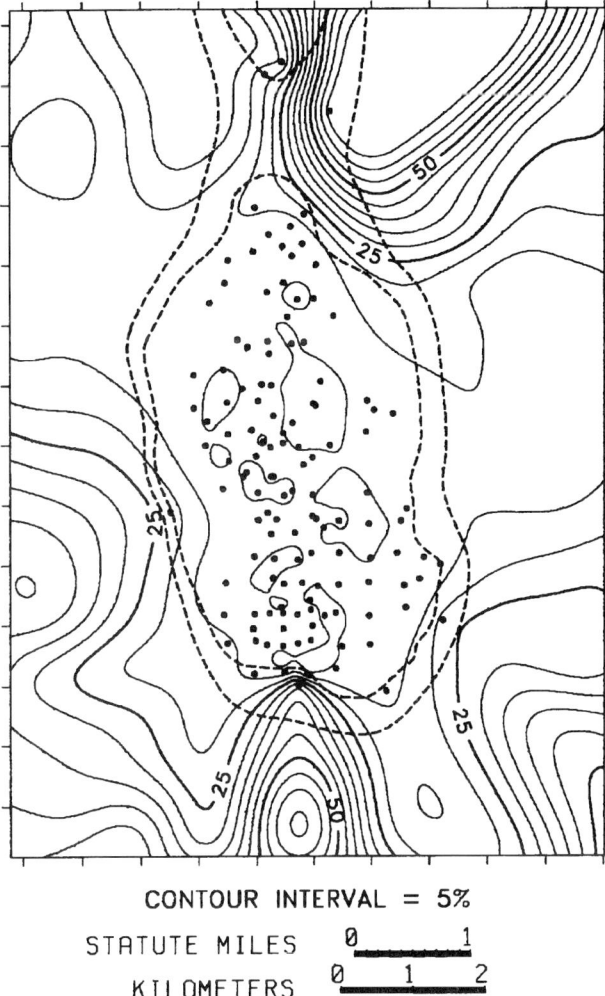

Figure 8. Contour map of water saturation built by gridding average water saturation data for the Tensleep Formation. The dashed lines represent the top of the transition zone (inner line) and the zero edge of the reservoir (outer line).

See Appendix 1 of this chapter for details of how the hydrocarbon pore thickness grids and volumes were constructed. Appendix 1 contains descriptions of: (1) the data used to build the component grids for this equation, (2) how the grids were constructed for the top and base structure, gross pay, net-to-gross ratio, and average porosity (methods for constructing water saturation grids were described above), and (3) the problems encountered when combining the component grids into an HPT grid and graphically portrays the steps in the process. Figure 9 shows final hydrocarbon-pore-thickness maps for the Tensleep and Phosphoria formations built using a water saturation grid derived from multiple water saturation curves.

Ten HPT grids were built, five for each reservoir. Each grid used either a different average-water-saturation model or a different fluid contact. Table 1 provides a list of names for the HPT grids and a summary of how they were constructed. HPT grid names are the same for both reservoirs.

VOLUMETRICS

Volumes were calculated within a polygon defined by the X-Y limits of the grid area. To estimate volumes, the algorithm calculates the volume for each grid cell within the polygon. The volumes of the cells are then summed to cumulate the total volume.

Five volume estimates were made for the Tensleep Formation and five for the Phosphoria Formation, each using an HPT grid built as described above and in Appendix 1. For comparison purposes, the OOIP estimates have been normalized. To do this, the volumes generated using the multiple saturation-curves method and the free-energy surface were set equivalent to 100 for both reservoirs. Volumes from the other HPT grids were then normalized to that same scale. Table 2 summarizes the results of volume calculations for both the Tensleep and Phosphoria formations.

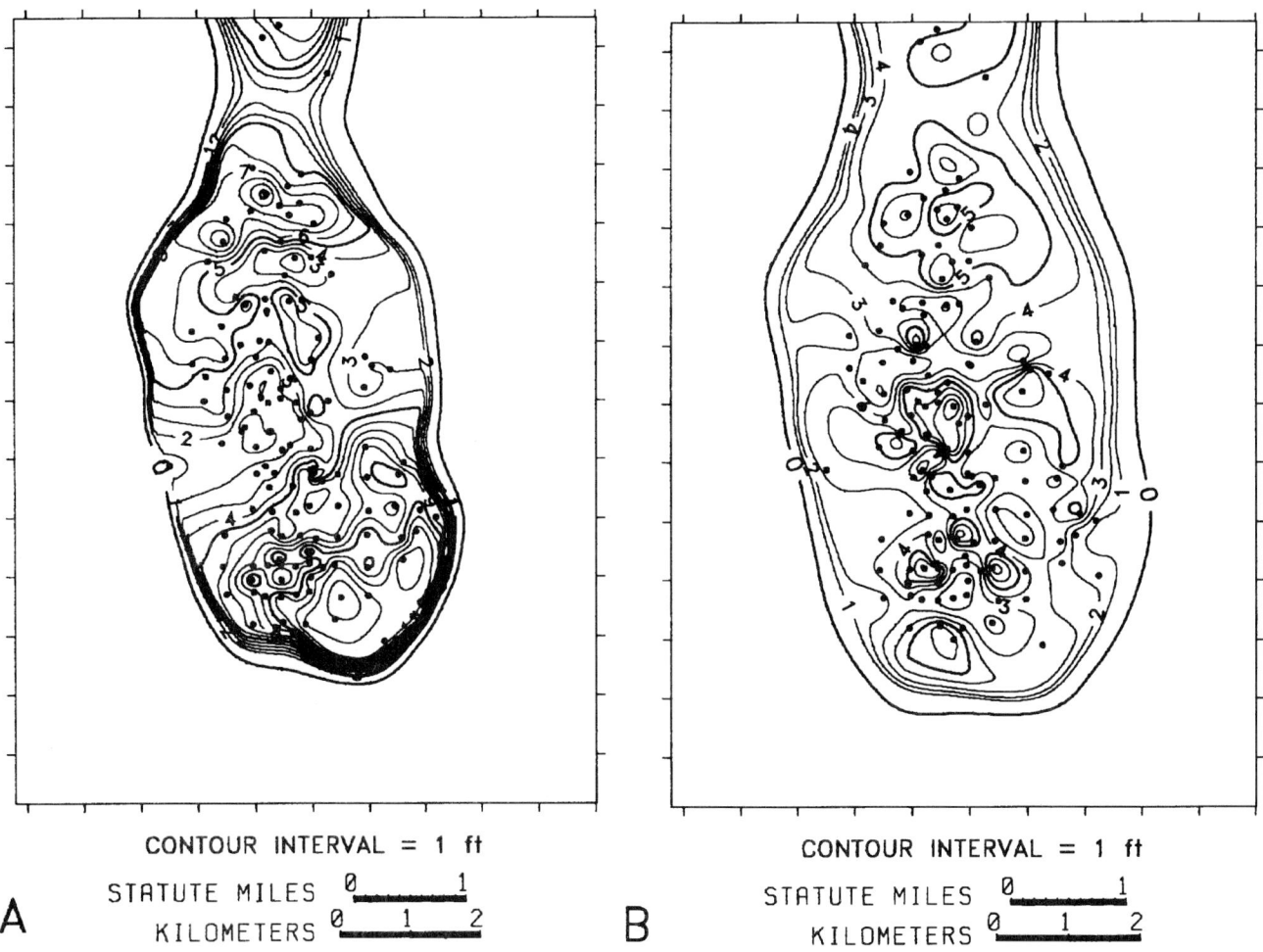

Figure 9. Contour maps of hydrocarbon pore thickness. (A) Tensleep Formation. (B) Phosphoria Formation.

As expected, in both reservoirs the major difference in OOIP estimates is associated with the fluid contact: free-energy surface (FE) or dry-oil production contact (DO). The change is much greater for the Phosphoria because vertical separation between the two fluid contacts is significantly greater than in the Tensleep. When the fluid contact is the same, there is still considerable volume variation between the methods for calculating water saturation. Maps of difference grids built by subtracting HPT grids (Figure 10) show where those variations occur and, even when volumes are nearly identical, show that the distribution of those volumes vary.

Differences between the HPT-Curves-FE grid (multiple saturation curves/free-energy surface) and the HPT-Curves-DO grid (multiple saturation curves/dry-oil production surface) are zero above the dry-oil production surface with the base of reservoir surface (Figure 10A). Below that surface, the differences increase rapidly. As mentioned above, hydrocarbons below the dry-oil production surface probably do not contribute significantly to production but do contribute to reservoir energy in the form of solution gas as reservoir pressure decreases.

Differences between the HPT-Curves-FE grid and the HPT-1Curve-FE grid (single saturation curve/free-energy surface) are small in the central portion of the reservoir (Figure 10B). This is because porosities in that area range from 12 to 16% and the reservoir is above the transition zone (Figure 4A; top of the transition zone is about 200 ft [60 m] above the free-energy surface). Since the water saturation curves from 12% to 16% are close together above 200 ft (60 m), results from a single water-saturation curve of 14% will produce results similar to a range of saturation curves. All major HPT variations occur in the transition zone where porosities drop from 13% to 12%. As seen in Figure 4A, water saturations between these two porosities differ by as much as 60% at a height of 65 ft (20 m) above the free-energy surface; major differences, although less, continue up to about 200 ft (60 m) above the free-energy surface. In the transition zone on the west edge of the reservoir and to some degree on the east edge, there is almost no difference in HPTs, although porosities are less than 13%. Both gross pay thickness and N:G ratio are small in these areas, result-

Table 1. Fluid Contact and Oil Saturation Grids Used for HPT

HPT Name	Fluid Contact	Average Oil Saturation Model
HPT-Curves-FE	P + 1110 T + 1110	Grid from capillary curves
HPT-1Curve-FE	P + 1110 T + 1110	Grid from one capillary curve
HPT-Constant-DO	P + 1450 T + 1200	Constant P 77.2% T 88.8%
HPT-Wells-DO	P + 1450 T + 1200	Grid from wells
HPT-Curves-DO	P + 1450 T + 1200	Grid from capillary curves

P = Phospohoria
T = Tensleep
FE = Free-Energy Surface
DO = Dry-Oil Production Surface

Table 2. Normalized Volume Calculations

HPT Grid	Phosphoria	Tensleep
HPT-Curves-FE	100.00	100.00
HPT-1Curve-FE	98.48	101.36
HPT-Constant-DO	77.31	93.83
HPT-Wells-DO	74.66	92.33
HPT-Curves-DO	74.67	91.87

ing in small HPTs and small HPT differences.

Differences between the HPT-Curves-DO grid and HPT-Constant-DO grid (constant saturation/dry-oil production contact) vary dramatically in the transition zone and in areas of low or high porosity (Figure 10C). Variations above the transition zone are small because water saturations for the Tensleep Formation show small variation across a large range of porosities (Figure 4A). The Phosphoria Formation has considerably more variation between these two HPT grids above the transition zone due to a larger range in water saturation with changing porosity.

Differences between the HPT-Curves-DO grid and HPT-Wells-DO grid (saturation from wells/dry-oil production contact) vary dramatically in and above the transition zone (Figure 10D). Variations above the transition zone relate to projections of low water saturation away from wells in the transition zone into areas that are clearly above the transition zone (north-central, south, and south-east areas). Within the transition zone, variations are similar to those seen for the constant-water-saturation HPT. That is, wells above the transition zone have projected low water saturations into and beyond the transition zone (west and east edge).

CONCLUSIONS

Four methods for modeling oil-saturation distribution and two types of fluid contacts have been compared through the use of computer modeling. These methods and contacts were applied to both the Tensleep and Phosphoria formations. Volumes were calculated for all methods for both formations and normalized so that comparisons could be made. Maps showing differences in hydrocarbon pore thickness between methods were made for the Tensleep Formation and compared. These comparisons indicate that, for these reservoirs:

(1) Computer mapping and volume estimation of original oil-in-place for a variety of scenarios can be done quickly. Building the initial grids took several weeks, and testing the scenarios presented here took only a few hours. Although volumes were calculated for only one lease, the same scenarios could have been run for hundreds of leases with little increase in time.

(2) The use of computers makes it practical to use saturation curves as a function of height and porosity to gain greater detail for lateral and vertical oil saturation distribution.

(3) Using a free-energy surface to define the reservoir limits resulted in 25% (Phosphoria) and 8% (Tensleep) higher estimates of original oil-in-place than estimates made using a dry-oil production contact. Although little of this volume is producible, it may contribute significantly to reservoir energy, in the form of solution gas, as reservoir pressure decreases.

(4) Comparison of the differences between maps of hydrocarbon pore thickness, whose saturations were built using different methods, indicates that large variations in hydrocarbon saturation and in resulting hydrocarbon thickness occur around the perimeter of these edge-water-drive reservoirs (in the transition zone). More subtle variations occur above the transition zone.

(5) The volume estimate of original oil-in-place from a grid of hydrocarbon pore thickness, built using a single hydrocarbon saturation curve, is similar to the volume estimate from a hydrocarbon pore thickness grid, built using multiple saturation curves. This indicates that an acceptable single saturation curve was selected (curve for average porosity of the reservoir); it does not indicate that the two hydrocarbon pore thickness grids are the same. If a different single hydrocarbon saturation curve were used, one not associated with the average porosity of the reservoir, it is unlikely that the volume results would be as close.

(6) Reservoir maps of saturation, built from multiple saturation curves tied to porosity and height above fluid contact, provide more detail of lateral and vertical hydrocarbon saturation variations than maps built using a single saturation curve tied to height. This detail provides more geologic control for reservoir simulators and more information to aid in reservoir planning.

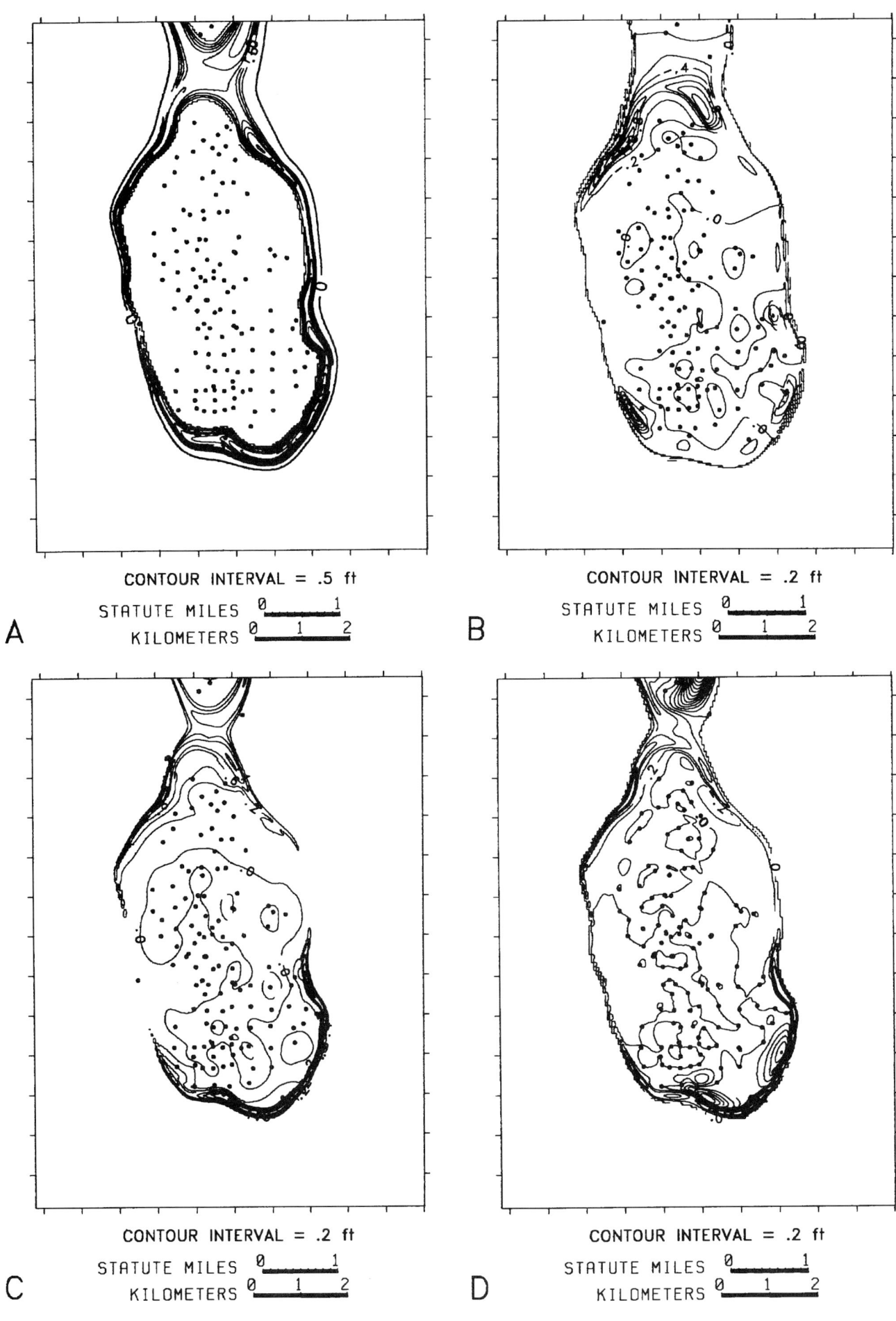

Table 3. Data Type and Range

Data Type	Range of Values	
	Phosphoria	Tensleep
X-coordinate (ft)	10,000 – 37,000	
Y-coordinate (ft)	11,000 – 42,750	
KB elevation (ft)	5,273 – 6,148	
Structure top elevation (ft)	–610 – +2,502	–843 – +2,246
Gross thickness (ft)	34–48	21–155
Net average porosity (%)	12–24	8–17
Net thickness (ft)	1.5–10	6–107
Net-to-gross ratio	0.14–0.96	0.27–0.98

(7) Volume estimates calculated using a constant value for water saturation are larger than volumes calculated when average water saturation grids built from wells or from multiple water saturation curves are used. These higher volumes are fictitious because oil saturation is not reduced in the transition zone.

(8) Volume estimates calculated using average-water-saturation from wells can be highly variable and appear to range between volumes calculated using a constant water saturation and volumes calculated using an average-water-saturation grid built from multiple water saturation curves. The variation is controlled by the presence and absence of wells penetrating the transition zone and by where those wells are positioned relative to other wells in the reservoir. This observation is generally true because the technique used to calculate saturations in the wells is similar to that used in the multiple curve saturation method.

ACKNOWLEDGMENTS

The authors thank Marathon Oil Company and Landmark/Zycor Incorporated for permission to participate in preparation and publication of this paper.

REFERENCES CITED

Campbell, C. V., 1962, Depositional environments of Phosphoria Formation (Permian) in southeastern Big Horn basin, Wyoming: AAPG Bulletin, v. 49, p. 1833-1846.

Curtis, B. F., J. W. Strickland, and R. C. Busby, 1958, Patterns of oil occurrence in the Powder River basin, in L. G. Weeks, ed., Habitat of oil: Tulsa, AAPG, p. 268-329.

Heymans, M. J., D. R. Petro, A. J. Kaltenback, and R. M. Just, 1990, South Dome Oregon basin field: An integrated study for improved oil recovery opportunities (abs.): AAPG Bulletin, v. 74, p. 673-674.

Hunt, J. M., 1953, Composition of crude oil and its relation to stratigraphy in Wyoming: AAPG Bulletin, v. 37, no. 8, p. 1837-1872.

Jones, T. A., D. E. Hamilton, and C. R. Johnson, 1986, Contouring geologic surfaces with the computer: New York, Van Nostrand Reinhold, 314 p.

Jones, T. A., and D. E. Hamilton, 1992, A philosophy of contour mapping with the computer, (this volume).

Partridge, J. F., 1958, Oil occurrence in Permian, Pennsylvanian, and Mississippian rocks, Big Horn basin, Wyoming, in L. G. Weeks, ed., Habitat of oil: Tulsa, AAPG, p. 293-306.

Stewart, F. M., D. L. Garthwaite, and F. K. Krebill, 1955, Pressure maintenance by inert gas injection in the high relief Elk basin field: Transaction A.I.M.E. Technical Note 84, Petroleum Technology, v. 192, p. 376-379.

Stone, D. S., 1967, Theory of Paleozoic oil and gas accumulation in Big Horn basin, Wyoming: AAPG Bulletin, v. 51, no. 10, p. 2056-2114.

Swanson, B. F., 1981, A simple correlation between permeabilities and mercury capillary pressures: Journal of Petroleum Technology, v. 33, p. 2498-2504.

Thompson, J. C., 1954, Resume of several fields in Wyoming and Montana south and west of the Elk basin field in Pryor Mountains—northern Big Horn basin, Montana, Billings Geological Society Guidebook: Billings, Billings Geological Society, p. 117-125.

Walton, P. T., 1947, Oregon basin oil and gas field, Park County, Wyoming: AAPG Bulletin, v. 31, no. 8, p. 1431-1453.

APPENDIX 1: DATA, GRID CONSTRUCTION, AND VOLUME CALCULATION

Data

Data from 348 wells penetrating the Phosphoria and Tensleep units were stored in a SAS data set. Tops and other values were entered and edited through a SAS editing screen tailored to this data base. All tops, recorded as measured depths, were converted to elevation and vertical thickness for mapping and volumetrics work.

System options allowed the user to request up to 13 maps containing contours and values for these data, as well as printouts of data statistics. This automated approach allowed users to make maps for evaluating consistency of top picks and to edit those maps (grids) to reflect realistic structural and stratigraphic interpretations.

Figure 10. Contour maps for the Tensleep reservoir built by subtracting HPT grids built using different water saturation grids and different fluid contacts. See Table 1 for a summary of how each HPT grid was built. (A) HPT-Curves-FE minus HPT-Curves-DO. (B) HPT-Curves-FE minus HPT-1curve-FE. (C) HPT-Constant-DO minus HPT-Curves-DO. (D) HPT-Wells-DO minus HPT-Curves-DO.

Table 3 summarizes the data types and their range of values. Net thickness was defined as the total thickness of all intervals in the Tensleep whose porosities were greater than 7%. Previous studies had shown that no oil is found in Tensleep rocks having porosities less than 7%. A 10% porosity cutoff was applied to the Phosphoria for the same reason.

Porosity values were recorded as percent and calculated based on well-log cross-plots of formation density and compensated neutron logs. The average porosity value was a thickness-weighted average of porosities in net reservoir rock above and below the fluid contact.

In this study, two types of fluid contact are of concern: the free-energy surface, and the dry-oil production contact. A free energy surface, as discussed above, was common to both formations at an elevation of +1110 ft (338 m). This surface is the lowest occurrence of oil in the reservoir.

A dry-oil production contact is an elevation below which both oil and water would be produced. This contact is commonly used to define the base of a reservoir volume. The contacts for the Tensleep and Phosphoria formations differ because of variations in pore types and their associated fluid flow characteristics. The dry-oil production contacts are +1200 ft (366 m) for the Tensleep and +1450 ft (442 m) for the Phosphoria.

Water saturation curves with height above the free energy surface on the ordinate and water saturation on the abscissa were provided. These curves were digitized as contours of porosity.

Construction of Attribute Grids

Top and Base Structure

Gridding parameters for the top and base structures of both reservoirs were the same, with the final grids having an increment of 150 ft (46 m) (249 rows and 181 columns). A least squares algorithm, biharmonic filter, and grid increment of 600 ft (182 m) were used to build an initial structure top grid. The grid increment was reduced to 300 ft (91 m), with new node values being automatically interpolated from surrounding nodes in the initial grid. A biharmonic filter was applied to this finer grid, using the data to ensure that both new and old nodes honored the data. This grid increment reduction and filtering process was repeated once more, resulting in a final grid increment of 150 ft (46 m). Starting with a grid increment of 600 ft (182 m) allowed the algorithm to project a smooth surface across large open areas between data and to project away from the data acceptably on the flanks. Gradually reducing the grid increment to 150 ft (46 m) while filtering to the data allowed the detail of the closely spaced points to be honored in the final grid.

Structural surfaces on the flanks of the reservoir and for distance beyond well control were interpreted to continue projecting deeper. However, because no wells penetrated the flanks, the least squares algorithm tended to "bend" the surface back to elevations similar in value to points on the perimeter of the data. Interpretive control points (dummy wells) were used around the flanks of the reservoir to force the structure grids to project deeper (Figure 11A).

The base of structure grid for each reservoir unit was constructed by subtracting the gross-rock-thickness grid from the top of unit structure grid. Gross-rock-thickness grids were built using the process described above. Because the reservoir units do not pinch out over the field, no special processing was needed. A few wells penetrated minor faults. These had a negligible effect on volumes, and were not used to build the top-structure grids. Also, well intervals affected by faulting were not used to build the thickness grids.

Final gross-rock-thickness grids did not project to negative values (Figure 11B), so they were used without modification to build the base-of-structure grid for each unit. The base-of-structure grids were filtered to the base-of-structure data to ensure that data were honored from wells not used due to faulted out thickness.

Gross Pay

The grid of gross-pay thickness was created by isolating the portion of the gross-rock-thickness grid above the free-energy surface. This was done by creating two grids, one defining the top of gross pay and the other defining its base, and then subtracting them (Jones et al., 1986). For these reservoirs, the top of gross pay was defined by only one grid: the structure top. The base of gross pay was defined by two grids: the base of structure, and the fluid contact. These grids were combined to form a new base-of-gross-pay grid which was the maximum of the two grids' values at each node location.

The base-of-gross-pay grid was subtracted from the top-of-structure grid, creating the gross-pay thickness grid. This grid is positive where reservoir rock exists above the free-energy surface and negative or zero everywhere else. Negatives in this grid were expected and were useful because they allowed a sharp zero edge of the reservoir to be defined for volume calculation and map display. On contour maps, only zero and positive-valued contours were displayed (Figure 12).

Net-to-Gross Ratio

A grid of net thickness was built using data both above and below the fluid contact. Before the least-squares algorithm and biharmonic filter were used to build the grid, an isopach preprocessing algorithm was applied to the data (Jones et al., 1986; Jones and Hamilton, 1992). This replaced zero-thickness values with a negative value that was determined by continuing the gradient of the positive-thickness surface adjacent to the zero edge. After zero values were replaced by negatives, the least-squares algorithm and biharmonic filters were applied just as for building the structure grids. Using the negative zero-replacement value allowed a more reasonable zero edge as the resulting net thickness grid projected negative. These negative values were replaced with zeros before creating the net-to-gross ratio grid.

The net thickness grid was divided by the gross-

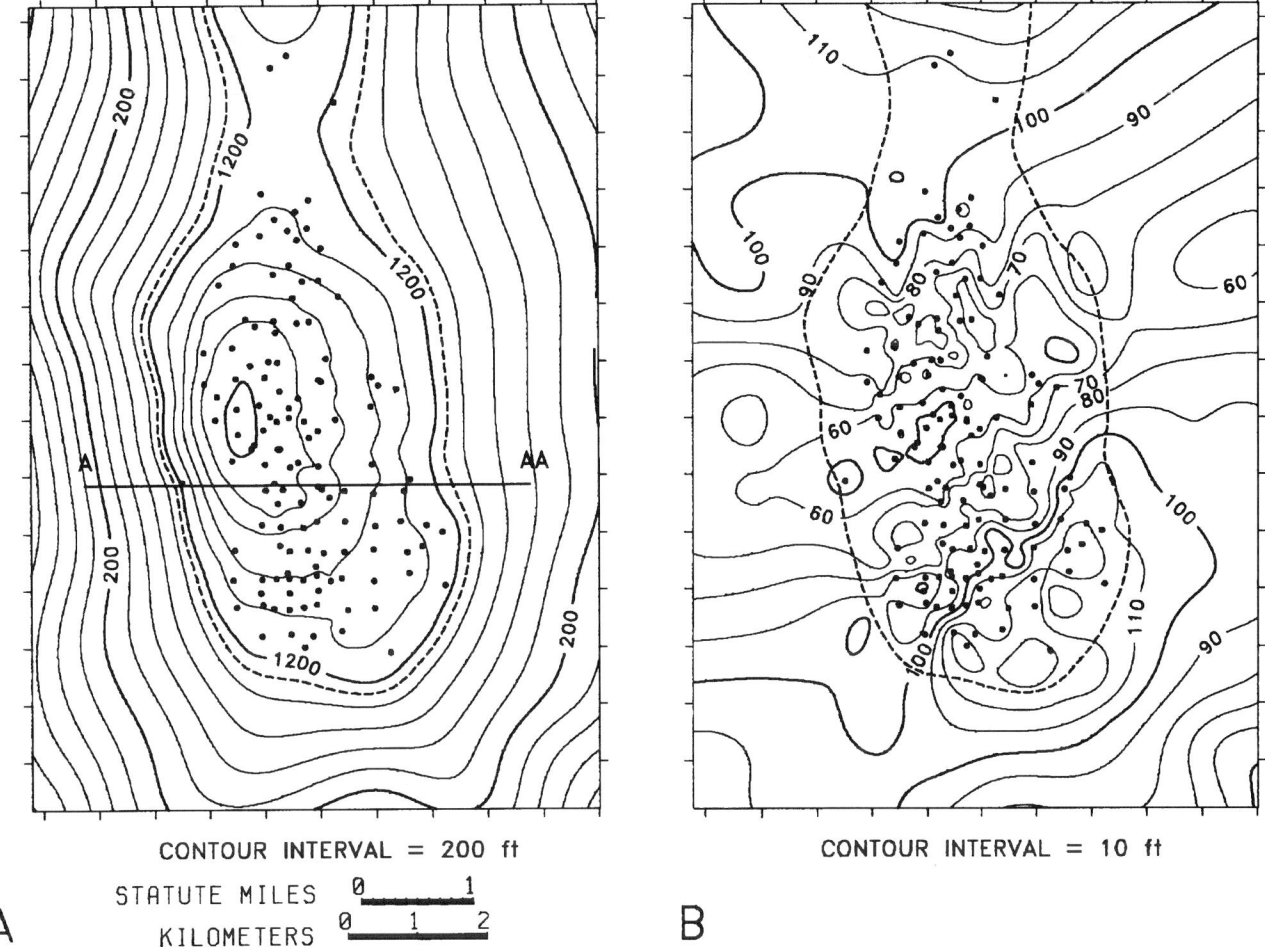

Figure 11. Contour maps for the Tensleep Formation. The dashed line represents the intersection of the structure top and the free-energy surface (fluid contact). (A) Structure top. (B) Gross-rock thickness.

rock-thickness grid, creating the net-to-gross ratio grid. As mentioned above, the gross-rock-thickness grid did not project negative, so the net-to-gross ratio did not go below zero. However, in areas where net was nearly equal to gross, the net thickness grid sometimes projected thicker than gross thickness. This resulted in a net-to-gross ratio greater than one and was corrected by clipping the net-to-gross grid to a maximum value of one (Figure 13A).

Creating the net-to-gross ratio grid in this way does not account for heterogeneities in the vertical distribution of net reservoir. These variations would cause the net-to-gross ratio for the hydrocarbon bearing portion of the reservoir to vary across the wedge zone (intersection of the fluid contact and the top and base of the reservoir). Correcting for this situation was not necessary because net reservoir rock is uniformly distributed vertically through both reservoirs.

Porosity

The porosity grids were built using the same procedures used to build the structure top grids. However, in place of least squares, a weighted-average gridding algorithm was used. This algorithm tends to bring the surface quickly back to a local average of neighboring data values when calculating nodes progressively farther from a data point. The biharmonic filter removed sharp high and low peaks at data locations, a characteristic of the weighted-average algorithm. As the surface projected away from data towards the structure flanks, its value approached the average of nearby data points so that extrapolation was well behaved. Values in the resulting porosity grids were well within the appropriate ranges, so no further work was required to handle negative or unrealistically high values. Fewer data (wells with adequate wireline logs) were available for porosity than for structure, resulting in somewhat smoother looking models (Figure 13B).

Building the Hydrocarbon-Pore-Thickness Grid

Our goal in preparing input for the volumetrics algorithm is to build a grid of Hydrocarbon Pore

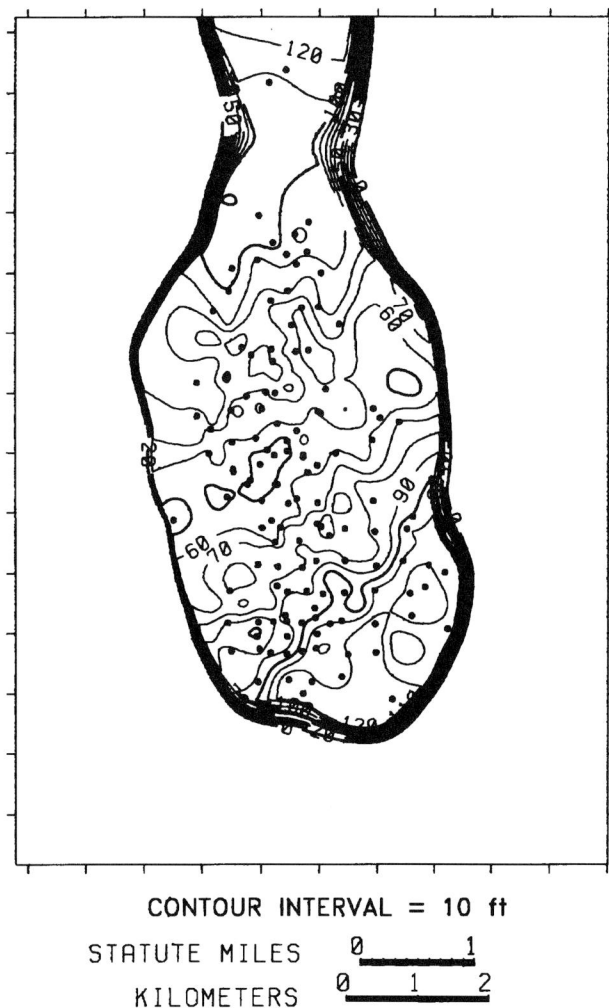

Figure 12. Contour map of gross-pay thickness for the Tensleep Formation. The thickness was built using the free-energy surface.

Thickness (HPT):

$$HPT = GP \times N{:}G \times POR \times S_o$$

After each component grid in the HPT equation is built, creating the HPT grid is a simple process of multiplying each of the component grids. However, each of those grids could project negatively. In fact, the gross-pay (GP) grid was deliberately built so that it would project to negative values. If two of the four grids projected negative at a node, their product (HPT) would be positive at that node. This can be prevented by simply clipping the N:G, POR, and S_o grid to a minimum value of zero, thus allowing only one grid to project negative. This often produces flat areas of all zeros in the final HPT grid, which creates a small band of extra thickness (up to one grid increment wide) at the zero edge (Jones et al., 1986). A special multiplication operator is available in some programs which returns a negative value when two negatives are multiplied, thus avoiding flat, all zero, areas in the HPT grid. We clipped the grids to zero because small volume variations at the zero edge (fraction of a percent of total volume) were not significant to this study.

Figure 14 helps visualize how the HPT grid is built by showing thickness profiles for each step in the process overlayed on one picture. The profiles are from the Tensleep Formation, cut west-to-east across the crest of the structure (see Figure 11A for cross-section baseline), and use the average water saturation grid built with the capillary-pressure-curves method. The thickest profile represents gross hydrocarbon rock thickness. The portion of gross-rock-thickness above the fluid contact is gross-pay, which rapidly wedges from full thickness to zero when the fluid contact is encountered. Multiplying the gross-pay grid by the net-to-gross ratio grid creates the net-pay thickness grid. Further reductions in thickness are seen with multiplication by porosity and oil saturation. Maps of HPT thickness for the Tensleep and Phosphoria formations are shown in Figure 9 in the body of this paper.

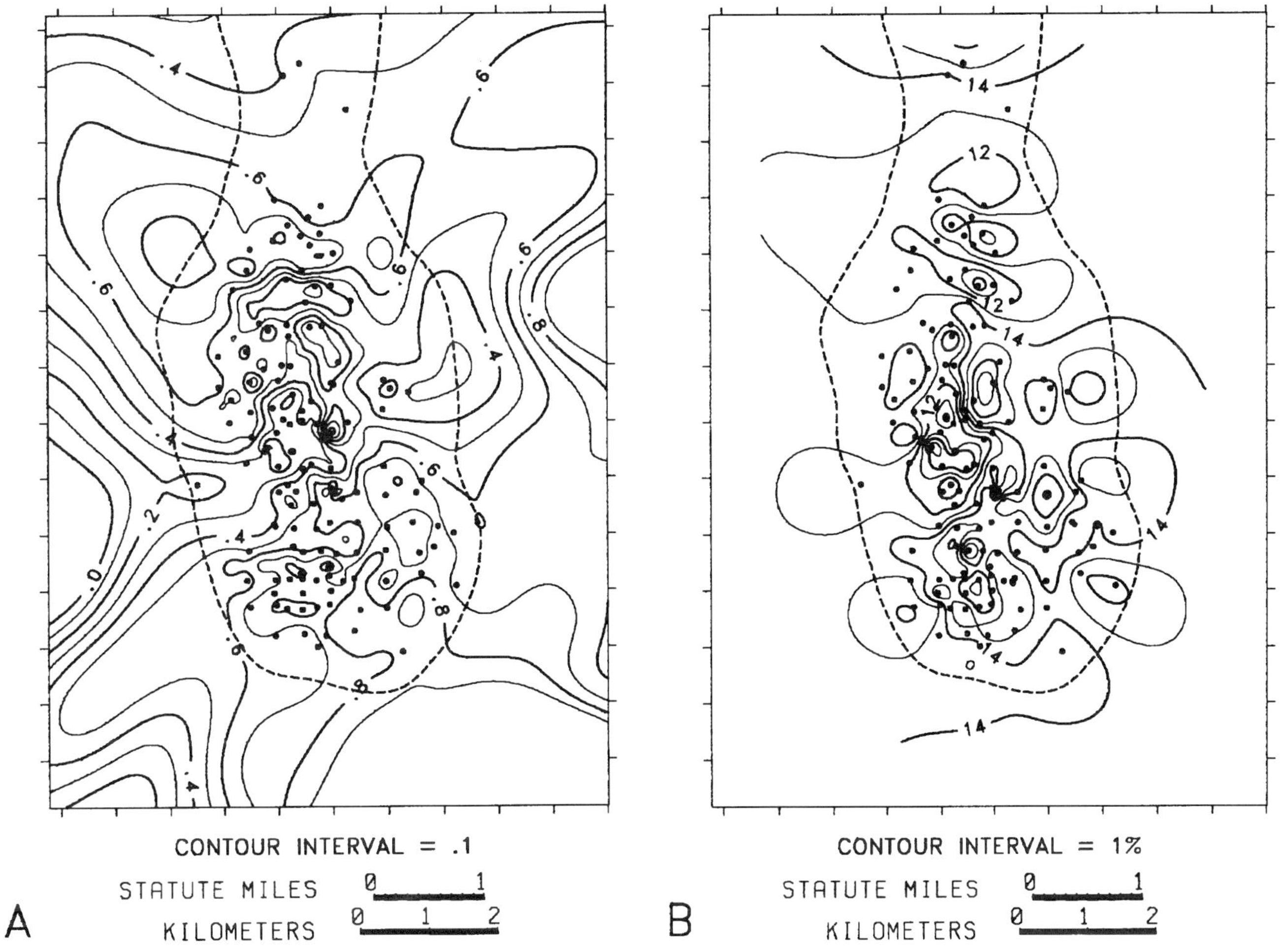

Figure 13. Contour maps of reservoir attributes for the Tensleep Formation. The dashed line represents the intersection of the structure top and the free-energy surface (fluid contact). (A) Net-to-gross ratio. (B) Average porosity.

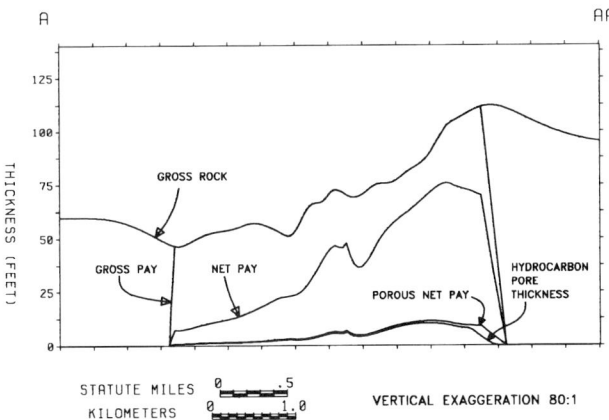

Figure 14. Series of thickness profiles displayed on the same section for a west-to-east line through the Tensleep Formation (See Figure 5A for the position of the cross section). The profiles show the change from gross rock thickness to hydrocarbon pore thickness as components of the hydrocarbon-pore-thickness equation are combined. Note how gross-pay thickness sharply wedges from full thickness to zero thickness when the fluid contact is encountered.

Chapter 10

Fault Representation in Automated Modeling of Geologic Structures and Geologic Units

Steven Zoraster
Landmark/Zycor, Inc.
Austin, Texas, U.S.A.

ABSTRACT

The standard petroleum industry technique of treating faults as opaque barriers in structure and unit modeling does not provide the accurate models and hardcopy maps required today for hydrocarbon exploration and production efforts. The shortcomings of these old methods is especially apparent when attempting to develop cost-effective models of small hydrocarbon reservoirs.

The process of incorporating fault surfaces into digital subsurface models is going through a period of rapid change as new techniques for incorporating fault geometry into structure and unit modeling applications have been introduced over the last few years. These methods, which are reviewed in this chapter, provide superior solutions to the problem of incorporating faults into subsurface modeling and mapping efforts than techniques which simply treat faults as opaque barriers. However, these methods represent only interim solutions.

Inevitably, full 3-D subsurface modeling in faulted environments will replace the two-dimensional programs now available. Experimental algorithms have been developed for modeling and editing structures which have multiple depth values for single (X,Y) coordinates, such as structures cut by reverse faults or structures that include over-turned folds. At the same time, geophysicists are exploring new methods for determining both structure shapes and rock properties from seismic data.

INTRODUCTION

Faults play a significant role in the creation and identification of many hydrocarbon traps and the production of hydrocarbons from reservoirs. Faults trap hydrocarbons in dipping geologic units, and faults are commonly associated with the structural geologic deformations and folds which create other types of hydrocarbon traps. Knowledge of the fault structure in a reservoir is an important factor in planning a production strategy from that reservoir. Therefore, accurate representation of faults is a critical step in generating the digital models and paper maps required for petroleum exploration and production.

The inherent difficulty of visualizing complex, intersecting faults and geologic structures in three-dimensional space results in errors appearing even in hand-contoured, faulted, subsurface maps. Typical errors in hand-contoured subsurface maps include incorrectly honoring throw instead of vertical separation when contouring across faults; missing contour levels; an odd number of contour traces terminating along an isolated fault; unjustified structural incompatibility in adjacent fault blocks on individual maps; and geologically incompatible interpretations on maps of adjacent structures (Tucker 1988; Tearpock and Bischke, 1990a, 1990b).

Given that incorrectly faulted contour maps are often created by manual contouring, it is not surprising that incorporating accurate fault-modeling capabilities into subsurface mapping and modeling software has proven to be a difficult task. In fact, inaccurate representation of faulted geologic structures and isochores is the most obvious deficiency of most mapping and modeling programs now used in the petroleum industry.

It has been more than 20 years since contouring programs were first used to support hydrocarbon exploration and production applications. Still, at many petroleum companies the standard operating procedure for computer-supported modeling of faulted structures first requires contouring the structure inside individual fault blocks by hand, followed by tedious and error-prone digitization of the paper interpretation for each block. Only then, after the interpretation is completed and the digital contour and fault data has been collected manually, are computer programs used to transform manual interpretations into merged digital models for further computer processing.

Automated fault modeling for petroleum applications is improving rapidly for both scientific and economic reasons. This chapter reviews problems associated with using computer programs to model faulted structures and units, and discusses the current state and likely future for integrated fault and surface modeling in industry software. The chapter is organized into three sections. The first reviews the history of past attempts to incorporate faults into contouring programs. The next section discusses some of the newer and more accurate methods for fault modeling. Finally, the last speculates about the future of integrated structure and fault modeling in petroleum industry software.

COMPUTER MODELING TECHNIQUES

The representation of structures or geologic units by rectangular grids is the computer modeling technique used most frequently in the petroleum industry. Grids can model very large data sets, have simple data structures, and are convenient for computational

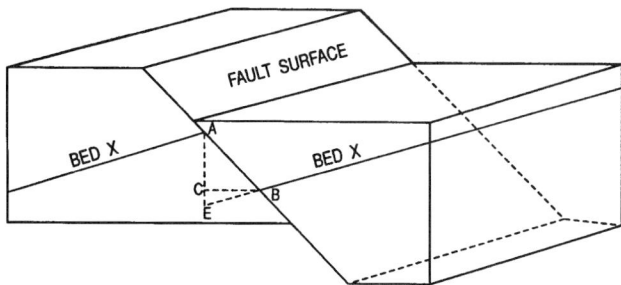

Figure 1. Fault terminology for a normal fault. Each fault component is measured in a plane perpendicular to fault strike. Fault strike is the trend of a horizontal line contained on the face of fault. AE is the fault vertical separation. AC is the fault throw. CB is the fault gap. <BAC is the fault dip angle. (Tearpock and Bischke, 1990. Reproduced with the permission of the authors.)

operations between structures. Triangulated irregular networks (TINs) provide an alternative to rectangular grids. TINs have an advantage over grids in that they model each data point with complete accuracy. On the other hand, efficient algorithms for creating TINs from large data sets with discontinuities have not existed until recently (Fortune, 1987; Cline and Renka, 1990). Building surface models using rectangular grids or using TINs are both described as gridding by many geoscientists, a convention followed in this chapter except when it is important to distinguish between the two techniques. Algorithms for building grid-based or TIN-based structure or unit models are reviewed by Sabin (1985).

While it is possible to find gridding and contouring programs which do not support fault modeling, nearly all software programs licensed to the petroleum industry incorporate some type of fault modeling capabilities. Often the fault modeling capabilities provided are too simplistic to support accurate structure and unit modeling. Today, the fault data input to most gridding programs consists of single-line vertical traces cutting a structure or a sequence of structures, or fault polygons or cut-outs marking the approximate intersection of the fault with the upthrown and the downthrown parts of a structure (Jones et al., 1986). Neither model adequately represents the three dimensional geometry of a fault, which is usually described by the terminology shown in Figure 1.

Faults as Opaque Barriers

Given simple vertical fault traces or fault polygons, most gridding programs treat faults as opaque barriers. As shown in Figure 2, well or seismic data on one side of a fault is not allowed to directly affect the value assigned to a grid node on the opposite side of the fault (Mallet, 1989a).

Structure models created by a program which treats faults as opaque barriers are likely to be inaccu-

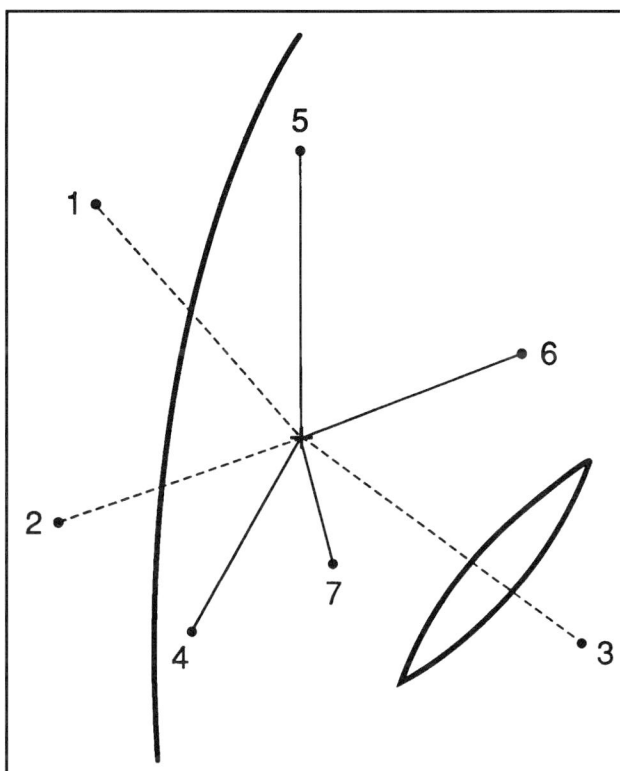

Figure 2. The opaque vertical fault trace on the left and the fault polygon on the right block the use of data points 1, 2, and 3 from influencing the initialization of the grid node indicated by the + symbol.

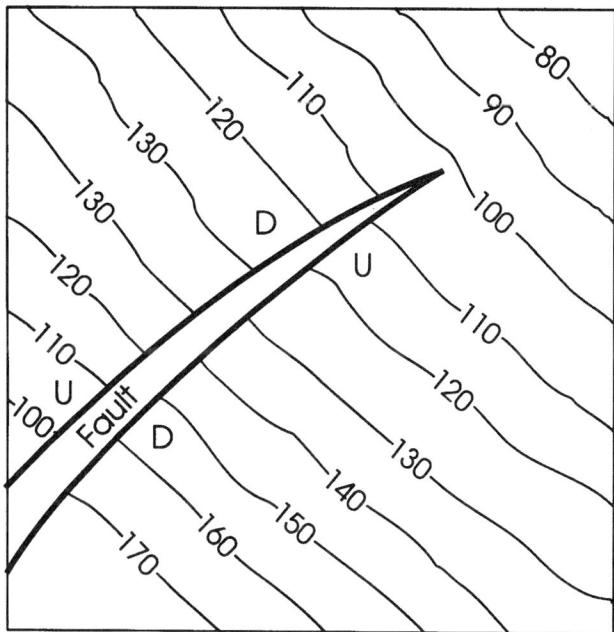

Figure 3. Subsurface structure contours showing inconsistent contouring across the fault. The fault "scissors" without narrowing of the fault gap.

rate or even geologically impossible near faults. With opaque fault barriers and no information about fault geometry, the user of a gridding algorithm must trust that algorithm to project the surface interpolated in each fault block up to the fault barriers, and hope that the combined structure model is accurate across the entire area modeled. Unless the input data are dense in every fault block, it is unlikely that the user's hopes will be fulfilled. Common anomalies in structure models created with opaque fault barriers include undefined grid nodes inside small fault blocks and in wedge zones defined by fault polygons; geologic shapes which fail to continue across simple faults; and, as shown in Figure 3, inconsistent representation of vertical separation along a fault.

The inaccuracies encountered in working with opaque fault barriers in modeling a single structure are compounded when faults are assumed to extend vertically through a sequence of horizons, even though they are known not to be vertical. However, such simplifications are often made because they are required to support reservoir simulators, because they expedite the completion of mapping projects performed under tight deadlines, or because the available technology allows the user no option.

Naturally, fault-related surface modeling inaccuracies are more likely to occur when working with sparse data sets, such as those produced from exploration well data or 2-D seismic surveys. Models of geologic structure and units which are based on 3-D surveys tend to be more accurate overall because of the large amounts of data available. However, because of the cost involved, 3-D surveys are not yet the major source of data available for geologic modeling. Even when 3-D survey data is available, if the fault data obtained from a 3-D seismic interpretation workstation consist of opaque vertical fault traces or fault polygons, highly faulted structures or units will still be poorly modeled across small fault blocks with little data.

Early Attempts at Improved Automated Fault Handling

Early efforts to improve fault representation in geologic modeling programs attempted to honor the correlation between data in one fault block and grid nodes in an adjacent fault block without attempting to model the intervening fault geometry explicitly. For example, in an early version of the CPS gridding program produced by Radian Inc., two separate ways of treating fault discontinuities were provided. In one version a fault discontinuity was treated as a local phenomenon and its contribution to the overall shape of the surface was assumed to diminish with distance from the fault; data at a distance greater than a user-defined tolerance from a fault trace could be used as if the fault did not exist to aid grid initialization in an adjacent fault block, but data near a fault trace could

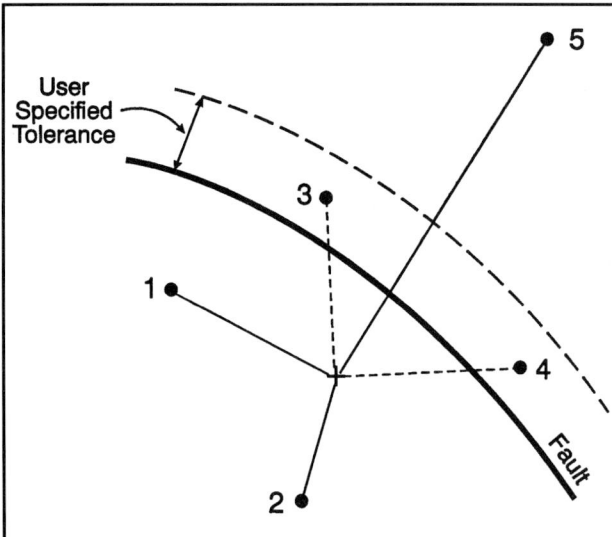

Figure 4. One method for deciding which data points could be used to initialize a grid node in an early version of CPS. Data points 1, 2, and 5 are used. Data points 3 and 4, which are on the opposite side of the fault and within the user specified tolerance of the fault are not used.

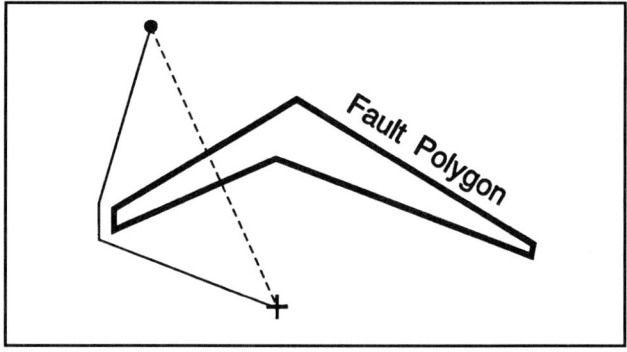

Figure 5. In the algorithm by Bolondi et al., the distance between a data point and a target grid node is measured by the shortest path which does not cross a fault. The dashed line indicates the shortest path ignoring the fault and the solid line is the shortest path which does not cross a fault.

not be used to grid in an adjacent fault block. The relationship between a vertical fault trace, a target grid node, data points, and the user specified tolerance are shown in Figure 4. Structures with many small faults were appropriate for this type of processing. The second option provided was to treat the faults uniformly as opaque barriers under the assumption that data on opposite sides of a fault trace were totally unrelated. This second option was designed for environments with large faults involving large vertical displacements and wrenching, where in fact the correlation across faults is poor and, of course, is the opaque fault barrier method used today in most gridding and contouring programs.

Another method for honoring the correlation between data in one fault block and grid nodes in an adjacent block was suggested by Bolondi et al. (1976). Their algorithm defined the distance between a data point and a grid node to be the length of the shortest polygonal path between them which did not cross a fault. An example of this distance measurement for one grid node, one data point, and a single fault polygon is shown in Figure 5.

Most gridding programs assign a weight to each data point used to determine the value of a target grid node, with the weight decreasing in proportion to the distance between the point and grid node. Therefore, using the shortest path which does not cut a fault is a conceptually simple method for reducing the influence of data in adjacent fault blocks without ignoring that data entirely. This technique seems to be difficult to implement and does not appear to have been used in practical applications. More recently, Marechal (1984) has suggested a way to use simplified fault models to help create accurate faulted geologic structure models when building those models using kriging algorithms. Again, there are no references to the use of this method in practical applications.

The Changing Environment for Fault Modeling

Algorithms for modeling faults in petroleum industry contouring programs are improving. The primary driving force for improvement is the cost of finding new hydrocarbon reservoirs, especially in parts of the world which have already been subject to intense exploration efforts. Most of the remaining onshore petroleum discoveries in the onshore United States will be in small fields, with less than 25 million barrels of reserves (Lindseth, 1990). Locating these smaller fields is expensive relative to the amount of oil they contain, and achieving profitable hydrocarbon production from them depends upon understanding their fault geometry.

Naturally, the increased value of good fault data to the petroleum industry has led both vendors and petroleum companies to significantly increase their investment in the hardware and software required to obtain such data. Fortunately, the state of the art in computer workstations and graphical user interfaces has advanced to the point where complex programs can be developed and used efficiently to collect fault geometry data. Interactive well-log correlation programs help make better fault picks in wells, more accurate estimation of missing section caused by faulting, and better correlation of faults between wells. Improved well-logging devices have made it less costly to obtain information about fault-dip directions from well bore data. At the same time, large amounts of accurate fault-geometry data are obtained from the seismic workstations which are now widely used in the petroleum industry.

Fault handling in industry software is also improving because research has increased our geologic understanding of faults and fault systems (Barnett et al., 1987; Gibson et al., 1989; Walsh et al., 1991). One significant discovery has been that the vertical displacement along an isolated fault tends to follow an elliptical pattern when projected against a vertical plane aligned along the fault strike. Other studies have produced equations which relate the average length of a fault to its maximum vertical displacement, at least in the particular geologic environments studied (Walsh et al., 1991). These discoveries have supported the development of interactive programs which can assist geoscientists in correlating fault picks from seismic data (Freeman et al., 1990).

INNOVATIONS IN FAULT MODELING

The most accurate way to incorporate faults into subsurface structure maps is to hand contour the significant fault surfaces and the targeted geologic structures in the area of interest, and to integrate the fault-contours models with each other and with the structure-contour models in the proper geologic sequence (Tearpock and Bischke, 1990b).

It may be impossible for any geologic modeling software to automatically replicate the complex simultaneous interpretation of faults and geologic surfaces as is done in hand contouring. However, this has not stopped vendors from creating new programs to approximate the hand-contouring process for subsurface mapping, and it has not prevented users of existing programs from trying to approximate the hand-contouring process through careful use of those programs.

In the following sections we discuss three different approaches to the incorporation of faults in horizon models. These different approaches attempt to strike a balance between the cost of implementing integrated fault and structure modeling in software and the geologic validity of the models and maps produced.

Rectangular Gridded Fault Surfaces

Because many fault surfaces are single-valued functions of their (X,Y) coordinates, it is often possible to model individual faults using algorithms provided by the gridding and contouring programs available today. Individual fault models can be created by gridding fault cuts in wells, or by gridding data from depth-converted fault cross sections interpreted on seismic workstations.

When sufficient fault data are available, it may be possible to create surface models for multiple faults, and to intersect those models with each other and with unfaulted structure models in a sequence of computer operations which approximate hand-contouring efforts. Compiled initially from the intersection of fault surface models and unfaulted structure models and edited appropriately, the fault polygons for each individual structure map are likely to be accurate representations of the intersection between a fault and the structure; the resulting digital structure models may approach the fidelity obtained through manual contouring of fault surfaces and geologic structures; and the structure maps with their contours and fault polygons may approach the accuracy of structure maps created entirely by hand.

In practice, using existing contouring programs to replicate manual contouring of fault surfaces and structures is a complex task. Each individual surface, whether it is a fault or a structural horizon, must be painstakingly modeled and validated. This can be particularly difficult when the data defining the surface is provided by deviated wells or a mixture of well and seismic data. Defining the extent of a particular fault surface may require analysis of intersecting faults and structures in an iterative process with multiple reinterpretations of many intersecting surfaces. Extrapolating a fault surface from regions in which it is well defined by well or seismic data into regions where it terminates or is intersected by another surface may require manual contouring, digitization, and another gridding step to get the extrapolated interpretation back into digital format. Finally, because this approach normally treats fault polygons as opaque barriers, incompatible contouring is likely to occur in adjacent fault blocks. Clarke (1992) discusses many of these problems and their resolution for a complex field involving 1200 wells and 30 significant faults.

The extra work involved in modeling faults as gridded structures is often worthwhile. For example, once a set of fault and structure grids is completed, the grids can be used to load simulation programs which provide full three-dimensional models of reservoir attributes by using the various input grids as limiting surfaces and further subdividing the intervals between surfaces based on well data piercing the surfaces. These 3-D models can be useful in planning production from a reservoir (Denver and Phillips, 1990).

Probably the most important reason that fault geometry has not been more frequently incorporated into geologic models through grid-based fault modeling is the difficulty involved in keeping track of the multiple surface models involved and their correct geologic relationships. Programs which allow a user to describe the correct geologic relationship between surfaces while providing for user intervention and editing can resolve some of these problems.

Hooper et al. (1992) describes a new program, the Full Fault Modeling System (FFMS), implemented by Radian Inc., for coordinating the grid-based modeling of multiple fault surfaces with geologic structural modeling. Fault models are derived from a sequence of digitized horizon fault traces for faults cutting a structure, either the upthrown or downthrown trace but not both. Along with these traces, which make up one side of the fault cut-outs used in many existing subsurface mapping programs, are recorded parameters describing the attitude and nature of the fault surface, including fault type (planar/listric), fault dip, strike, elevation (calculated from the input horizon structure map), throw, priority level and fault identification. Some fault properties, such as dip may change from horizon to horizon, while others, such as fault type will stay constant for all traces belonging to a single fault. Initial fault models are generated from data stored along one horizon's single sided fault traces. When enough single sided fault traces and fault parameters are available from multiple horizons, final fault surface models are generated by a gridding algorithm. Each fault model is stored is as a separate grid, allowing overlapping and intersecting faults to be represented.

In the FFMS program, the first targeted structure, usually the shallowest surface, is initially modeled as though the available upthrown or downthrown fault trace is a vertical fault. Then each fault surface is intersected with the structure model to produce the missing side of each fault cut-out. After some editing, complete fault cut-outs are available the target horizon. If significant editing is required to complete the missing side of certain fault cut-out, then the edited traces may be used to update the existing fault surface models.

After the first structure model is completed, adjacent structure models are created by a sequence of operations involving repeatedly gridding unfaulted isochores, adding those isochores to an adjacent structure model and then intersecting the new structure model with current fault surfaces. This results in new fault traces for the downthrown blocks on the new structure, and the downthrown fault traces are migrated upwards to complete the fault cut-outs for the upthrown blocks in the new structure. Finally, the depth values from the associated fault plane grids can be merged into the cut-outs of the structure model of the new horizon. (Figures which illustrate the use of the FFMS program are included with the paper by Hooper et al., 1992).

The FFMS approach produces structurally consistent models of multi-layered reservoirs cut by non-vertical faults. In addition, it can produce properly faulted models of isochores for use in volumetric calculations. However, the various steps required to develop the fault polygons and the fault plane models are not completely automated. Significant interactive editing apparently is required to achieve the full potential of the program.

Single Horizon Methods

At least two vendors of subsurface contouring software have modified the traditional single surface grid-based modeling algorithms now used in the industry to honor the geometry of normal faults. One program of this type, developed by Landmark/Zycor, accepts as input fault data with vertical displacement encoded at vertices of the fault trace (Zoraster and Ebisch, 1990). Data points on one side of a fault which have been adjusted for fault displacement can directly affect the value assigned to grid nodes on the other side of the fault, guaranteeing compatible contouring in adjacent fault blocks. To make it easy to tell one side of a fault from the other, faults in the form of single "center-line" traces are used during grid initialization and these single line traces are automatically expanded into fault polygons by the gridding program. Center-line faults and expanded fault polygons with horizon picks are shown in Figures 6 and 7. The expanded fault polygons with contours of the structure and the fault face traced through a grid produced in the Landmark/Zycor program are shown in Figure 8. A similar program which uses fault geometry data for single structure modeling has been introduced by Dynamic Graphics Inc. (S. Pack, 1991, personal communication).

For the Landmark/Zycor program, the data near a target grid node are selected to initialize the value of that node without consideration of the fault traces. Only after the data points are collected is the possibility that faults may cross between them and the grid node investigated. If one or more faults cross between a data point and the target grid node, then the value of the data point is adjusted in proportion to the total displacement of all those faults. In effect, the time or depth value of each data point used to initialize a grid node is automatically normalized to the time or depth of the fault block in which the node is located.

This technique produces grid-based structure models defined up to the center-line fault traces as though they were vertical faults. However, the use of constantly changing sets of data points to initialize each node can lead to general roughness and small anomalies in any digital model based on a regular grid and created from randomly spaced data. Many gridding algorithms reduce this problem by smoothing the grid with a mathematical filter. In the Landmark/Zycor program a biharmonic filter is used to smooth the center-line grid after it is initialized (Briggs, 1974). The biharmonic filter uses 12 nearby grid nodes to help reduce the roughness at each target grid node. Special smoothing operators are defined near data points to tie the smoothed grid to the data.

Smoothing grid models that honor opaque faults can be difficult since the quality of the grid smoothing may deteriorate when all 12 nodes required by the standard biharmonic operator are not available. Of course this happens frequently near opaque faults because some of the 12 nodes are hidden on the opposite side of an opaque fault barrier. This problem is avoided in the Landmark/Zycor program because the vertical separation along center-line faults is honored during smoothing. Grid nodes which are required by the biharmonic operator, but are on the opposite side of a center-line fault from the node being smoothed have their values temporarily normalized to the fault block of the target grid node. In the Landmark/Zycor program the only grid nodes which are regularly smoothed with a biharmonic operator using less than 12 neighboring nodes are nodes along the sides of the grid where it is easy to develop mathematically correct operators using less than 12 neighboring nodes.

After smoothing, the center-line faults are then automatically expanded into properly bifurcating

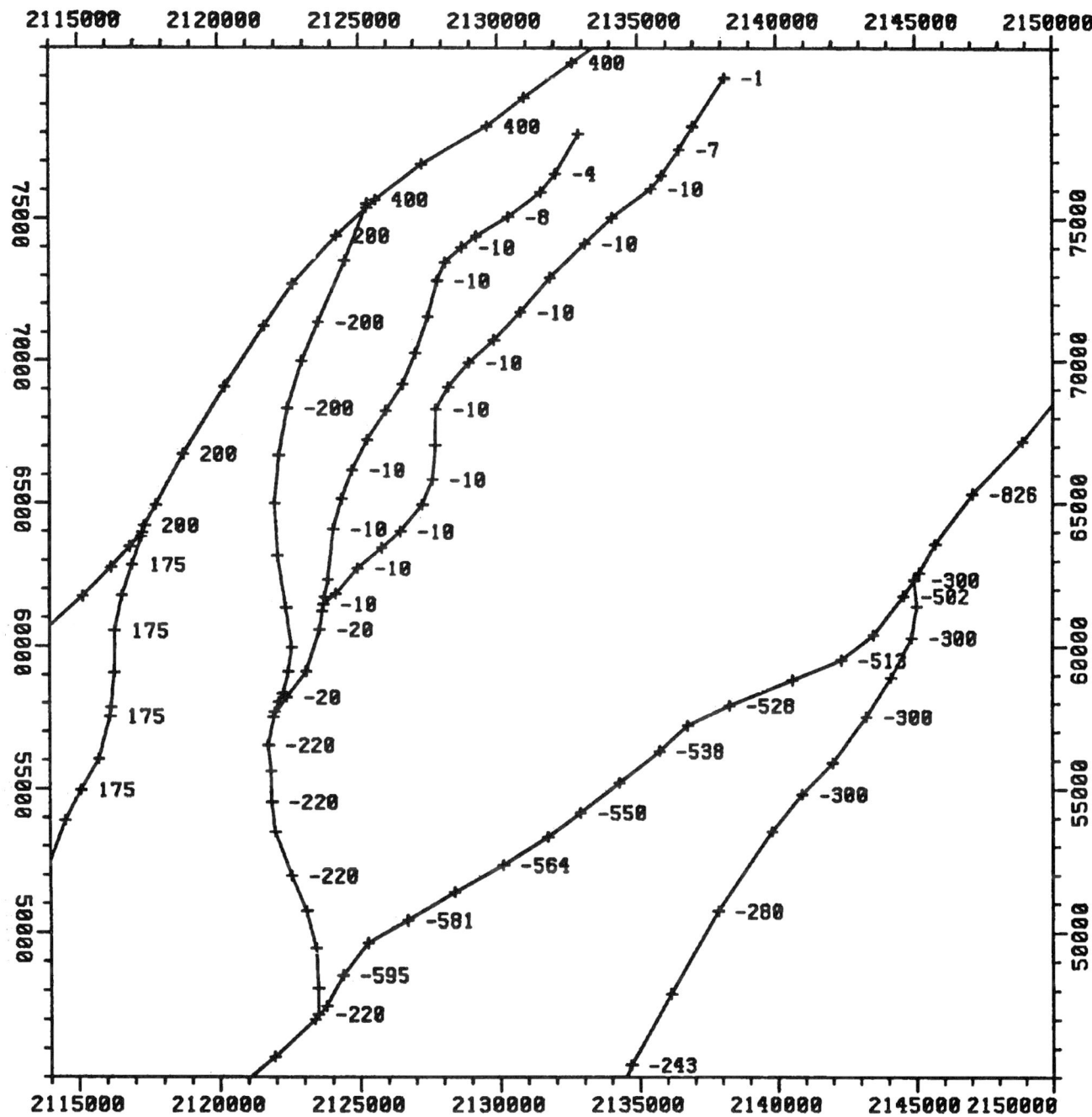

Figure 6. Center-line fault traces as used in the Landmark/Zycor gridding algorithm. Vertical separation is posted at every other fault vertex. Vertical separation is measured from left to right across a fault, with left and right determined relative to the direction in which the fault is digitized. For this data set most faults are digitized in order from lower left to upper right, and the predominance of negative separation values indicates that the horizon, measured in negative units, is downthrown to the right and below the fault traces.

fault polygons using the dip and vertical separation of the faults. These fault polygons define the intersections of fault faces with the structure. Grid nodes inside the fault polygons are assigned values which model the faults as they cut the horizon. This is accomplished by obtaining values along the fault-structure intersection (fault polygons or cut-outs) from the center-line grid. A second gridding step is then performed using as data only the horizon values sampled along the fault polygons. Connecting these structural surfaces at fault vertices across the fault gap produces a good combined representation of both fault displacement and dip.

The use of center-line faults enables isochores

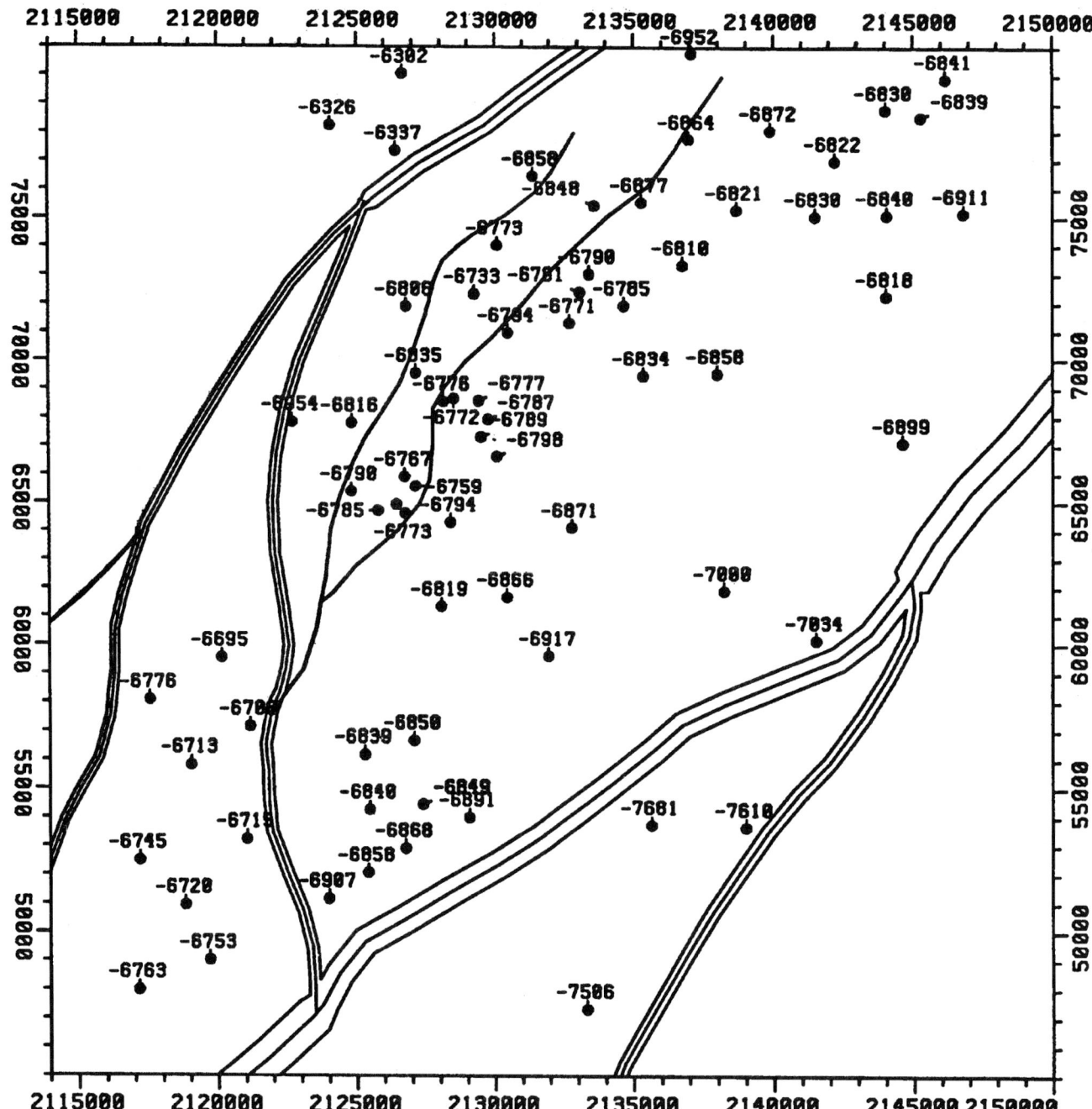

Figure 7. Expanded fault polygons and the data picks which define the structure that will be modeled using the Landmark/Zycor algorithm. Note the complete absence of data in the fault block at the lower right side of the figure.

which then cut across faults to be built directly from thickness data, thus allowing sequences of conformable structure models to be built in a faulted environment. To directly model thickness requires fault traces for both the top and base of a unit. Unlike FFMS, which uses one of the structure grids, an unfaulted isochore grid, and the fault grids to create traces for the other structure surface, the Landmark/Zycor program uses traces from one structure, fault dip, and unfaulted isochore to migrate the fault traces to the adjacent structure. As with FFMS, the character of a fault may be listric or planar.

All gridding algorithms which incorporate fault vertical separation must accurately represent vertical separation at points where bifurcating or compensating faults meet. When a fault bifurcates, the sum of the vertical separations, of the two splitting faults near the bifurcation should be equivalent to the vertical separation of the original fault just before the bifurcation. A current disadvantage of the Landmark/Zycor pro-

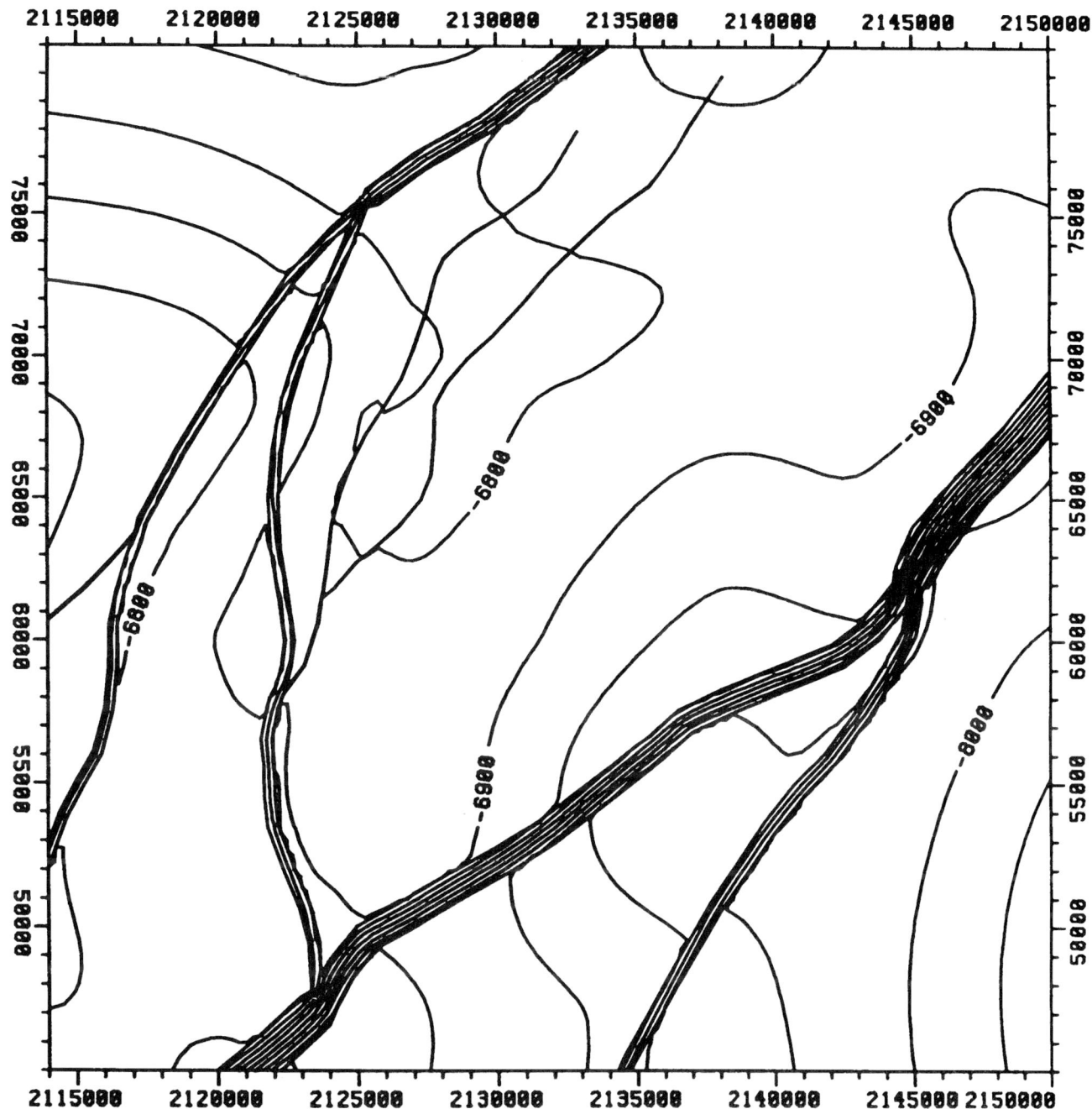

Figure 8. Expanded fault polygons and contours drawn across the structure and the fault face cutting that structure as generated by the Landmark/Zycor algorithm. By honoring vertical separation along the bounding center-line faults, it is possible to generate a reasonable structure interpretation in fault blocks without data.

gram is that this step is not fully automated. Balancing and fine tuning of vertical separation can require multiple iterations of fault interpretation, structure modeling, and contour map generation to get the correct fault interpretations and accurate structure models. Fortunately, once a correct interpretation is achieved for one structure, that balanced interpretation can be automatically migrated with the center-line fault models to adjacent structure surfaces, and are easily fine tuned with a few edit/modeling iterations.

MULTIPLE SURFACE TRIANGULATED MODELS

The approaches to incorporating faults into structural and unit models discussed above are grid-based and therefore can not be guaranteed to honor every single input data point exactly. A more important complaint about them is that their operations for modeling multiple structures are essentially sequential in nature, with the various individual fault surfaces, fault cut-outs, isochores, and models for each structure or unit created one at a time with limited automated checks on the validity of the combined interpretation. Usually the checks which exist are visual (reviewing screen or hardcopy contour maps), and when errors are discovered significant interactive editing and repeated user directed execution of major parts of each program may be required.

The Mapping-Contouring System (MCS) by Scientific Computer Applications, tries to avoid both problems by using TIN-based surface modeling techniques (Banks, 1991), and by attempting to automatically replicate the manual contouring procedures recommended by experts such as Tearpock and Bischke (1990b). For example, MCS starts the way Tearpock and Bischke say manual subsurface mapping should start: by modeling faults and the vertical separation associated with each fault.

MCS uses a four step approach to model faulted structure surfaces:
(1) Build models of fault surfaces and their associated displacement,
(2) Move top picks to their prefaulted positions,
(3) Build unfaulted structural surface models, and
(4) Displace the surface models by shifting them to their faulted positions.

Fault data required by MCS include correlated fault cuts and fault displacements (vertical separation). These data come from wells, seismic, or can be a user's interpretation. The fault interpretation required by MCS also includes a list of faults which border each fault block and their relative age relationship. When two faults intersect, the fault that terminates is considered younger and the one that continues is considered older.

Figure 9 is a contour map of a surface cut by two faults and modeled by MCS. The general fault modeling process will be described using this surface as an example. Figure 10 is a west-to-east cross section showing the fault configuration. The faults are antithetic with fault A being dominant and therefore older. Three fault blocks are seen in the cross section, block number 1 above A and B (youngest), block number 2 above A and below B (intermediate), and block number 3 below A (oldest).

Two surface models are built for each fault: the fault surface and its displacement. The sign of the displacement tells the program whether the fault is normal or reverse (positive = normal, negative = reverse). Displacements that vary from positive to negative are treated as scissors faults and displacements that are zero define areas where the fault does not exist (fadeouts). Figures 11 is a contour map of fault A.

Restoring the structural surfaces back to their prefault position is done one fault block at a time. The user provides a fault interpretation to MCS by using a simple ordering convention when listing faults that bound a fault block. The youngest fault is listed first with progressively older faults following. For each of the older faults, the position of the youngest fault relative to that older fault is specified. For example, for the youngest fault block (above faults B and A) fault B is youngest and it is ABOVE fault A. This syntax tells MCS to use the portion of fault B's surface that is above fault A. The vertical displacement of the youngest fault bounding that fault block (fault B) is used to shift all picks within the block to their prefault B position. Figure 12 is a contour map of the surface after restoration of the first fault block, note that fault B no longer exists.

Once the fault block has been restored, the youngest fault of that block no longer exists and thus is not part of the fault block description for older fault blocks. Since fault A is the only fault remaining, the next and final fault block description would indicate that fault A is the youngest and only fault that bounds the block. The displacement of fault A is added to all picks above fault A, restoring them to their prefault positions. Figure 13 is a contour map of the restored (unfaulted) surface.

The MCS operations we have described for one structure can be applied to multiple structures during the execution of the program. Once top picks for all structures have been restored to their prefault positions, surface models are built. This is done by either modeling the data directly (for unconformities or isolated depositional surfaces) or using MCS's multiple surface stacking procedure (for sequences of conformable surfaces). Once modeled, all surfaces are broken and shifted back to their faulted positions. This step is the inverse of the restoration process, with the oldest fault block being shifted first and so on. For each fault block, the unfaulted restored surface is shifted by the vertical displacement experienced by that fault block and all older fault blocks. This shifts the restored surface to that fault block's present-day position. Contours are generated using this shifted surface but are prevented from crossing any faults that bound that fault block. The fault trace defining the edge of that fault block is created by intersecting the shifted surface with the faults that bound that fault block.

Three points should be noted about this method: 1) because unfaulted surface models are built and then faulted, continuity in surface form carries across faults, 2) because a smooth model of vertical displacement is used, separation is consistent along faults, and 3) because displacements of each fault block are progressively added together, displacements always balance at bifurcations.

MCS blends aspects of modeling fault surfaces and of using vertical separation to carry structural form across faults. Therefore, it has components similar to both of the modeling techniques previously described, however, unlike the techniques described above, MCS can handle thrust faults as a standard part of the program operation.

To use MCS correctly a geoscientist must understand the sequence of faulting events which took place in order to correctly guide the restoration of a

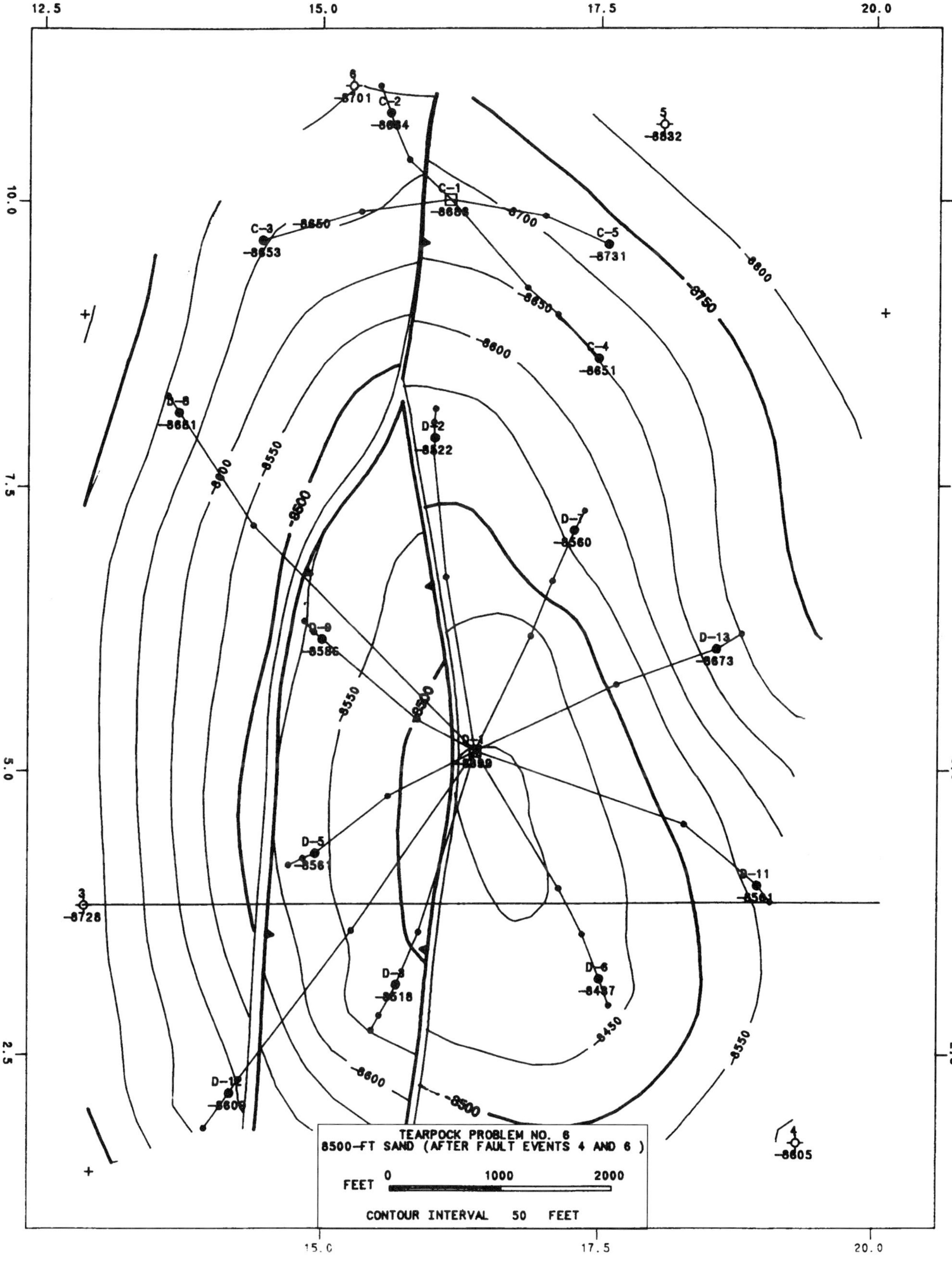

Figure 9. Top of the 8500-ft sand as modeled by MCS. Example taken from Tearpock and Bischke (1990). (Figure provided by Scientific Computer Applications)

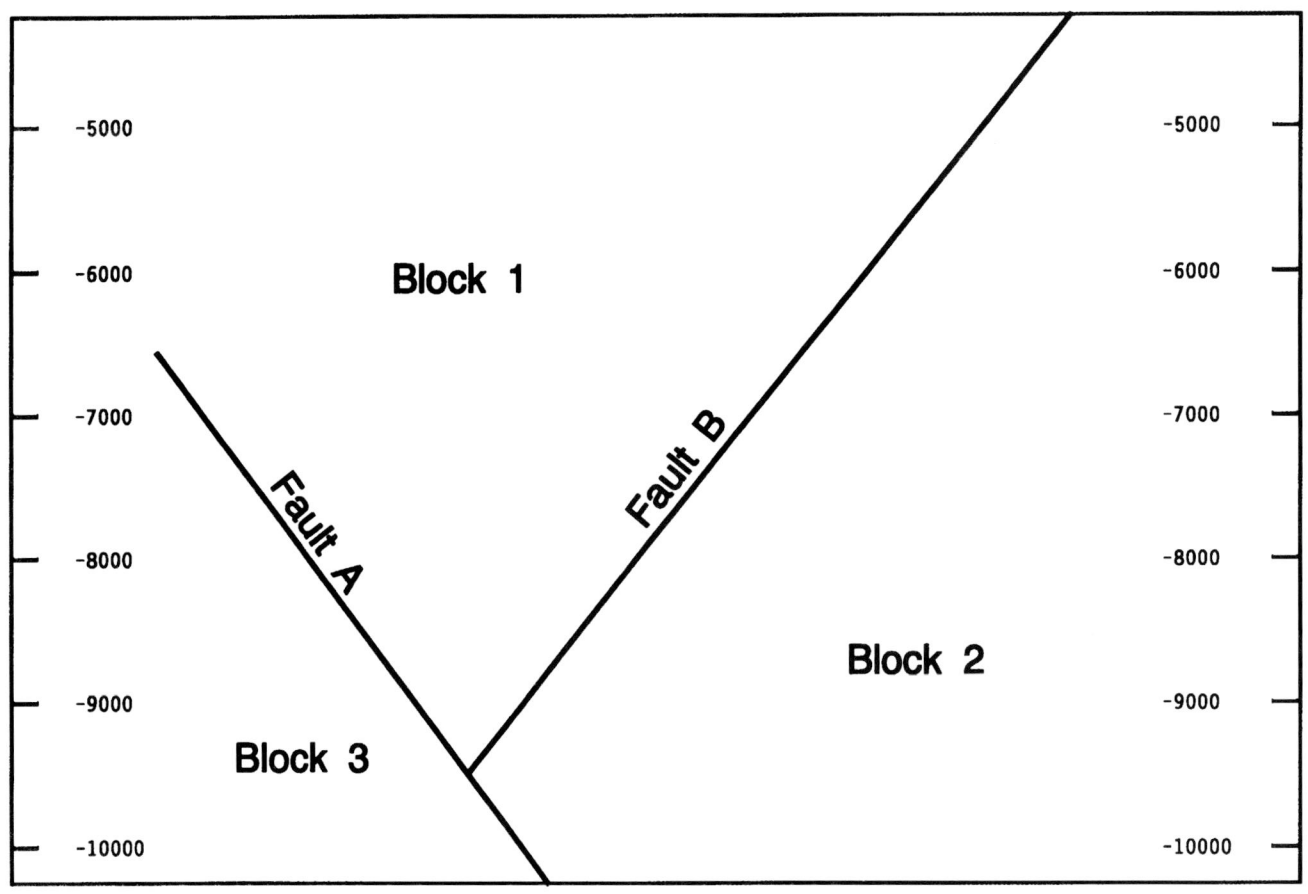

Figure 10. West-to-east cross section showing the fault configuration affecting the structure model displayed in Figure 8. (Figure provided by Scientific Computer Applications)

sequence of faulted structures to their unfaulted condition. This can require a fairly long learning curve even though the users' manual for the program provides detailed guidelines with several examples. Fortunately, the program itself contains error checking software which can detect incorrect reconstruction sequences and inform the user which sequences are causing trouble.

The algorithms described above, and in this section all can require user intervention to get the desired geologic interpretation into the final surface models. These algorithms described previously are parts of programs which employ interactive contour, grid and fault editing tools which allow the user to incorporate this interpretation. Integrating interpretations about structural environments into the MCS models is achieved through the use of dummy data points, a less interactive but equally effective approach.

THE FUTURE

The future of fault modeling belongs in the domain of 3-D modeling and 3-D visualization. Although the programs described in the previous section provide better models of faulted subsurface structures and units than older programs, they should be considered temporary solutions which will be replaced by more sophisticated software within the next five years.

There are at least four reasons to expect improved methods for modeling of faulted structures to appear within the next few years. First, the same economic pressures which led to the development and acceptance of the programs described above will continue to encourage the creation of more versatile programs. Certainly all the major vendors of subsurface modeling software are interested in 3-D techniques and petroleum companies are encouraging that interest. Second, the amount of data available to support enhanced modeling efforts will increase significantly as 3-D seismic workstations become the dominant source of data for both exploration and production work. Third, the ongoing development of 3-D modeling and visualization programs for other domains, such as medicine, fluid dynamics, and biological applications, will provide valuable new ideas to the developers of subsurface modeling programs (Helman and Hesselink, 1991; Nelson et al., 1991). Finally, none of the existing programs provide practical methods for handling the extremely complex, multi-valued

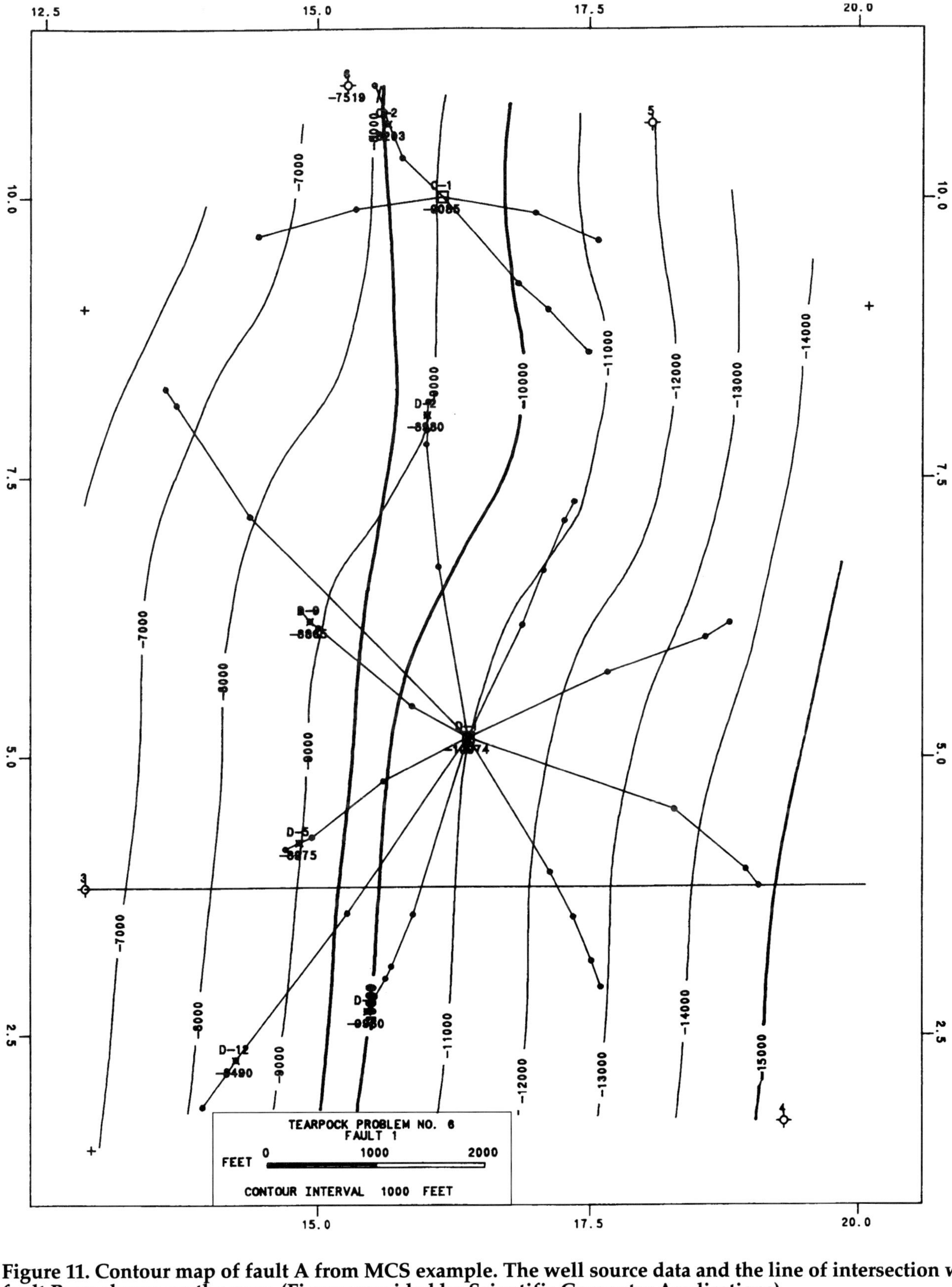

Figure 11. Contour map of fault A from MCS example. The well source data and the line of intersection with fault B are shown on the map. (Figure provided by Scientific Computer Applications)

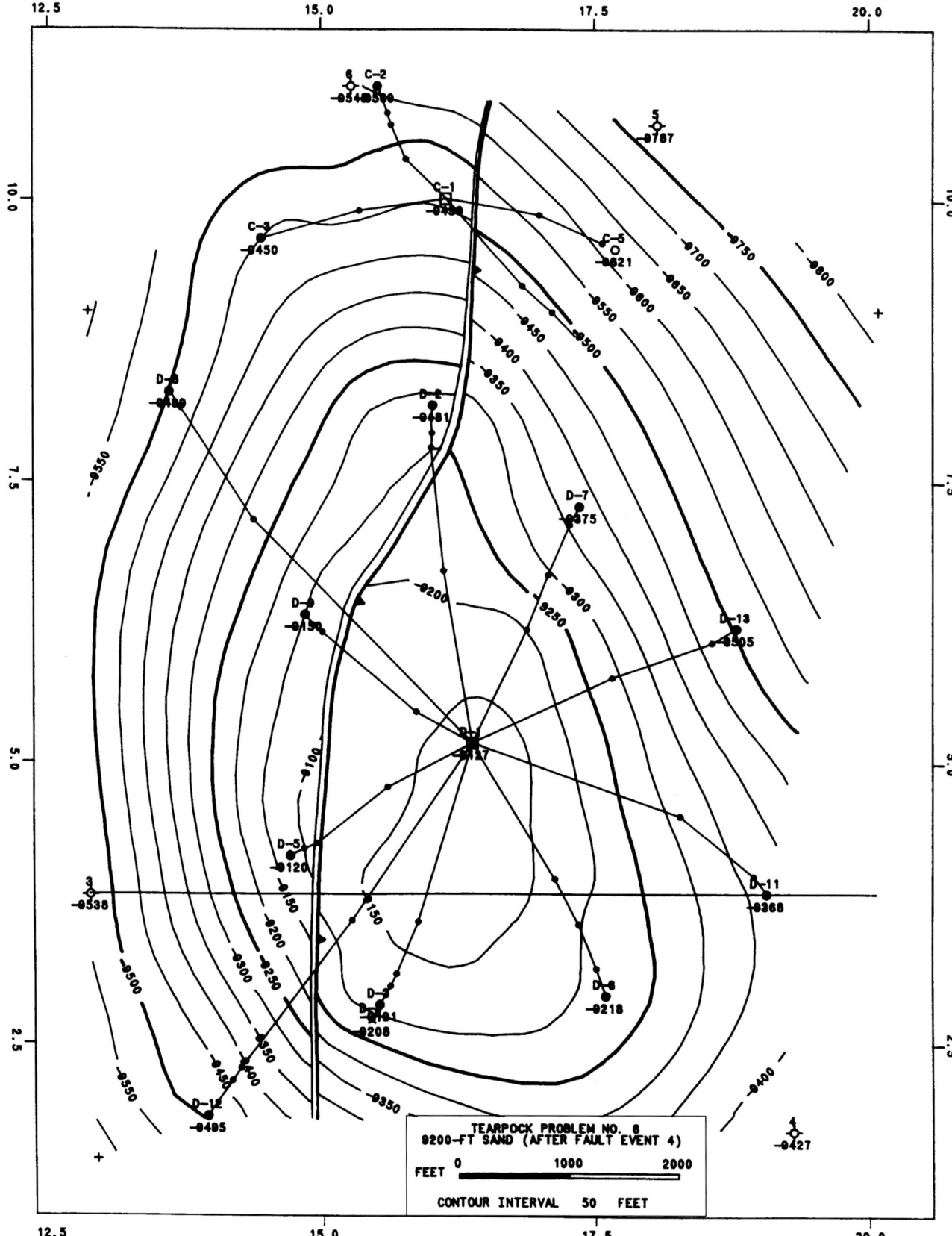

Figure 12. Top of the 8500-ft sand after restoration of fault B. (Figure provided by Scientific Computer Applications)

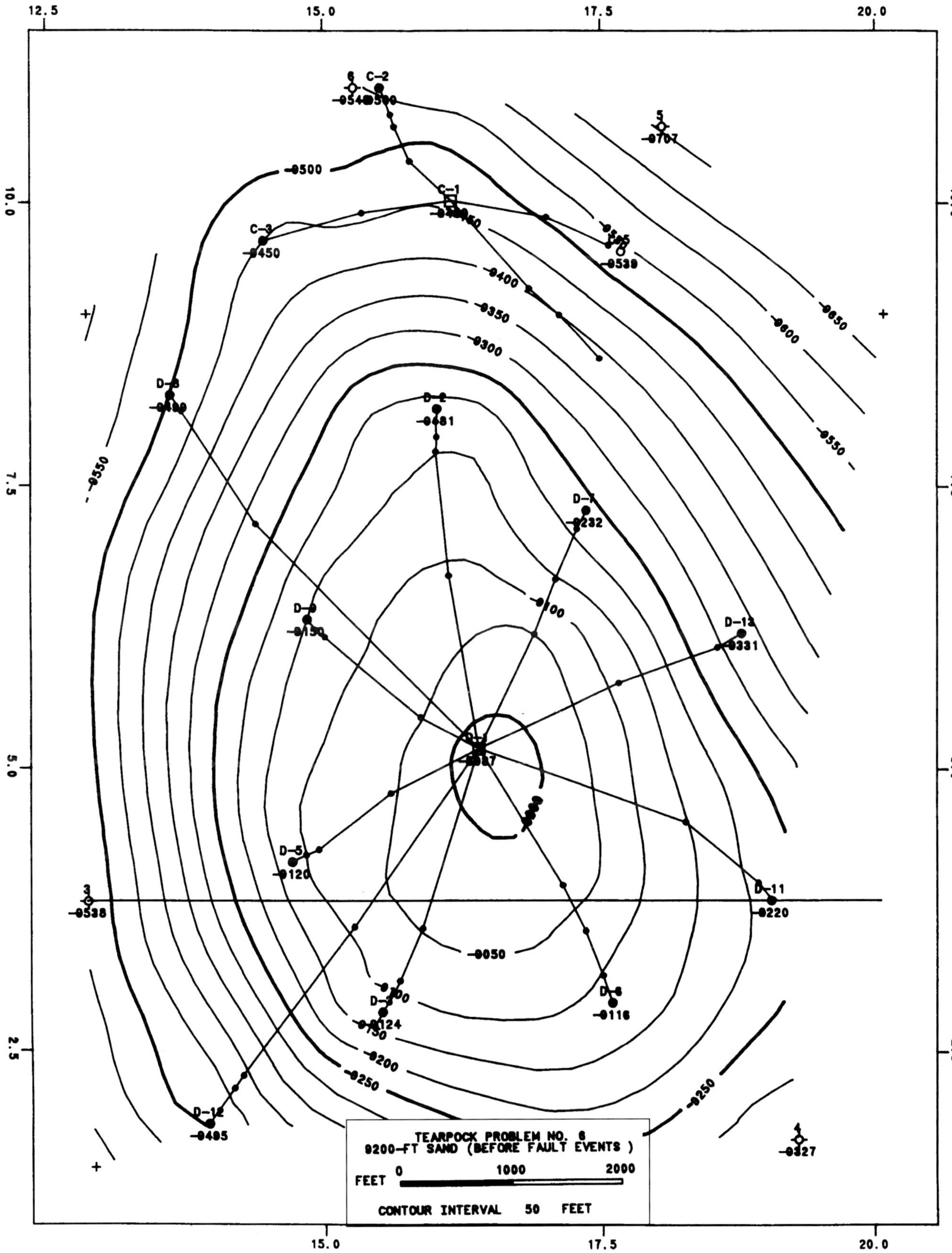

Figure 13. Contour of the 8500-ft sand after restoration of both faults A and B. (Figure provided by Scientific Computer Applications)

geologic features such as salt domes and roll-over structures which are often associated with hydrocarbon reservoirs.

Improved integrated 3-D subsurface modeling incorporating faults is being studied from two directions. One approach relies on geologically oriented CAD-like programs which allow any surface, whether a fault or a structure boundary, to bend over itself in three dimensions. GOCAD is probably the best-known example of this type of program (Mallet, 1989c). GOCAD provides sophisticated interactive editing capabilities which allow the geoscientist to shape any surfaces to fit specific geologic models while still honoring the available data. However, according to available literature, GOCAD remains a research project and has not reached the point where it is used in practical applications.

A second approach to full 3-D volume modeling is based on improvements to seismic data analysis. These methods combine enhancements to traditional 3-D seismic signal processing techniques, cross-well borehole seismic methods, traditional well logging, and geostatistics to produce accurate 3-D models of bedding patterns, depositional environments, petrophysical properties, and the fault structures of reservoirs. Several reports of the use of these combined methods in practical applications have appeared in recent geophysical literature (DeBuyl, 1989; Nolen-Hoeksema, 1990).

The true future of fault representation in automated subsurface mapping applications will almost certainly involve a holistic solution which incorporates multiple data sources and multiple data views along with interactive editing capabilities. An important factor in the acceptance of holistic, computer-based solutions to fault modeling problems will be achieving significant improvement in the speed with which interpretation efforts can be completed at a workstation compared to the time now required to obtain manually compiled interpretations.

Data sources for a single, integrated fault and structure modeling effort may include some or all of the following: fault picks in wells with missing sections and, sometimes, fault dip direction and dip angle; depth-converted fault trace cross-sections from 2-D or 3-D seismic workstations with estimates of vertical separation and fault gap, and confidence factors on vertical separations and fault gaps based on the angle at which seismic lines cross the strike of the fault surface; fault trace cross sections from well-log correlation programs, again incorporating confidence factors on fault separation and gap; map view fault polygons with depth values and other fault attribute data; partial fault information not sufficient to fully define a fault structure, such as the fact that a particular fault cuts only two wells; and empirical models of how faults grow in different tectonic environments.

A mixture of user intervention, graphic display of different fault attributes, and sophisticated algorithms will be required to blend these various sources of fault data. Fault 3-D display tools which show true perspective views of interpreted fault surfaces and display on the fault surfaces contours of fault vertical separation would help the geoscientist decide whether a particular fault interpretation is valid. Algorithms which automatically adjust existing interpretations to account for the incorporation of new data, or reported on localized inconsistencies between old interpretations and new data, would be extremely useful, as would algorithms which automatically balance vertical separation at fault bifurcations. And of course, the geoscientist must have the ability to impose his or her own geologic knowledge into any interpretation system.

The form in which full 3-D fault modeling will be achieved can not be predicted. Many different systems, each with its own strengths and weaknesses, will appear. Some will disappear quickly. Others will evolve and coexist with each other over long periods of time, just as the existing mapping and modeling programs which exist today have managed to coexist, often with two or more systems used simultaneously at the same petroleum company.

ACKNOWLEDGMENTS

Comments on an earlier version of this paper by Tom Jones of Exxon Production Research, David Hamilton of Zycor/Landmark, and Dick Banks and Joe Sukkar of Scientific Computer Applications are greatly appreciated.

REFERENCES CITED

Banks, R. B., 1991, Contouring algorithms, Geobyte, v. 6, p. 15–22.

Barnett, J. A. M., J. Mortimer, J. H. Rippon, J. J. Walsh, and J. Watterson, 1987, Displacement geometry in the volume containing a single normal fault: AAPG Bulletin, v. 71, no. 8, p. 925–937.

Bolondi, G., F. Rocca, and S. Zanoletti, 1976, Automatic contouring of faulted subsurfaces: Geophysics, v. 41, no. 6, p. 1377–1393.

Briggs, I., 1974, Contouring using minimum curvature: Geophysics, v. 41, no. 6, p. 1377–1393.

Cline, A., and R. J. Renka, 1990, A constrained two-dimensional triangulation and the solution of closest node problems in the presence of barriers: SIAM Journal of Numerical Analysis, v. 27, no. 5, p. 1305–1321.

DeBuyl, M. D., 1989, Optimum field development with seismic reflection data: Geophysics: The Leading Edge of Exploration, v. 8, no. 4, p. 14–20.

Denver, L. E., and D. C. Phillips, 1990, Stratigraphic geocellular modeling: Geobyte, v. 5, p. 45–47.

Fortune, S., 1987, A sweepline algorithm for Voronoi diagrams: Algorithmica, v. 2, no. 2, p. 153–174.

Freeman, B., G. Yielding, and M. Badley, 1990, Fault correlation during seismic interpretation: Geophysics: The Leading Edge of Exploration, v. 8, no. 3, p. 87–93.

Gibson, J. R., J. J. Walsh, and J. Watterson, 1989, Modelling of bed contours and cross-sections adjacent to planar normal faults: Journal of Structural Geology, v. 11, no. 3, p. 317–328.

Helman, J. L., and H. Lambertus, 1991, Visualizing topology in fluid flows: IEEE Computer Graphics and Applications, v. 11, no. 3, p. 36–46.

Jones, A. J., D. E. Hamilton, and C. R. Johnson, 1986, Contouring geologic surfaces with the computer: New York, Van Nostrand Reinhold, p. 151–157.

Lindseth, R. O., 1990, The next wave of exploration: Geophysics, v. 55, no. 12, p. 9–15.

Mallet, J. L., 1989a, Discrete smooth interpolation: ACM Transactions on Graphics, v. 8, no. 2, p. 121–144.

Mallet, J. L., 1989c, Geometric modelling and geostatistics, in M. Armstrong, ed., Geostatistics, v. 2: Boston, Kluwer Academic Publishers, p. 737–747.

Marechal, A., 1984, Kriging seismic data in the presence of faults, in G. Verly et al., eds., Geostatistics for Natural Resources Characterization, Vol 1: Boston, D. Reidel Publishing Company, p. 271–294.

Nelson, G. M., B. H. Foley, B. Hammann, and D. Lane, 1991, Visualizing and modeling scattered multivariate data: IEEE Computer Graphics and Applications, v. 11, no. 3, p. 47–54.

Nolen-Hoeksema, R. C., 1990, The future role of geophysics in reservoir engineering: Geophysics: The Leading Edge of Exploration, v. 9, no. 12, p. 89 - 97.

Sabin, M. A., 1985, Contouring—the state of the art, in R. A. Earnshaw, ed., Fundamental algorithms of computer graphics, NATO ANSI Series, v. F17, New York, Springer-Verlag, p. 413–482.

Tearpock, D. J., and R. E. Bischke, 1990a, Mapping throw in place of vertical separation: A costly subsurface mapping misconception: Oil and Gas Journal, v. 88, no. 29, p. 74–78.

Tearpock, D. J., and R. E. Bischke, 1990b, Applied subsurface geological mapping: Englewood Cliffs, Prentice Hall, p. 195–399.

Tucker, P. M., 1988, Seismic contouring: A unique skill: Geophysics, v. 53, no. 6, p. 741–749.

Walsh, J., J. Watterson, and G. Yielding, 1991, The importance of small-scale faulting in regional extension: Nature, v. 351, p. 391–393.

Zoraster, S., and K. Ebisch, 1990, Incorporating fault geometry into geologic horizon models, Geobyte, v. 5, no. 2, p. 30–36.

Chapter 11

The Gridded Fault Surface

Don Clarke
Department of Oil Properties
City of Long Beach
Long Beach, California, U.S.A.

ABSTRACT

Faulted structural surfaces present difficult problems for the computer mapper, though several solutions are possible. In this paper, I present a method whereby the fault plane is treated as if it were a structural surface. The method uses fault picks, oil/water contact data, directional survey data, and reservoir data. The data are manipulated and mapped with a commercially available mapping package using standard gridding, editing, and data-to-grid manipulations. The resulting fault grids can be integrated with grids of structural surfaces, used as boundary surfaces, or used to develop fault gaps on structure maps.

Fault-surface modeling is most applicable to large oil fields with many wells. Several advantages have been realized, including accurate fault locations at all horizons and accurate predicted fault penetrations for new well locations. When new penetrations are encountered, they can be used to update the fault plane and then the revision can be employed to correct the structure maps. Multiple interpretations for the fault surface can be tested for validity. The techniques and products described were developed for the Wilmington oil field in Long Beach, California.

BACKGROUND

The method described in this paper for gridding individual fault surfaces was developed for the Long Beach Unit of the East Wilmington oil field in Long Beach, California (Figure 1). Individual oil properties are consolidated into a unit so that an oil field can be developed in an efficient and cost effective manner. One task of the unit operator is to determine a fair equity for all of the unit participants. The Wilmington oil field has been divided into several operating units. The City of Long Beach is the Unit Operator for the Tidelands Unit and the Long Beach Unit. The Long Beach Unit is composed of the eastern five miles of the Wilmington anticline. The unit is operated by the City of Long Beach for the State of California and approximately 10,000 individuals, and THUMS Long Beach Company is the field contractor.

The Wilmington anticline is an asymmetrical, northwest-trending, doubly-plunging, faulted anticline, 16 mi (25 km) long and 3 to 5 mi (5 to 8 km) wide. Production is from Pliocene- and Miocene-age sands that occur between 2000 and 10,000 ft (600 and 3050 m) subsea depth (Figure 2). The cumulative production is more than 1.3 billion barrels of oil. Cumulative Long Beach Unit production is greater than 700 million barrels of oil.

The Long Beach Unit was developed from offshore facilities, even though 10% of the field is onshore. More than 1200 wells have been drilled to date, and

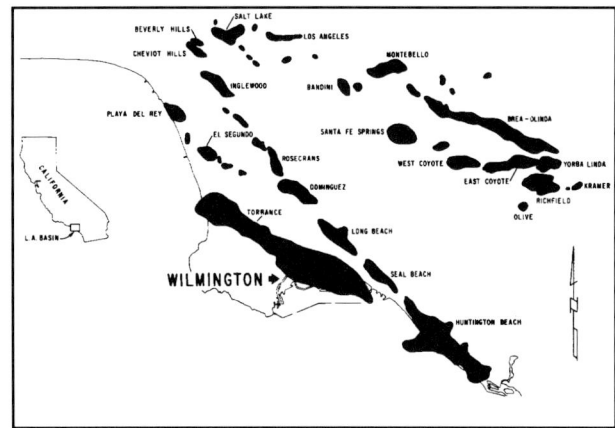

Figure 1. Index map showing the Wilmington oil field location relative to the other oil fields in the Los Angeles basin. (Published by permission of the Department of Oil Properties, City of Long Beach.)

our data base includes an additional 300 wells from adjacent leases and units (Figure 3). The Long Beach Unit has been divided into five economic blocks bounded by faults. Each of the blocks have smaller faults associated with them (Figure 3). The entire Long Beach Unit has 30 faults that are large enough to be modeled numerically by computer. These faults have vertical separations that vary from zero to three hundred feet. The economic blocks are divided stratigraphically into seven zones (Tar, Ranger, Terminal, Union, Pacific, Ford, and 237); see Figure 2. The zones are subdivided into fifty subzones and the upper thirteen subzones are further subdivided into forty sub-subzones. All of the faults, subzones, and sub-subzones have been digitally mapped as structural surfaces.

The Long Beach Unit was formed in 1965, long before computers played a significant part in subsurface geological mapping. The types of maps to be produced and the general procedures for map construction were designed for hand work. In the middle 1980's, the decision was made to computerize the Long Beach Unit. About the same time, it was discovered that all the directional data were in error because magnetic declination corrections were incorrectly applied. As a result, all geological maps had to be reconstructed. The equity determination required repeatable geological procedures. Because we were legally bound to satisfy all of the original unit procedures for map construction, new methods and proce-

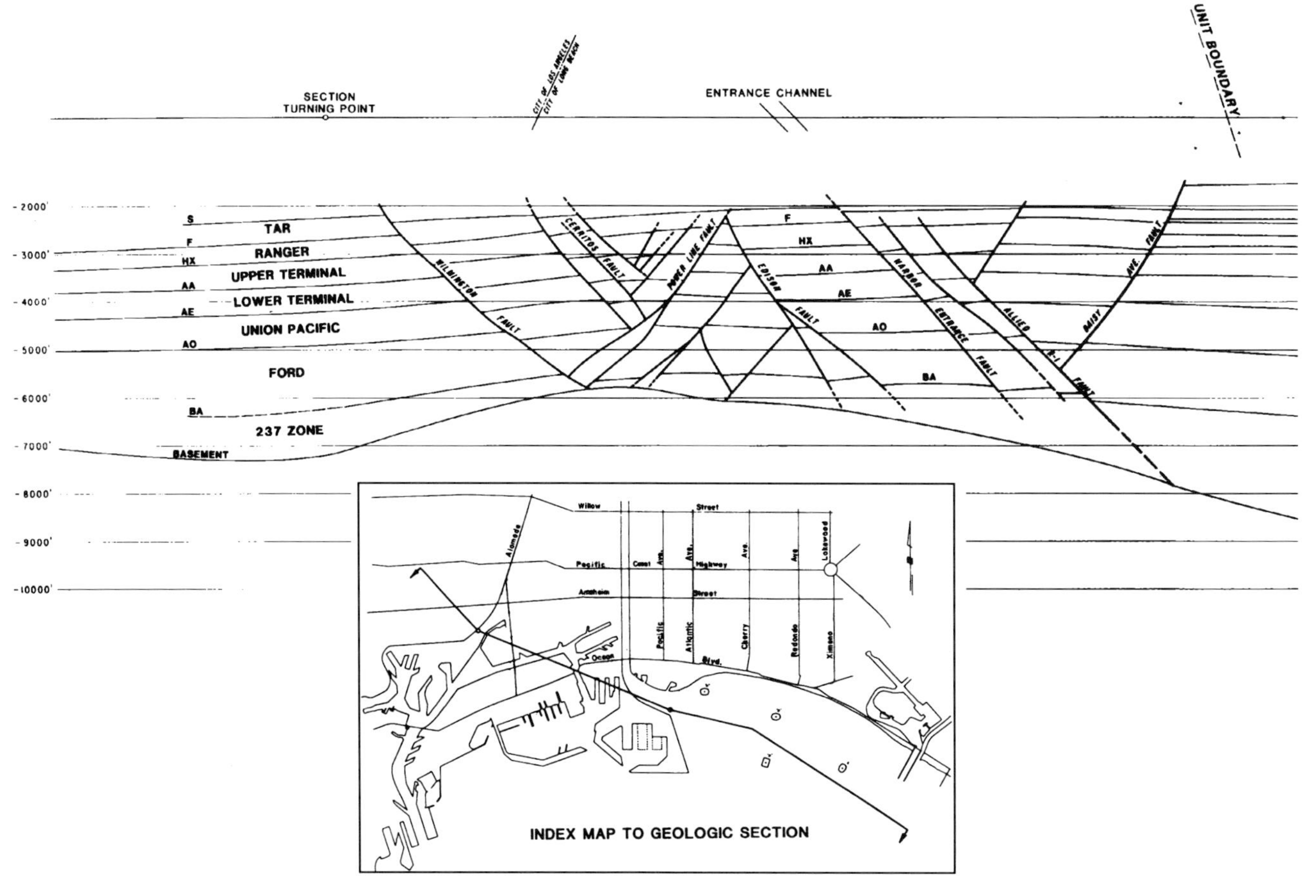

Figure 2. Axial cross-section of the Wilmington oil field showing the seven producing zones and the major faults. Most oil production comes from the Ranger zone.

dures were developed for computer mapping that complemented the original procedures without violating the original requirements.

These methods were applied to the equity effort. The equity for the Long Beach Unit was finalized on April 1, 1990. Hundreds of fault surface maps, structure maps, and isochore maps were produced. The geological description portion of the equity determination was far more detailed and accurate than originally anticipated.

The geological description used for the equity effort is now being expanded and applied to development work. Applications have been found for the detailed geological work in reservoir description and reservoir modeling. These results are now used as a powerful development tool. Current completion practices now selectively target small intervals or areas that were delineated during the equity work.

FAULT SURFACE MAPS

Accurate subsurface mapping techniques are necessary for proper exploration and exploitation of hydrocarbons in faulted structures. "A reasonable structural interpretation, in most faulted areas, must begin with an accurate fault interpretation resulting from the construction of fault surface maps and the proper integration of these fault maps with structure maps." (Tearpock and Bischke, 1991, p. 195.) No background information is provided for structural geology or structural styles; texts such as Dennis (1972), Billings (1972), Harding and Lowell (1979), Suppe (1985), and Tearpock and Bischke (1991) provide an excellent background.

The construction of fault surface maps is fundamental to providing accurate and sound geological interpretation in faulted areas. Not only does the integration of fault surface maps into structure maps provide three dimensional validity but it allows more accurate volumetric calculations. Erroneous geologic interpretations can result if the geologist attempts to reconstruct a complicated structural surface using fault cut data from oil well electric logs without constructing a fault surface map.

The integration of the fault surface map into the structure map provides (1) accurate delineation of the fault locations (fault gap) for all cut horizons (precise location of upthrown and downthrown fault traces), (2) the correct width of the fault gap or overlap, and (3) proper projection of structure contours across the fault.

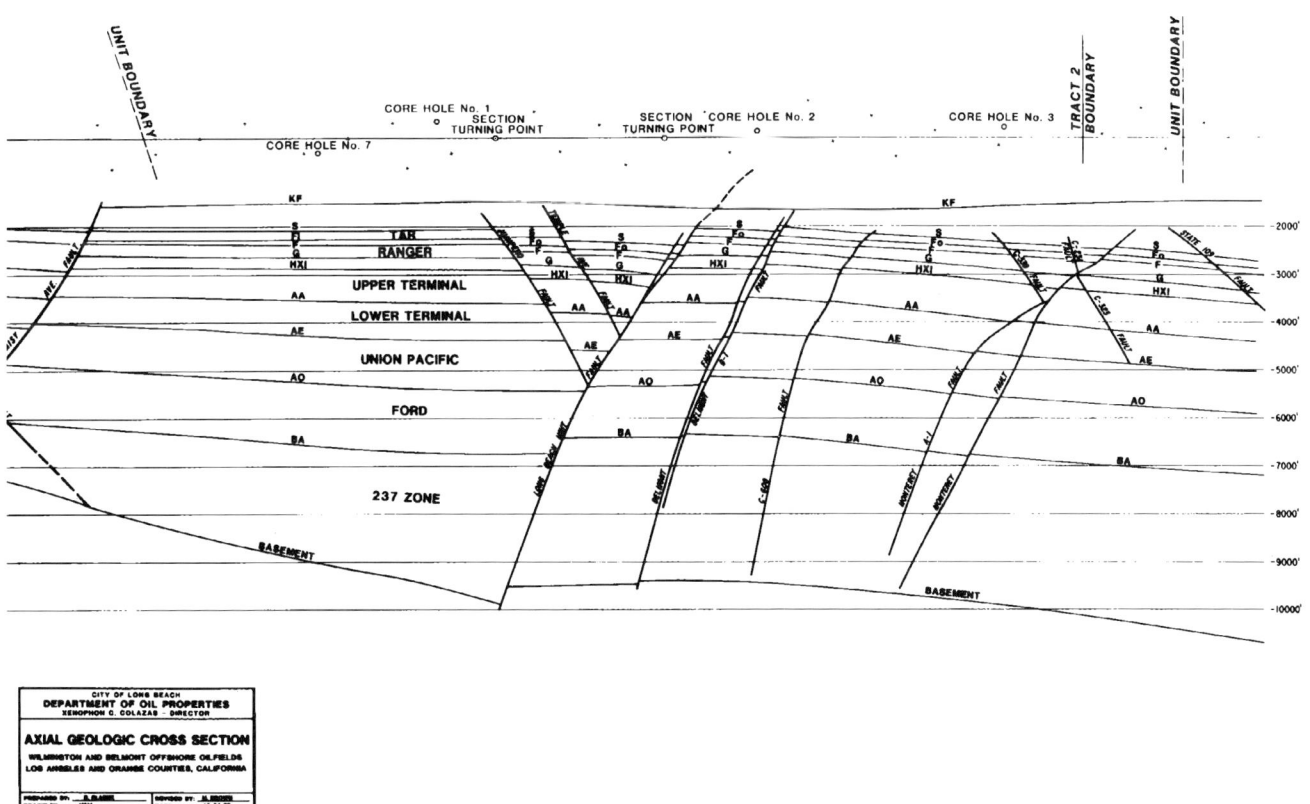

Figure 2—Continued.

DEFINITIONS

Fault terms are not defined consistently in the literature. I have attempted to use the terminology of Tearpock and Bischke (1991, p. 197-200). The following three definitions will help the reader.

Fault cut (fault penetration) is a location along the well bore where stratigraphic section is missing or repeated as the direct result of a fault cutting through the well bore.

Vertical separation is the vertical component of bed displacement. It is measured as the vertical distance between a line or plane (such as a formation top) projected from one fault block across a fault to a point where the projection is vertically over or under the same line of plane in the opposite fault block. Vertical separation, seen in vertical well bores, shafts, and vertical cross sections (Dennis 1972) is that distance.

Fault gap is an enclosed area (fault polygon) that represents the intersection of the fault surface with a structural surface. Usually there is displacement along the surface. The width of the fault polygon depends upon the relationship between the angle and direction of the fault formation dip and vertical separation. The width of the polygon on a structure map can not be intuitively determined.

SPECIAL CONDITIONS

Several problems must be solved and special conditions satisfied in order for a fault surface to be successfully constructed. The location, shape, and orientation of the fault surface is determined by analyzing four types of control:

1. Fault penetrations and non-penetrations along a well course.
2. Stratigraphic irregularities.
3. Apparent structural irregularities.
4. Reservoir properties.

Fault Penetrations, Non-Penetrations and Stratigraphic Irregularities

The well course provides a locus of known data points through the rocks. This is the basic control. The wells originate from five drilling areas (Figure 3). Ninety-five percent of the wells are directionally drilled, most with hole angles exceeding 45°, and some as high as 89°. The well courses range from sigmoidal to helical spirals.

The well directional information is stored in a data base called the borehole database. Typical data include well name, east and north coordinates, and measured and vertical subsea depths for each directional survey station along a well course. Locations for information such as fault intercepts are interpolated between the known directional survey station data.

All well penetrations of a fault surface must be accounted for. Most fault penetrations are initially determined from electric logs (e-logs). The coordinates for each well penetration are calculated and used as raw fault data. The fault surface is constructed so as to not penetrate the well bore a second time. If it does, a second penetration should be found on the e-log showing the opposite displacement. If the first penetration resulted in lost section, then the second penetration should have repeated section (Figure 4). This is usually observed as a shortening or lengthening of known section. Occasionally a paradox will arise, e.g., a channel or lenticular layer will add length to the section and scouring can cause loss of section. These also must be resolved. All nearby wells are checked for penetration by the fault surface.

The process is iterative: Develop a surface with these intercepts, then check the well courses for penetrations. Make adjustments to the surface as a result of the penetration check. Make another penetration check. Repeat the process until a satisfactory surface is achieved.

Apparent Structural Irregularities

The full extent of a fault normally is not defined by well-course penetration and well-course avoidance. Information from structural anomalies and reservoir data must also be used. Structure maps of the zones cut by a fault should be checked for irregular contour spacing or odd bends. If these features are observed on several horizons, they strongly suggest that the fault extends into or through the area of irregularity (Figure 5).

These structure maps are only work copies because the final structure maps are dependent on the integration of the fault surface data.

Reservoir Properties

Reservoir properties often provide information about fault position and should be considered when constructing fault surfaces. This is because fault surfaces often act as barriers that separate oil pools within the reservoir. Properties such as permeability, oil gravity, cut, and gas/oil ratios are helpful. Differences in oil/water contacts provide striking fault evidence. In Wilmington, we have found offsetting oil/water contacts in adjacent wells to be very useful for extending fault surfaces where we lack fault penetrations (Figure 6). The projected oil/water contacts for the Mn sand and the Mn_1 sand restrict the fault trace to an area between wells A-204I and A-167IS. Mn and Mn_1 are geological markers designating the tops of productive sands. The strike of the fault surface as defined south of the figure projects nicely between the wells. Oil gravities were significantly different in Fault Block VIIA (Figure 7), more detailed work showed these differences to be good indicators of a new fault.

Pressure data provide strong faulting evidence since large differences in measured pressure between adjacent wells are difficult to explain without a barri-

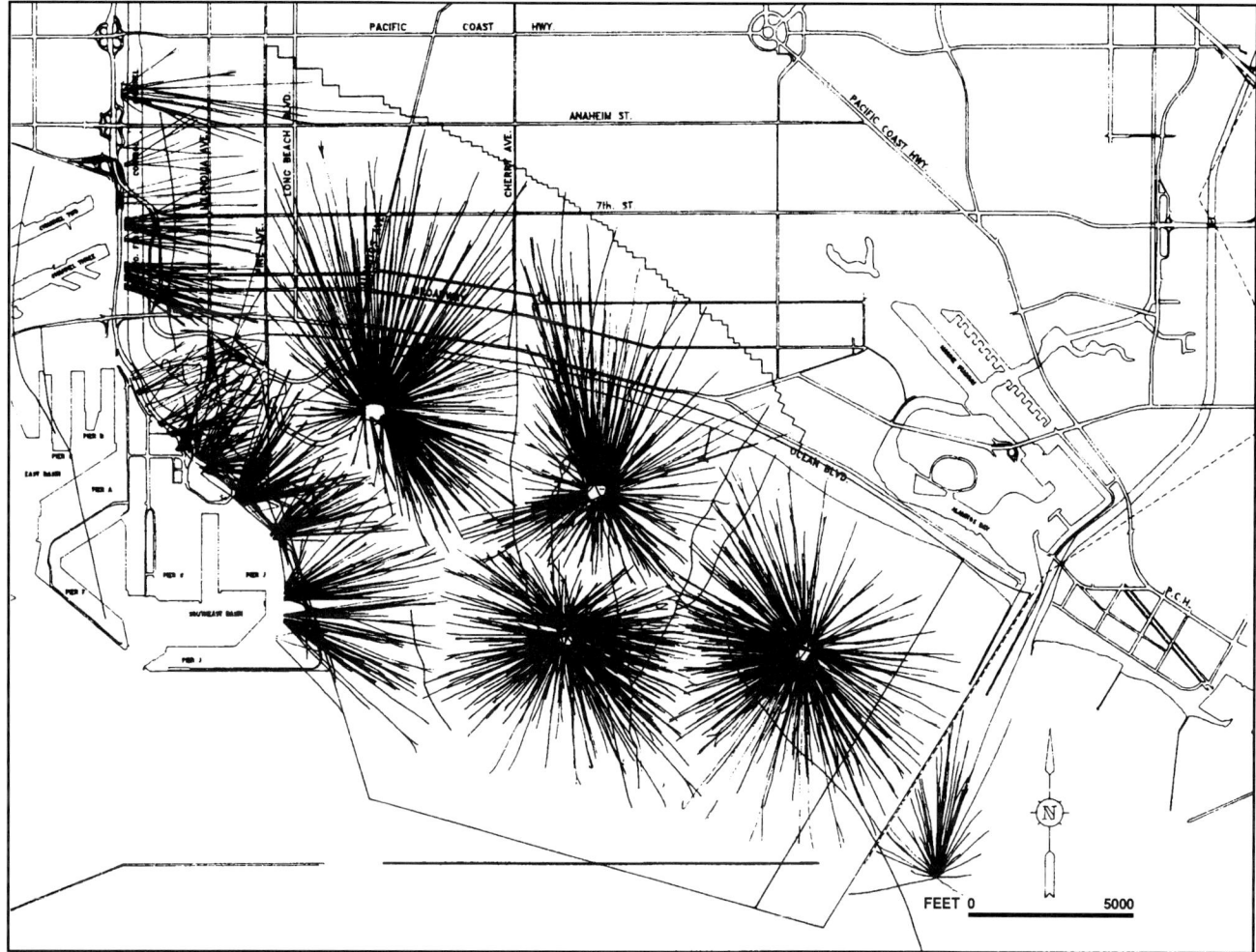

Figure 3. Horizontal trace map of the East Wilmington oil field. This map shows approximately 1500 directionally drilled wells. The major subsurface faults are shown. (Published by permission of the Department of Oil Properties, City of Long Beach.)

er. Well production history was also used. Many more factors can be used but we found the aforementioned to be the most useful. The more variables that are considered for each decision, the greater is the mapper's confidence in the interpretation. Figures 8 and 9 are examples of how reservoir data have been applied to fault delineation.

FAULT-TO-FAULT INTERCEPTS

Faults often bifurcate or merge into a single fault. The intercepts must be handled correctly; otherwise, the resulting grid surface will have erroneous values. Each splay must be modeled independently, and it is possible for two or more faults to come together at a given loci of points. A well may penetrate at or near the intercept of two faults. It is difficult to recognize closely spaced fault breaks on electric logs. Also, small fracturing associated with the fault creates a confused e-log response. With these limitations in mind, it becomes possible to use one fault penetration for two fault surfaces.

The fault surface must match the geological horizons. If calculated intercepts occur in the wrong place, it is likely that the fault interpretation is not correct. Usually this is seen as a sharp change in the strike of structural surface on one side of a fault, which occurs when one well has been mapped on the wrong side of the fault. The fault or structure surface must be altered to accommodate the offending well. The new surface still must satisfy all of the other conditions.

Determining actual fault movement or displacement may prove troublesome. It has been our observation that added or lost log sections vary from well to well. This can be due to well plot, field geometry, stratigraphic variations, and faulting during deposition and folding (Figures 4, 10, and 11). As mentioned above, the fault may actually be a combination of

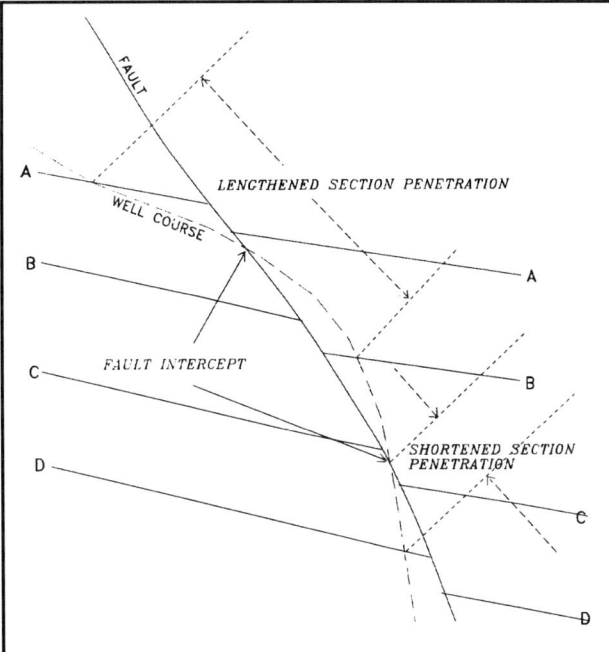

Figure 4. Diagrammatic cross section across the C-608 fault in the Wilmington oil field. The well direction passes from the footwall to the hanging wall and then recrosses the same fault farther down hole. The first crossing causes a lengthening of section because the same stratigraphic interval is crossed on both sides of the fault resulting in a repeat of section, and a high hole angle also results in lengthened section. The second crossing resulted in a shortened section since some of the section was faulted out and low hole angle shortens the section. The two factors are additive and the differences can be dramatic.

many smaller breaks (anastomosing fault surface). This causes the apparent separation to vary (Figure 10). Post-depositional compaction may also be a factor here.

The relative magnitude of the added or lost log section is still useful as a guide when defining the fault surface, or when differentiating between two closely spaced fault surfaces. For example, two nearby wells with similar hole angles will be located near two closely spaced fault surfaces in the Wilmington oil field. One well crosses a fault high in the section and results in a vertical separation value of 50 ft (15 m). The second well crosses a fault deeper in the section and has a smaller vertical separation value. It is most likely that the wells have penetrated different fault surfaces. If the well that penetrated the deeper horizon exhibited greater vertical separation, then it may have penetrated the same fault surface as the first well since many of our faults are growth faults that increase in vertical separation with depth. The vertical separation information will be useful later when we work out the structural history of the oil field.

DESCRIPTION OF FAULT MODELING METHOD

This technique for modeling individual fault planes involves a combination of numerical modeling and geological decision making. Hereafter I will use the term *grid* to refer to the numerical model of the fault surface. The Temple Avenue B-1 fault will be used as a working example. This fault was recently discovered during a stratigraphic analysis. When we reviewed past geological and engineering data, we found that the data supported the fault interpretation and allowed us to extend the fault trace. The figures that correspond to the following procedure are all in rectangular coordinates (state plane coordinate system, California Zone 7).

Procedure

Step 1—Start Up

Marker discontinuities on electric logs are compiled. The vertical subsea depth, and north and east coordinates are calculated for each discontinuity value. The discontinuities are usually the result of faulting or stratigraphic change.

The data fields for each point are defined as follows:

East Coordinate: 4243247
North Coordinate: 4019247
Well name: A-2471A
Measured depth: 2319
Vertical subsea: –1962
Back Interpolation: –1962
Subtract: 0
Apparent Separation: –20

The fields and values for the east and north coordinates, well name, measured depth and vertical subsea depth need no further definition. *Back interpolation* and *subtract* are data fields used for fault-surface checking; their use will be explained later. We use missing or added log section for apparent separation. A positive value indicates repeated section, a negative value indicates lost section. Vertical holes will yield vertical separation values but directional wells yield only apparent vertical separation. The added or missing log section values in the apparent separation field can be corrected to vertical separation.

A data set is created for each fault surface to be mapped. We do not store the fault name with the data points.

Step 2—Fault Groupings

The discontinuities are collected by a geologist into spatially related groups, each representing a fault. These discontinuities are now considered fault picks. When two or more faults are closely spaced, grouping of the fault picks by fault can be difficult. Incorrectly placed fault picks will become apparent at Step 5.

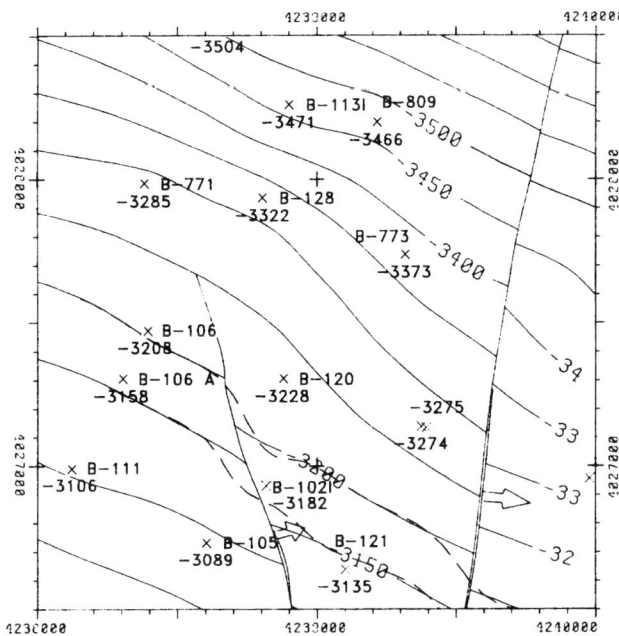

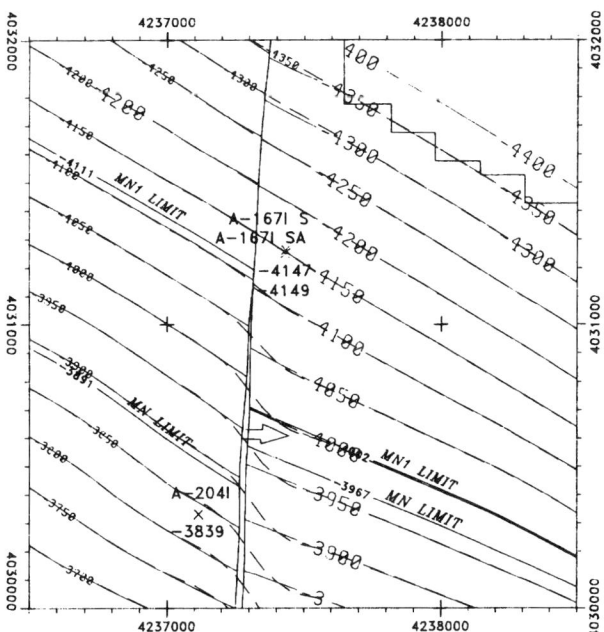

Figure 5. Temple B-1 fault in Fault Block VII of the Wilmington oil field. The −3150 and −3200 contour lines with faults (solid) and without faults (dashed) show the irregular contour spacing and strike change that is strong evidence to extend a fault trace. The northernmost fault penetration occurs in well B-102I. Contour interval 50 ft (15 m).

Figure 6. The northern most fault penetration of the Junipero fault is 100 ft (30 m) south of the area illustrated. The fault location was extended because of contour variations and limit differences on either side of the fault.

Some faults have been mapped with no fault picks. These surfaces are not as accurately located as fault surfaces with fault picks but they still must satisfy all of the tests below.

Step 3—Selection

Those discontinuity picks that cannot be placed with a group are set aside for further study.

Step 4—First Plot—Initial Gridding

Next a grid is constructed for each grouping. Little or no manual adjustment is necessary for this grid. Make sure that the grid is large enough to encompass the entire expected areal extent of the fault. Contour the grid and post the data points for the first check plot (Figure 12).

Step 5—Bull's Eye Contour Patterns

A *bull's eye* is a set of concentric contour lines (Figure 12). Bull's eyes typically indicate areas that deviate from the average surface as either a pit or a hill. When a bull's eye has no points associated with it, it is caused by poor extrapolation or interpretation. The mapper can either move the surface to the bull's eye level by editing the grid or remove the bull's eye by editing it down to the grid level. If the bull's eye is centered on a data point, the point must be evaluated. If the point is correct, then the grid is altered to match the point. If the point was mispicked, then it is corrected. If the point does not belong to this fault surface, it is removed from the data file and the grid is corrected. Remember that the point may belong to a nearby fault.

Sometimes a bull's eye is produced by a point on its edge, usually the result of two closely spaced points with depths that differ by more than 5%. The gridding algorithms will develop a slope based on the two points and project the slope away from the control. If this is a recurring problem, the gridding algorithm or parameters must be changed. If it is not common, edit the proper changes where necessary. It should be noted that problems with a directional survey can cause a well point to be improperly placed. If the survey can not be checked or fixed, you may have to remove the point from the fault pick data set.

Step 6—Second Plot—Contour and Data Check

Post wells with oil/water contact data, locations of any structural anomalies, fault picks, and contours of the fault grid onto a paper plot. Also plot the total depth and the top of the logged interval for each well (Figure 13).

These extra data will be used to guide the fault surface where there are no penetration data. This is also the time to incorporate any useful reservoir and production data.

The map at this stage appears very disorderly. This is because fault plane data usually occurs in clusters or groups of points with an uneven distribution.

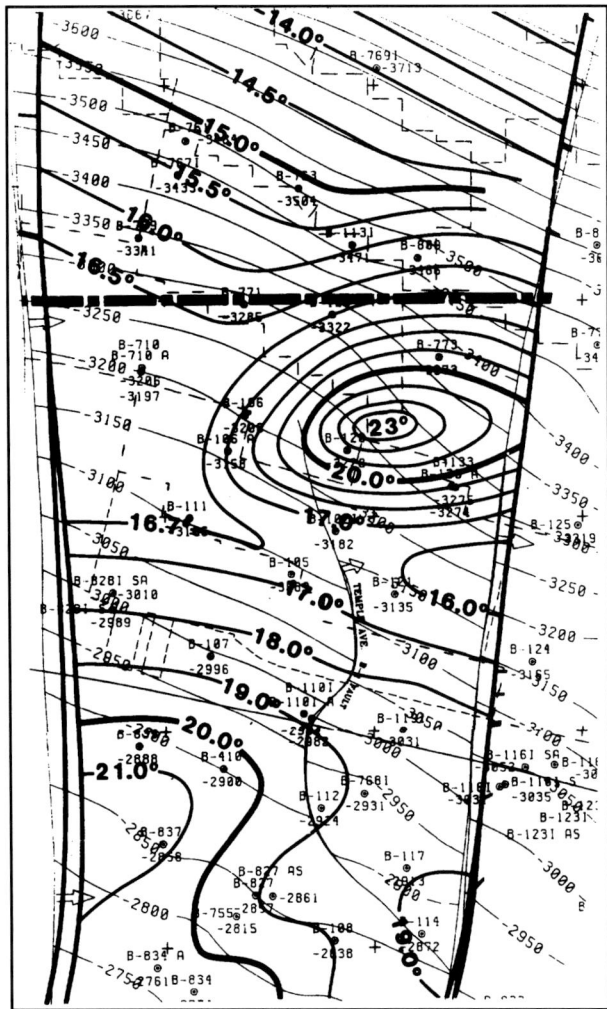

Figure 7. Overlay of oil gravity (bold contours) and computer structure (fine contours) in Fault Block VII A. Oil gravity data were contoured for equity prior to discovery of the Temple Avenue B-1 fault. The structure map contours were constructed using the Temple Avenue B-1 fault surface. The Temple B-1 fault surface was extended further north based on the oil gravity interpretation. The oil gravity data were subsequently re-interpreted using the Temple Avenue B-1 fault as a barrier. The fine dashed lines represent property boundaries.

Step 7—Interpretation

The mapper manually extends selected contour lines between the clusters of fault picks into sparse and no data areas. This extension is done so as to minimize curvature. Every contour line need not be extended. Contour lines where there is a dip change should be extended (Figure 14). Points that deviate significantly from the surface should be analyzed and removed if necessary. If there were no fault picks or fault grid, the mapper must place probable contour lines to represent a surface through the posted data.

Step 8—Edit and Regrid

The fault plane interpretations must now be digitized for entry to the computer. This can be done in several ways. We typically edit the original digital fault plane contour files so that they match the interpretation. The fault surface is then regridded using both the edited fault points and edited contours. Care must be taken to remove extraneous contours or parts of contours because they will provide erroneous control.

A second method is to digitize the significant hand altered contours and then grid anew using these contours and the edited fault points. These two methods produce similar results. Extra contours should be added outside of the data area, both up= and down-dip to control extrapolation.

Step 8 results in a second approximation of the fault surface. It is important to note that the modeled fault surface should extend beyond any intercepts with other faults. This extension will help later when fault intercepts are determined and incorporated.

Step 9—Penetration Check File

The fault surface must be checked for well penetrations that were originally missed by the geologist. Tearpock and Bischke (1991, page 270) explain the problems encountered with directional wells. Also, errors in the shape of the fault surface can result in erroneous fault penetrations. This check is important in oil fields with large numbers of directional wells because their geometry is difficult to visualize. A quick visual check will often be sufficient in areas with straight vertical wells.

The fault grid model is now ready to check against the borehole database (well courses) for well penetrations of the modeled fault surface (Figure 15). Values for the north coordinate, east coordinate, well name, measured depth, and vertical subsea depth for each survey station occurring within a given x-y range are attached as a file. The range should be large enough to include any logical extension of the fault surface. This is somewhat subjective since the amount of non-penetration data contributes greatly to the original grid size. For example, in the Wilmington oil field, faults typically extend along strike well beyond the fault pick control. Oil/water contact data will indicate that the fault extends well down the flanks of the anticline. Without these data, the fault surfaces would be several thousand feet shorter.

A value at every directional survey station location is calculated (back interpolated) from the fault surface and placed in a new Z-field called back. The new appended file is named *back interpolation*.

Since the object of this task is to determine how far the survey station points are from the fault, we must subtract the values in the *back* field from the values in the vertical subsea field and append this new Z-value to the back interpolation file as a new Z-field called *subtract*. The subtract field now indicates the vertical distance between the actual survey station and the fault surface.

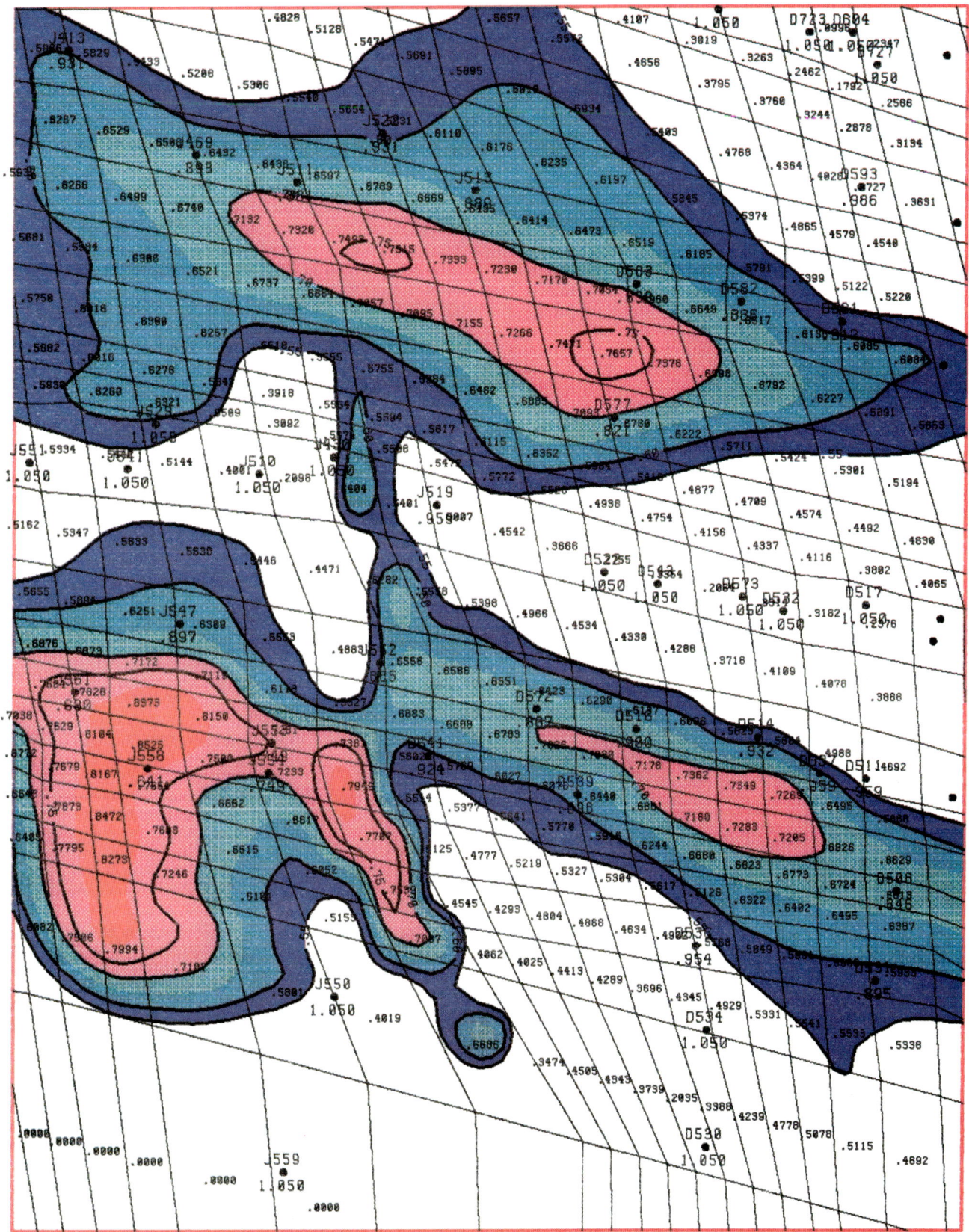

Figure 8. Moveable oil saturation for the F-interval of the Ranger zone was mapped as part of a reservoir simulation in the south flank of Fault Block VI. The reservoir model blocks are displayed as the uneven polygon grid. Average values for moveable oil saturation are posted and the wells are posted at the F-marker. The contour interval is 5%. The grid nodes from the reservoir model were entered as data points and gridded and contoured using a commercial mapping package. The linear feature just left of center that trends generally north corresponds to the D-523 fault. Note the movement of the moveable oil to the north on the east side of the fault. This map was constructed by T. Brix, Department of Oil Properties and J. Skinner, Petresim Integrated Technologies Inc. and published by permission of the Department of Oil Properties, City of Long Beach.

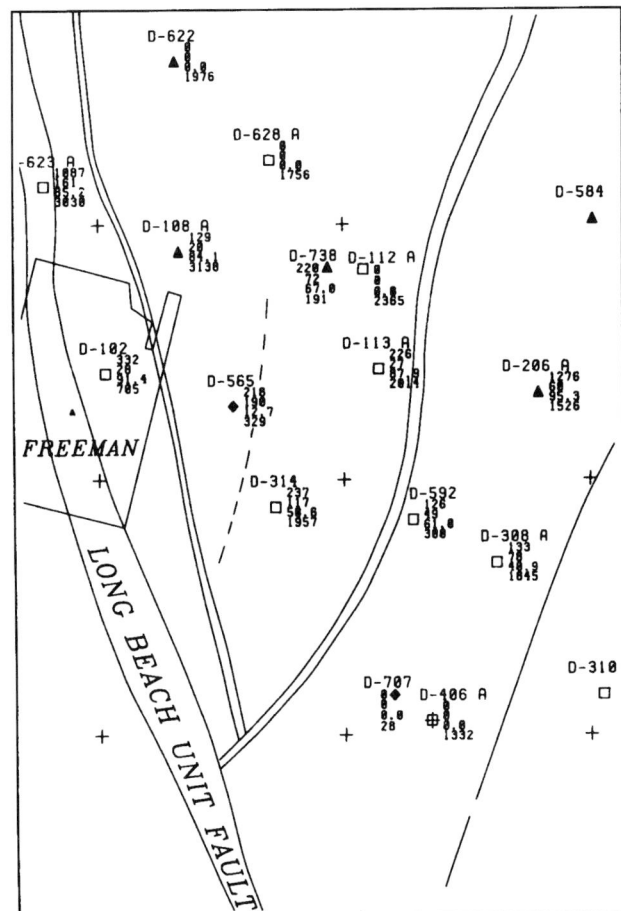

Figure 9. Union Pacific-Ford zone production map with current production (bbl), net production (bbl), water cut (%) and cumulative production (Kbbl) posted at each well. Note that the cut in D-565 is dramatically lower than surrounding wells. D-314 is up structure from D-565 yet shows a higher water cut. This data is currently being used to help delineate a small displacement fault between D-565 and the wells to the east as shown by the dashed fault line. (Published by permission of THUMS Long Beach Company.)

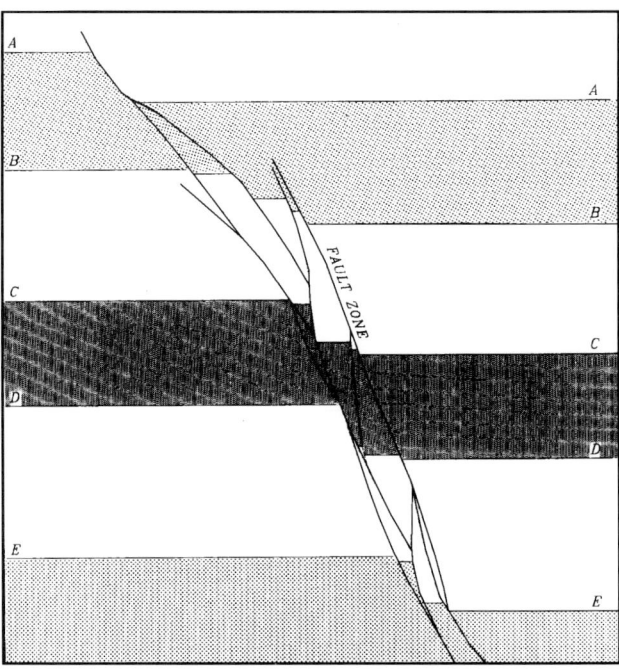

Figure 10. Anastomosing fault in diagrammatic cross section. Modeling surfaces to represent this fault zone is impossible when using well data. The geologist maps the zone as a single surface that represents all of the breaks. The size of the brecciation must be small compared to the extent of the zone. If individual fractures can be resolved in several wells, then the mapper should consider modeling each separately. In the Wilmington oil field, this problem is greatest along the Long Beach Unit fault.

Figure 11. Growth fault in diagrammatic cross section. This fault has had repeated or continuous movement over time. Layer H shows greater displacement than layer B and layer A shows no displacement. Close study of these offsets give geologists a good idea of the faulting history. The faulting can be episodic or continuous. If the events are episodic, the timing of the events can be determined. Growth faults often change dip due to compaction of the footwall and can die out down dip. In the Wilmington oil field the sand and shale layers alternate which should produce a wavy fault surface but, in practice, we do not see this.

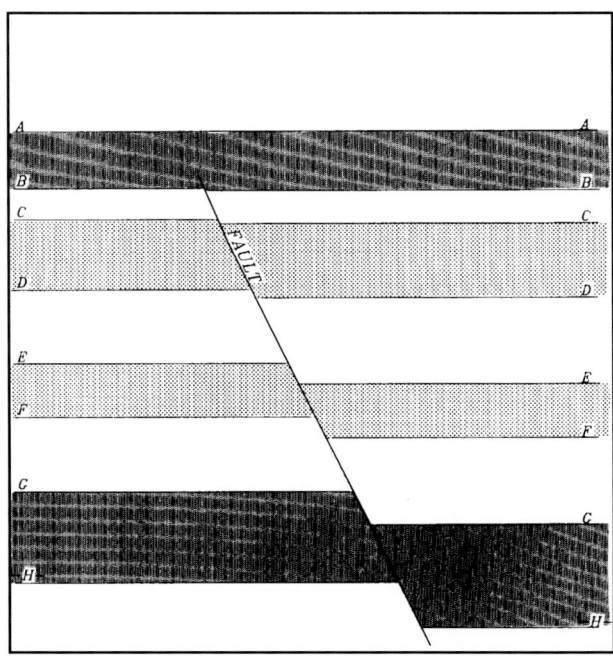

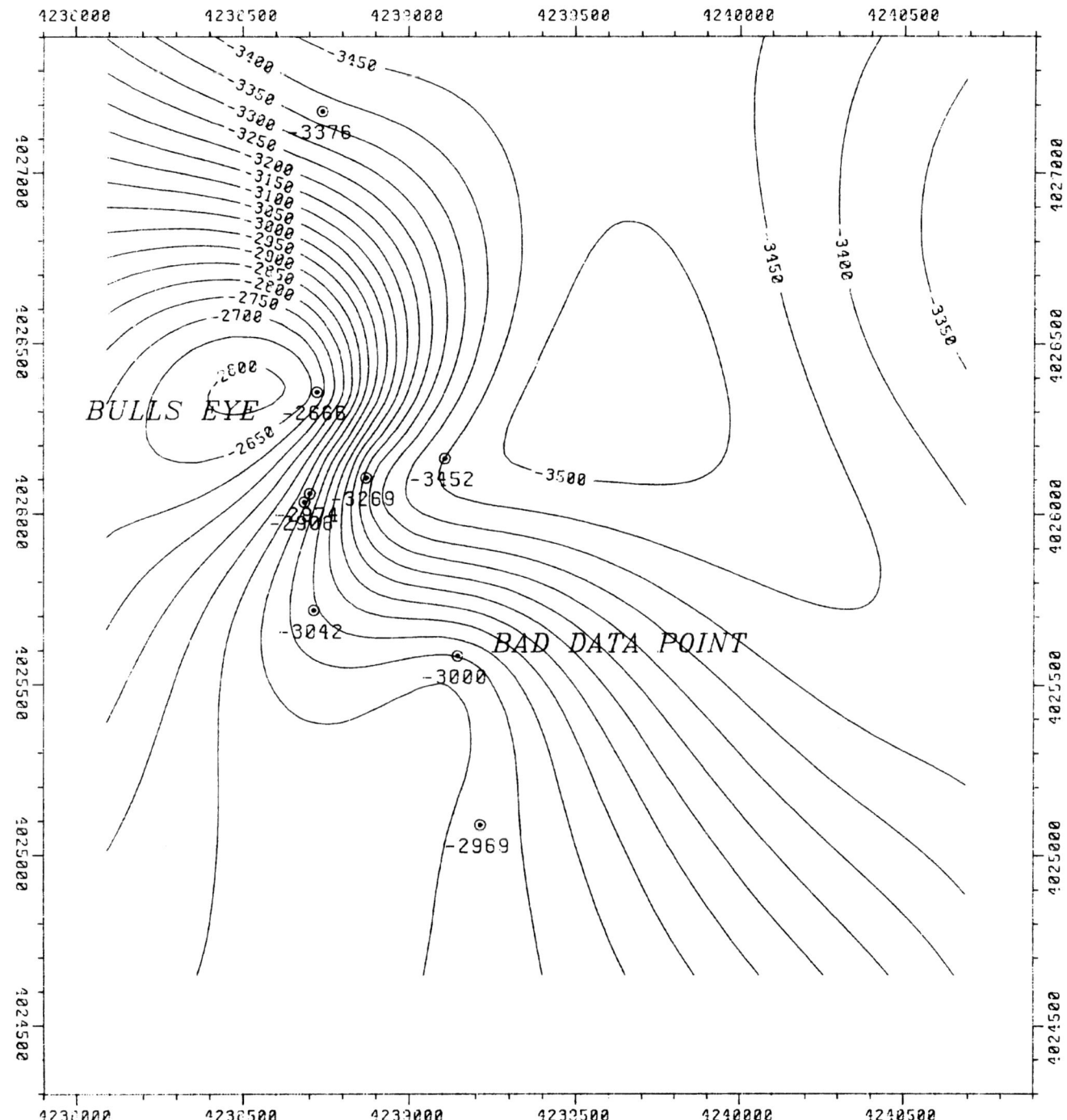

Figure 12. Gridded and contoured raw fault data. Data are typically sparse, therefore algorithms often produce bull's eyes divergent contour lines and other bends. Bad data points further complicate the first model. The actual fault surface strikes roughly north-south and dips steeply to the east.

This data set may still be too large to analyze easily because the station locations, that are large distances above or below a fault, are unlikely to be in error. Therefore, we perform another edit to remove all data beyond a vertical distance of plus or minus 200 ft (60 m). We call this vertical range-editing. It should be noted that the vertical range-editing is another subjective edit. High angle faults require a much larger vertical edit window. Overwrite the subtract file with the new range edited points. This will reduce the size of the working file. The only files to be saved permanently are the original station file and the vertical range-edited file. The range-edited file is sorted by well name and measured depth and printed out for

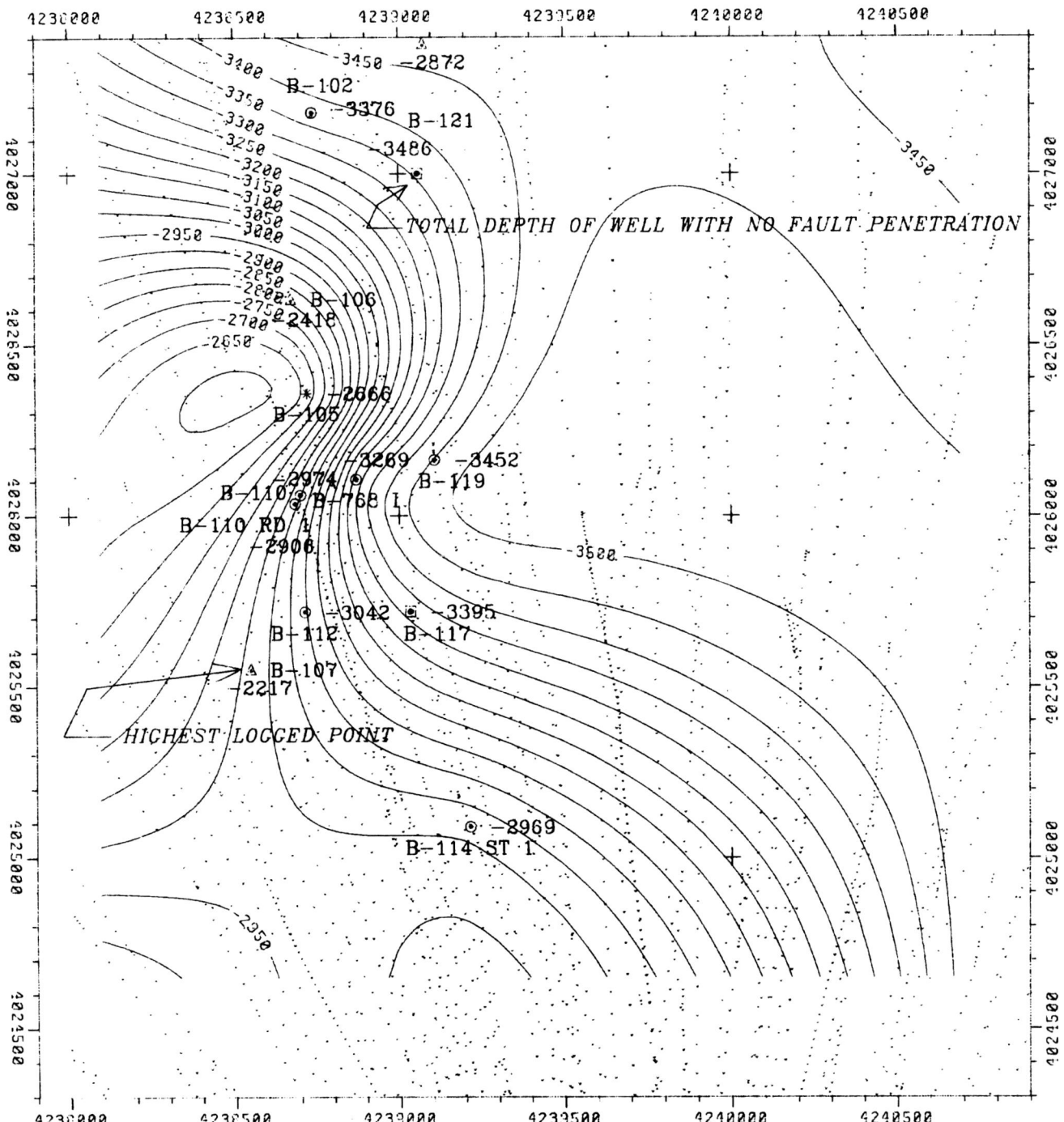

Figure 13. Second generation contours of the fault surface with additional data for error checking. Directional survey station points (dots), top of logged interval, and total depth of the well are displayed. Bad data points have been removed. The resulting surface still does not represent the fault adequately. Operations comparing the grid of the contoured surface to the directional survey stations are performed. All penetrations of the grid by well courses must be reconciled, either by adjusting the grid or making another fault pick. Penetrations shallower than the top of the logged interval are impossible to verify.

geological review. The data are also plotted on the fault surface map. This file represents the actual proximity of the fault surface (grid) to each well course in the study area.

Step 10—Fault Penetration Check

Fault penetrations should agree with the subtracted values (subtract z-field). Changes from negative to

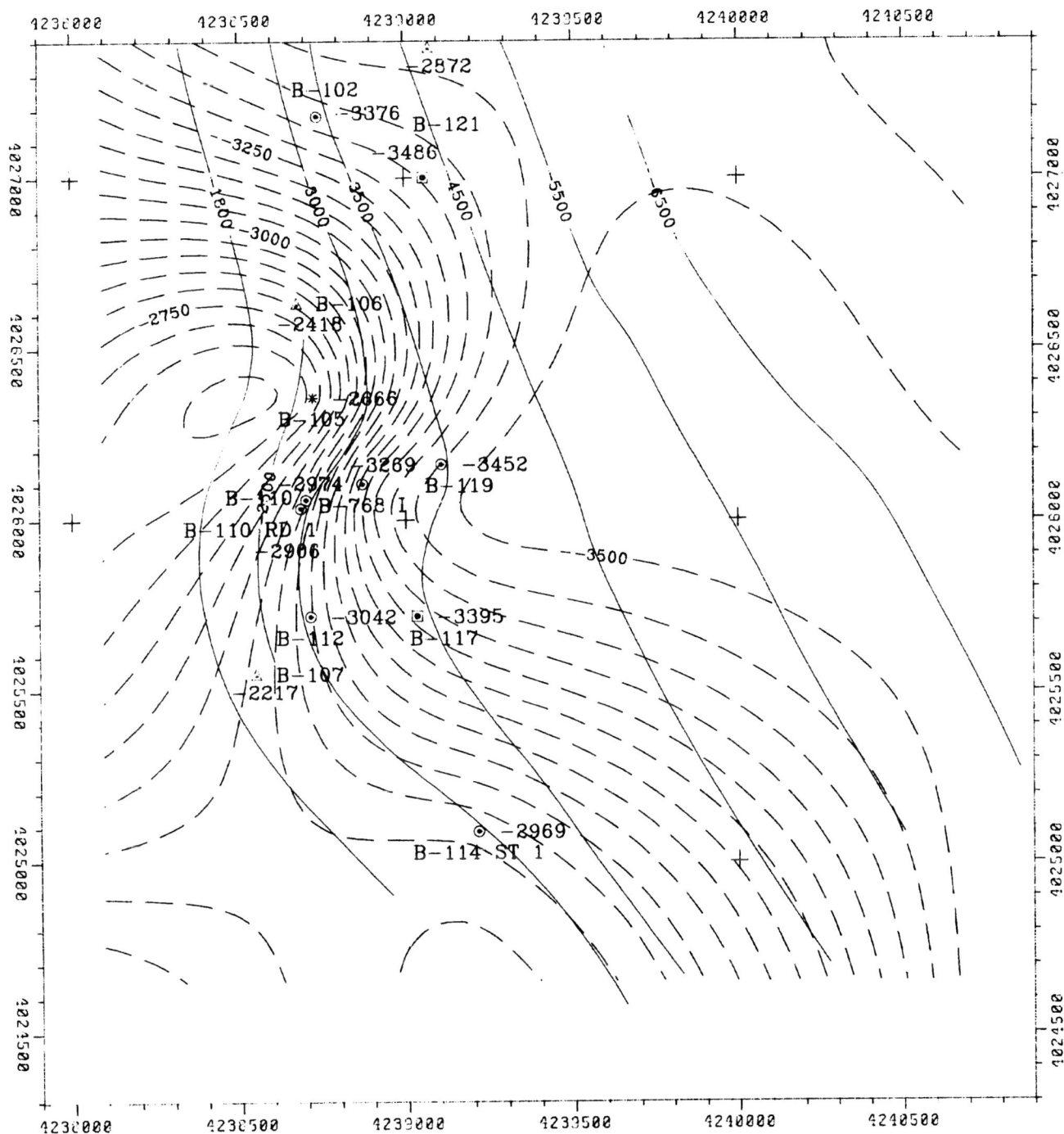

Figure 14. Digitize contours to force control. Only a few carefully selected contours need to be digitized. The digitized contours can be used along with the point data to rebuild the fault grid. This process will probably have to be repeated once or twice more to get an acceptable fault surface.

positive or vice versa along the locus of points indicates the penetration of the fault as modeled. The vertical distance from the fault surface is indicated by the value in the column labeled *subtract*. Some well plots will have multiple penetrations, other well plots will have one penetration, while still others will have no penetrations. If a well course shows one or more penetrations but no penetrations are known, the fault model must be moved or corrected so that no penetration occurs. If there is one verified penetration for a multiple-penetration well plot, the model must be adjusted to miss any offending penetrations. If there is a known penetration and the well course indicates none, then the model must be adjusted to include a

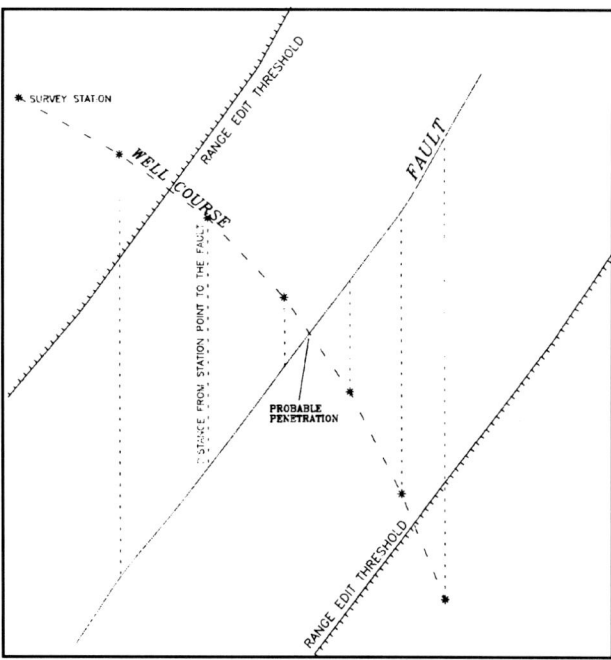

Figure 15. Diagrammatic cross-section along a well direction showing the well penetrating a fault. The well course is shown by large dashed lines. The well survey stations are shown as asterisks. The survey stations are back interpolated to the fault. This value is subtracted from the vertical subsea value of the station point. The result is the vertical distance from the fault surface. The range of these distances is limited to about plus or minus 200 ft (60 m). Anything greater is deleted. This distance is represented as the range edit threshold.

penetration. Each well course in the fault proximity must be checked in a similar manner.

Special problems can occur that must be considered when checking data against the modeled fault surface (grid). Fault penetrations can only be verified where e-log or other geophysical data exists. Some faults may not extend beyond unconformities because no fault activity occurred on the fault after the geologic discontinuity. In the Wilmington oil field, most faults do not extend across the Pliocene-Pleistocene unconformity. If the unconformity is mapped, it can be used to truncate the fault. There is no point in checking data above a known unconformity that truncates a fault.

The required changes are incorporated into the model in much the same way as the original geological interpretation was in Steps 7 and 8. Contour lines are edited and or digitized and a new grid is built using the edited contours and fault penetration points as control.

The resulting third-pass grid is again tested against the directional survey station data for each well. The fault penetration checking process described above is repeated as many times as is necessary to get acceptable results. In practice, we have found that twice is usually enough. An edit for one well occasionally causes a problem in another place. Obviously this problem is greatest where well density is greatest. We have found cases of unexpected curves or irregular contour spacing. These irregularities have proven to be surprisingly accurate fault representations when new wells were drilled through the fault.

Step 11—Multiple Fault Surfaces

The final step is to determine intercepts with other fault surfaces. One fault grid can be clipped against another and then contoured or the geologist can contour the grids and manually pick their intercept and digitize the intercept information into the computer. This should be done for each intercept. Some faults in the Wilmington oil field intercept three or more other faults. Since each fault has its own grid, this can get complex. We treat the intercept information as a "fault trace" when contouring the grid of the fault surface, that is contours extend up to the intercept but not past it. We determine the actual intercept line by overposting contours from two separate fault surfaces and then digitize the locus of points that represent the intersection of respective equal-value contours as a fault trace. We have found that grid operations such as subtracting two fault grids and plotting the zero line do not yield satisfactory intercept lines.

Typically, we find multiple fault intercepts in the same area. The most through-going fault with the greatest amount of displacement is mapped (modeled) first. The secondary faults usually terminate against the larger fault. It is easiest to visualize these as conjugate fault sets. We have not seen faults crossing in the Wilmington oil field, so we do not consider the offsetting of one fault by another as found in an intersecting fault situation. The faulted fault surface should be modeled with care. If one grid is used, an overlap of surfaces could result in multiple grid values for the same location. Therefore, it may be a better idea to map the fault as two separate faults. Some of the newer 3-D mapping systems are designed to handle this. Tearpock and Bischke (1991, p. 224) give a detailed explanation on the geometry and recognition of intersecting fault patterns.

Figure 16. Temple Avenue B-1 fault. The final fault map includes all necessary cartographic information along with contours of the fault surface and posted penetration depths. Note the intercept with the Temple Avenue Fault and the dashes in the contours where the fault surface has been extended. (Published by permission of the Department of Oil Properties, City of Long Beach.)

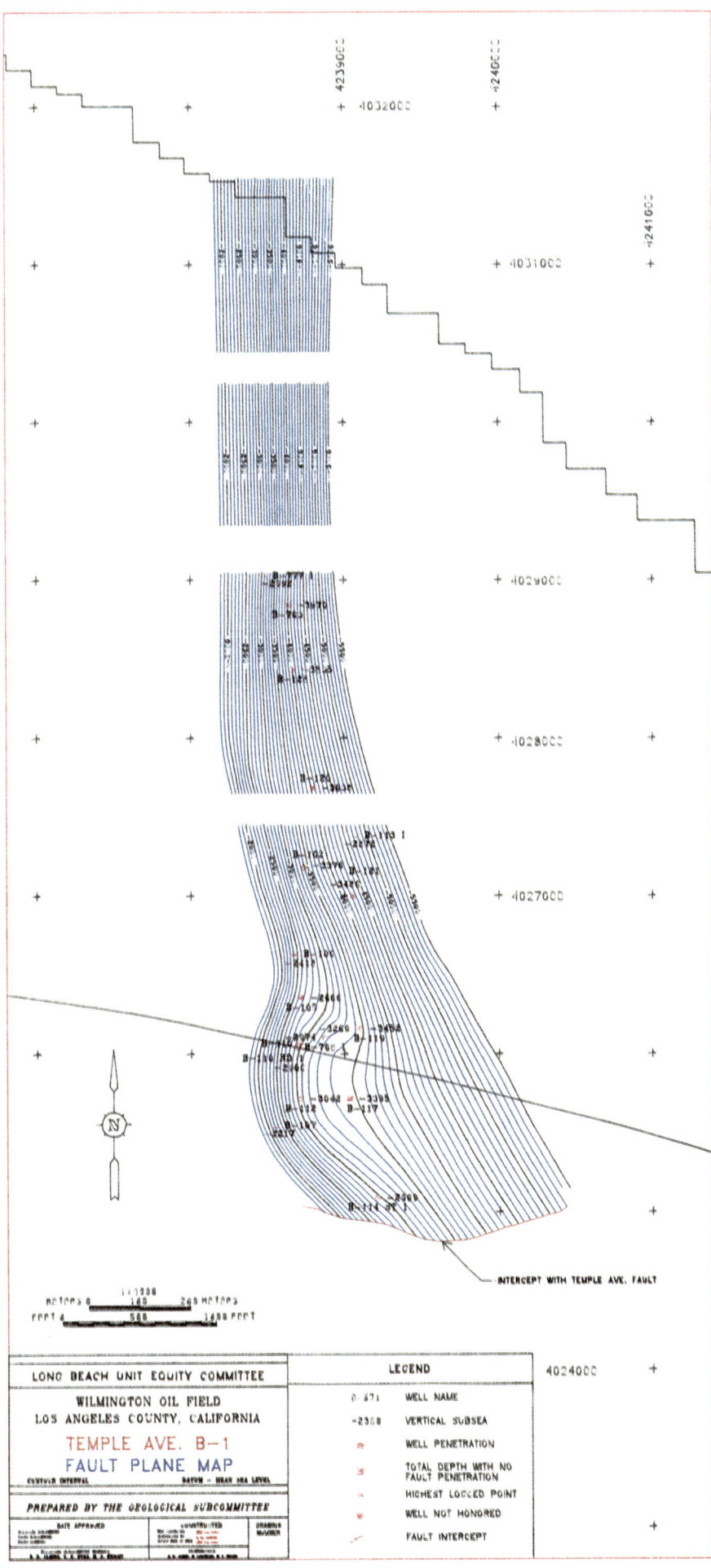

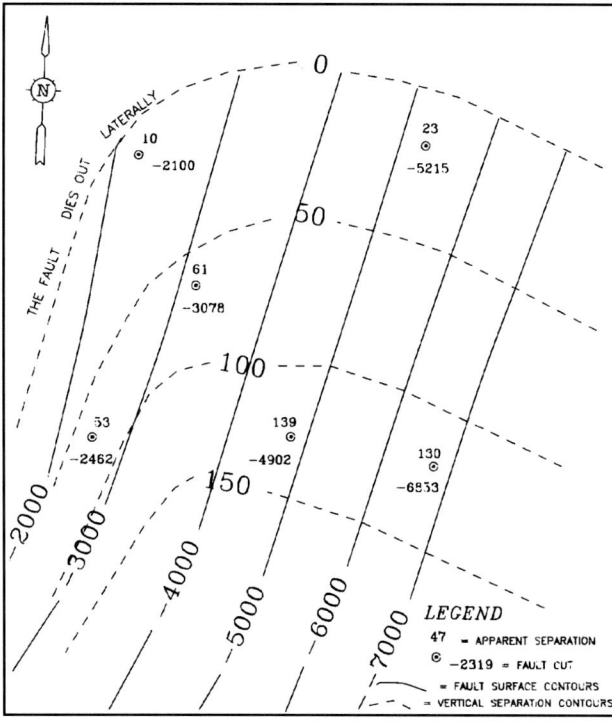

Figure 17. Vertical separation contours are overlain on the fault surface contours. This fault surface is a post-depositional fault. The fault was active from the time represented by the depths –2000 ft to –3500 ft (610 m to 1066 m). The intercepts of the fault structure contour with stratigraphic structure contours will indicate the actual time that faulting occurred. (Modified from Tearpock and Bischke, 1991).

Step 12—Final Map (barrier)

The final presentation should include all pertinent information for the fault surface (Figure 16), including:

1. Well penetrations.
2. Near occurrences—Total-depth points of wells not penetrating the fault. Top-of-log point or unconformity point for wells that penetrate the fault where we cannot determine the intercept along the well course. Directional survey station points for near grazes.
3. Contours of the fault surface.
4. Intercepts with other faults.
5. Limit and reservoir data where used.
6. Cartographic information—Scale, cultural data coordinates, North arrow, title block, etc.

VERTICAL SEPARATION, GROWTH FAULTS, AND TIMING OF FAULT ACTIVITY

Vertical separation can be mapped when sufficient data are available. The contours for vertical separation are drawn directly onto the fault contour map. The missing or added section for fault cuts in vertical holes represents vertical separation but directional wells yield exaggerated section, which can be corrected to vertical separation. A detailed description of the correction process can be found in Tearpock and Bischke (1991). If the vertical separation contours are perpendicular to the fault-surface contours, the fault is post-depositional. Deviations from the perpendicular trend indicate the time fault activity (Figure 17). Increase in vertical separation with depth indicates a growth fault.

Faults may die laterally up section or down section. When this occurs, the grid should contain null values in the nodes that extend beyond the die-out. Further analysis of the fault may be necessary for development work. The Restored Top Method as described in Tearpock and Bischke (1991) for determining the timing and activity for growth faults and Thorsen's Expansion Index (Thorsen, 1963) for estimating the size of growth faults can be easily applied to digital data.

INCORPORATION OF FAULT GRID INTO STRUCTURAL SURFACE GRIDS

The geologist has now spent a considerable amount of time defining a set of faults. These faults must be incorporated into the structure maps of the many geologic surfaces cut by the faults. We have done this in two ways. (1) Treat the intersections of the fault surface with the geological surface as a discrete boundary at the geological surface. The result is an envelope that itself is treated as a vertical fault on structure maps and in the grids. (2) Incorporate the fault surface into the geological surface. The result is a surface that is continuous with no fault breaks, only sharp changes where the fault is crossed.

Faults Treated as Boundaries

Structural surface grids are subtracted to create an isochore grid which is then integrated by the computer to determine the enclosed volume. Since most faults are non-vertical, volumetric errors will arise in fault areas because the fault gap moves and changes shape from horizon to horizon. The fault envelope or fault gap represents the face of the fault as it crosses a geologic surface. Typically, this envelope has no values assigned to its enclosure. The computer geologist also must consider one other condition prior to using this method. The fault surface and the structural surface are represented by a grid with a resolution defined by the grid cell size. This grid cell defines the resolution of the mapped surface. If the width of the fault gap is narrow compared to the grid cell size, then there is no point in combining the geological surface with the fault surface since the resulting grid will produce only averaged results. The occasional grid value from the fault surface that falls within the fault gap will not improve the results significantly.

The locus of points representing the intersection of

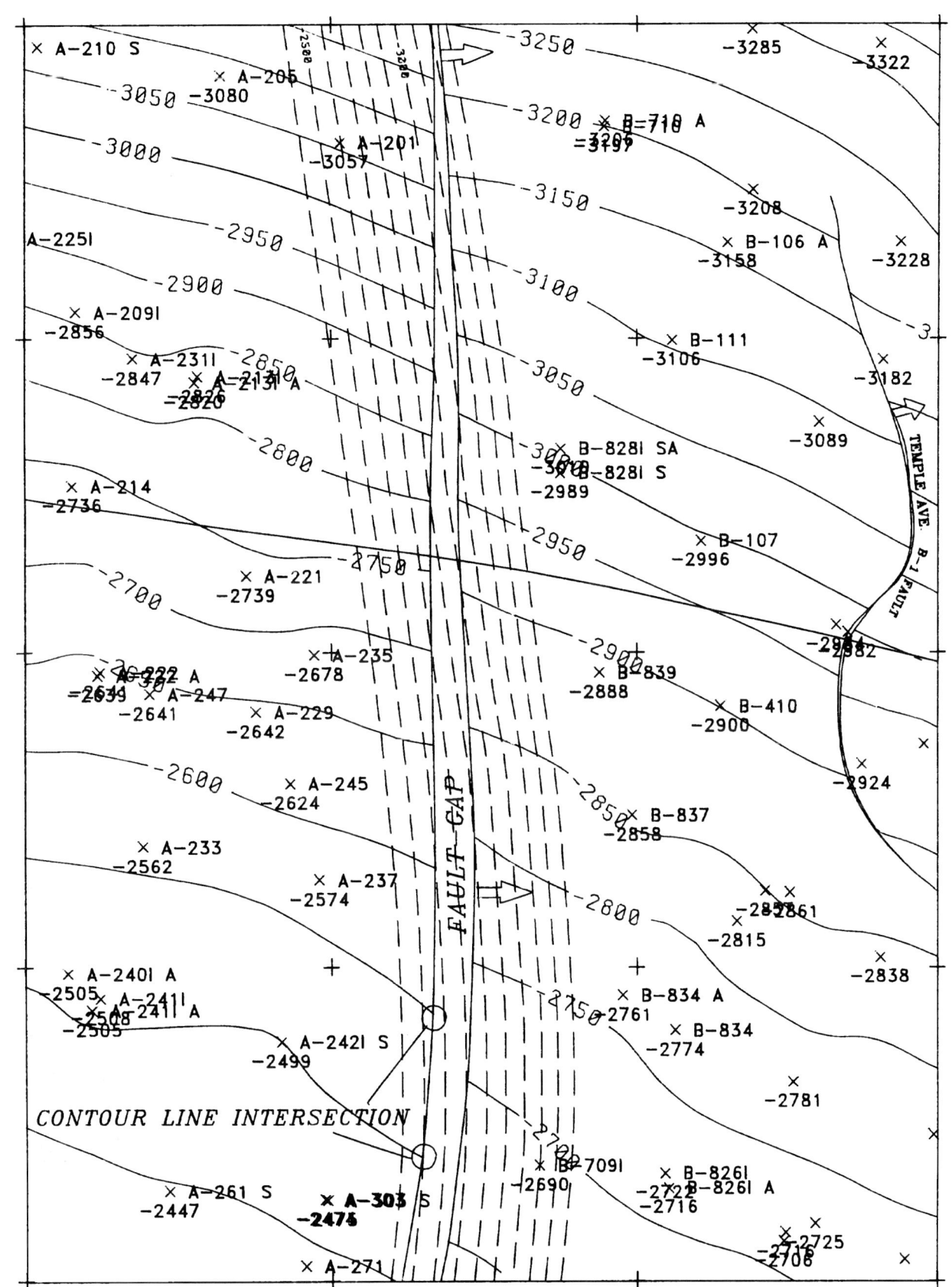

Figure 18. Final structure contours of the F-marker showing Junipero fault contours superimposed. The combination of the two surfaces will result in a complete enclosing surface (fault gap). The fault gap is created by connecting the intersections of the structural contours from the F-surface and the Junipero surface. Notice the displacements along the fault. (175 ft [54 m] at –2500 vss and 60 ft [18 m] at –3200 vss).

Coarse Grid with Narrow Non-vertical Fault Polygon

Base Map with Nodes and Polygon

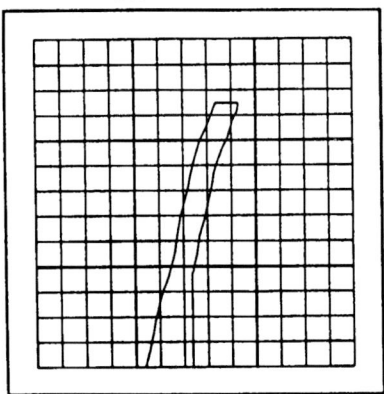

Steps Occurring in Narrow Polygon with Too Few Grid Nodes

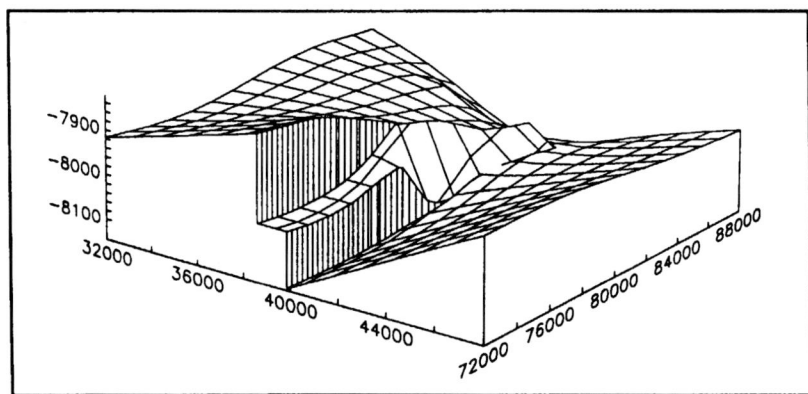

Fine Grid

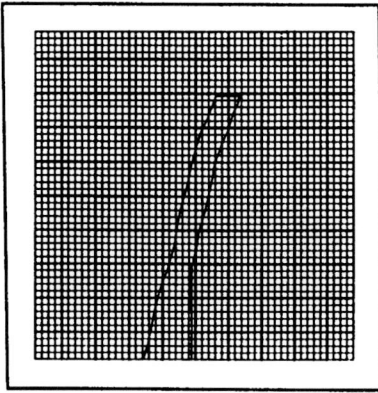

Fine Grid Perspective

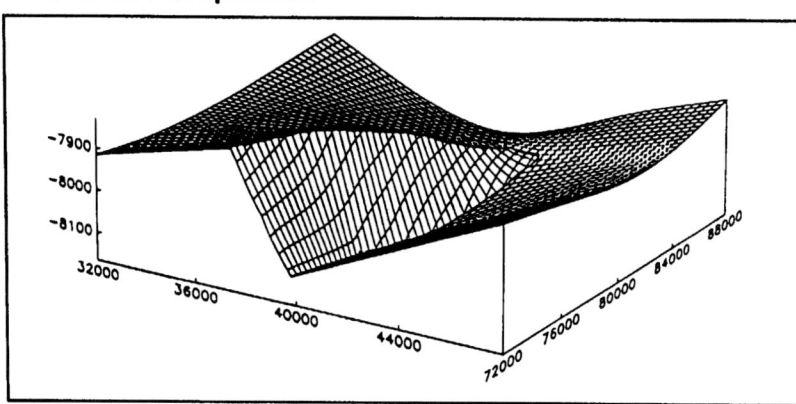

Figure 19. The grid cell size is critical to the accurate modeling of a faulted surface, especially if volumetrics are to be determined accurately. The optimal grid cell size for the structure surface is usually different from the optimal grid cell size for the fault surface. Compromises must be considered if the grids are large. (Dynamic Graphics, Inc., 1990)

the fault surface with the geological surface must be determined in order to establish the fault gap. There are several ways of doing this, including: (1) Subtract the fault surface from the structural surface and contour the zero line. We have found this to be unreliable since we have many faults and the structural surface tends to average across the fault location, resulting in erroneous intercepts; (2) Plot fault contours onto a paper plot of the structure contours; (3) Use the edit features of the mapping program to graphically overlay the fault contours onto the working structural contours. The working structural contours are generated from a grid that was constructed with approximate fault locations. A fault-gap file is built in a manner similar to the manual method by connecting contour intersections (Figure 18). We used Method 3 and it has proven to be very accurate and rapid. It also allows the geologist to make revisions as data are entered. The fault gaps are constructed in descending order, with the largest and most persistent fault first and the smallest, and most limited fault, last.

The final structure map is constructed with the fault gaps. The fault gaps are treated as vertical faults and grid cells within the fault gap should contain a null value. They also replace the approximate fault location files that were used for the preliminary geologic structure grid construction. This method produces accurate structure maps. Many of the faults have a varying amount of displacement and some die-out completely. Computer structure mapping with the fault envelopes has not produced any significant problems because of varying displacement or fault die-out. Occasionally, some minor grid editing is needed. These structure maps are used for most planning and development work in the Wilmington oil field. These maps are not used for volumetrics.

Stratigraphic Horizon and Non-Vertical Faults as a Surface

The second method of integrating faults and structural surfaces involves incorporating the fault into the actual geological surface. This is by far the most accurate way to define a geological surface. But, significant difficulties are encountered when implementing it.

The structural surface that represents each geological marker is gridded along the surface of all intersecting faults. The enclosed area of the fault gap has grid values that represent the fault surface. Integrating the fault surface into the structural surfaces above and below a zone of interest is an excellent method to calculate volumetrics of the enclosed zone. But if the grid cell resolution is too coarse, this method will not produce accurate results.

Grid Cell Resolution

The grid cell resolution of both the geological surface and the fault surface must be fine enough to permit this type of map construction. In the Long Beach Unit, we have found that a grid spacing for structure maps of 200 ft (60 m) is about right for our well density and spacing. Our fault surface maps are modeled with a 100-ft (30-m) spacing because of the high dip angles. Figure 19 illustrates the density of grid nodes at 200 ft and 50 ft (60 m and 15 m). The most detailed maps are gridded to a 20-ft (6-m) grid in order to retain the true shape of the fault in the narrow fault gap.

SUMMARY AND CONCLUSION

The Wilmington oil field is an excellent proving ground for fault modeling techniques. Several difficult faulting problems have been solved. These problems include: non-vertical faults, intersecting faults, faults with varying displacement, poorly distributed fault penetration data, no fault penetration by wells, reservoir anomalies, and gridding incompatibilities. Four types of data are used to define the locations, shape, and orientation of each fault surface:

1. Fault penetrations and non-penetrations along a well course.
2. Stratigraphic irregularities.
3. Apparent structural irregularities.
4. Reservoir properties.

Fault penetration data are collected and stored in a data base. For each fault the penetration data are gridded, contoured and analyzed with respect to the structure, stratigraphy, oil/water contacts and reservoir properties. The contour and the grid files are altered to honor all available data. The corrected files are then compared to the directional survey data for each well in the area. Any well crossings are corrected and the grid is recompared. This is done until an acceptable fault surface is modeled.

The fault surface is then used to adjust the structural surfaces for each marker. The fault surface can be represented as an envelope on the structural surface or it can become part of the surface by patching the fault grid into the structural surface. Grid cell size is a significant factor for this operation. Accurate volumetrics or accurate surface representation will require that special care be given to the selection of grid cell size in areas of faults on structural surfaces.

Software vendors have provided excellent tools to model surfaces. Faulted surfaces have always proven to be difficult to model. I propose that the grid cell size be variable. This can be accomplished by assigning a null value to the grid nodes within the fault gap. The null value should be unique from the null value that is used to indicate no data. A finer grid can then be constructed for each of the areas. The smaller fine grids can be stored behind the regular grid data in groupings that are matched to the null defined areas. The result will be a patchwork of pieces. Contouring and editing would require a double pass in areas of multiple grid cell size. The final grid would be equivalent to taking the fine grid nodes from the lower picture on Figure 19 and inserting them into the fault envelope for the coarse grid.

Each fault surface should be defined if possible. But if insufficient data are available or the faults are part of a fault zone, the data should be mapped as an average surface representing the data as closely as possible.

These methods have been successfully applied to the East Wilmington oil field. Faults with as few as two penetrations as well as faults with 100 penetrations have been mapped. The mapped faults have been combined with the structural surfaces to create accurate structural representations and very accurate volumetrics.

The methods described here were illustrated through the construction of one small fault, the Temple Avenue B-1 fault. These techniques can be applied to any oil field with a large number of wells.

ACKNOWLEDGMENTS

I especially thank Curtis Henderson for working with me to develop these mapping techniques. My appreciation goes to the City of Long Beach and to the Director of the Department of Oil Properties, Xenophon C. Colazas, for permitting the techniques to be shared. I thank Sheryl Gallup for her word-processing and editorial help in preparing this manuscript. And also a thank you to the staff of Landmark-Zycor Inc. for their technical support and software that enabled us to perform the fault mapping functions.

REFERENCES CITED

Billings, M. P., 1972, Structural geology, 3rd ed.: Englewood Cliffs, Prentice-Hall.

Dennis, J. G., 1972, Structural geology: New York, Ronald Press.
Dynamic Graphics, Inc., 1990, Interactive surface modeling user guide: Berkeley, Dynamic Graphics.
Harding, T. P., and J. D. Lowell, 1979, Structural styles, their plate-tectonic habitats, and hydrocarbon traps in petroleum provinces: AAPG Bulletin, v. 63, no. 7, p. 1016–058.

Suppe, J., 1985, Principles of structural geology: Englewood Cliffs, Prentice-Hall.
Tearpock, D. J., and R. E. Bischke, 1991, Applied subsurface geological mapping: Englewood Cliffs, Prentice-Hall.
Thorsen, C. E., 1963, Age of growth faulting in southeast Louisiana: Transactions Gulf Coast Association Geological Society, v. 13, p. 103–110.

Chapter 12

Computer Modeling of Multiple Surfaces With Faults: The Ivanhoe Field, Outer Moray Firth Basin, U.K. North Sea

N. J. Hooper
Amerada Hess Limited
London, U.K.

J. G. M. Raven
Shell (U.K.) Exploration & Production
London, U.K.

M. J. Kilpatrick
Radian Corporation
Austin, Texas, U.S.A.

ABSTRACT

Largely due to inadequacies of mapping software and complexities of data management, faults are traditionally treated as vertical planes for each separate reservoir zone or throughout the total reservoir within volumetric field models. This simplified approach results in errors in geometry and volume calculations.

We discuss a software system which automates the volumetric modeling of multiple horizons exhibiting nonvertical faulting. The modeling is based upon the construction of fault plane grids from digitized fault traces and a set of attributes to describe the attitude of the fault. The fault plane grids are intersected with structural and isochore grids for each reservoir zone, producing a realistic set of migrated fault traces for each structural horizon. This greatly enhances the structural definition for volumetric analysis.

Extensional fault models, incorporating normal, listric, and strike-slip faults can currently be built. The resulting horizon and fault grids may form the input for more advanced three-dimensional modeling software, providing a better structural framework. The Full Fault Modeling System (FFMS) is a stand-alone module. It has been designed to work closely with Radian's CPS-3 Advanced Mapping Software, but can equally operate with any computer mapping package which uses rectangular grids. The FFMS software provides a structurally consistent approach to computer-based modeling of multiple surfaces affected by nonvertical faulting.

The Ivanhoe field, situated in the Outer Moray Firth of the U.K. North Sea, has been used as a case history to demonstrate the application of the software and its methodology.

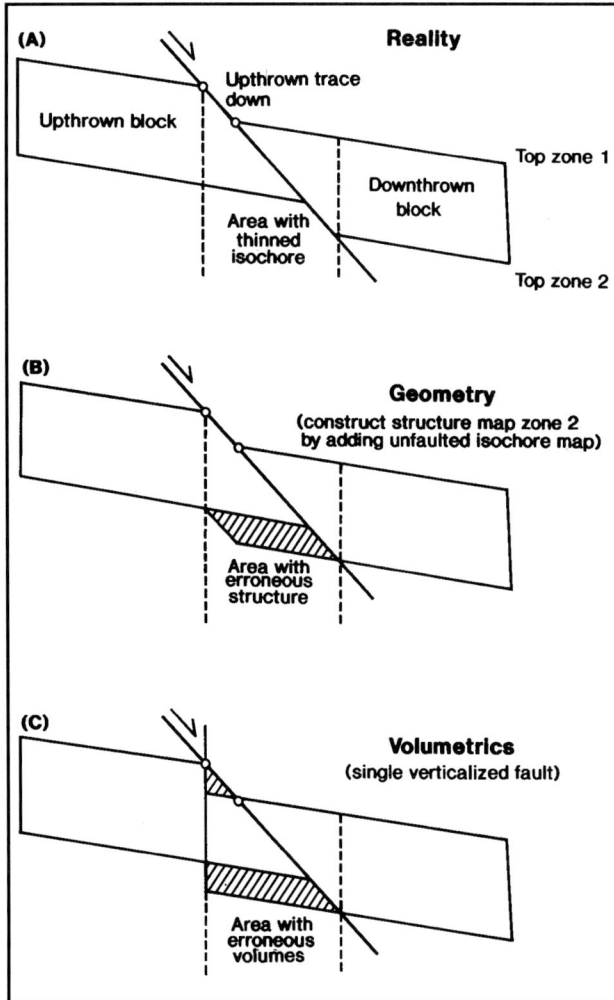

Figure 1. Demonstration of problems inherent in computer mapping without detailed fault modeling. (A) Reality, (B) Geometry, (C) Volumetrics.

INTRODUCTION

Computer mapping of faulted surfaces for reservoir modeling and volumetric studies has long been restricted to the simplified vertical fault model. This is largely due to a combination of three factors:

- inadequacy of available computer mapping software,
- complex data management requirements, and
- processing performance of available computers.

Nonvertical fault models for single surfaces have been possible for several years, but these may be very complex and do not incorporate the geometric details required for producing a consistent model of a multiple-layer reservoir.

Modeling multiple surfaces exhibiting nonvertical faults has also been possible to a limited extent using a basic fault plane gridding methodology (Newton, 1983; Hooper, 1990a). The problem is maintaining integrity of the model as fault intersections migrate from one surface to another. The procedures are both complex and laborious, and are impractical for more than a few major faults.

The Full Fault Modeling System (FFMS) has been developed by Radian Corporation to provide a solution to modeling faulted multi-layer reservoirs for structural and volumetric analysis.

The system handles the complex data management, leaving the user free to work interactively with the data to build a reliable geological model.

PROBLEM DEFINITION

Field mapping of North Sea reservoirs is commonly done in the form of a layer-cake model: the structure is mapped using interpreted seismic horizons and well control at or near the top of the reservoir, and possibly at or near the base of the reservoir. Seismic definition is often not available to allow mapping of any of the intra-reservoir units. Successive horizons can vary from parallel to sub-parallel, and may truncate or pinch-out as modeled by the unit isochores. Well data are used to map true vertical thickness isochores for each reservoir unit. Since well control is too sparse to provide adequate control for modeling isochore thinning in the fault zone, well thicknesses, where faulted, are restored to their unfaulted condition (or omitted). These adjusted well data are used to produce an unfaulted isochore for the reservoir unit. The isochore is added to the top reservoir depth map (or subtracted from base reservoir), to produce a depth structure map for the relevant intra-reservoir unit.

Figure 1 demonstrates the potential errors in both geometry and subsequent volumes following this operation with an unfaulted isochore. In terms of the geometry (Figure 1b), an error is apparent in the base reservoir unit structure, with the fault section appearing vertically below the faulted in the upper structure, and not being migrated to honor the dip of the fault plane. In terms of the volumetric model (Figure 1c), whether verticalized to the upthrown fault trace (as is commonly the case) or utilizing both upthrown and downthrown traces, significant errors are realized. The result is an over-estimation of bulk volumes when both downthrown and upthrown blocks are considered, but of perhaps greater significance is the reduction in bulk volumes attributable to the upthrown block within the fault wedge. This reduction in volume originates from the verticalization of the faults and from the incorrect geometry as a result of not migrating the fault traces to produce an accurate model of the structures within the fault zone.

These errors are exacerbated with numerous intra-reservoir units if not corrected, and are sensitive to fault angle, reservoir unit thickness, and fault throw.

The goal is to be able to use an automatic procedure

which correctly models multiple horizon geometry in the vicinity of faults, producing a structurally consistent model with more accurate and reliable volumetrics. Such a system should be able to correctly model horizon and fault intersections, migrate those intersections from horizon to horizon, and handle intersections between faults (bifurcations). The procedure should have a system for managing data, and provide a clear audit trail. It should also be interactive, automating repetitive tasks and allowing manual intervention for editing and interpretation.

FAULT MODELING METHODS

Several fault modeling methods have been used to date, all with certain limitations or compromises. These have arisen for a number of reasons, either due to limitations in available software, insufficient or poor data, inadequate computer resources, or lack of expertise. Most of these methods maintain a simple approach and are equally simple to invoke when considering a theoretical or limited model. Numerous practical complications arise in reality which restrict their use to simple structures, small areas, or a few major faults within a field. Jones et al. (1986, p. 141–173) cover a number of fault modeling techniques. A few comments on some of the more widely used techniques are considered relevant at this stage to highlight the principal concerns:

(1) Continuous Structural Surfaces—This technique uses fault traces for individual horizons and assigns elevation values to these polygons. These horizons are then gridded, allowing interpolation across the fault polygon. The results can be reasonable for individual horizons, but consistency is not automatically maintained between horizons, allowing an unreliable structural model to be built.

(2) Regrid within Fault Polygons—This is a derivation of the methodology discussed above. The structures are created first, retaining blank or null areas within the fault polygons. These fault polygons can then be regridded using the upthrown and downthrown surfaces as control. This method can be extended to multiple horizon models (Newton, 1983), but it is not possible to model the fault plane itself, leading to some inconsistencies.

(3) Model Faults as Separate Surfaces—This can be done either on a segmented basis considering individual horizons (Kemp and Grinstad, 1984), or can be considered on a field-wide basis in an attempt to maintain a consistent fault model for the entire field and all horizons (Hooper, 1986, 1990a, 1990b; Clarke, 1992). Either way has limitations: large numbers of faults and fault intersections render these techniques impractical.

(4) Verticalize Faults per Horizon—This is perhaps the most commonly used method, either verticalizing the faults for the entire reservoir, or migrating the faults for each horizon and verticalizing per zone. Migrating the faults per zone can be cumbersome, requiring manual operations. The resulting model, whether verticalized by zone or for the entire reservoir, will exhibit incorrect geometry and result in inaccurate volumetrics (Figure 1).

(5) Restore to Pre-Faulting Conditions—This is an extension of manual mapping techniques (Tearpock and Bischke, 1991, p. 195–279), and allows maximum flexibility in terms of varying fault throw and intersections. This approach maps the fault planes and throw from well control and restores horizons to their unfaulted condition. The isochores are sub-sequentially stacked to form the layer-cake model. Once built, the model is then re-faulted (Banks, 1990; Jones et al., 1986). This method is acceptable where there is sufficient well control to describe the fault geometry and displacement, but is implausible with limited well control or in areas of complex geological history exhibiting syndepositional (growth) faulting or structural inversion.

6) Fault Throw Gridding—This technique uses additional information about the fault, namely throw (vertical separation), to adjust grid node values across a fault during initial gridding (Zoraster and Ebisch, 1990; Zoraster, 1992). This technique can also create contours and surface values within the fault polygons. As with other techniques, this method only models horizons individually, and does not produce a consistent model for multiple horizons and intersecting fault planes.

The choice of fault modeling method depends upon a number of factors, including:

- available data,
- complexity of field to be modeled,
- purpose of modeling, and
- computing capacity available.

It goes without saying that more and better-quality data available to describe the fault plane will lead to a better ultimate model, producing more reliable volumetrics. Equally, if the model is only for a quick-look study or rough estimation of volumes, it is unnecessary to employ detailed modeling techniques. By inference, any detailed modeling involving numerous elements, such as individual faults and fault blocks, is likely to be a computer intensive process and therefore will require significant resources.

In field development and equity work, reliable and accurate models are required to define the structure and calculate volumes of the reservoir. It is also useful to run sensitivities in order to gauge the impact of certain key parameters or unknown/uncertain quantities on final volumes.

Fault models are one of these unknowns: it is only necessary to look at the life cycle of a field from prospect through appraisal and on to mature development to see how fault interpretations change. This is due to the varying amount and quality of data available throughout the life of the field.

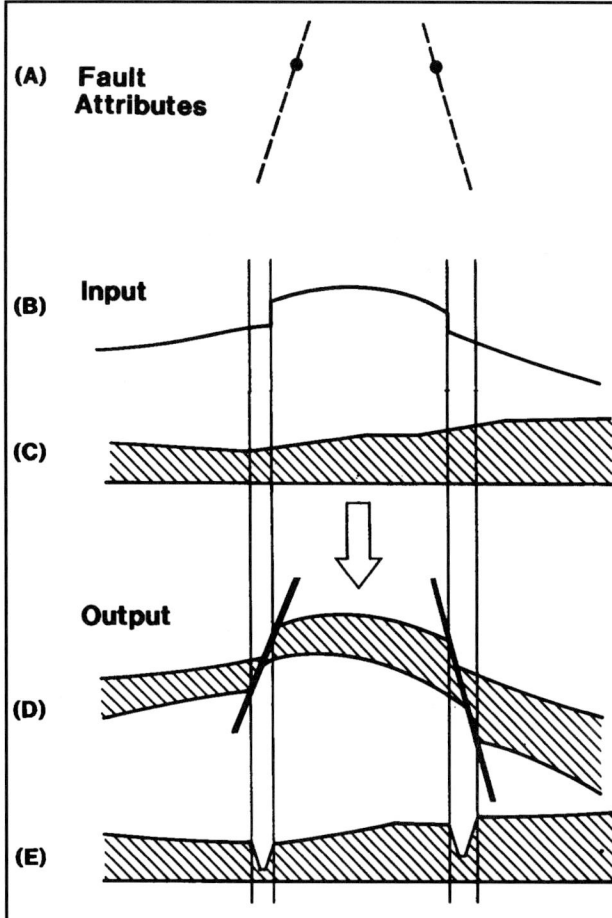

Figure 2. Full Fault Modeling System (FFMS) method in cross section: (A) calculate fault plane map from digitized fault traces and fault attributes, (B) input structure map for top of zone 1 with vertical fault traces, (C) input unfaulted isochore map for zone 1, (D) output structure maps for top of zones 1 and 2 with nonvertical faults, (E) output fault-corrected isochore showing thinning in fault zones.

Typical input data available for fault modeling include:

- seismic horizon picks,
- well horizon picks,
- seismic fault cuts, traces, and polygons,
- fault cuts in wells, and
- geological isochore maps.

In addition to these more traditional data sources, greater use of seismic workstations and other interpretive and predictive tools continue to provide better data and in greater quantities.

METHODOLOGY USED IN FFMS

The method used in FFMS is demonstrated schematically in Figure 2. It is analogous to manual methods involving the manipulation and intersection of separate planes representing horizon tops and fault planes (Tearpock and Bischke, 1991, p. 195–279). It is based upon modeling faults as separate surfaces and manually intersecting these with horizon tops (Kemp and Grinstad, 1984; Hooper, 1990a). FFMS automates these manual processes, allowing easy modification and reconstruction, providing good data management, and a solid audit trail.

Data used by FFMS include:

- digitized fault traces,
- fault characteristics (dip, strike, throw, heave),
- fault cuts in wells,
- faulted horizon structure map, and
- unfaulted isochore map(s).

The first step is to define and describe the individual faults to be modeled. This typically involves digitizing a series of fault traces, currently either the upthrown or down thrown trace only. Included with these traces are a set of parameters describing the attitude and nature of the fault plane (dip, strike, elevation, throw, etc.). These parameters can vary along the length of the trace, or can be assumed constant for each fault. Fault parameters can also be varied from one horizon to the next if required. Obviously, the more parameters specified for each fault trace, and the ability to vary these along the fault to match variations observed/measured in the geological setting (Barnett et al., 1987), ultimately leads to a better structural model and more accurate volumetric calculations.

The minimum specification for each fault is the digitized trace and dip angle (elevations can be calculated from the input horizon structure map), the maximum level of control is afforded from the following parameters:

Fault Identification (i.e., name or number)
Trace Aspect– (upthrown/downthrown)
Model Type–(planar/listric)
Planar Properties
- trace elevation
- dip of fault plane
- strike of fault plane
- throw (vertical separation)
- heave (horizontal separation)

Priority Level

Once the fault traces and parameters are loaded, a fault gridding algorithm computes grid node values for each fault. Each fault plane is stored separately, allowing overlapping and intersecting faults to be modeled efficiently, allowing multiple Z-values at any given location. Well cuts can be used to control fault plane projection, automatically modifying any general dip parameters where necessary, thus ensuring that well data are honored where available during the fault plane gridding process. The horizon structure map is then loaded to the system as a gridded surface: this has been verticalized to the upthrown fault trace (foot-

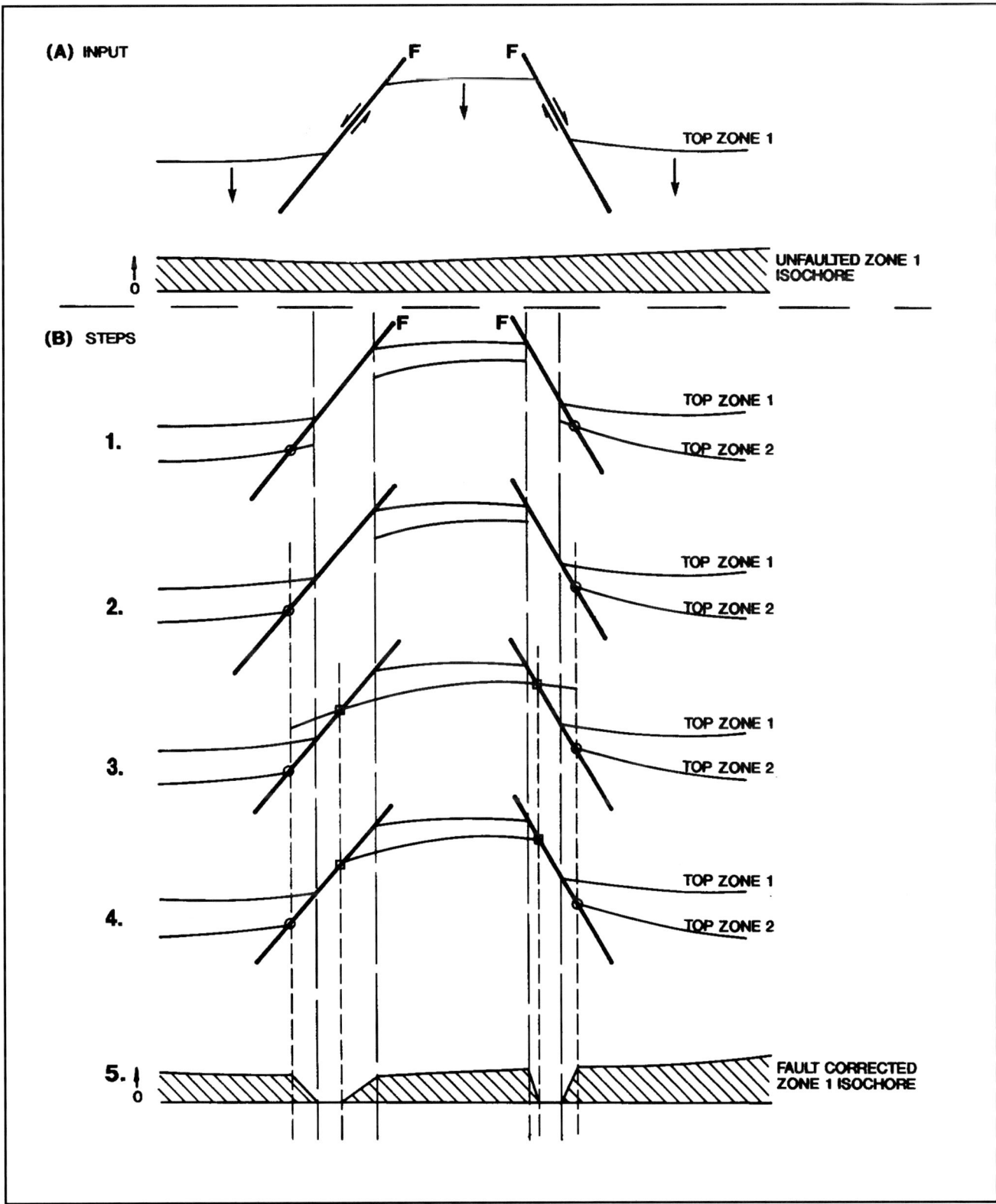

Figure 3. FFMS methodology: stacking (adding) isochore from Top Zone 1 to produce faulted Top Zone 2, migrated fault traces, and fault corrected Zone 1 Isochore. (A) Input: nonvertically faulted structure on Top Zone 1, and unfaulted isochore for Zone 1. (B) Steps: 1. Add unfaulted isochore to Top Zone 1 and intersect resultant Top Zone 2 with fault plane to calculate downthrown fault traces for Top Zone 2; 2. Clip Top Zone 2 at downthrown traces; 3. Extrapolate Top Zone 2 within upthrown blocks and intersect with fault planes to calculate upthrown traces; 4. Clip upthrown blocks at upthrown traces producing correctly migrated and faulted Top Zone 2; 5. Calculate fault corrected isochore for Zone 1 showing thinning in fault zones.

wall) and gridded using conventional methods. Normal quality control procedures must be followed to confirm structural consistency of the input horizon and the fault displacement (Tearpock and Bischke, 1991). Fault throw gridding may be used to assist in the generation of this surface (Zoraster and Ebisch, 1990; Zoraster, 1992). During the next stage the system calculates all of the intersections of the fault planes (from Stage 1) with the verticalized horizon structure grid. These intersections are then filtered and checked for truncations/bifurcation. The result of this intersection/filtering step is a set of downthrown block (hanging-wall) fault traces for the input horizon.

Editing of the downthrown fault traces is possible if required. If edited, the system can pick up the edited fault traces and continue with the modeling of the faulted input horizon structure grid honoring the new downthrown traces. At this stage we have the faulted horizon structure map with both upthrown and downthrown fault traces defined. The surface between the fault traces is blank, although the software will fill these in from the relevant fault plane grids.

The next step is to add the unfaulted isochore (Figure 3). An unfaulted isochore grid is generated using only well data which are not cut by any faults. Repeat sections or missing sections are accounted for, or the wells are omitted. If the isochore is eroded or pinches out, the grid should be clipped to zero before starting the stacking procedure. The resulting initial structure map on zone 2 will have the faults vertically below those of zone 1. The system will correct this structure with the following automated sequence of operations:

1) Determine the outer limits of the fault zone on Top Zone 2 by computing the locus of the downthrown traces (downward fault migration step). This is the same basic process as used to define the downthrown fault traces for the input horizon.
2) Reconstruct the upthrown blocks in the fault zones, effectively verticalizing the Top Zone 2 structure to the downthrown traces just computed for Top Zone 2. Note that an external data-tying step can be introduced at this point, if desired and if data are available.
3) Determine the locus of the upthrown traces for the Top Zone 2 structure (upward fault migration using the fault/horizon intersection procedure as per step 1).
4) Truncate the upthrown fault blocks at the upthrown traces by blanking the surface between the computed loci of upthrown and downthrown fault traces.
5) Merge the depth values from the associated fault plane grids into the Top Zone 2 structure at the appropriate fault zone locations determined by the computed upthrown and downthrown traces. This results in a complete faulted Top Zone 2 structure with its associated fault traces.

A faulted isochore map for the unit is produced by subtracting the two faulted horizon structure maps. This isochore shows thinning within the fault zone (Figure 2). The process can be interrupted at any stage, typically where the fault traces are calculated by intersecting horizon and fault planes. The traces can be edited, if required, and the process re-started.

The output calculated by FFMS includes:

- fault plane maps for each fault
- upthrown and downthrown fault traces with depth values
- structure maps with nonvertical faults
- structure maps with verticalized faults to either the upthrown or downthrown trace
- fault corrected isochores.

The approach just described ensures that a structurally consistent computer model of multi-layered reservoirs displaced by nonvertical faulting is created. The methodology has been tested on a number of fields by Shell U.K. Expro and Amerada Hess. The procedures have shown consistent and reliable results, producing high quality maps, with significant impact on volumetrics.

Grid resolution is not a significant issue in the application of the FFMS procedure. This is predominantly due to the use of consistent fault plane grids throughout the model building process, rather than restricting the extent of fault surfaces to within the constraints of the individual fault trace pairs. In this way, grid resolution criteria can be considered the same as for normal computer mapping exercises. Nevertheless, although the overall model will be structurally correct, individual horizons would be enhanced if a more refined grid cell were to be selected; this would be true both for contouring and volumetrics. This is because the contouring and volumetrics algorithms have not yet been modified to use the additional pieces of the fault surfaces where they project beyond the relevant fault traces. This capability is currently being added as an enhancement to allow the algorithms to use this additional control, resulting in a true grid resolution dependent only upon the horizons to be mapped.

The procedures continue to be developed and enhanced. Among others, some of the current developments are:

- input upthrown and downthrown traces, or multiple horizons, and automatically calculate fault plane dip/azimuth,
- automatic fault plane cross sections for checking communication of sand bodies across faults, and
- enhanced volumetric algorithm to utilize additional structural information available in fault grids.

A CASE HISTORY: THE IVANHOE FIELD, OUTER MORAY FIRTH BASIN, U.K. NORTH SEA

The Ivanhoe field was discovered in 1975, and is located in block 15/21a of the U.K. North Sea, within the Outer Moray Firth basin (Figure 4). The field is oriented NW–SE and covers an area approximately 2 miles by 1 mile (3,200 × 1,600 meters) in size. The reserves occur in tilted fault-block traps of the Upper Jurassic, Piper Sandstone Formation (Parker, 1991).

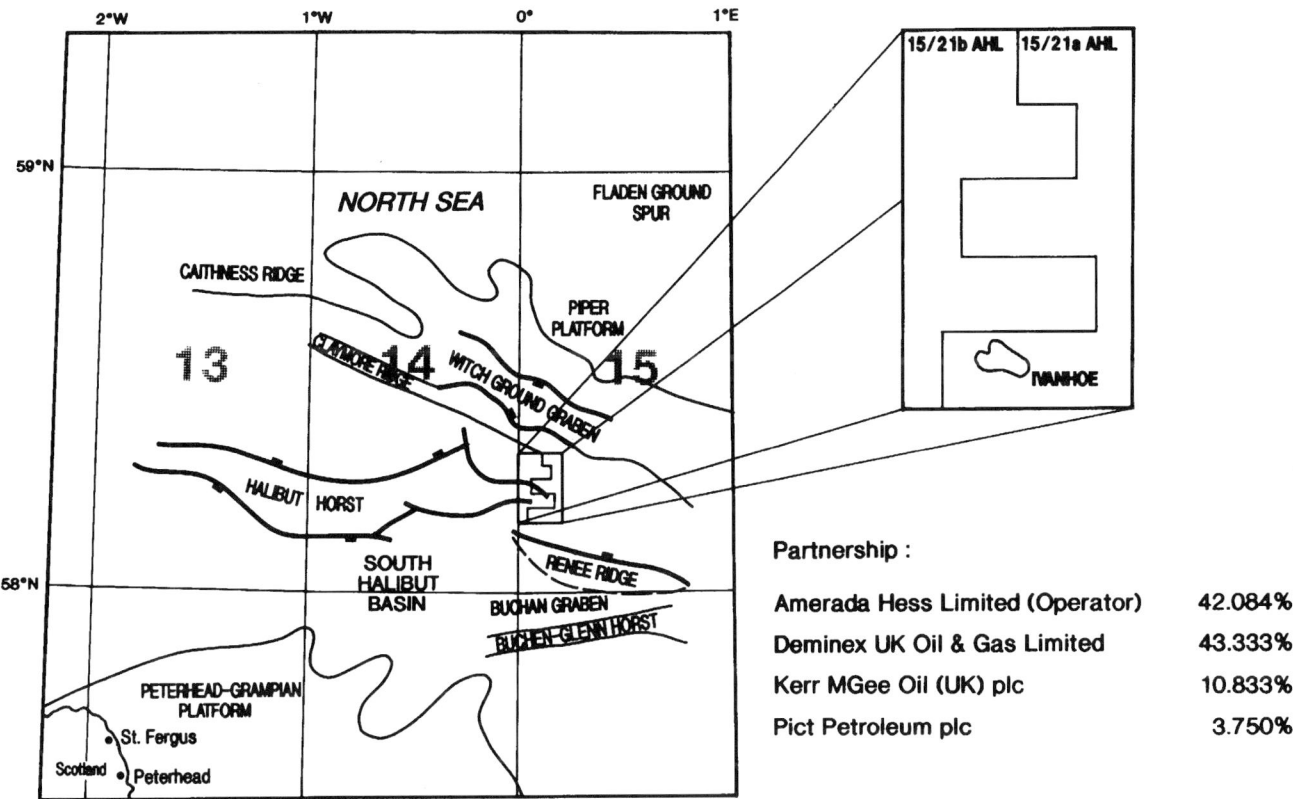

Figure 4. Location map of Ivanhoe field, U.K. North Sea, showing the major structural elements (after Parker, 1991).

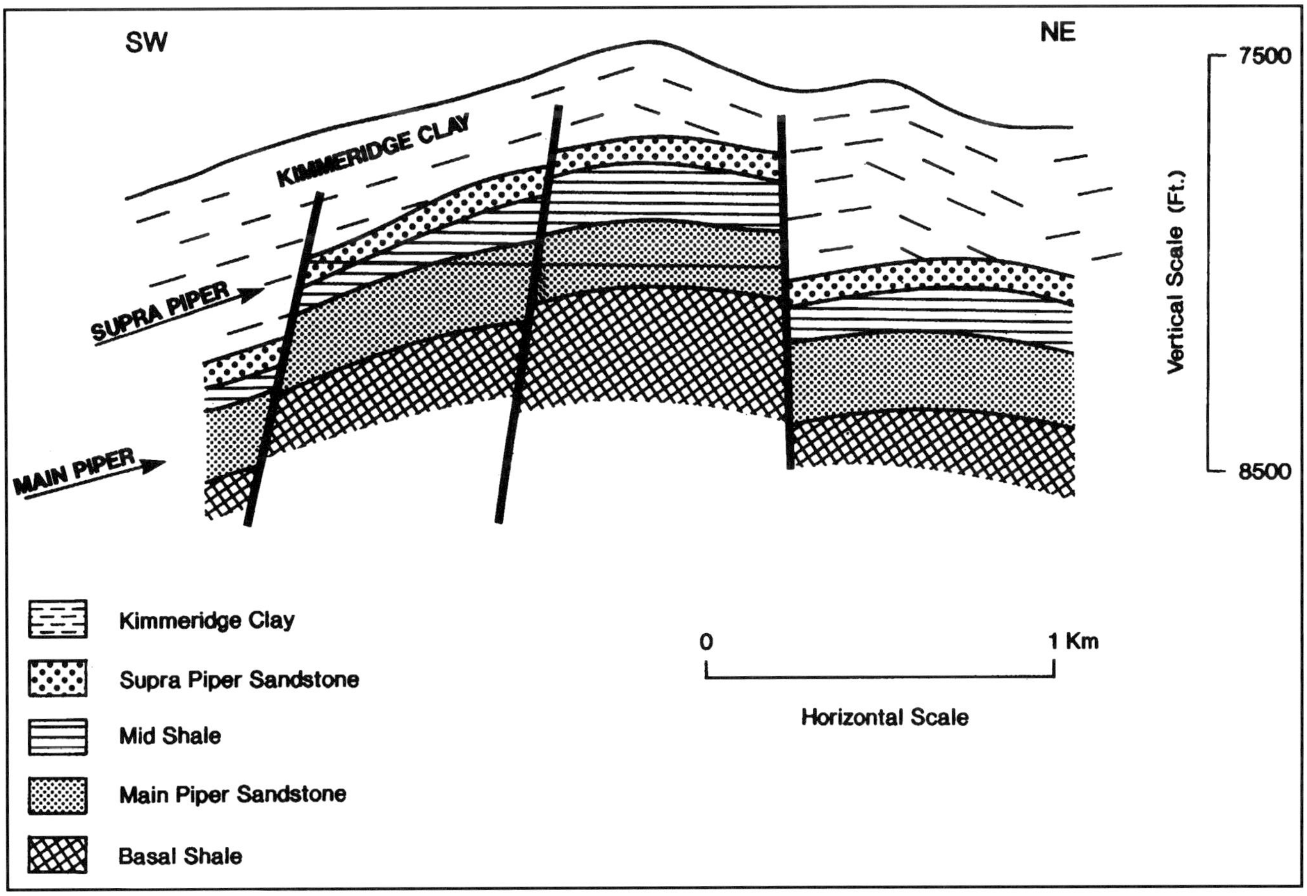

Figure 5. Schematic structural cross section across Ivanhoe field (after Parker, 1991).

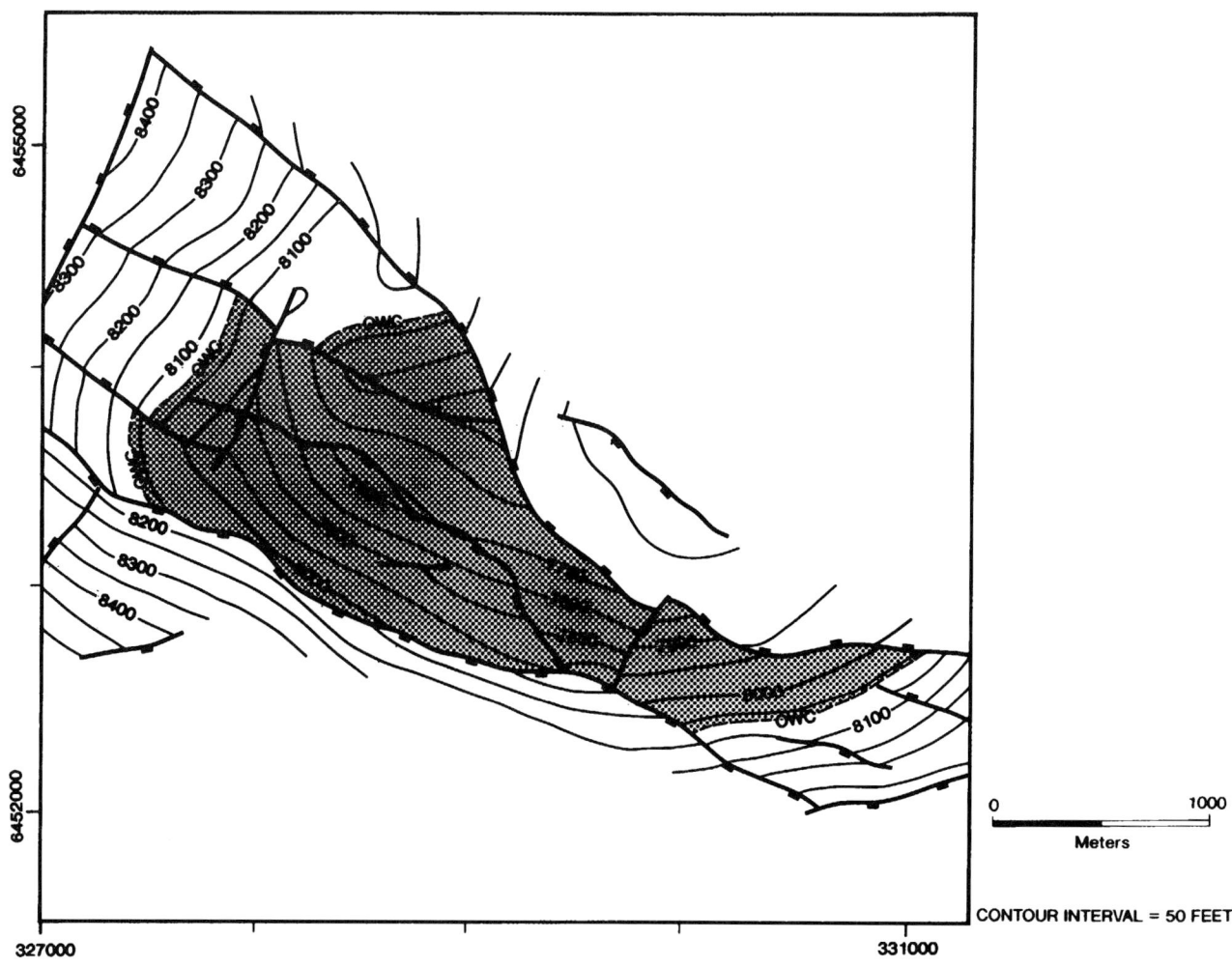

Figure 6. Top Supra Piper depth map for the Ivanhoe field verticalized to the upthrown fault traces.

The field contains two reservoir sandstone units, the Supra Piper and Main Piper sandstones, which are surrounded by shale units (Figure 5). The Main Piper Sandstone unit is of late Oxfordian age. A significant unconformity separates the Piper Formation from the overlying Kimmeridge Clay Formation. Major extensional tectonic activity occurred in the early Volgian, giving rise to the tilted fault-block and basin terrain which developed in the Outer Moray Firth, and are typified by the Ivanhoe field structure.

Previous computer mapping of the Ivanhoe field had been carried out in preparation for field development. This earlier mapping project, taken as the starting point for mapping of the Ivanhoe field using FFMS, provides a useful comparison between the FFMS model and the more traditional approach based upon a verticalized fault model. Individual structure maps for the entire reservoir, verticalized to the upthrown trace, were available for input to FFMS; these horizons are:

- Top Supra Piper Sandstone
- Top Mid Shale
- Top Main Piper Sandstone
- Top Basal Shale.

Only limited well control was available during the original mapping, and this was used to tie the maps of the horizon structures and to create unit isochore maps. No wells contained fault cuts, and so were not used during the fault manipulation and modeling procedures of FFMS. The Top Supra Piper Sandstone structure map and associated vertical fault traces, along with the individual unfaulted unit isochore maps were used to start modeling with the FFMS software.

Figure 6 shows the original structure map for the Top Supra Piper Sandstone verticalized to the upthrown trace: some 17 fault traces are represented, many showing intersections with other faults. The field is fault bounded to the northeast by a major NW–SE-trending fault with a maximum throw of approximately 450 ft (140 m) as mapped, downthrown to the northeast.

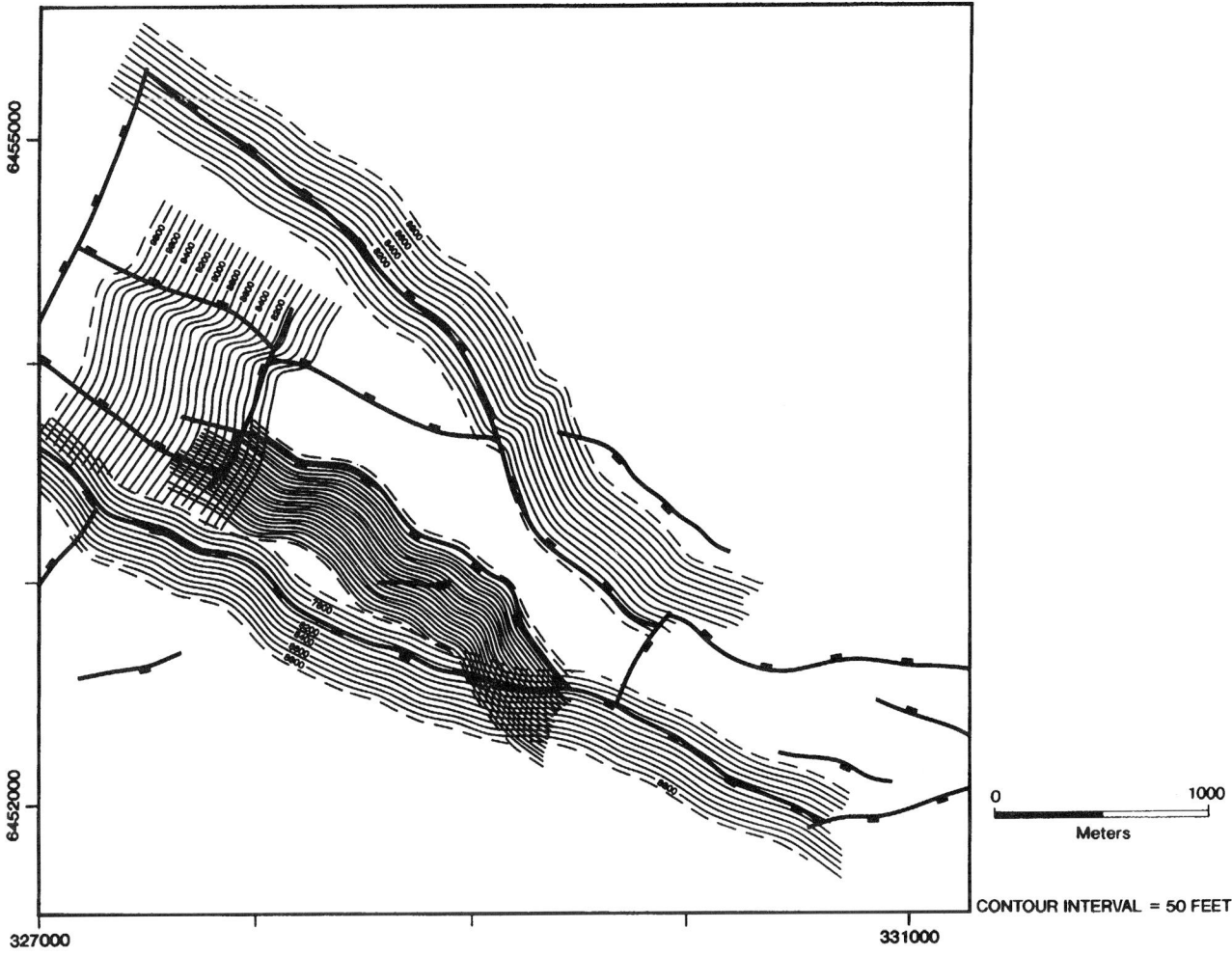

Figure 7. Contoured fault planes of selected faults, demonstrating projection of fault planes and overlapping faults.

Dips were estimated from seismic mapping and assumed to be constant for each fault plane. The faults are all fairly high angle, ranging between approximately 70° and vertical. Elevations for the fault traces were calculated by back interpolation from the Top Supra Piper Sandstone map. This was the only control given to the FFMS process for this exercise.

Fault attributes were loaded and the individual fault planes projected (gridded). Figure 7 shows the result of fault plane projection for selected faults. Projection can be both up- and down-dip and can be varied either horizontally or vertically for each fault. This figure shows several fault intersections, demonstrating the ability to define multiple values at any given point.

The fault planes were subsequently intersected with the verticalized Top Supra Piper Sandstone map, producing a nonvertically faulted Top Supra Piper Sandstone map (Figure 8). Minor amounts of editing of the projected fault traces were required during the construction of this map, predominantly associated with some of the intersecting high angle faults. During the editing process, the original unclipped calculated fault traces could be displayed and modified interactively, either digitizing on screen or via a digitizing tablet.

The next step was the addition of the unfaulted Supra Piper Sandstone isochore to generate the Top Mid Shale. The isochore had originally been mapped from well data, and shows a significant thickening from 10 ft (3 m) in the northwest to 160 ft (50 m) in the southeast. The isochore was added to the Top Supra Piper Sandstone map, taking into account the fault plane information, producing a Top Mid Shale structure map with correctly migrated fault polygons and fault plane segments. The resultant isochore calculated from the two faulted structure maps (Figure 9) exhibits thinning within the fault polygons caused by displacement of the Supra Piper Sandstone by these faults. This is further demonstrated by the cross sections in Figure 10: the first cross section is of the Top Supra Piper Sandstone and Top Mid Shale along A–A' indicated in Figure 9, while the second cross section is through the faulted isochore along the same base line.

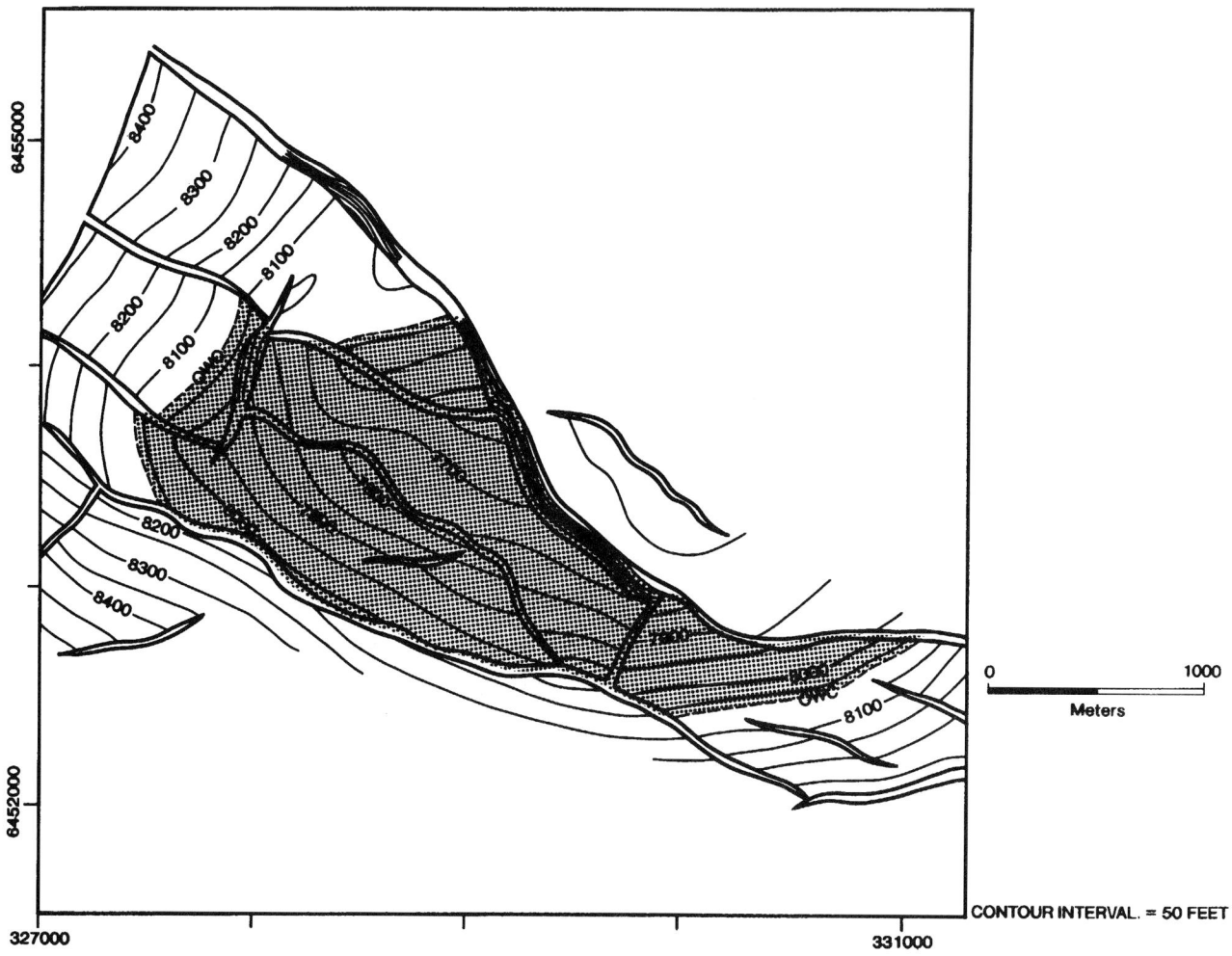

Figure 8. Nonvertically faulted Top Supra Piper depth map showing contours within fault polygons.

Stacking (adding) the Mid Shale isochore in the same way produces the Top Main Piper Sandstone. This process can be repeated for as many reservoir units as required. Figure 11 shows another cross section, this one through the Top Supra Piper Sandstone, Top Mid Shale, and Top Main Piper Sandstone. This section demonstrates the overall integrity of the model produced by FFMS. The fault planes have been projected from the Top Supra Piper Sandstone and intersected with deeper reservoir horizons. These horizons in turn have been corrected by the FFMS process and the fault traces migrated to honor the projected fault plane, producing a rigorous and stable model. Varying the fault attributes and creating a new fault model can be used to run sensitivities and test the impact of varying the model through areas of uncertainty.

As discussed earlier, one of the main reasons for detailed fault modeling is to obtain more reliable and accurate volumetrics. Gross pay volumes have been calculated for the Ivanhoe field using the FFMS model. These are compared with two alternate models: (1) the original vertical fault model (verticalized throughout the reservoir), and (2) a vertical fault model with fault traces migrated for each horizon (also output by FFMS). The results of these volumetrics are tabulated below (Table 1). Contrary to what was expected based on the earlier discussions of modeling geometric and volumetric errors (Figure 1), volumes for the Ivanhoe field have increased for both reservoir zones. These increases are reasonable and can be explained. Whereas the theoretical case carried volumes in both the upthrown and downthrown blocks, the Ivanhoe field is predominantly fault-bounded, consisting of tilted fault-blocks. The main fault to the north seals the reservoir. The downthrown block is below the contact and is non-reservoir. The net result of fault modeling is to re-allocate additional reserves into the upthrown block.

The FFMS volumes show an overall 9% increase compared with the original verticalized model, and a 5% increase compared with the migrated verticalized model.

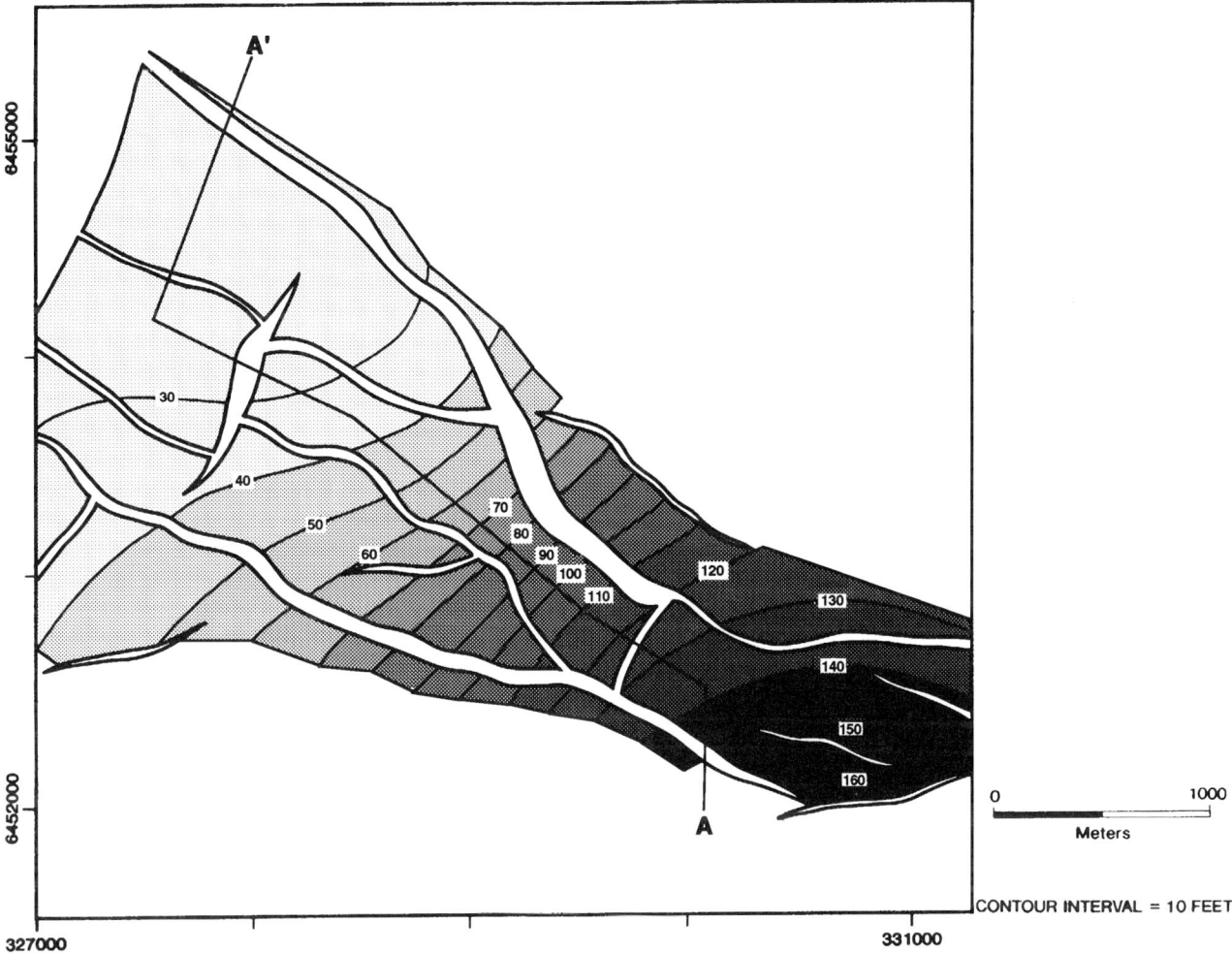

Figure 9. FFMS Output: Fault corrected isochore map for Supra Piper Sandstone showing thinning in fault polygons.

Very little variation is seen in the gross pay volumes in the Supra Piper Sandstone. This is probably due to the combination of a number of factors: high angle faults; relatively thin reservoir unit; and most of the larger faults are above the oil-water contact, therefore fault modeling has merely re-apportioning volumes between up- and down-thrown blocks. Any increases are probably a result of this shift of volume into the up-thrown blocks under the fault wedge.

More significant variations are seen in the deeper Main Piper Sandstone. The volumes obtained using FFMS show a 16% increase compared with the original vertical fault model, and a 7% increase compared with the migrated vertical fault model. This unit is considerably thicker, and the base of the Sandstone does not come above the oil-water contact. The significant increases are accounted for by migration of fault traces downdip along the fault planes, encompassing a larger area than the other two models, and resulting in an increase in size of the upthrown fault blocks above the oil-water contact.

CONCLUSIONS

Computer modeling of multiple surfaces using non-vertical faulting producers produces more reliable structural models. This has been demonstrated using the Ivanhoe field. The result is a more accurate model both in terms of geometry and volumetrics. Existing fault modeling methods have been briefly discussed, and for the most part shown to be either impractical in terms of detailed modeling or to result in geometry and volumetric errors.

FFMS provides a powerful and flexible tool for the automated modeling of multiple surfaces with faults. It can be used for both structural and volumetric analysis of complex, multi-layer reservoirs, and to run volumetric sensitivities based upon variations in the fault model. Moreover, the results are precise, consistent, repeatable, and allow clear auditing of the process.

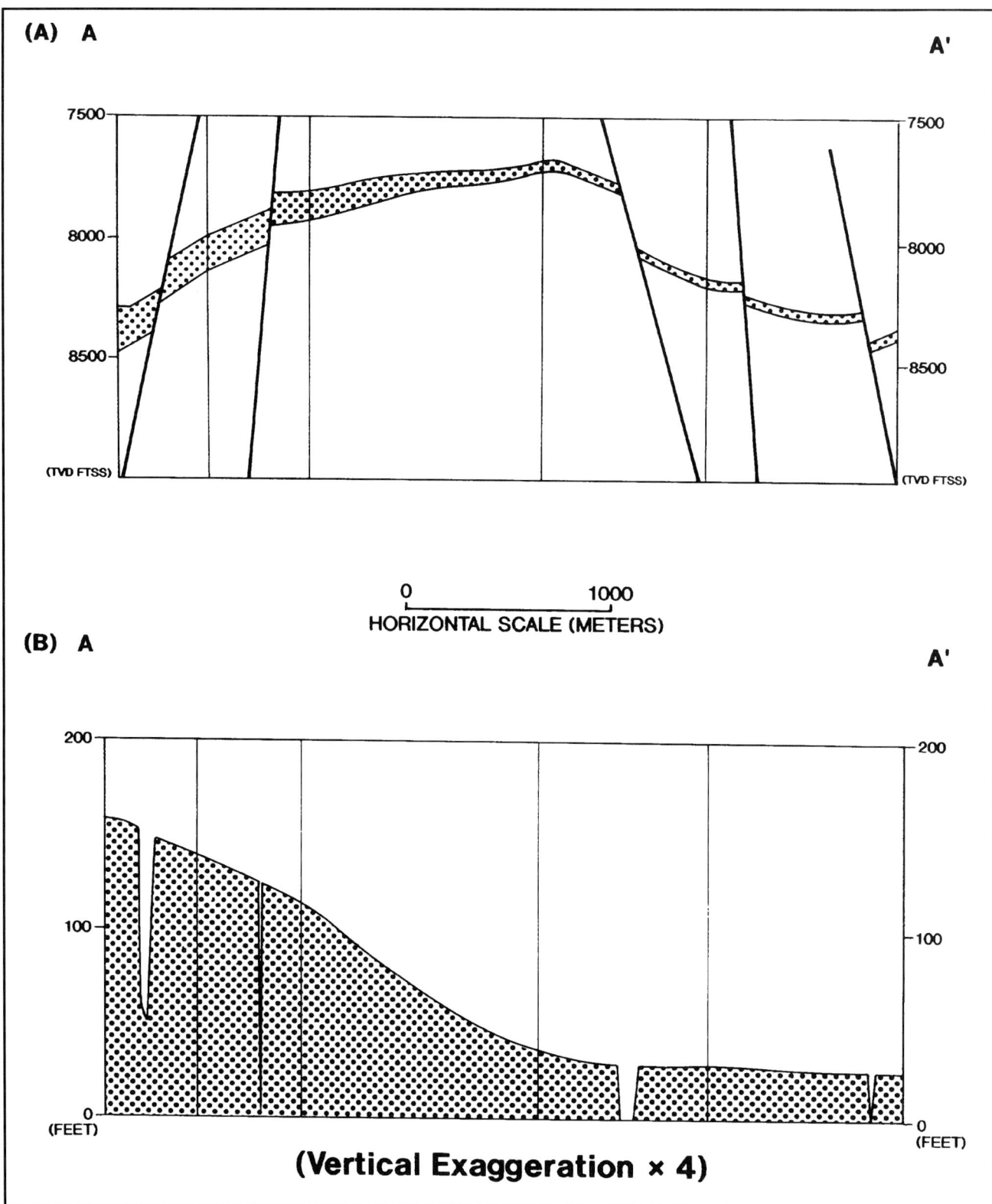

Figure 10. Cross section A-A' showing: (A) Top Supra Piper Sandstone and Top Mid Shale, (B) fault corrected Supra Piper isochore. See Figure 9 for location of cross section.

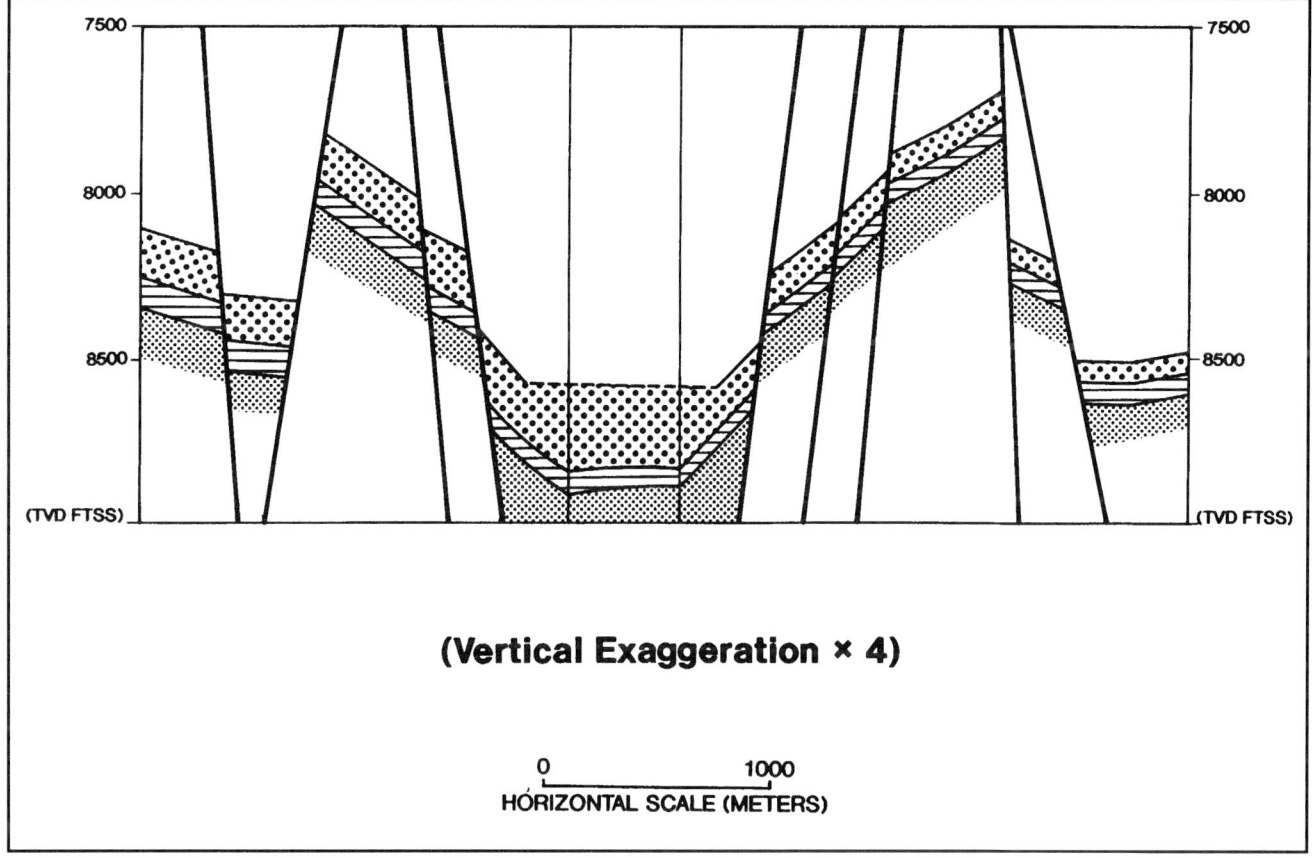

Figure 11. Cross section through Top Supra Piper, Top Mid Shale, and Top Main Piper, demonstrating structural consistency of all surfaces and relationship with fault planes.

Table 1. Ivanhoe field: Gross pay volumetrics. Volumes are in acre-feet. Oil-water contact: 8052 ft (2454 m).

	FFMS Nonvertical Model	Original Vertical Model	Migrated Vertical Model
Supra Piper Sandstone	38,600	38,500	38,500
Main Piper Sandstone	62,200	53,800	57,500
Total	100,800	92,300	96,000

ACKNOWLEDGMENT

The authors wish to thank Amerada Hess Limited, Deminex U.K. Oil & Gas Limited, Kerr-McGee Oil (U.K.) plc, and Pict Petroleum for their permission to present data from the Ivanhoe field in this paper. CPS-3 is a registered trademark of the Radian Corporation.

REFERENCES CITED

Banks, R., 1990, Modeling geological and geophysical surfaces: Geobyte, v. 5, no. 5, p. 20–23.

Barnett, J. A. M., J. Mortimer, J. H. Rippon, J. J. Walsh, and J. Watterson, 1987, Displacement geometry in the volume containing a single normal fault: AAPG Bulletin, v. 71, p. 925–937.

Clarke, D., 1992, The gridded fault surface, (this volume).

Hooper, N. J., 1986, Computer mapping and reservoir modeling, in Proceedings of Annual European CPS Users' Conference: Austin, Radian Corporation.

Hooper, N. J., 1990a, Layer-cake modeling, in CPS-3 training manual: Austin, Radian Corporation, p. 82–88.

Hooper, N. J., 1990b, Advanced surface modeling techniques, in CPS Update: Austin, Radian Corporation, p. 4–7.

Jones, A. J., D. E. Hamilton, and C. R. Johnson, 1986, Contouring geological surfaces with the computer: New York, Van Nostrand Reinhold, 314 p.

Kemp, A. C., and K. C. Grinstad, 1984, Computer modeling of a faulted reservoir, *in* Proceedings of Annual European CPS Users' Conference: Austin, Radian Corporation.

Newton, I., 1983, Field study structure maps with nonvertical faults, *in* Proceedings of Annual European CPS Users' Conference: Austin, Radian Corporation.

Parker, R. H., 1991, The Ivanhoe and Rob Roy fields, Block 15/21a-b, U.K. North Sea, *in* U.K. Oil and Gas Fields 25 Years Commemorative Volume, The Geological Society Memoir No. 14: London, The Geological Society of London, p. 331–338.

Tearpock, D. J., and R. E. Bischke, 1991, Fault maps, *in* Applied subsurface geological mapping: Englewood Cliffs, Prentice-Hall, 648 p.

Zoraster, S., 1992, Fault representation in automated modeling of geologic structures and geologic units, (this volume).

Zoraster, S., and K. Ebisch, 1990, Incorporating fault geometry into geologic horizon models: Geobyte, v. 5, no. 2, p. 30–36.

Chapter 13

Extensions to Three Dimensions: Introduction to the Section On 3-D Geologic Block Modeling

Thomas A. Jones
Exxon Production Research Company
Houston, Texas, U.S.A.

ABSTRACT

Geological concepts invariably involve spatial relationships, so most exploration and reservoir analysis should be three dimensional. Significant benefits can be gained by working directly in three dimensions, rather than using a series of two-dimensional analyses to gain three-dimensional insights. These benefits include use of appropriate averaging, better analysis of reservoir continuity, ability to interface data with other programs, and increased geologist's productivity.

Because of spatial complexity and the need to process large amounts of geologic data, computer programs are used to map and model rock properties. The papers in this section deal with what we shall call 3-D geologic block models. In these models, we divide the geological volume to be studied into blocks, and use observed data to assign geological properties to the blocks to make a complete representation of the existing geology. The set of blocks with assigned geologic attributes thus constitute 3-D geologic block models.

Geological interpretation is important with 2-D computer mapping, and interpretation similarly should be used during the modeling process. The assignment of values to the model blocks should be made in a way that incorporates stratigraphic and structural information. This paper summarizes benefits of 3-D geologic block modeling over 2-D analyses, and then describes some current approaches to this type of modeling.

INTRODUCTION

Geologists and geophysicists work in the world of three dimensions. Virtually all interpretation requires that relationships be considered in 3-D space. Of course, it is difficult to visualize three dimensions graphically, so earth scientists use two-dimensional tools—geologic maps, contour maps, and cross sections. These displays represent projections of information from the difficult three-space to a more manageable two-space.

Maps and cross sections were used by earth scientists long before the computer existed. The computer allows these 2-D displays to be generated more rapidly and from larger datasets than can be managed manually, and computer mapping has become a valuable tool in the solution of geologic problems. In addition to the papers in this book, several authors have discussed the benefits of computer mapping, but more importantly, they have outlined methods to incorporate geological interpretation so as to generate more meaningful maps and other displays.

Use of the computer has not substantively changed the way these displays are created, interpreted, and used. It is unfortunate, therefore, that some geologic problems cannot be solved adequately through mapping, whether manually or with the computer. Because mapping is inherently a two-dimensional tool, certain three-dimensional problems can be solved only in an approximate manner. Other problems are so strongly tied to three-dimensional processing that mapping cannot approach them at all.

This paper summarizes some difficulties that can arise when dealing with 3-D problems in two dimensions. Then, as an introduction to the following papers in this book, it presents some basic concepts that are involved in moving toward a modeling solution. The papers here are not primarily concerned with the valuable interactive display capabilities now available with modern computers (e.g., Lee et al., 1990), but with methods to build models to make most effective and accurate use of data, known geology, and interpretation.

DIFFICULTIES WITH MAPPING

Mapped geologic variables generally can be divided into two groups: geometric or attribute. Geometric mapping involves working with structural or stratigraphic relationships, and most problems of this type can be handled by 2-D mapping, although perhaps with difficulty. The bulk of the material in Jones et al. (1986) applies to this type of problem. On the other hand, mapping rock attributes (for instance, ore grade, porosity, permeability, or percent sand) is likely to lead to difficulties if approached from a two-dimensional viewpoint. Attribute analysis (and ultimately mapping) is best done in three dimensions.

This section summarizes some shortcomings of using two-dimensional analysis to solve problems that are inherently three-dimensional.

INAPPROPRIATE AVERAGING

Mapping often uses data that have undergone significant averaging. Heterogeneities are smoothed if we map an average of some property over a reservoir interval. If the property within the reservoir is homogeneous, the map may be reasonably accurate in depicting the distribution of the attribute. However, the map will be less useful if the attribute varies, particularly if we are dealing with distinct properties such as pay and non-pay. This averaging can result in a map that does not represent the interval at all.

For instance, assume a constant-thickness interval that contains homogeneous, high-permeability rocks, and also includes a thin, low-permeability shale layer equidistant between the top and base of the interval. A contour map of average permeability will show a relatively simple distribution of medium permeability. If we are interested in lateral flow in the reservoir, however, the high-permeability portions will allow more rapid movement than mapped horizontal permeability would indicate. On the other hand, the shale barrier would prevent vertical fluid movement, again not indicated by a map of average vertical permeability. Looking at the interval only in two dimensions thus gives a misleading picture.

Denver and Phillips (1992) show examples of how generating grids over excessively thick intervals can influence volumetric calculations. Further, they show that volumetrics, and implicitly a group of grids and contour maps, may be inconsistent because dependencies with other variables are not taken into account. For instance, if we calculate porosity thickness and net thickness at each well and draw two maps independently, differences in extrapolations and projections can vary between sand thickness and average porosity information to create contradictions.

These problems can be reduced through calculation and analysis in three dimensions in which geologic properties are not averaged prematurely. Several variables or geologic entities are handled individually, allowing each to reflect spatially the specific property of interest. The individual variables and entities can then be combined at the last stage of processing (for example, combine facies and porosity to create a map of pore thickness), rather than being combined before spatial processing begins. Further, this combination can be made at a fine (e.g., foot by foot), rather than coarse (zone by zone), resolution.

Reservoir Continuity

A severe problem with simple mapping is that two-dimensional displays, even many of them, are not adequate to answer some questions. For instance, an important part of production work involves determination of connectivity or continuity of porous bodies. We may need to know if a given well is in communication with another well.

Two tools used to analyze continuity are cross sections and contour maps. To determine if two wells are in communication, a common method is to draw a cross section between the wells. However, a single cross section may not answer the question; the wells may be in contact via sands that lie off the line of section. Figure 1 shows a cross section that seems to indicate that two wells are not in communication. However, the block diagram shows contact off the line of section. The single cross section would cause us to

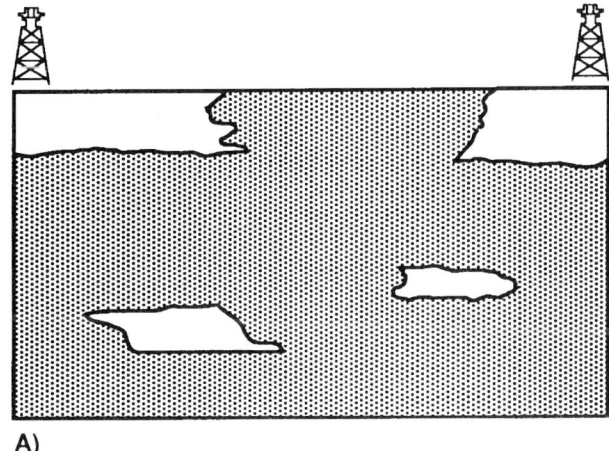

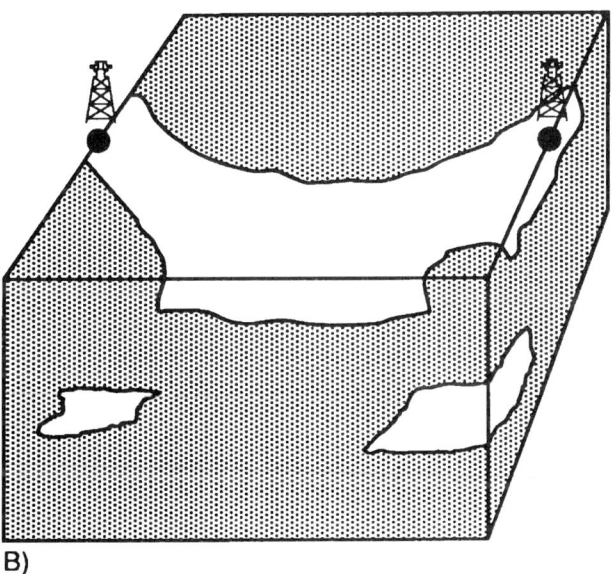

Figure 1. A simple cross section seems to indicate that the two wells are not in fluid communication, but the block diagram shows that they are connected.

misinterpret the reservoir's performance.

The second commonly used method for determination of continuity is a contour map over an interval of interest, say the reservoir sand. If the map shows non-zero thickness at the two wells, and if a path can be followed with non-zero thickness from one well to the other, it is generally assumed that the wells are connected hydraulically. However, it is possible that several isolated sands occur in the interval; if they overlap, projection onto a 2-D map may falsely imply that they are connected.

Cross sections and contour maps are inadequate for solving most problems of continuity. Construction of many maps and sections may help determine whether the two wells are in communication, but this often requires more time than is available, and we still may not know if all pathways were evaluated. With the properties of all points in a volume stored in the computer, and with use of appropriate software, however, all possible spatial relationships are available. The computer can find pathways that lead from one well to another, even though these pathways may be circuitous. Hence, three-dimensional analysis leads to a more accurate reservoir description.

Interfaces to Other Programs

A number of programs exist in the mining and petroleum industry that require extensive geologic information for operation. For instance, a mine planning system must have detailed information on the spatial distribution of ore and waste. Geologists and engineers must interpret lithology and ore grade throughout the area of the mine, and then input this information to the program. The detail required to create a mine plan is so great that simple mapping is not adequate to feed the planning system, so an extra interpretive (perhaps manual) step is required.

Similarly, petroleum reservoir simulators require extensive information. Reservoir engineers have developed complex programs to simulate the movement of fluids and changes in physical properties (e.g., pressure) in the reservoir over time in response to petroleum production and fluid injection. A modern 3-D simulation model might contain more than 20,000 blocks, each requiring porosity and block-interface permeabilities. When simulators and computers were limited to small models, these data were obtained from hand-drawn maps and cross sections.

Because mine planning and reservoir simulation systems are so complex and allow such detailed analyses, substantial information is required, more than can be generated manually. If we turn to the computer for mapping or construction of cross sections, more detail is available but we are still limited. Only through a full 3-D analysis can we handle the vast amount of information required by advanced programs and be assured that the data are processed in a spatially consistent manner (e.g., Johnson and Jones, 1988).

Geologist's Productivity

Another advantage of 3-D processing is that it increases the productivity of the geologist. When working only with maps and cross sections, the geologist must translate continually between two and three dimensions instead of concentrating on geological concepts. If the computer has processed the data consistently in three dimensions, the geologist need not be so distracted. All of his time can be spent on geological interpretation.

In addition to removing distractions, the geologist's interpretative process is enhanced when working in three dimensions. Spatial relationships that are processed in three dimensions are more accurate and easier to interpret than if they are pieced together from a series of 2-D representations. The geologist can more readily detect and recognize implications of such relationships, thereby speeding the work. Further, the ability to see the spatial effects of alternative interpretations and hypotheses increases interest in following up new ideas.

3-D MODELING

The problems with mapping discussed above can be severe, and errors can result if these problems are ignored. Accordingly, a number of programs have been developed to allow earth scientists to extend their work from 2-D mapping to 3-D modeling. The following chapters in this book describe the use of computing to solve specific geologic problems. As an introduction to the remainder of the book, this chapter summarizes 3-D geologic block modeling (Jones, 1988a,b).

This approach to modeling may be thought of as creating a static or descriptive computer-based model of some portion of the earth. The intent here is to use all available data from various sources, plus geologic knowledge and interpretation, to replicate the area's geology. The intent is also to gain the ability to predict away from data locations. Other approaches to modeling with similar goals are discussed, for example, by Fried and Leonard (1990), Fisher and Wales (1990), Lasseter (1990), and papers in the book edited by Raper (1989).

Simulation modeling has been done for a number of years, with applications in such diverse fields as geophysics, tectonics, geochemistry, hydrology, and geomorphology (c.f., Harbaugh and Bonham-Carter, 1970). Many of these simulation models are dynamic, showing the development of geologic characteristics as they change through time. These models normally do not have the goal of honoring and interpolating large amounts of data, but more often are concerned with learning how geologic processes operate.

Most descriptive or static 3-D models built for mining or petroleum applications are in one of two forms: a three-dimensional array of blocks or points, or a series of layers. If in the form of a 3-D array of blocks or points, the entire set of array locations that have been assigned geologic attributes constitutes the model. Dimensions of blocks or spacing of points should be chosen so that the rock attributes are relatively homogeneous within a block or between points, without creating an excessive number of either. We have used blocks in a regional study that were 2 mi (3.2 km) on a side and 500 ft (152 m) thick, and have also modeled mineral deposits with blocks 20 m wide and 0.1 m thick.

There is a difference between a three-dimensional array of blocks and an array of points. During model construction of either, the interpolation algorithm assigns a value to a single point in space. However, the block approach assumes that the value of the block's attribute is homogeneous over the entire block. In the point approach, the 3-D array is treated in a manner analogous to a 2-D grid and the attribute's value is assumed to be smoothly varying from point to point in the array. Once constructed the 3-D array of points requires different methods for display generation, volume calculation, and vertical averaging than those used typically for a 3-D array of blocks. Paradis and Belcher (1990) discuss an example of this type of system.

We may think of a geologic volume as being made up of a series of layers, and then use the concept that each layer of rocks may be modeled by a 2-D array with attributes assigned to each node. Even if the model's data structure is not in terms of layers, the concept can aid understanding. This layer-based approach leads to a model with variable-thickness, wedge-shaped blocks. Denver and Phillips (1990; 1992) and Badiozamani et al. (1992) describe such systems. Here the geologic interval is divided into a number of strata that correspond to layers of the variable-thickness blocks.

There are advantages to each type of data structure, whether a 3-D array or layer-based, and either can be considered to make up a block model. Regardless of which type of system may be available, the comments that follow generally apply.

Required Input

Three types of information are required for 3-D geologic block modeling. First, we need data describing rock properties down the well. The continuous data string in the hole is divided into a series of intervals, and an attribute is assigned to each interval. Typical variables include lithology, ore grade, porosity, and permeability, but digitized values from wireline logs are also common. Intervals typically are 1 ft thick.

The second type of input information for modeling is a means to control correlations between well and model. This information might be as simple as rules that correlations follow mine benches, but more accurate models result if stratigraphic-structural correlations are used. These correlations can be defined by a set of surfaces in the form of computer-grids (c.f., Jones et al., 1986) that make up a stratigraphic framework; Figure 2 shows a cross section through a typical framework. Rocks associated with horizons A, B, and C were deposited conformably, and a later period of erosion after folding created unconformity D. Following this, horizons E and F lapped onto the unconformity, with a final period of erosion. The set of computer grids should incorporate all significant stratigraphic and structural features because these relationships can have a substantial impact on correlations within the final model.

This leads to the third type of input information. Just as in mapping with the computer, geologic knowledge or interpretation provides important control for modeling. Interpretation includes definition of stratigraphic frameworks and geologic history, plus

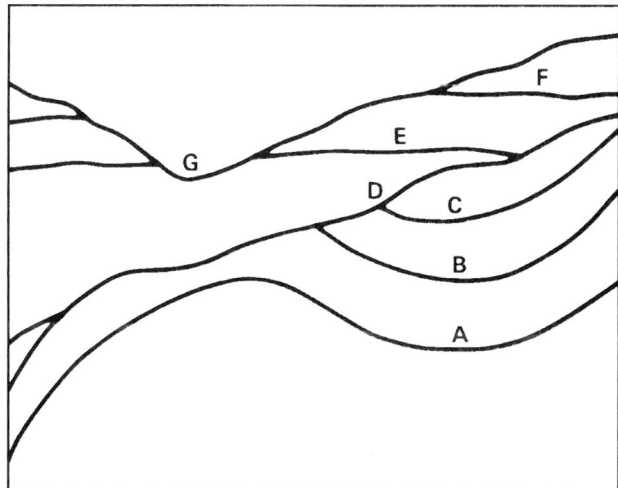

Figure 2. Diagrammatic cross section showing typical stratigraphic framework containing truncation, baselap, and conformity. Stratigraphic horizons indicated by letters. (After Jones et al., 1986)

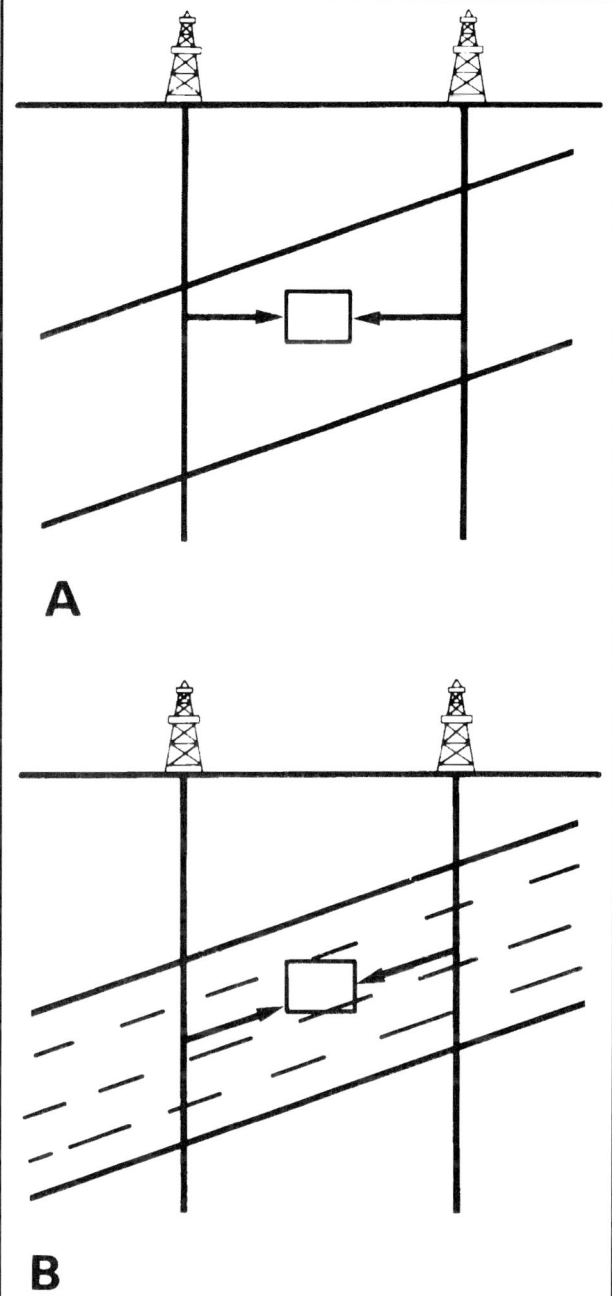

Figure 3. Correlation of data from borehole to block. (A) Horizontal, without regard to geology. (B) Following stratigraphy depicted by chronostrata (dashed lines). (After Jones, 1988a)

other considerations discussed below.

In order to make a good model for an economic application, it is important that the geology of the area is well understood, including structural and depositional history. The model will be a poor description if fundamental geological elements are violated. This is not to say that one cannot develop alternative interpretations from such models, but best results are obtained when all information is used in the modeling process. This differs from simulation modeling where a significant goal is to learn about geologic processes and to compare results with alternative hypotheses.

Modeling Process

The procedure in modeling is to define the set of blocks or layers, and then assign a geologic property to each. The assignment of a value to a given block or layer-node consists of three steps:
(1) Determine which boreholes are influential for that block. These typically are nearest to the block. To remove effects of clustering, the holes are usually selected in such a way that several sectors around the block contain data points, although other selection criteria may be used.
(2) Determine which portions of the selected holes are to be used for calculating the value. Many mining programs project information horizontally from the data to the block as if along a mine bench (Figure 3A). The horizons dipping from right to left are ignored, and it is possible that incorrect information may be put into the block because the portions of the wells providing the data do not correlate stratigraphically with the block.
Structural dip is taken into account by projecting the well information to the block in question through use of computer grids that depict stratigraphic correlations. In many cases the gridded surfaces represent geologic time lines, so we define a chronostratum (indicated by dashed lines in Figure 3B) to be a subinterval that correlates a block to nearby holes and conforms to the geologic interval being modeled. As we move laterally over the entire model, the computer grids indicate the posi-

tion of this chronostratum so correct data intervals from the various wells are correlated to the block. Use of chronostrata ensures that information assigned to the block is more likely to reflect the geology, preventing such errors as sandstone porosity being placed into a shale zone.

(3) Assign an attribute to the block by using those data values that are near enough to the block to have influence and that are within the chronostratum passing through the block. For discrete or qualitative data (e.g., lithology), this can be done by simply finding the nearest data point to the block and using that value, that is, nearest-neighbor or polygonal modeling. For a quantitative variable, such as porosity, the values may be combined arithmetically through simple averaging, kriging, spline fitting, or other methods. Information about continuity or heterogeneity, such as gained from variogram analysis, should be used when selecting calculation methods.

For each block or layer in the model, follow this three-step process of selecting wells, finding those data values to be used for the block or point, and calculating a value. After the model is built, it may be used to provide geologic maps, contour maps, cross sections, or calculations.

Other Considerations

Even given a program to perform these modeling steps, the task may not be as simple as it sounds. Geologic interpretation is an important aspect of computer mapping and 3-D geologic block modeling, and complete analysis involves special considerations that must be taken into account.

Stratigraphic relationships between horizons are important considerations. Even though grids that make up the stratigraphic framework define the exact forms of the correlation surfaces, geologic interpretation is also required for specifying how the model is to be built. Figure 4 shows three versions of the same information. In Figure 4A, the two horizons are conformable to each other. In this case we assume that sands were also deposited conformably, thinning as the interval between horizons thins. This leads to correlating the sands between wells with both horizons as control.

Figure 4B assumes the upper horizon laps onto the older surface, so the sands presumably do also and correlations should parallel the upper horizon. In Figure 4C, the upper surface truncates sands which were deposited parallel to the base; therefore, the sands should be correlated parallel to the basal horizon. The same considerations also apply to non-sedimentary deposits, as found in mining applications, because the same geometry can generally be found and modeled. A related concept is the use of phantom horizons for correlations that are not controlled by unit boundaries.

A second type of interpretation deals with trends. Just as trends may be incorporated into gridded surfaces, so also should trends be modeled in attributes. For instance, we may feel that porosity in rocks

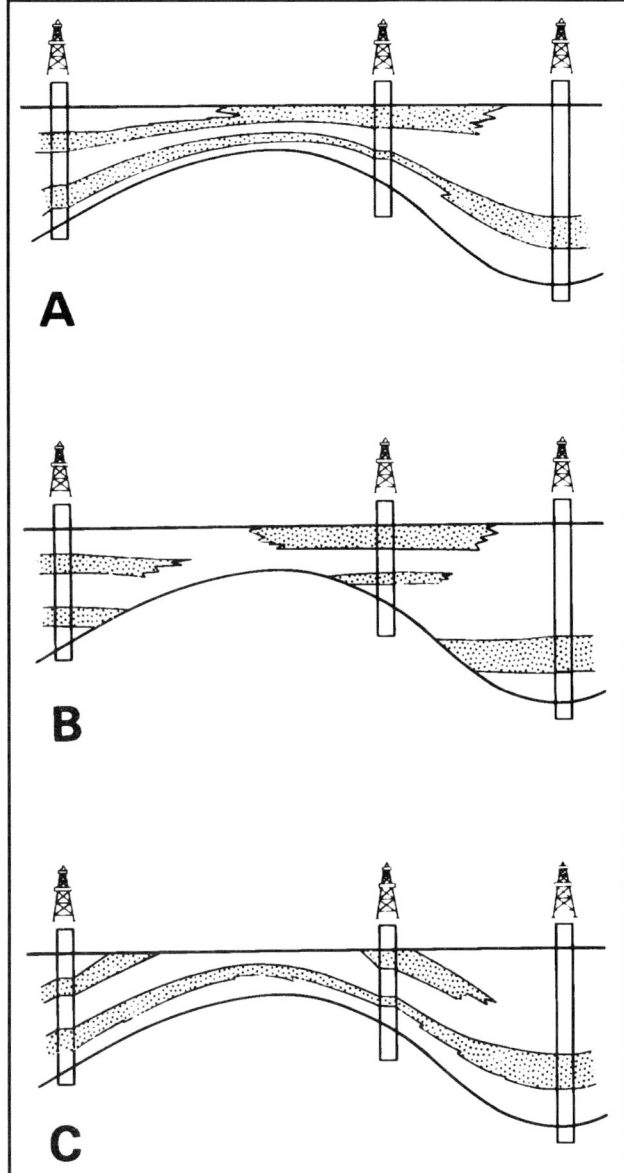

Figure 4. Three types of stratigraphic correlation, and their effects on the model. (A) Conformable. (B) Baselap. (C) Truncation. (After Jones, 1988a)

deposited from a river system can be predicted more accurately downstream than across stream. The interpreted trend can be important to the model, and it can be introduced during calculation of the block values (c.f., Jones, 1988a, figure 6).

Other types of interpretation that potentially can have a significant impact on a model involve geologic boundaries and heterogeneity. Porosity across a boundary between wells containing sand and shale may be gradational or may change abruptly (e.g., sediment filling at a stream channel). Further, even for gradational boundaries, reservoirs are usually more heterogeneous than simple modeling would imply. If we need only broad average values, simple modeling may be adequate. However, permeabilities to be sent to reservoir simulators must accurately represent the actual spatial distribution in the reservoir (Johnson and Jones, 1988).

A related aspect is modeling a variable (say, porosity) which has substantially different values or properties in different lithologies (e.g., shale versus sandstone) or facies. Simple modeling might mix populations of the attribute. Badiozamani et al. (1992) discuss making calculations separately for each facies and then combining the results into a final model.

Of course, faults and complex structures are important interpretive aspects of a project. This area is still unresolved for mapping, let alone for 3-D problems, although special approaches for selected projects have been effective.

AN ALTERNATIVE APPROACH TO MODELING

Modeling programs are not yet widely available to earth scientists. However, some approaches to 3-D geologic block modeling can actually be viewed as advanced mapping applications. In a layer-based system, we can regard each layer as being equivalent to a map of an attribute. This section outlines how a general-purpose mapping system may be used to emulate the 3-D block-modeling process.

The procedure for simulating 3-D geologic block modeling consists of two steps. The first step is to define a large number of thin layers or subzones in the earth over some specified region; the entire set of layers would encompass the stratigraphic interval to be modeled. Chronostrata are used in the modeling context, and also are used here. Now they represent both correlative intervals and storage media for the attributes, replacing blocks. Because a given chronostratum represents an interval in which attributes are correlative, grids of its top and base should be constructed with stratigraphic relationships taken into account.

The second step is to assign attributes to the chronostrata. All data that fall into a single chronostratum are extracted from the wells, and treated as individual data points which are gridded using standard algorithms. This grid represents the spatial distribution of the attribute within this chronostratum. The same process is followed for each chronostratum, leading to an attribute grid for each. It is not necessary to be concerned with extrapolation into areas where the chronostratum pinches out; comparison of the chronostratum thickness with the attribute grid allows us to blank attribute nodes corresponding to zero or negative thickness. All information in the model thus resides in the geometric grids that define the chronostrata and the attribute grids associated with the chronostrata. If another geologic attribute is to be modeled, a second set of attribute grids is made.

After the grids are generated, they can be used to draw attribute maps along chronostrata, to create contour maps of attributes averaged or cumulated over several chronostrata, or to compute volumetrics. This alternative system, however, will not allow doing further, complex 3-D processing or creating such displays as cross sections. These tasks would require additional coding or special capabilities.

APPLICATIONS

Although 3-D computer modeling of the type described here is relatively new, a few papers have described applications. Exxon's cellular modeling system has been used for several types of applications. Delaney and Tsang (1982) studied continuity of high- and low-permeability zones in a carbonate reservoir. Lindsay et al. (1984) studied the submarine development of the Mississippi delta over the past century. This application was unusual in that the model's vertical dimension was time rather than depth, allowing temporal changes in channel depth (erosion and deposition) to be tracked. Hamilton et al. (1985) discussed petroleum and minerals applications. Johnson and Jones (1988) discussed the tie between geologic models and reservoir simulators.

Other systems have also been described (c.f., Fried and Leonard, 1990). Denver and Phillips (1990) discuss SGM, a system that is based on mappable layers (chronostrata). Paradis and Belcher (1990) discuss IVM, a system that uses a three-dimensional grid system analogous to an array of blocks.

Six papers in this volume deal with 3-D geologic block model construction and application. Lindsay (1992) discusses the use of block models for understanding the surficial geology in the Canning River delta of Alaska. Hamilton and Didur (1992) present a study of the Kutcho Creek sulphide mineral deposit in Canada, and show that geologic and economic information can be combined during the modeling process. Denver and Phillips (1992) point out the advantages of using 3-D geologic block modeling and discuss a layer-based modeling approach to calculating volumes in the Hugoton gas field and Delaware Basin. Badiozamani et al. (1992) extend layered modeling to allow more complexity in the layer definitions. They stress the use of geologic information in their study of Canadian oil sands. Barrett and Bailey (1992) demonstrate the usefulness of 3-D models for characterizing reservoir heterogeneities, understanding their affects on a steam flood, and applying that understanding to increase production in the South Belridge field. Mayoraz et al. (1992) identify complex structures (reverse faults and recumbent folds) not easily modeled using currently available 3-D modeling software and demonstrate the effectiveness of newly developed software for modeling structures of this type in the Swiss Alps.

REFERENCES CITED

Badiozamani, K., F. Roghani, and G. Hawes, 1992, Computer modeling of geologic surfaces and volumes, (this volume).

Barrett, R. A., and J. Bailey, 1992, Three-dimensional modeling techniques in the analysis of a mature steam drive, (this volume).

Delaney, R. P., and P. B. Tsang, 1982, Computer reservoir continuity study at Judy Creek: Journal of Canadian Petroleum Technology, v. 21, no. 1, p. 38–44.

Denver, L. E., and D. C. Phillips, 1990, Stratigraphic geocellular modeling: Geobyte, v. 5, n. 1, p. 45–47.

Denver, L. E., and D. C. Phillips, 1992, The impact of vertical averaging on hydrocarbon volumetric calculations: A case study, (this volume).

Fisher, T. R., and R. Q. Wales, 1990, A case study of Noonan Ranch Field, Denver Basin, Colorado: Geobyte, v. 5, no. 1, p. 39–41.

Fried, C. C., and J. E. Leonard, 1990, Petroleum 3-D models come in many flavors: Geobyte, v. 5, no. 1, p. 27–30.

Hamilton, D. E., J. L. Marie, G. M. Moon, F. J. Moretti, and W. P. Ryman, 1985, Application of three-dimensional computer modeling for reservoir and orebody analysis (abs.): AAPG Bulletin, v. 69, no. 2, p. 262.

Hamilton, D. E., and R. S. Didur, 1992, Computer modeling of geologic surfaces and volumes, (this volume).

Harbaugh, J. W., and G. Bonham-Carter, 1970, Computer simulation in geology: New York, John Wiley and Sons, 575 p.

Johnson, C. R., and T. A. Jones, 1988, Putting geology in reservoir simulations: a three-dimensional approach: SPE Preprint 18321, 63rd Annual Technical Conference, p. 585–594.

Jones, T. A., 1988a, Modeling geology in three dimensions: Geobyte, v. 3, no. 1, p. 14–20.

Jones, T. A., 1988b, Geostatistical models with stratigraphic control: Computers & Geosciences, v. 14, p. 135–138.

Jones, T. A., D. E. Hamilton, and C. R. Johnson, 1986, Contouring geologic surfaces with the computer: New York, Van Nostrand Reinhold, 314 p.

Lasseter, T. J., 1990, An interactive 3-D reservoir modeling system: Geobyte, v. 5, no. 1, p. 48–49.

Lee, Y.-H., P. A. Martinez, and J. W. Harbaugh, 1990, Dynamic 3-D graphics critical element in Stanford's SEDSIM project: Geobyte, v. 5, no. 1, p. 37–38.

Lindsay, J. F., 1992, Three-dimensional geologic block model of a polar fan-delta complex, Canning River, North Slope, Alaska, (this volume).

Lindsay, J. F., D. B. Prior, and J. M. Coleman, 1984, Distributary-mouth bar development and role of submarine landslides in delta growth, South Pass, Mississippi Delta: AAPG Bulletin, v. 68. p. 1732–1743.

Mayoraz, R., C. E. Mann, and A. Parriaux, 1992, Three- dimensional modeling of complex geological structures: New development tools for creating 3-D volumes, (this volume).

Paradis, A., and B. Belcher, 1990, Interactive volume modeling: Geobyte, v. 5, no. 1, p. 42–44.

Raper, J. (ed.), 1989, Three-dimensional applications in GIS: Philadelphia, Taylor and Francis, 189 p.

Chapter 14

Three-Dimensional Geologic Block Model of a Polar Fan-Delta Complex, Canning River, North Slope, Alaska

John F. Lindsay
Bureau of Mineral Resources
Canberra, ACT, Australia

ABSTRACT

The Canning River delta on the Beaufort Sea coast of Alaska has been modeled in three dimensions as a typical example of a polar fan-delta complex. Polar deltas result from severely inhibited sedimentary processes. Free water is abundant only during the short summer when the carrying capacity of the stream is greatly exceeded so that most deposition occurs subaerially in braided-stream channels. Consequently, polar fan-deltas prograde seaward only minimally.

Three-dimensional geologic block modeling with the computer was highly effective in assimilating the large volumes of data available from shallow borings and has provided a new and unexpected insight into the structure and origin of this complex sedimentological body. The model shows that at least five depositional sequences are present beneath the delta surface. Each sequence began with the subaerial deposition of a massive, poorly sorted gravel unit in braided-stream channels. The massive gravels are generally followed by thick units of dark-colored, poorly sorted silt or silty sand. These reflect a fresh or brackish water swamp environment in the lower section and suggest shallow-water nearshore marine deposition in the upper sections. The shallow-water marine sediments show signs of ice scouring and some sequences are not complete because erosion has occurred during extreme sea-level lows. Thus some apparently massive gravel units are complexes representing more than one sea level cycle.

Barrier islands are common along the Beaufort Sea coast and their origin has been frequently discussed in the literature. The Canning River delta model shows the islands to be ephemeral structures built around cores of earlier gravels or marine sediments left behind from the Woronzofian transgression. These islands and their associated lagoons may not be typical of polar fan-deltas but may be a fortuitous occurrence resulting from the fact that the present sea level is slightly lower than during the Woronzofian sea-level maximum. The main effect of the lack of a lagoon would be reduction of the width of the zone of shore-fast sea ice, which would result in more intense scouring of the shallow-water marine sediments by sea ice.

INTRODUCTION

River deltas formed in temperate latitudes have received considerable attention from geologists in the 20th century, and consequently, three-dimensional geologic block modeling of these large-scale sedimentary bodies is to some degree possible. However, few models have been built for deltas formed in the highly stressed polar environment where temperature extremes vary from well below to well above freezing. In these regions, water, the most essential active ingredient in delta sedimentation, is unavailable most of the year. We thus could expect polar rivers to produce significantly different deltas from their counterparts in the more temperate regions of the earth. This paper is intended as a step towards filling this information gap through a detailed three-dimensional computer modeling study of the Canning River delta.

CANNING RIVER DELTA

The Canning River delta, a small fan-delta (Holmes, 1965), lies well inside the Arctic Circle (70°N, 146°30'E) on the Beaufort Sea coast between Prudhoe Bay and the Alaska-Canada border (Figure 1). The fan-delta extends 20 mi (30 km) inland and at its maximum is approximately 35 mi (60 km) wide (Figure 2). It is small compared with the Mississippi River or other well-known temperate deltas but is comparable to similar bodies built at the mouth of other arctic rivers.

The Canning River drainage basin lies totally in a permafrost region and discharges into a polar sea (Naidu and Mowatt, 1975). It heads in the Brooks Range 100 mi (160 km) from the Beaufort Sea and much of its drainage area lies above 4000 ft (1200 m) in rugged terrain. The Brooks Range is the source for most of the stream's detrital materials, which are composed mainly of Paleozoic and Mesozoic sedimentary rocks, some low to intermediate-grade metamorphic rocks, and a minor component of igneous materials. The stream breaks out of the mountains at an altitude of about 400 ft (120 m) above sea level about 20 mi (30 km) from the coast. Most of its load is deposited before it reaches the sea and forms the delta (Figure 3). The Canning is one of several large rivers entering the Beaufort Sea along this section of coastline (e.g., Colville, Kuparuk, Sagavanirktok). All form similar deltas that coalesce laterally into a major depositional complex along the coastal plain.

Pleistocene History

The Pleistocene history of the North Slope is complex (O'Sullivan, 1961; Black, 1964) involving at least eight marine transgressions (Hopkins, 1967). The last major marine transgression along the Alaskan Coast, before the onset of the late Wisconsin and Holocene transgression, was the Woronzofian transgression (25,000 and 48,000 years ago). This transgression is believed to have brought sea level to within a few feet

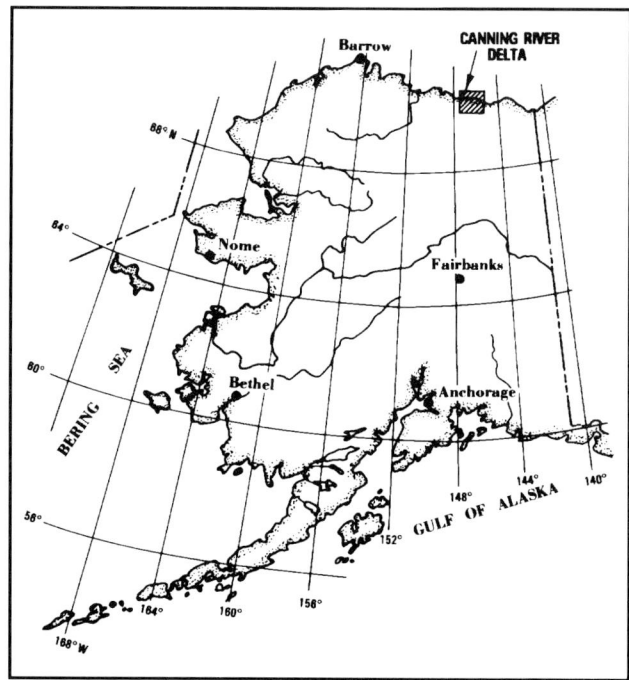

Figure 1. Alaska and the location of the Canning River delta.

of its present position and any delta building at that time was close to the present Canning River delta. The last major sea-level change occurred between 15,000 and 25,000 years ago when sea level fell an estimated 300 to 450 ft (90 to 140 m) exposing the area covered by the present Canning River delta and adjacent shelf area to subaerial erosion. Delta building must have occurred to the north of the study area.

Sea level has been rising during most of the last 15,000 years and longer. This rise has not been continuous and minor fluctuations have caused stillstands and short periods of accelerated rise and fall. Most important to the present study is a major stillstand between 105 and 130 ft (32 to 40 m), which occurred between 11,500 and 12,500 years ago (Creager and McManus, 1967; Curray, 1960, 1961; Fairbridge, 1961). A second minor fluctuation has been recognized between 65 and 130 ft (20 to 40 m) and is believed to have occurred 8500 to 11,000 years ago. Since then sea level appears to have risen steadily, reaching its present level between 3000 and 4000 years ago.

Climate and the Hydrologic Cycle

The climate is dominated by severe winters lasting from 8 to 9 months, during which there is essentially no hydrologic activity. Precipitation is held as snow and all stream flow ceases. Annual precipitation is low, about 5 in. (12.7 cm), making the region semiarid, and the mean annual temperature is –12°C.

Hydrologic activity is also considerably modified offshore by sea-ice formation. Early in the winter,

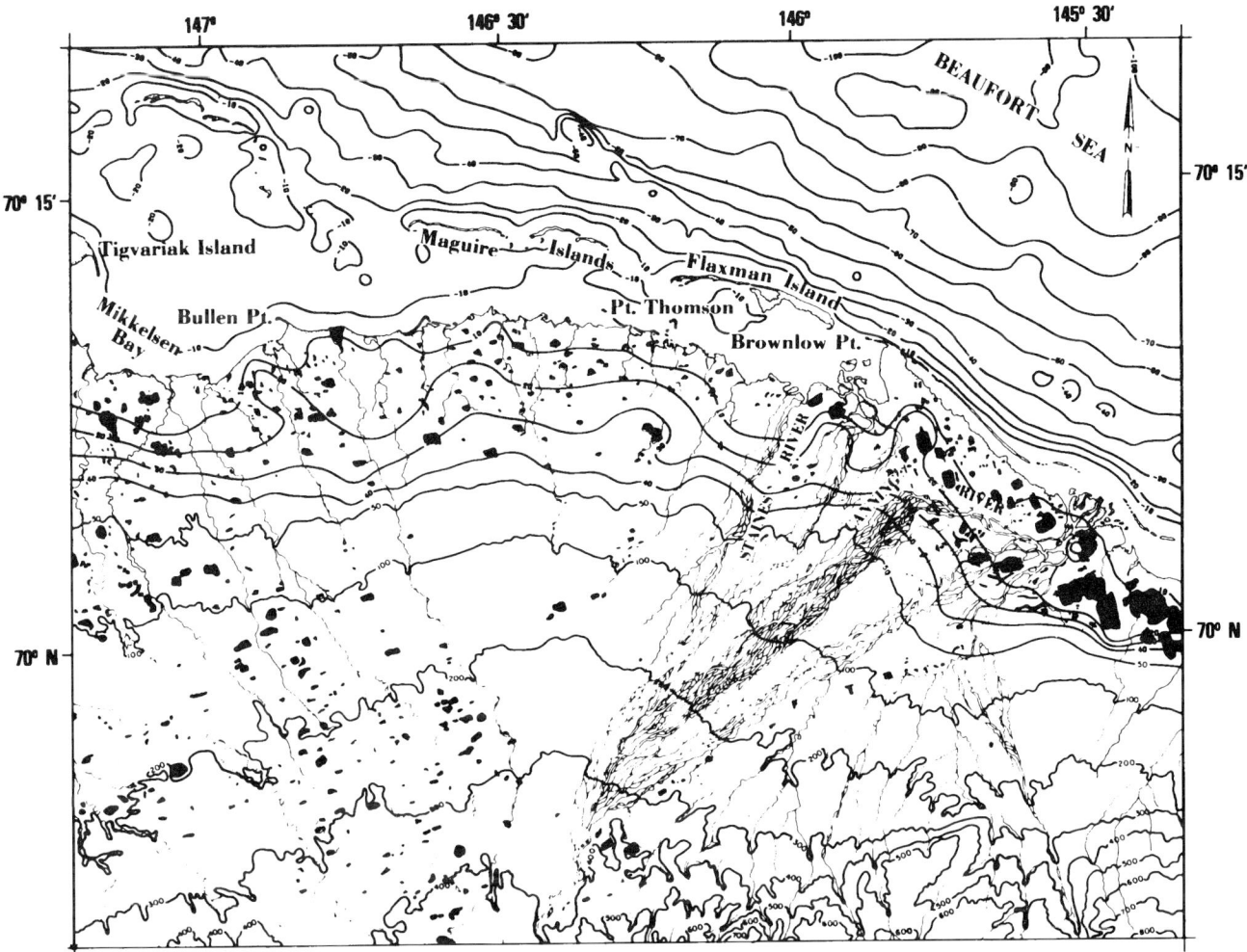

Figure 2. Topographic map of the Canning River delta. Thermokarst lakes are shown solid black.

pack ice is pushed onshore by the prevailing northeasterly winds while, at the same time, sea ice begins to develop. As the sea ice thickens, stream flow also declines so that by early winter the flow of fresh water beneath the sea ice ceases and salt water moves toward the delta front (Walker, 1974). Ice growth in shallow lagoonal areas behind the barrier islands restricts the flow of water (see Figure 4) and increases its salinity to twice that of normal sea water (Schell and Hall, 1974). By March sea-ice thickness has surpassed 5 ft (1.5 m) in lagoonal areas (Figure 4A) and the saline waters beneath have been reduced to complexly interconnected lenses from 5 to 15 ft (1.5 to 4.5 m) in thickness (Figure 4B). Current activity can reasonably be assumed to be minimal because of these sub-ice waterways.

In the spring, large areas of the delta are flooded. At the coast, these floodwaters advance across the as-yet-unbroken sea ice, slowing and dumping their sediment load before draining through the sea ice and mixing with the sea water beneath (Walker, 1974). The period of over-ice flow lasts less than 2 weeks. Some of the floodwaters move beneath the shore-fast sea ice in the lagoons; however, flow must be severely impeded by the complex pathways (Figure 4B).

Because the mean annual air temperature is about –12°C (Naidu and Mowatt, 1974), permafrost is general throughout the area onshore and probably for some distance offshore. The distribution, depth, and character of the permafrost is determined largely by the nature of clastic materials and drainage. Permafrost bonds the loose sedimentary material, making mechanical erosion difficult except for two or three months of the year.

Contemporary Depositional Environments

The Canning River delta region can be divided into three environmental zones: (1) an onshore environment dominated by braided stream deposition, (2) a near-shore shallow lagoonal environment separated from the open Beaufort Sea by an arcuate chain of barrier islands, and (3) the offshore environment beyond the barrier islands (Figure 2).

Onshore Environments

The Canning River sediments are derived from the mountainous terrain of the Brooks Range. The stream valley is narrow and relatively straight with a gradient of 250:1. When the stream breaks out of the mountains the average gradient of the stream remains about the same but the stream bed broadens from about 0.5 mile (1 km) to more than 5 mi wide (8 km). The stream rapidly loses its ability to transport coarse materials and its channel braids. The present Canning River flows northeast, entering the sea near Brownlow Point. However, over some unspecified time period in the past, the channel has swept through an arc of about 95°, reaching as far west as Mikkelsen Bay and producing a large-scale fan-delta complex.

Small lakes are created on the delta and adjoining coastal plain (Figure 2) as ice-rich sediment thaws. Once initiated the thaw lakes grow rapidly and coalesce (Black and Barksdale, 1949). The lakes form in fine-grained sediments which have higher ice content than coarser materials. Fine-grained sediments tend to accumulate in lower energy, water-rich environments of the delta in areas with low slopes (less than 500 to 1). Thus lakes tend to form in finer sediments that accumulate on the lower slopes around the seaward margin of the delta and are indicators of contemporaneous dewatering and compaction of the finer sediments. They also play a significant role in coastal erosion (Hopkins and Hartz, 1978) as the coalescing lakes ultimately break through to the sea, creating large embayments such as Prudhoe Bay.

The role played by lakes as sites for sediment accumulation is not so clear. Many lakes along the seaward margin of the delta are regularly flooded each spring and must accumulate fines ultimately becoming completely sedimented. A test boring made in the center of a small lake lying 1 mi (1.7 km) from the coast near Point Hopson consisted of alternate units of gravel and black clay (Figure 5), suggesting that a succession of lakes may have developed in the past and may have been disrupted regularly by braided channel deposition. Naidu and Mowatt (1975) found that the lakes on the Colville delta were almost entirely filled with peaty tundra vegetal detritus. The role of thermokarst lakes in the evolution of the delta may thus be more important than is immediately obvious and is discussed further in relation to the three-dimensional model.

Nearshore Environment

The coast of the mainland is characterized by narrow low-lying beaches backed by low bluffs. The mainland beaches are seldom wider than 60 ft (18 m), no more than a few inches thick and are almost entirely gravel which has an average size of 1 in. (25 mm) and a maximum of 2 in. (50 mm) (Hopkins and Hartz, 1978; Rex, 1964). The mainland beaches appear sediment-starved. Because of the rapidly decreasing competence of the overloaded river channel, little gravel reaches the shore directly (Arnborg et al., 1966; Barnes and Reimnitz, 1974). Beach gravels thus are derived almost entirely by erosion of the coastal bluffs. Since sea ice eliminates wave action during winter and the pack ice dampens wave motion much of the rest of the year, movement of beach materials is limited (Wiseman et al., 1973). The mainland beaches of the Canning River delta are further protected from wave action by the offshore islands.

Coastal erosion and beach formation is also restricted by permafrost, which acts as an effective cementing medium. Consequently, thermal erosion is as important, if not more important, than the mechanical action of waves in the formation of the beaches. Erosion generally proceeds stepwise as undercutting of the bluffs causes slumping of large blocks of permafrost, which are then reworked by waves on the beach (Naidu and Mowatt, 1974). Thermal erosion is further impeded by draped vegetal mats and by drifts of winter snow that must melt first.

A string of elongate low-relief islands begins at Brownlow Point on the western seaward edge of the delta and extends along the edge of the shelf. Behind the islands lies a shallow narrow lagoon about 7 mi (12 km) at its widest and extending for 20 mi (33 km) along the delta front. To the east the lagoon pinches out as the island chain merges with the coast. Westward the lagoon extends almost indefinitely although constricted by Tigvariak Island near the western end of Mikkelsen Bay. The lagoon is very shallow, averaging 10 ft (3 m) and only exceeding 20 ft (6 m) at its western end. Bottom relief is limited and bottom slopes do not exceed 1 in 500 except along the beach fronts. At least nine small closed depressions with low relief occur on the lagoon floor; the largest is about 2 mi (3.5 km) in diameter.

The waters of the lagoon are in continuous communication with the open sea by two deep channels and several smaller passages between the islands. However, because of its shallowness the lagoon is entirely in the fast-ice zone so that wave action is prohibited 8 to 9 months of the year. Consequently, for much of the year water circulation almost ceases and all fines, even clay-sized particles, have an opportunity to settle out.

Flooding in the spring carries sediment out onto the surface of the sea ice and the sea ice breaks up. Although the Canning River does not empty into the lagoon today, it has done so in the past. Studies of the nearby Colville River delta which does exit into a lagoon indicate that after the floodwaters have drained through the sea ice, a layer 0.5 to 7 in. (1.3 to 18 cm) thick consisting mainly of fine sand and silt with a large organic admixture remains on the surface of the ice (Walker, 1974).

As the sea ice breaks up the sediments deposited on the sea ice are gradually dumped into the lagoon and wind- and current-driven ice flows gouge the soft sediments on the sea floor. Initially, fast-ice thins and lifts off the bottom. However, not until about mid-July does the fast ice weaken enough to break up under the influence of winds and currents (Reimnitz and Barnes, 1974). This activity reworks the sediments to depths from 6.5 ft to 18 ft (2 to 5.5 m), mixing new and old

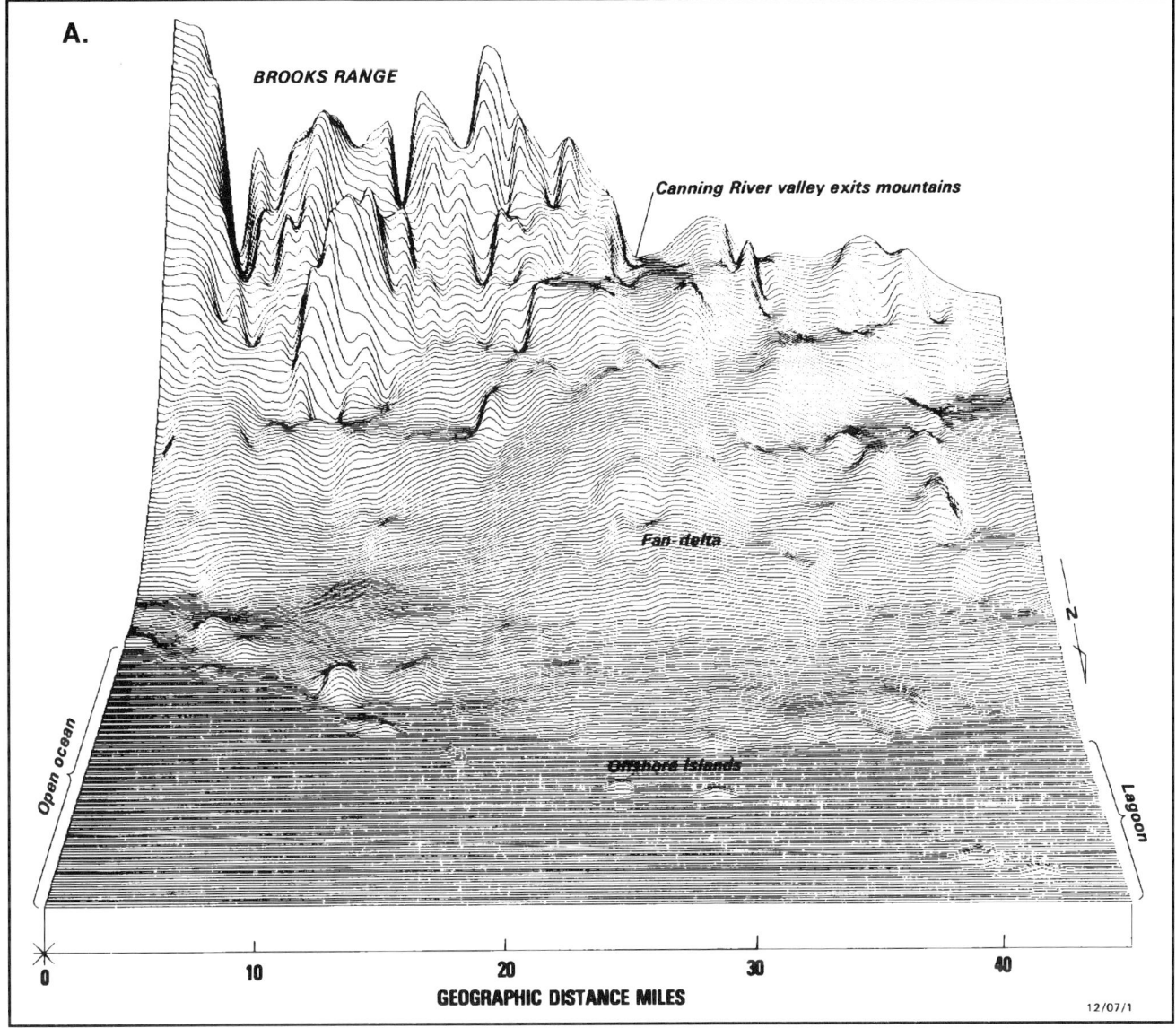

Figure 3. Block diagram of the Canning River delta as viewed directly south. The vertical exaggeration is 150 to 1. A. Showing the ocean surface. (Figure 3B on following page).

lagoon sediments and creating relatively homogeneous, massive, poorly sorted silts and sands with disrupted, highly contorted bedding throughout. The environment is also hostile to any sessile benthic organism, although shell beds do occur in the shallows close to the barrier islands, presumably in areas protected from wind-driven ice.

Most sediment deposited on the ice by the spring flood is dumped nearby as the sea ice melts (Barnes and Hopkins, 1978). Thus, the role of ice-rafting is minor (Reimnitz and Barnes, 1974) with ice being more important in mixing the spring flood deposits than in transporting the sediments laterally.

Nine major and many minor islands form a curvilinear chain immediately offshore from the Canning River delta. All the barrier islands are elongate, the longest being 4.4 mi (7.3 km). None of the islands exceeds 16 ft (5 m) above sea level and few are more than a few hundred feet wide.

Most authors favor the view that the islands are largely constructional features. In some cases the islands consist of earlier materials left as erosion remnants (for example, Barnes and Hopkins, 1978; Hopkins and Hartz, 1978). Without doubt, both points of view are correct. The islands superficially consist largely of sands and gravels in the form of beaches and accretionary spits. These constructional areas are seldom more than 9 ft (2.7 m) above sea level and are cored by older sediments. Autumn storm surges are the dominant modifying agent (Barnes and Hopkins, 1978).

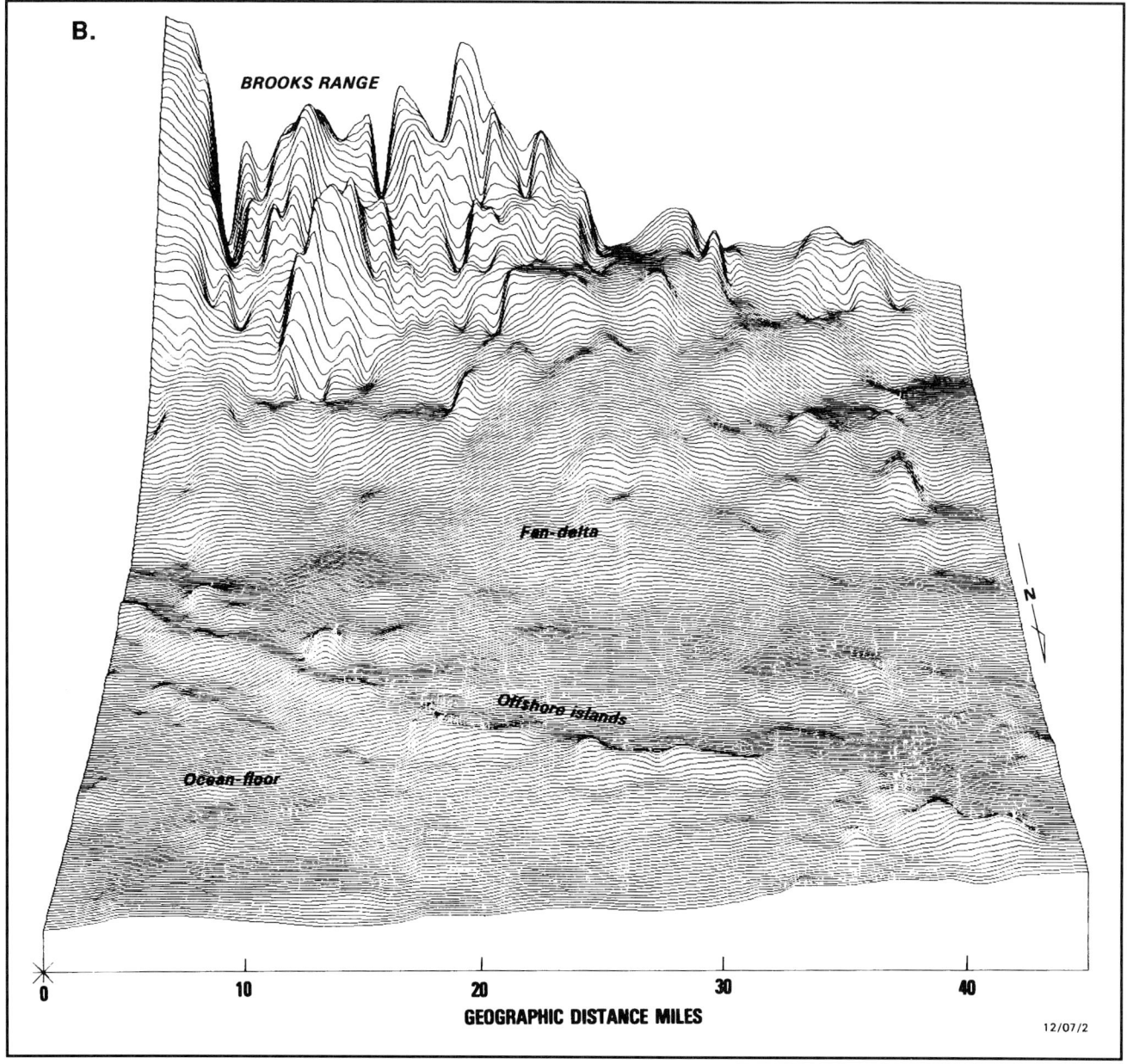

Figure 3B. With the ocean removed to show that the offshore islands form a continuous ridge.

Clastic materials of the barrier-island beaches are derived from a source separate from the mainland beaches. The mainland beach material is derived from the Brooks Range, whereas the pebbles forming the barrier-island beaches are derived locally from the older Flaxman Formation which contains a different suite of lithologies (Hopkins and Hartz, 1978). Gravels derived from the Flaxman Formation are transported in a westerly direction by the prevailing longshore current system to form the barrier islands which migrate westward from 43 to 100 ft (13 to 30 m) per year and shoreward from 9 to 21 ft (2.7 to 6.4 m) per year (Hopkins and Hartz, 1978; Lewellen, 1970; Wiseman et al., 1973).

Offshore Environment

Beyond the barrier islands, the sea floor descends sharply to depths of 100 ft or more (30+ m), with bottom slopes from 1 in 250 near the beach front to 1 in 125 farther seaward. Little river sediment has reached into this environment because, in its immediate past history, the Canning River emptied into the lagoon, which trapped most of the fines.

Ice scouring is more intense beyond the barrier islands. The shear zone between shore-fast ice and the pack ice of the open sea lies just seaward of the islands. Consequently, significant ice scouring is found to water depths greater than 150 feet (46 m)

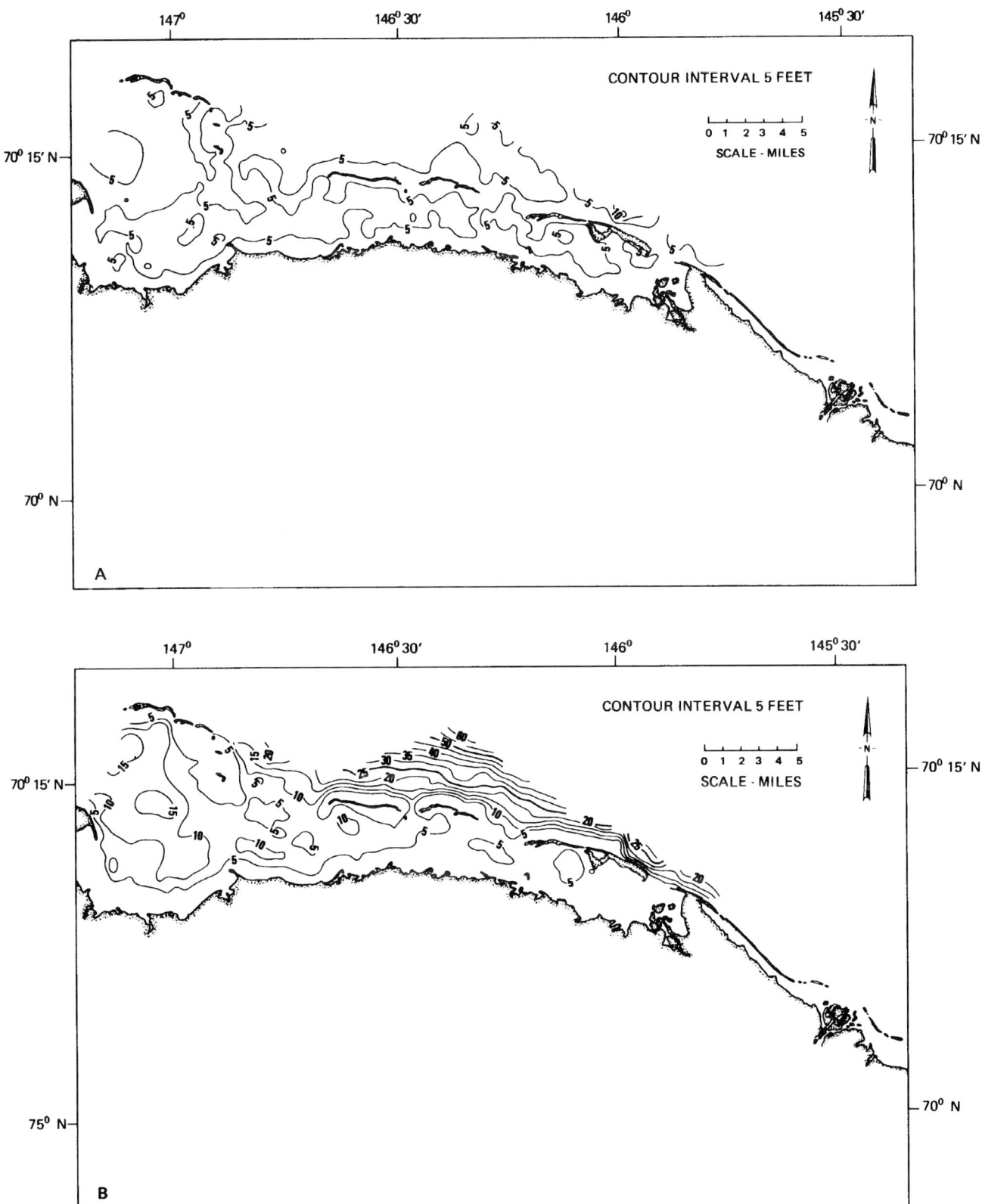

Figure 4. A. Sea-ice thickness offshore from the Canning River delta at the height of winter. B. Free water beneath the winter sea ice. The complexly interconnected waterways make current activity virtually impossible beneath the sea ice in the shallow lagoon (modern shoreline stippled).

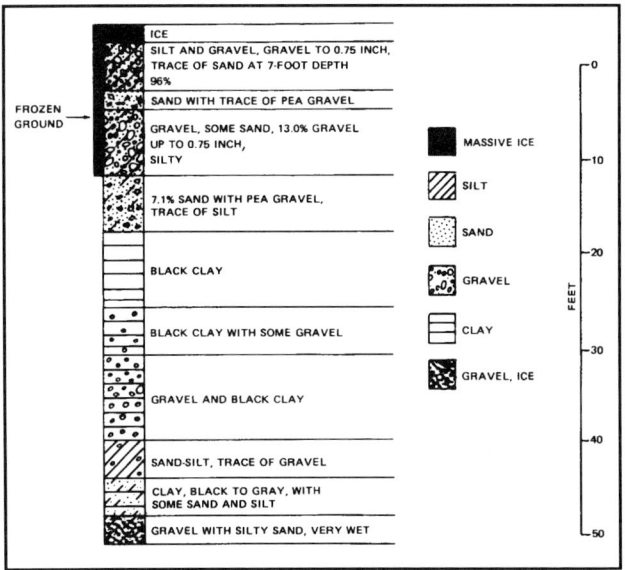

Figure 5. Stratigraphic sequence beneath a small thermokarst lake lying 1.5 mi (2.5 km) inland from Point Sweeney on the Canning River delta. The alternating sequence of clays and gravels may be due to deposition in preexisting thermokarst lakes.

(Reimnitz and Barnes, 1974), which is far beyond the limits of the present study area. Reimnitz and Barnes (1974) found that bedding becomes more uniform in sediments taken in waters deeper than 150 ft (46 m).

MODELING IN THREE DIMENSIONS

Large volumes of data available for the study of the Canning River fan-delta made computer modeling necessary. The delta was modeled in three dimensions using proprietary software developed by Exxon Production Research Company. The broad outlines of the major algorithms used in this software package are discussed in Jones et al. (1986) and Jones (1988, 1992). The model produced by the software is in essence a three-dimensional matrix consisting of discrete blocks or cells containing information about the physical properties of the sediments. The blocks are 2500 × 2500 ft (762 × 762 m) in horizontal dimension and 2 ft (61 cm) thick. This model resolution was selected by the spacing of the borings and the available detail of the lithologic descriptions. Lateral lithologic variability (facies change) is much less than vertical change, hence the large difference between horizontal and vertical resolution used for the model. The objective in modeling is to effectively interpolate the properties of cells within the matrix where no direct observational data are available.

Data

Modeling in three dimensions requires the input of two types of data: (1) rock properties, usually in the form of well information, and (2) a three-dimensional framework (cf. stratigraphic-structural surfaces) to control lateral interpolation. Finally, to bring these two data types together as a three-dimensional model requires the input of geological knowledge about the body being modeled and knowledge about the limitations of the software used in modeling (Jones, 1988).

The model uses 1207 borings (Figure 6). The majority of the data came from shot holes made while the area was surveyed seismically during the 1970 and 1971 winter field seasons. The holes were drilled at the height of winter during March and April when the surface soils were frozen solid and the sea ice was close to its maximum thickness. Thus offshore holes were drilled from the surface of the sea ice, providing details of sea-ice thickness and water depth as well as lithology. The holes vary from 15 ft to 105 ft (4.6 to 32 m) deep; the most common depth is 75 ft (23 m).

Twelve borings drilled along the delta front during 1975 and data from velocity survey holes on Flaxman Island provided additional data between seismic lines. Some of these holes extended deeper than 100 ft (30+ m). Four well-documented holes drilled from the frozen surface of four small thermokarst lakes in the Point Thomson area to 50 ft (15 m) below the lake bottom were also included.

Finally, the most important of the supplemental data sources was the five holes drilled during 1979 for the US Geological Survey (Bruggers and England, 1979). The holes are important because they are deep, 140 to 340 ft (43 to 104 m) and strategically located in deeper water areas where shot hole data are few or nonexistent.

The modeling software used in the present study allowed the physical properties of the sediments to be handled in one of two ways: discrete or continuous (Jones, 1988). The physical properties of the sediments could thus be handled either qualitatively as a discrete data form (e.g., lithology), or as a continuous variable (e.g. porosity). Either type of model could be manipulated with logical and arithmetic operations to create hybrid models. In their study of the Colville delta, Naidu and Mowatt (1975) found that mean grain size effectively distinguished the facies of polar deltas. Since the present study was based on relatively recent unconsolidated sediments and data were recorded in terms of mean grain size, it was decided that the data could be most effectively manipulated in a continuous form and that they should be encoded as mean grain size to allow continuous modeling between borings. Most sediments have grain sizes that are log-normally distributed and most sedimentary processes act on detrital particles so that grain size tends to increase or decrease exponentially toward or away from the source. Thus a grain size unit with a logarithmic component (phi units) could be manipulated by simple linear approximations to produce a model more or less in conformity with nature. Consequently, lithologies were entered as codes representing phi grain size increments of 1 (Figure 7). The coarsest sediments encountered were

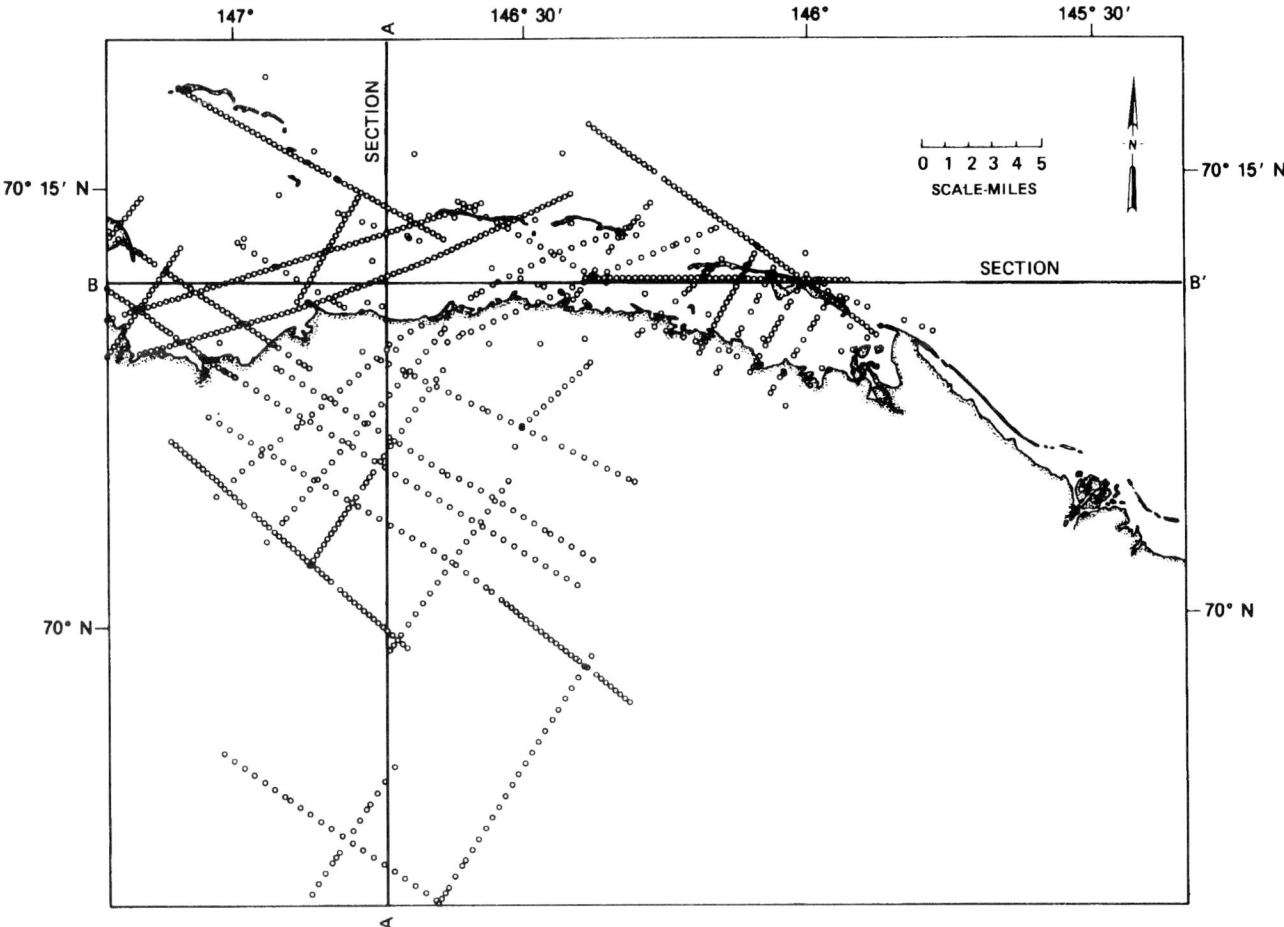

Figure 6. Location of boreholes used in the three-dimensional reconstruction of the Canning River delta (modern shoreline stippled). Data are lacking on the eastern edge of the delta where it extends into the Arctic National Wildlife Rangelands.

boulders with a mean grain size of −10ϕ; the finest sediments were clays with a mean grain size of 12ϕ. This approach assumes smooth lateral change in the value of the modeled attribute (phi) and any sharp lateral discontinuity that did exist would be smoothly interpolated between drill holes.

Typical data for a drill hole included (Figure 7):

General descriptive information
• Hole ID
• X-coordinate
• Y-coordinate
• Starting elevation
• Total depth

Unit boundaries (tops)
• Depth to interpreted top
• Identification of top

Detailed interval information
• Beginning depth of attribute
• Ending depth of attribute
• Value of attribute (Mean grain size of sediments in phi units and special codes to identify sea water, and sea ice.) Missing data were identified with special characters so the program could handle them properly.

The Correlation Framework

In a more typical three-dimensional geologic block modeling study, such as the evaluation of a petroleum prospect, the physical framework used in modeling would normally be based upon a series of time stratigraphic surfaces such as depositional sequence boundaries derived from the interpretation of seismic data. The necessary high resolution seismic data were not, however, available for the surficial deposits forming the Canning River delta and an alternative approach had to be devised.

Lateral control of the model interpolation was maintained using a series of 15 computer grids to form a stratigraphic framework to correlate from boring to boring. The framework was initiated by digitiz-

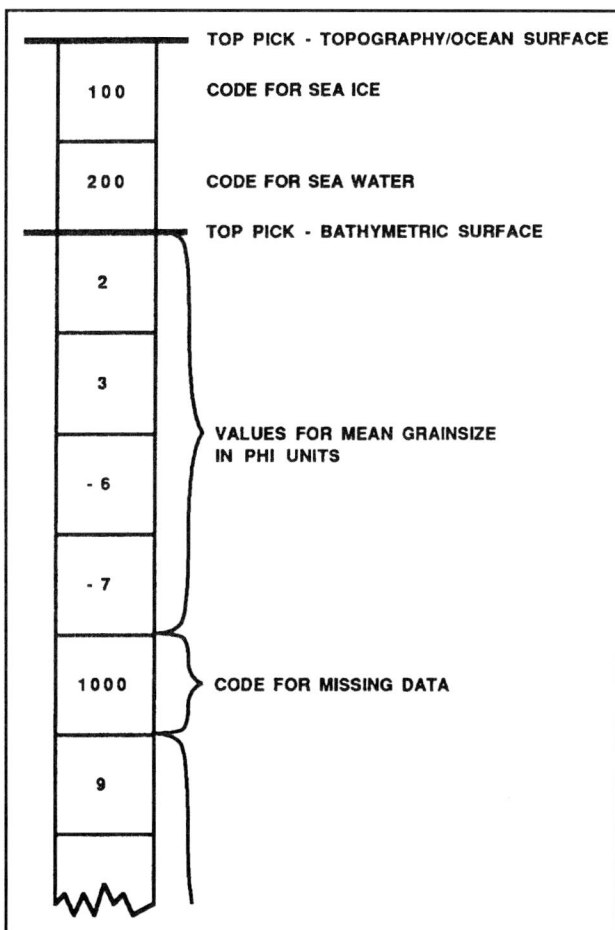

Figure 7. Distribution of information in a well, shown in simplified form.

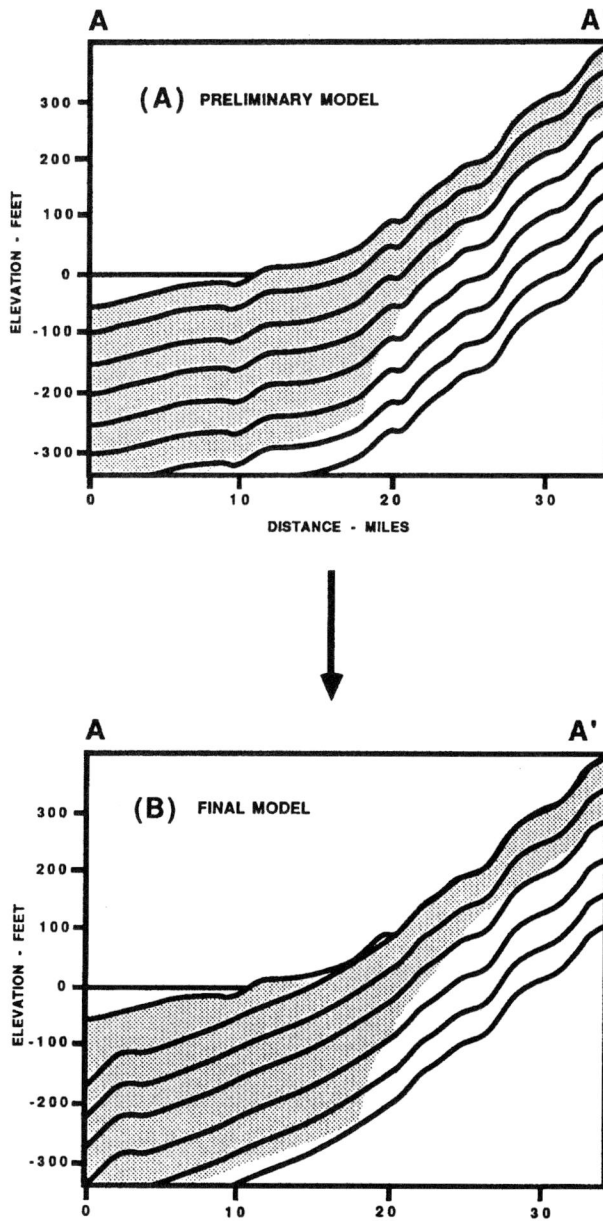

Figure 8. Simplified diagrammatic illustration of correlation controls for the two step modeling process. (A) Preliminary model in which the topography of the fan delta was used to control subsurface data and (B) Final model using a severely smoothed upper surface for the first major conglomerate unit as a controlling surface which was then extended to depth in several steps (see text). Stippled areas indicate the zone in which subsurface data density is adequate for modeling.

ing bathymetric and topographic maps covering the delta area. The uppermost grid conformed to the landward topography and the ocean surface to enable the sea water and seasonal sea ice to be incorporated in the model (Figure 4). The second grid follows the topographic and bathymetric surface. The third grid was constructed 2 ft (61 cm) below and parallel to the second grid to allow incorporation of data from offshore bottom samples. Below this point the correlation controls were not obvious. To establish these controls, a preliminary model was constructed and used to determine the gross distribution of rock bodies. This was done by shifting the second grid (topography and bathymetry) down through several steps to a maximum of 350 ft (106 m) below the topographic/bathymetric surface and thus below all bore hole data (Figure 8A). These grids were then used as the correlation framework for building a preliminary model.

Cross sections drawn through this preliminary model showed that a surface roughly parallel with the tops of gravel would reasonably control the correlations of the lower units. To generate this surface the elevation of the uppermost cell in each column of cells that had a phi grain size less than or equal to 0 phi (coarse sand) was found and stored in a grid. This grid was then severely smoothed to remove the irregularities resulting from the simple assumptions used.

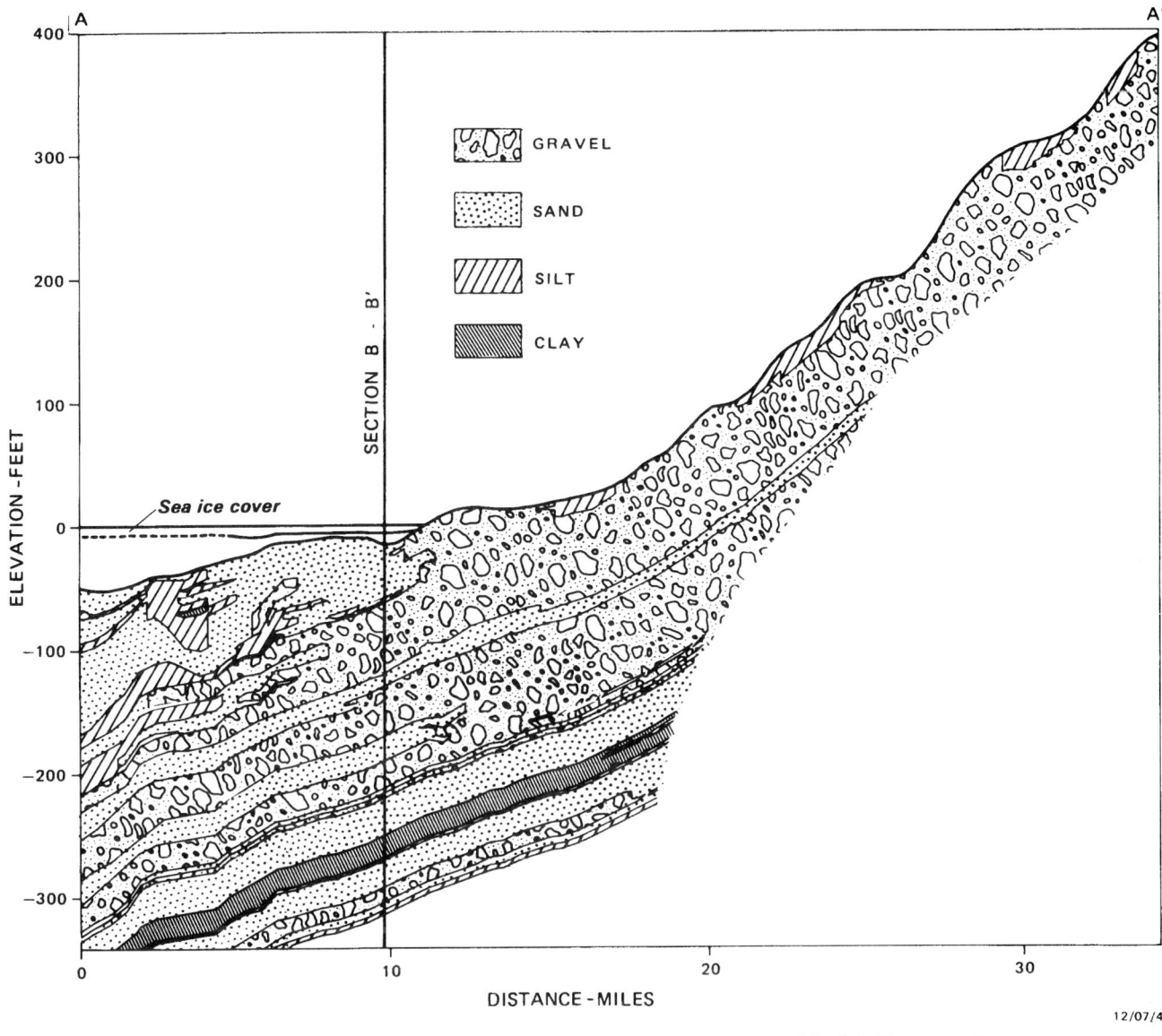

Figure 9. Section AA', a north-south profile of the Canning River delta. Note the thinning of the major gravel units in a seaward direction and their interfingering with finer sands and silts. The vertical exaggeration is 200 to 1. (See figure 6 for section location.)

This fourth, smoothed grid was shifted vertically to 11 more positions that were approximately 15, 30, 45, 60, 75, 100, 156, 200, 250, 300 and 350 ft (4.6, 9, 14, 18, 23, 30, 48, 61, 76, 91, and 107 m) below the topography. The spacings of these shifted grids were chosen to provide maximum support for the borehole information. The holes had been drilled to a series of regular depths of 15, 30, 45, 60, 75, 156, 300 and 340 ft (4.6, 9, 14, 18, 23, 48, 91, and 104 m). The spacings of the surfaces thus help define zones of decreasing data density at increasing depth. These 11 grids plus the original four comprised the correlation controls for building the final 3-dimensional geologic block model (Figure 8).

Model Construction

As mentioned in the discussion of data, a continuous (inverse distance squared) method of interpolation was used to build the final model. To assign values to the cells of the model (Jones, 1988), the program:
(1) Determined the position of the cell vertically with respect to the correlation framework.
(2) Identified nearby bore holes that had interval data in that correlative position.
(3) Selected the interval or intervals of data from the bore hole that covered the same correlative zone as the cell and converted them to one value using an appropriate interval thickness weighting technique.

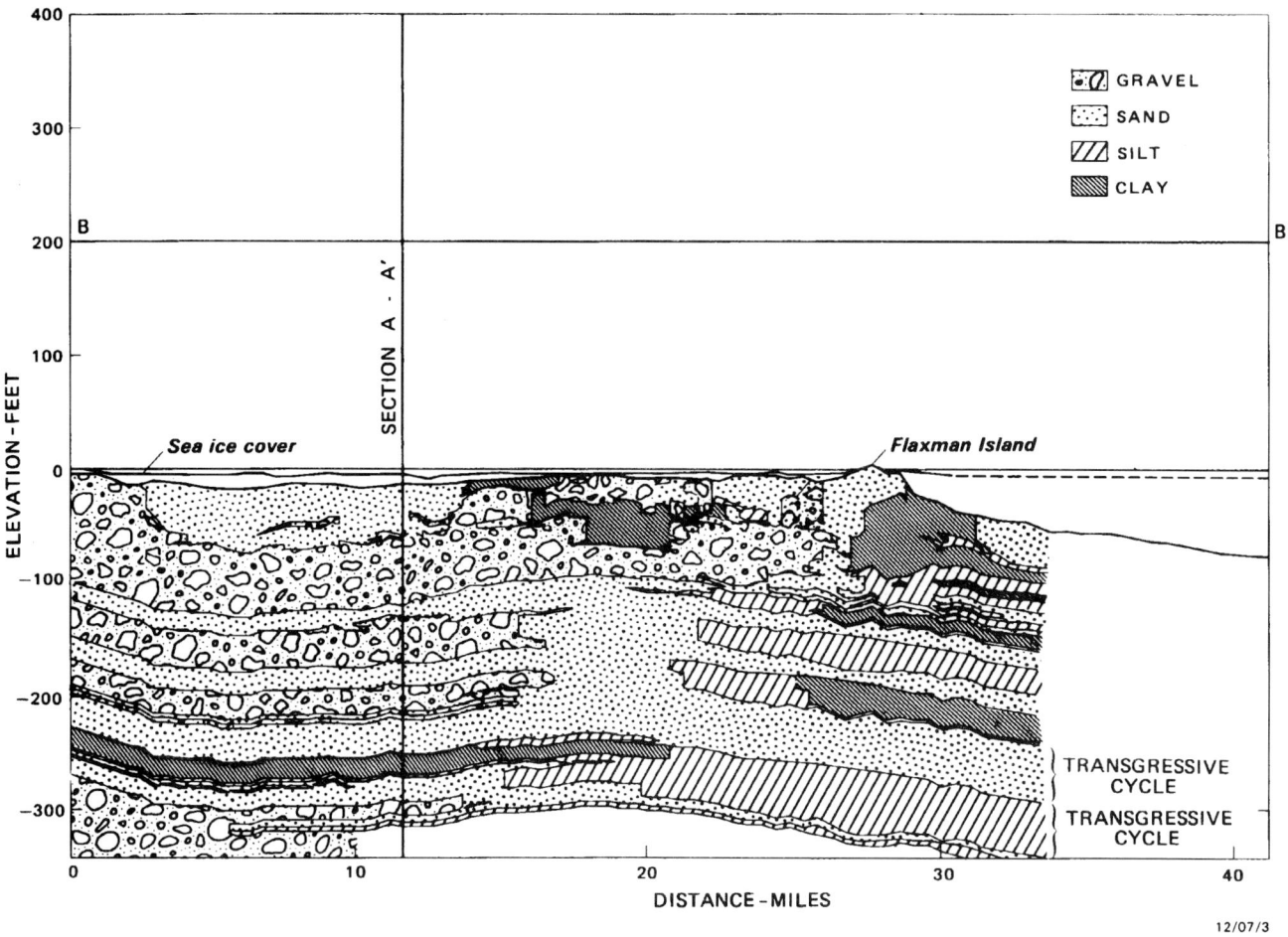

Figure 10. Section BB', an east-west profile through the Canning River delta close to the present shoreline. Note that the earlier major gravel units are not necessarily symmetrically disposed about the axis of the modern Canning delta suggesting a shift with time. The discontinuous erosional remnants of the Flaxman Formation are readily apparent beneath the barrier islands. The vertical exaggeration is 200 to 1.(See Figure 6 for section location).

(4) Weighted the converted data value for each bore hole by the distance the borehole was from the cell.
(5) Summed the weighted data values and assigned the resultant value to the cell. Any instance of missing interval data in a bore hole was treated as transparent and the program looked through that well to the next closest bore hole with data. Standard search sector and search radius methods were used by the algorithm to reduce the effects of clustering and restricted extrapolations.

As with most computer and hand modeling or contouring techniques, areas with more data produced more detail in the model. Areas with less data produced smoother, more continuous features in the model. No special effort or techniques were used to force the heterogeneity in the sparse data area of the model to match that of the dense data areas. For this study the volume of data was sufficient even at depth to produce an acceptable model for interpreting the delta's character.

Model Interrogation and Display

Once the model was developed, it was a relatively simple matter to return to it to derive intricate graphics which were either too time consuming or, in some cases, almost impossible to develop by hand methods. Relatively simple and obvious uses of the model, such as block diagrams, well location maps, or cross sections, are shown in Figures 3, 6, 9 and 10. These displays could be generated directly from the grids and model using standard display programs.

Less obvious and more sophisticated applications are illustrated in Figures 11, 12 and 13 where major continuous lithologic bodies that are not obvious otherwise have been identified by the modeling and structure contours and isopachs generated for them. To generate the surface on the top of the gravel and thickness of the Flaxman Formation required special functions of the program. Once the final model was built the program was instructed to identify pathways or determine connectivity of similar rock attributes. For example, cells with phi sizes less than a set value might be selected for this operation. Grids can then be generated on top and bottom of these continuous bodies and used to generate maps of their structural positions (Figure 11) or subtracted to show their distribution and thickness (Figures 12 and 13).

The interpretation of what each of these continuous bodies represents obviously is not done by the computer. Many of the bodies will be difficult or impossible to interpret geologically. However, often these bodies are identifiable as particular rock units such as the Flaxman Formation of Figure 12, and the insights gained from the knowledge of its shape and distribution are invaluable to the understanding of the geology of the area.

DISCUSSION

Major Gravel Units

Massive gravels form the main depositional units beneath the delta itself and some distance seaward from the modern delta. Where they have been sampled, the units consist of gray or gray-green, poorly sorted, sandy gravels or gravelly sands with mean grain sizes from -2 to -4ϕ, or 4 to 16 mm (Bruggers and England, 1979). Texturally, the gravels are identical to those currently deposited in the active braided channel of the Canning River. We can thus reasonably assume that the major subsurface gravel units were all subaerially deposited and that the distribution of most of the gravels is tied closely to Pleistocene sea-level changes. These gravel units are typically 100 ft (30 m) thick and areally extensive, covering at least 450 mi^2 (1250 km^2).

Beneath the modern delta, the gravel appears as a single morphologic unit (Figure 9). However, the gravels interfinger seaward with fine sands, silts and, to a lesser extent, clays; and the gravels lose some of their massiveness (Figure 10). Farther seaward, at least one of the units disappears; the others thin and isolated lenses of gravel begin to appear. This gravel complex onshore has been called the Gubik Formation, a Quaternary unit overlying the Tertiary sediments of the North Slope (Black, 1964; O'Sullivan, 1961).

All the spatial relations of the Gubik Formation suggest its deposition during a sea-level rise with its gravels as part of a single sequence. The overloaded braided stream simply stopped at the transgressing coastline, depositing all its coarse load and most of the fines. However, the sequence of events appears to be more complex than this. A structure-contour map of the surface of the uppermost gravel generated from the three-dimensional model reveals numerous depositional complexities (Figure 11). The main gravel mass forming the contemporary delta lies at or close to the surface. A short distance seaward from the present shoreline, the gravel surface steps down abruptly from about -10 ft (3 m) to -40 or -50 ft (12 to 15 m). This sudden drop represents the contemporary depositional limit for gravel-sized materials. In cross section, these most recent gravels are seen to rest on an earlier gravel sloping seaward. The contemporary gravel grades seaward into a tongue of sands and silts while the earlier gravels extend farther seaward.

About 80 ft (24 m) below sea level, the surface again steps down abruptly to a lower gravel unit, possibly a tongue of the earlier gravel unit discussed above. This abrupt step in the gravel surface may reflect late Pleistocene conditions when sea level stood between 105 and 130 ft (32 to 40 m) below present. However, in the following discussion, we see that this gravel tongue probably belongs to an earlier depositional episode, perhaps earlier than mid-Wisconsin.

The most distinctive feature in the top of the gravel surface immediately seaward from the contemporary delta (centered about 146°30'W) is a deep incision extending from the edge of the gravel unit landward for 10 mi (17 km) and terminating abruptly against the margin of the contemporary delta. The incision is 5 mi (8 km) wide at its seaward extreme and more than 80 ft (24 m) deep. All indications show that this was a major stream channel during the lowest phase of the Pleistocene sea-level decline. The dimensions of the channel are more than adequate to carry the total stream flow now carried by the Canning River.

Apart from the large buried channel, the most obvious feature of the top of gravel surface is its roughness and the presence of numerous knolls or hills projecting above present sea level. Offshore these highs form the nuclei for several islands. Similarly, gravel mounds and ridges rise above the general level of the present delta at Bullen Point and Point Thomson. Significantly, many of these same gravel highs have distinctive boulders associated with the Flaxman Formation (described below). Thus the top-of-gravel surface is defined by a complex body consisting of contemporary gravel and at least one older gravel unit deeply dissected during a Pleistocene sea-level low and then partly buried by the present delta at the end of the Pleistocene transgression. The offshore gravels may in part represent the lowstand systems tract developed prior to the final sea-level rise.

Flaxman Formation

Next to the major gravel bodies, the Flaxman Formation is perhaps the most distinctive and informative unit in the Canning River delta succession. The formation (Leffingwell, 1908, 1919) is a till-like deposit consisting of interbedded sands and clay with scattered pebbles and boulders, some as much as 9 ft (2.7 m) in diameter. The formation was first discovered on Flaxman Island and remnants have been found on several islands and along the coast (Reimnitz and Toimil, 1977; Leffingwell, 1908, 1919) and most later authors seem to agree that the Flaxman Formation is

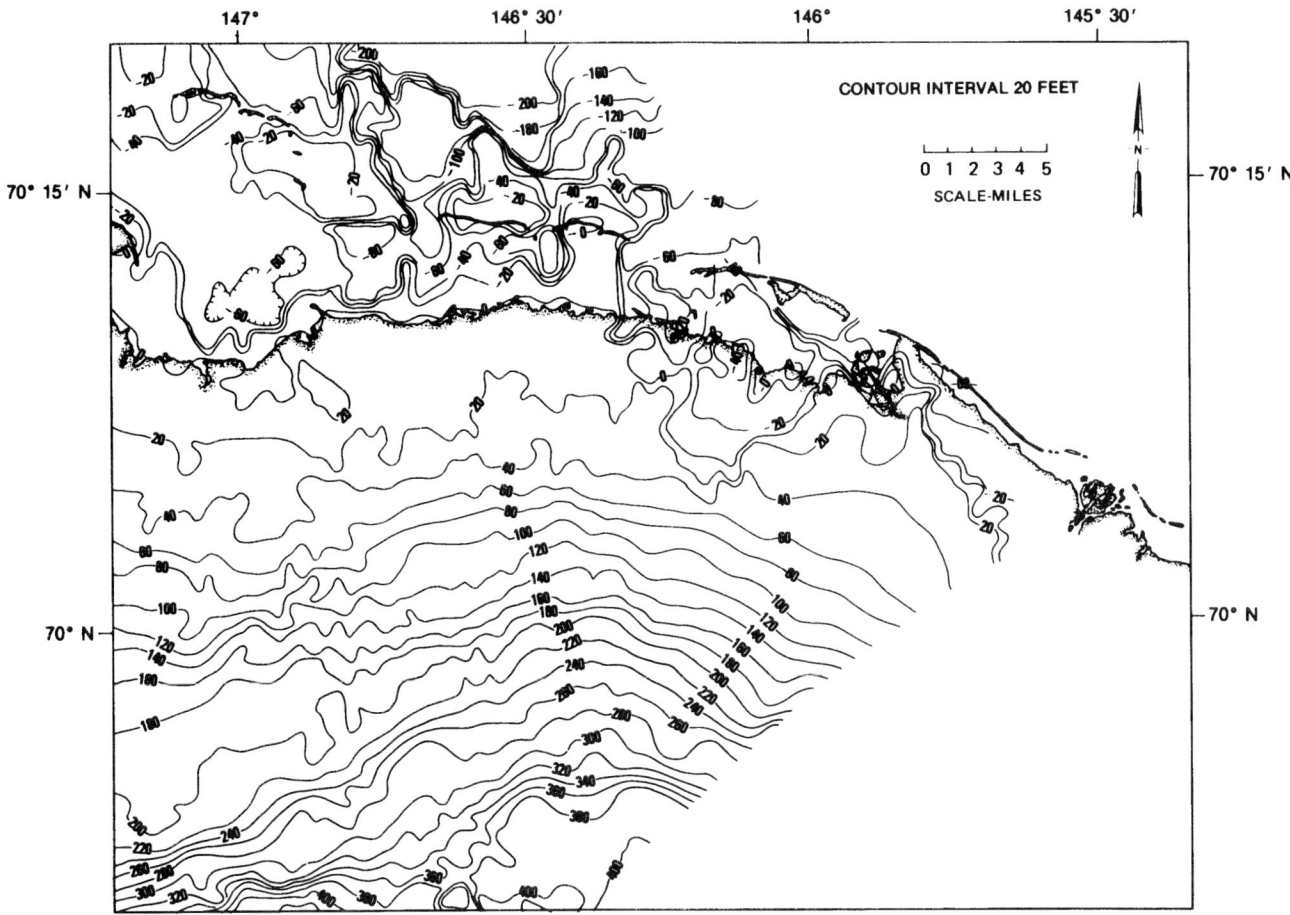

Figure 11. Structure-contour map of the upper surface of the first major gravel unit beneath the Canning River delta. The surface drops abruptly offshore from the delta indicating the rapid changes in the depositional environment as the stream reaches the sea (modern shoreline stippled).

glaciomarine in origin which is further supported by the fact that the clasts have a different composition from those contained in the main deltaic gravels.

The present distribution of the Flaxman Formation is discontinuous in the Canning River delta vicinity (Figure 12). The formation consists of numerous large irregularly shaped lenticular bodies with extremely sharp boundaries. These sedimentary bodies seldom exceed 60 ft (18 m) in thickness and appear on the Pleistocene-Holocene structure surface as a series of hill-like mounds. All indications show that they represent erosional remnants of a much more extensive sedimentary unit, suggesting that the offshore islands may in part be an erosion-resistant ridges of the Flaxman Formation. The 3-D model indicates that this is not normally the case. An erosional remnant of the Flaxman Formation underlies Flaxman Island and Flaxman remnants are the underlying reason for the localized topographic highs at Point Thomson and Brownlow Point. However, other topographic highs are underlain by older gravels.

The Flaxman Formation is now so completely dissected that its primary lateral relationships are difficult to define. Sections through the succession in the vicinity of Flaxman Island and Point Thomson provide evidence that the Flaxman Formation interfingers laterally into deltaic braided stream gravels; westward from Flaxman Island beyond the influence of the delta, Flaxman lithologies may dominate (Figure 10). The Flaxman Formation thus appears to be contemporaneous with the earlier gravels of the Canning River delta. Exotic boulders with a distinctive composition are widely but discontinuously distributed along the coast well beyond the Canning River delta. Leffingwell (1919) and MacCarthy (1970) identified these boulders with the Flaxman Formation. Hopkins (1967) and McCulloch (1967) interpreted the boulders as the result of ice-rafting during the Woronzofian or mid-Wisconsin transgression, which further supports the idea that the boulders are related to the Flaxman Formation. McCulloch (1967) connected the boulders with an ancient shoreline occurring along the North Slope about 25 ft (7.5 m) above the present sea level. Whether the shoreline is a product of eustatic sea-level change or local tectonism is not clear yet.

The present modeling is consistent with the obser-

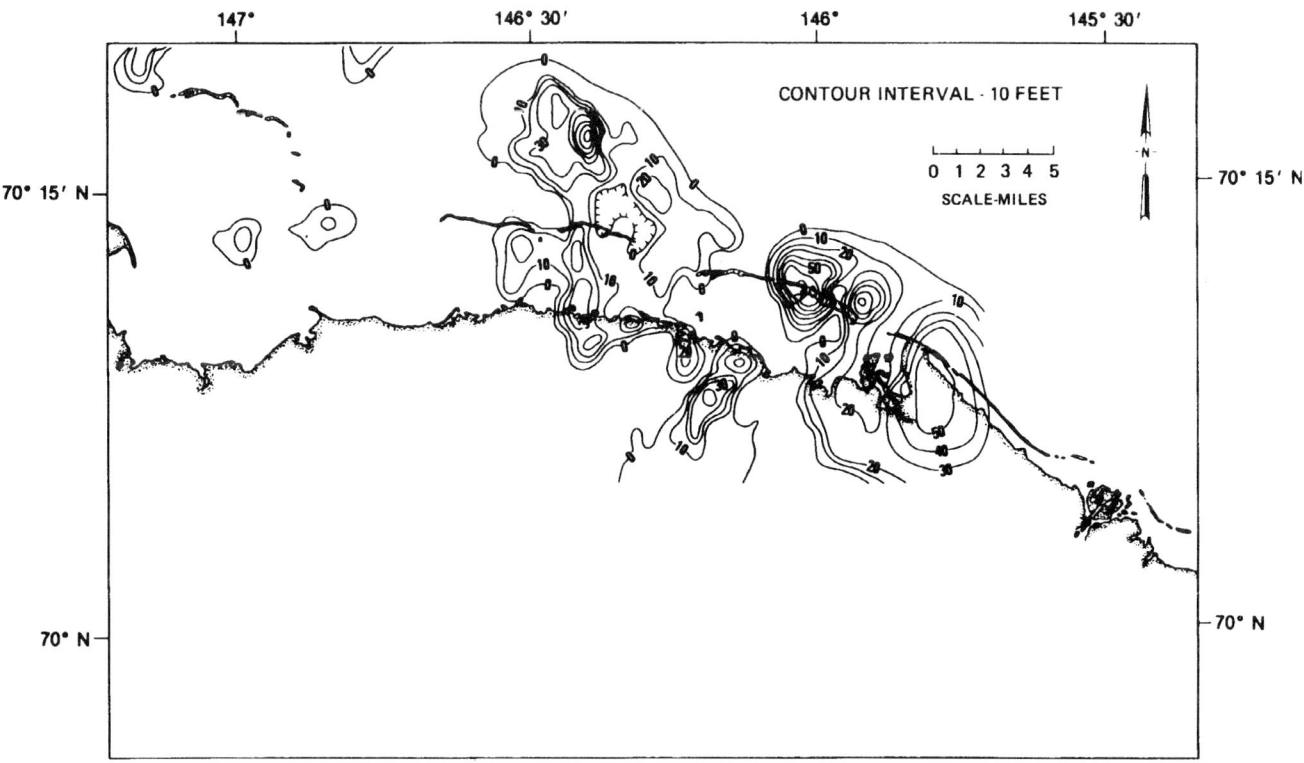

Figure 12. Isopach of Flaxman Formation lithologies interbedded with Canning River deltaic sediments. The formation appears as a series of discontinuous erosion remnants up to 80 feet (24 m) thick. Note in particular the large erosional remnant of the formation beneath Flaxman Island, an important factor in the existence of some of the barrier islands (modern shoreline stippled).

vations of the earlier authors; however, the improved details of the model add further complexity to the origin of the boulders. Figure 12 shows clearly that the boulders in the vicinity of Point Thomson, Brownlow Point, and Flaxman Island all relate to erosional remnants of the Flaxman Formation. However, boulder concentrations have also been found in exposures at such places as Bullen Point and Tigvariak Island. Both these areas are underlain by sandy gravels that are probably braided stream channel deposits. In all cases, the exotic boulders are found on topographic highs rising well above the general level of the contemporary deltaic gravels. All the data suggest that the Flaxman Formation, the older gravel component forming the mounds in the present delta, and the cores to the offshore islands are contemporaneous and that they relate to the mid-Wisconsin transgression represented by the 25-foot strandline. According to the chronology of Hopkins (1967), these data place the age of the early deposits somewhere between 25,000 and 48,000 years ago.

Sands And Silts

The major gravel units underlying the Canning River delta are interbedded with sands and silts. Shoreward these units tend to consist of medium sands and at times gravelly sand, although this particular facies is generally limited in extent and appears to represent the seaward fringe of the braided channel deposits. The majority of the units between the gravels consist of silty sand or sandy silt or, less frequently, clayey silt with occasional thin layers of clean medium sand or, less often, gravelly layers. These units are almost always gray or gray black and shell and wood fragments are common. Less frequent are layers of fibrous peat with abundant wood fragments.

At least five major silty or sandy units can be identified in the 300 ft (91 m) of section examined beneath the Canning River delta. All units are much the same and all vary considerably within the major unit so that a gray sand may be underlain by a gray or black clayey silt. However, of these units, only one (the uppermost) is well enough defined by the available data and the computer model to show its three-dimensional form (Figure 13).

The uppermost unit is extremely variable in thickness and conforms closely to lows in the antecedent topography resulting from the late Pleistocene sea-level minimum. Behind the barrier islands, the unit is typically 20 ft (6 m) thick. However, immediately offshore Bullen Point between the islands, the sediments increase to 40 and then 60 ft (12 to 18 m) in thickness where the late Wisconsin stream channel cuts deeply

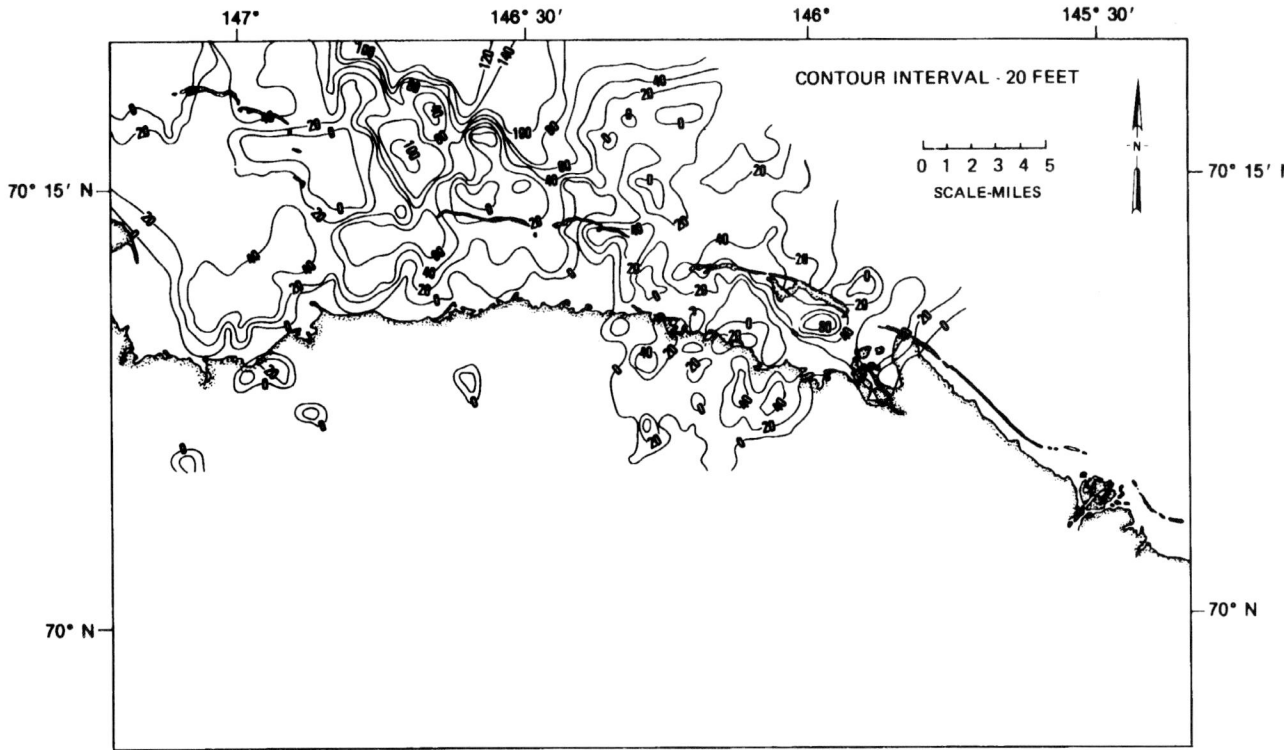

Figure 13. Isopach map of Holocene sands and silty sands overlying the youngest major gravel unit of the Canning River delta. In contrast to the major gravel units the Holocene sands increase abruptly in thickness seaward of the shoreline indicating a major change in depositional environment (modern shoreline stippled). Coarser sediments are not transported any distance offshore. Small volumes of sand are trapped behind and around the barrier islands but are scoured clean, presumably by tidal scour, from the channels between the islands.

into the underlying gravels. Immediately beyond the barrier islands, the channel fill increases to more than 100 ft (30 m). Laterally from the channel, the wedge of sediment spreads and thins to a mean thickness of about 40 ft (12 m).

Perhaps as significant as the local thickening of the sands and silts is the area over which they are absent or thin (Figure 13). To some extent, the thinning coincides with local highs forming the nuclei of the barrier islands. However, large areas exist where sands and silts have not been deposited inside the lagoon, for example, behind Flaxman Island. These bare areas are difficult to explain because in two areas they occur as closed depressions, whereas a third area appears on a broad ridge extending out to the Maguire Islands. Whether they are areas of nondeposition or areas scoured by tidal currents is unclear. Even more enigmatic are bare areas seaward from the Maguire Islands. They are unlikely to be due to current action because the trend runs at right angles to the major direction of longshore drift. Possibly the bare areas are areas of nondeposition in a protected zone behind the Maguire Islands. This idea is further supported by the convexity of the bathymetric contours over the thick sediment fill in the Challenge Channel (west of the Maguire Islands), suggesting that recently the channel has been a major conduit for sands and silts migrating from the lagoon into the deeper ocean.

Since the area was so deeply dissected during the late Wisconsin sea-level minimum, this uppermost sand and silt unit likely represents the sum total of sediment deposited during the late Wisconsin and Holocene marine transgression. The presence of wood fragments with marine shell fragments suggests that the sediment was deposited entirely in a shallow-water nearshore marine environment; this observation is consistent with the present lagoon and nearshore environment.

The alternating succession of massive gravel units and silt and sand units is typical only of the western half of the Canning River delta area. Eastward, the deeper older gravel units disappear, which suggests that the Pleistocene Canning River has shifted much farther eastward more recently. However, despite the lack of the braided stream gravels, the transgressive cycles are still recognizable (Figure 10). Seams of fibrous peat occur in at least two levels in US Geological Survey core #18 drilled about 1000 ft (300 m) north of Flaxman Island. The deepest of the fibrous peat layers occurs between 327 and 338 ft (99.7 and 103 m) below sea level at the base of a massive, gray-brown sandy silt that is about 57 ft (17 m) thick. The accumulation of peat requires a fresh or, at worst, brackish water swamp environment. However, toward the top of that same silt unit, shell fragments become numerous, which suggests a transition to marine conditions. Directly above this silt unit lies a thin, gray sand immediately overlain by a gray silty sand. At the base of the silty sand 358 ft (109 m) below sea level, numerous horizontal seams of peat were again encountered. Only 11.5 ft (3.5 m) above the peat, many shell fragments again appear. Each of these units perhaps represents transgressive cycles with peat deposition occurring in the onshore swampy areas laterally adjacent to deltaic braided stream deposits. The number of boreholes penetrating to this depth is limited, but the profiles through the delta suggest a lateral continuity of the peat-bearing horizons with thin gravel units farther west. With this continuity, as many as five transgressive cycles may be present in the more than 300 ft (91 m) of section examined beneath the Canning River delta. How many of these cycles represent eustatic sea-level change compared to local tectonism is not known.

Origin of the Barrier Islands

The chain of low-lying islands directly offshore from the Canning River delta appear to be geologically important because they significantly affect the movement of water and sediments in the immediate vicinity of the present day delta.

The previous description of the major gravel units and the Flaxman Formation shows that the majority of the islands have cores of older materials that were probably Woronzofian and that these cores are remnants left from an erosional episode during the last Wisconsin sea-level low. The general morphology of the islands suggests that they are in part constructional features.

Using gravel clast composition data, Hopkins and Hartz (1978) interpreted the constructional portions of the islands to be lag gravels derived directly from the core materials forming the islands. They also pointed out that the island chain does not form a unified sediment transport system because the major passes between the islands are so deep that they act as barriers to sediment transport. From this observation they concluded that the gravels forming the constructional part of each island must be derived locally from the island's core of older materials. These conclusions seem reasonable in part. However, Flaxman Formation clasts dominate beaches at localities obviously cored by delta (Brooks Range) derived gravels. This finding implies that either Hopkins and Hartz (1978) are wrong in their assertion that the gravels cannot cross the channels between the islands, or the gravels crossed at an earlier time when the channels were configured differently.

The evidence suggests that the offshore islands are ephemeral features that will almost certainly succumb to erosion in a geologically short time. This conclusion has important bearing on the delta model, as it suggests that such offshore islands are not necessarily an integral part of any generalized model for fan-deltas. They are the random product of sea-level change, which by chance returned the present sea level to a level almost coincident with the Woronzofian sea-level maximum. Thus, while offshore islands and lagoonal environments are common along the Beaufort Sea Coast today, they were not necessarily so in the past. Without the barrier islands, the depositional environment would be changed somewhat in that the shore-fast-ice zone would be much closer to the delta front and the silts and sands deposited immediately seaward of the delta would be much more intensely reworked by ice scouring. Also, without the protection of the barrier islands, the sands and silts would be more generally dispersed into the deeper ocean seaward from the delta. Finally, since the islands are such ephemeral structures, involvement by man in the vicinity could result in erosion and, perhaps ultimately, in the total disappearance of the islands because no clear evidence exists that the gravels forming their constructional parts are continually replaced.

Fan-Deltas in the Ancient Record

Fan-delta deposits should be recognized readily in cores taken through an ancient succession by the predictable depositional sequence. The most obvious lithology is the poorly sorted sandy gravel of the main delta unit, which is a typical braided stream deposit. However, the gravels in all cases should be overlain by poorly sorted sandy silts and silty sands that may be organic-rich swamp deposits or perhaps beach deposits at their base but grade rapidly upward to marine sands and silts with little or no change in sedimentary texture. These sands and silts should all show the effects of ice scouring, first by being poorly sorted but more by having highly disturbed, contorted bedding. In a transgressive sequence, the units overlying the gravel should all show a generally upward fining in response to the apparent retreat of the shoreline and hence the river mouth. Also distinctive from these sediments is the abundance of organic fragments that result from the spring flood flushing terrigenous materials some distance out to sea on the sea ice.

CONCLUSIONS

Polar deltas are different from deltas formed in more temperate climates because the severe climate of the region limits hydrologic activity for much of the year and thus affects depositional processes by:

1. locking up precipitation in snow or ice for much of the year,
2. forming erosion-resistant permafrost, and
3. producing sea ice which almost eliminates wave action.

Consequently, most hydrologic activity occurs for a short period in spring and early summer. The result is that the sediment-carrying capacity of the stream is overloaded and most deposition occurs in braided channels. Fan deltas, therefore, tend essentially to be large subaerial fans with little progradation.

Depositional environments within the modern delta include braided stream channels, thermokarst lakes, mainland beaches, and the open ocean, and may also include barrier islands and associated lagoons.

Three-dimensional geologic block modeling with the computer provided a unique insight into the structure of a typical polar delta that formed on the Canning River. The model shows that the delta is underlain by a succession of sediments representing at least five sea level cycles. Generally, the succession consists of a series of depositional sequences that begin with a massive, poorly sorted gravel unit deposited subaerially in braided stream channels, which is followed by a unit of dark-colored, poorly sorted silts or silty sands that may contain peat layers at its base and a marine shelly fauna toward its top. The marine portions of these silt or sand units generally show signs of considerable soft sediment deformation due to ice gouging. However, not all sequences in a stacked series will be so simple because, during sea-level lowstands, so much erosion is possible that gravel units may be complex deposits representing more than one sea level cycle.

The offshore barrier islands occurring in front of the Canning River delta and along much of the Beaufort Sea coast all appear to be ephemeral structures built about a nucleus either of an older deltaic gravel or of the Flaxman Formation. The nuclei of the islands thus all appear to be erosional remnants of Woronzofian age left after the late Wisconsin sea-level low. The existence of these islands and the lagoons landward of them thus fortuitously results from the present sea level being a few feet lower than the Woronzofian sea-level maximum. The lagoon environment therefore is not typical of fan deltas; the result is that the shore-fast-ice zone will, in most cases, be narrower and that deformation of nearshore marine sediments will be more intense around the margin of most fan deltas.

A polar fan-delta sequence should be recognized readily in the ancient record, particularly from the highly deformed nature of the ice-scoured silty sands and sandy silts overlying the main gravel units. The geochemical methods proposed by Naidu and Mowatt (1974, 1975), when tested on ancient sediments, may in combination with the model discussed here, aid in eliminating any doubt about the origins of such sediments.

REFERENCES CITED

Arnborg, L., H. J. Walker, and V. Peippo, 1966, Water discharge in the Colville River, 1962: Geografiska Annaler, v. 48, p. 195–210.

Barnes, P. W., and D. M. Hopkins, eds., 1978, Geological sciences, in Beaufort/Chukchi-National Oceanic and Atmospheric Administration/Bureau of Land Management, Environmental assessment of the Alaskan Continental Shelf, interim synthesis: Boulder, Colorado, p. 101–133.

Barnes, P. W., and E. Reimnitz, 1974, Sedimentary processes on Arctic shelves off the northern coast of Alaska, in J. C. Reed and J. E. Slater, eds., The coast and shelf of the Beaufort Sea: Proceedings of the Symposium on Beaufort Sea Coast and Shelf Research: Arlington, Virginia, Arctic Institute of North America, p. 439–476.

Black, R. F., 1964, Gubic Formation of Quaternary age in northern Alaska, in Exploration of Naval Petroleum Reserve no. 4 and adjacent areas, northern Alaska, 1944-53, U.S. Geological Survey Professional Paper 302-C, p. 59–91.

Black, R. F., and W. L. Barkdale, 1949, Oriented lakes of northern Alaska: Journal of Geology, v. 57, p. 105–118.

Bruggers, D. E., and J. M. England, 1979, USGS geotechnical investigation Beaufort Sea, Alaska—1979: Houston, Harding-Lawson Associates report, Job 9619, p. 005.08.

Creager, J. S., and D. A. McManus, 1967, Geology of the floor of Bering and Chukchi Seas—American Studies, in D. M. Hopkins, ed., The Bering Land Bridge: Stanford, Stanford University Press, p. 7–31.

Curray, J. R., 1960, Sediments and history of Holocene transgressions, continental shelf, northwest Gulf of Mexico, in F. P. Shepard, F. B. Phleger, and T. H. van Handel, eds., Recent sediments, northwest Gulf of Mexico: Tulsa, AAPG, p. 221–266.

Curray, J. R., 1961, Late Quaternary sea level: a discussion: GSA Bulletin, v. 72, p. 1707–1712.

Fairbridge, R. W., 1961, Eustatic changes in sea level: Physics and Chemistry of the Earth, v. 4, p. 99–185.

Holmes, A., 1965, Principles of Physical Geology: London, Thomas Nelson and Sons, 1288 p.

Hopkins, D. M., 1967, Quaternary marine transgression in Alaska, in D. M. Hopkins, ed., The Bering Land Bridge: Stanford, Stanford University Press, p. 47–90.

Hopkins, D. M., and R. W. Hartz, 1978, Coastal morphology, coastal erosion and barrier islands of the Beaufort Sea, Alaska: USGS Open File Report, no. 78-1063, 54 p.

Jones, T. A., 1988, Modeling geology in three dimensions: Geobyte, v. 3, no. 1, p. 14–20.

Jones, T. A., 1992, Extensions to three dimensions: Introduction to the section on 3-D geologic block modeling, (this volume).

Jones, T. A., D. E. Hamilton, and C. R. Johnson, 1986, Contouring geologic surfaces with the computer: New York, Van Nostrand Reinhold, 314 p.

Leffingwell, E. K., 1908, Flaxman Island, a glacial remnant: Journal of Geology, v. 16, p. 56–63.

Leffingwell, E. K., 1919, The Canning River region, northern Alaska, U.S.: USGS Professional Paper 109, 251 p.

Lewellen, R. I., 1970, Permafrost erosion along the

Beaufort Sea Coast: Denver, Geography and Geology Department, University of Denver, 25 p.

MacCarthy, G. R., 1970, Glacial boulders on the Arctic coast of Alaska: Arctic, v. 11, p. 71–85.

McCulloch, D. S., 1967, Quaternary geology of the Alaskan shore of Chukchi Sea, *in* D. M. Hopkins, ed., The Bering Land Bridge: Stanford, Stanford University Press, p. 91–120.

Naidu, A. S., and T. C. Mowatt, 1974, Clay mineralogy and geochemistry of continental shelf sediments of the Beaufort Sea, *in* J. C. Reed and J. E. Sater, eds., The coast and shelf of the Beaufort Sea: Proceedings symposium on Beaufort Sea coast and shelf research, p. 493–510.

Naidu, A. S., and T. C. Mowatt, 1975, Depositional environments and sediment characteristics of the Colville and adjacent deltas, northern arctic Alaska, *in* M. L. Broussard, ed., Deltas, models for exploration: Houston, Houston Geological Society, p. 283–309.

O'Sullivan, J. B., 1961, Quaternary geology of the Arctic coastal plain, northern Alaska: Ph.D. dissertation, Ames, Iowa State University of Science and Technology, 204 p.

Reimnitz, E., and P. W. Barnes, 1974, Sea-ice as a geologic agent on the Beaufort Sea shield of Alaska, *in* J. C. Reed and J. E. Slater, eds., The coast and shelf of the Beaufort Sea: Proceedings symposium of Beaufort Sea coast and shelf research: San Francisco, Arctic Institute of North America, p. 301–351.

Reimnitz, E., and L. Toimil, 1977, Diving notes from three Beaufort Sea sites, *in* P. W. Barnes et al., eds., Marine environmental problems in the ice-covered Beaufort Sea shelf and coastal regions: Natl. Oceanic and Atmospheric Administration, Environmental assessment of Alaskan Continental Shelf, principal investigator's reports for year ending March 1977, v. 17, p. J1–J7.

Rex, R. W., 1964, Arctic beaches, Barrow, Alaska, *in* R. L. Miller, ed., Papers in marine geology: Shepherd Commemorative Volume: New York, Macmillan, p. 384–400.

Schell, D., and G. Hall, 1972, Water chemistry and nutrient regeneration process studies, *in* Baseline data study of the Alaskan Arctic aquatic environment: Fairbanks, Institute of Marine Science, University of Alaska, Report R72-3, p.3–28.

Walker, H. J., 1974, The Colville River and the Beaufort Sea, *in* J. C. Reed and J. E. Slater, eds., The coast and shelf of the Beaufort Sea: Proceedings Symposium of Beaufort Sea Coast and Shelf Research, San Francisco, Arctic Institute of North America, p. 513–540.

Wiseman, W. J., J. M. Coleman, A. Gregory, S. A. Hsu, A. O. Short, J. N. Suhayda, C. D. Walters, and L. D. Wright, 1973, Alaskan Arctic coastal processes and morphology: Baton Rouge, Louisiana State University, Coastal Studies Institute, Technical Report no. 149, 171 p.

Chapter 15

Three-Dimensional Geologic Block Modeling Of The Kutcho Creek Massive Sulfide Deposit, British Columbia

David E. Hamilton
Landmark/Zycor Inc.
Austin, Texas, U.S.A.

Robert S. Didur
Fernie, British Columbia
Canada

ABSTRACT

Three-dimensional geologic block models were used to evaluate the Kutcho Creek volcanogenic massive sulfide deposit in north-central British Columbia. This stratiform deposit dips 45° and contains significant amounts of copper, zinc, silver, and gold mineralization. Over 16,000 m of drill core information and 1000 m of assays for each of the four metals and for specific gravity were used. Geologic contacts from directionally drilled holes were available for each of nine interpreted horizons.

Separate three-dimensional models were built for each metal and for specific gravity. Each model contained approximately 3.1 million cells, with cell dimensions of 10 m by 10 m (horizontal) by 2 m (vertical). Grids were used to define rock-unit boundaries and the correlations within those rock units. These grids controlled three-dimensional interpolation of assay values from the drill holes to the model cells.

Individual metal models were combined into a single copper-equivalent model using current milling and market costs for each metal. Sections through the model were used to quality check the model and to design several open-pit mine configurations. Geologic and mine-recoverable reserves, including average grade, tonnage, and waste/ore ratios, were determined for the entire deposit and on a bench-by-bench basis for the final pit. The calculations were used to fine-tune the final pit design and for mine planning.

INTRODUCTION

The Kutcho Creek volcanogenic massive sulfide deposit is situated in the Stikine Range of the Cassiar Mountains, north-central British Columbia (Figure 1). In 1967, a group of companies, including Imperial Oil Limited (IOL), located a significant copper and zinc stream-sediment anomaly near the deposit. From 1967 through 1974, IOL and Sumac Mines Limited (SML) ran soil and stream geochemical surveys, sampled outcrops, completed surface mapping, and flew several thousand miles of airborne electro-magnetic surveys. In 1974, both IOL and SML separately began drilling programs on adjacent leases of the Kutcho Creek deposit. This drilling defined several mineralized zones, the largest being the Kutcho Creek zone. In 1978, data from both companies were pooled to allow better understanding of the geology and a complete evaluation of the deposit's potential. By 1979, when this 3-D modeling project commenced, these companies had drilled 140 holes (Figure 1), logged over 16,000 m of core, and assayed more than 1000 samples for copper, zinc, silver, and gold. This paper describes the methods used to build three-dimensional models of this deposit (the Kutcho Creek zone), and how those models aided in understanding the geology and economics of the deposit.

Evaluation of the economic potential of the Kutcho Creek deposit required estimates of average ore grade, tonnage, and waste/ore ratios for the entire deposit, individual leases, and within a variety of open-pit mine designs. Manual techniques were ineffective in generating these estimates due to the quantity and complexity of the data and required output. Computer modeling efficiently used all the data and enabled testing of deposit sensitivity to variations in ore grade cutoff, dilution factors, bench heights, open-pit configuration, and to fluctuations in milling process and market value. Computer modeling also allowed rapid updating of these estimates when new bore-hole information became available.

Testing of the parameters described above was done using computer-generated 3-D geologic block models (Jones, 1988a). As discussed by Jones (1992) these models are static in time and attempt to use geologic interpretation with top picks and attribute data (e.g., ore grade, specific gravity) to assign attribute values to the cells of the block model. The result is a three-dimensional volume that is an estimate of the modeled attribute's distribution through the rock units. The model was built in 1979 and 1980 using Exxon's proprietary program.

Three-dimensional geologic block models were built for each metal and for specific gravity. The metal models were combined into a single copper-equivalent model using current milling and market costs for each metal. Cross sections and maps allowed quality checks of the model. Computer-assisted section-by-section design and optimization of the open-pit mine produced the final pit design. Average grade, tonnage, and waste/ore ratios were determined for the

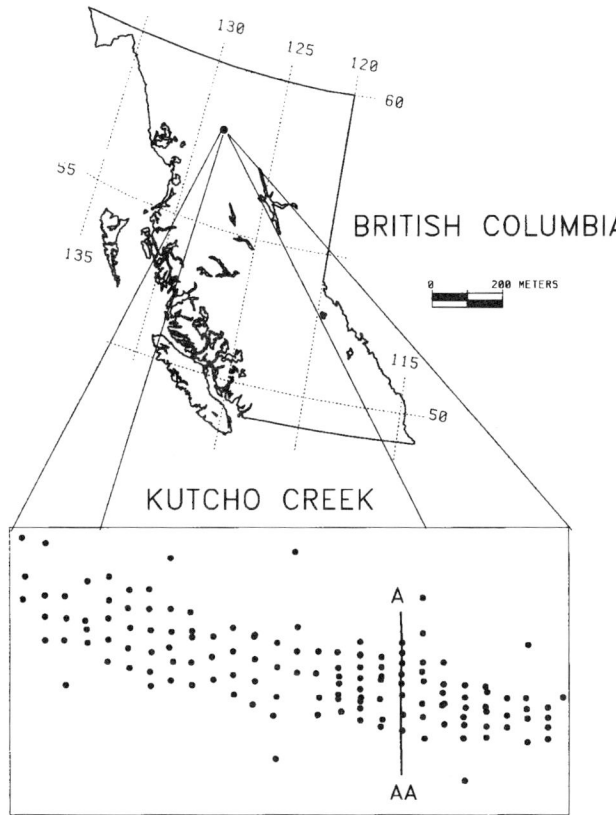

Figure 1. Map of British Columbia showing the location of Kutcho Creek and an enlargement showing the drill holes.

entire deposit, for each pit and on a bench-by-bench basis for the final pit. Steps to model and evaluate this deposit were fully automated. Those procedures have subsequently been used to update the model based on new data and require about three days for each update.

GEOLOGY AND DATA

Geology

The Kutcho Creek volcanogenic massive sulfide deposit occurs in a belt of predominantly silicic pyroclastic and flow rocks of Triassic age. The volcanic sequence, which includes some basic tuffs and flows, is interbedded with argillites at its top and is overlain by various sedimentary rocks of Lower Jurassic age (Figure 2). The overall sequence strikes east-west, dips 45 degrees to the north and has been foliated, folded, and metamorphosed to lower greenschist facies.

The ore deposit occurs at or near the top of the felsic volcanic pile in a 300-m thick Sericite Schist unit interpreted to be a metamorphosed rhyolite lapilli tuff. The upper 50–60 m of schist which underlie the deposit contain up to 50 percent pyrite near the zone,

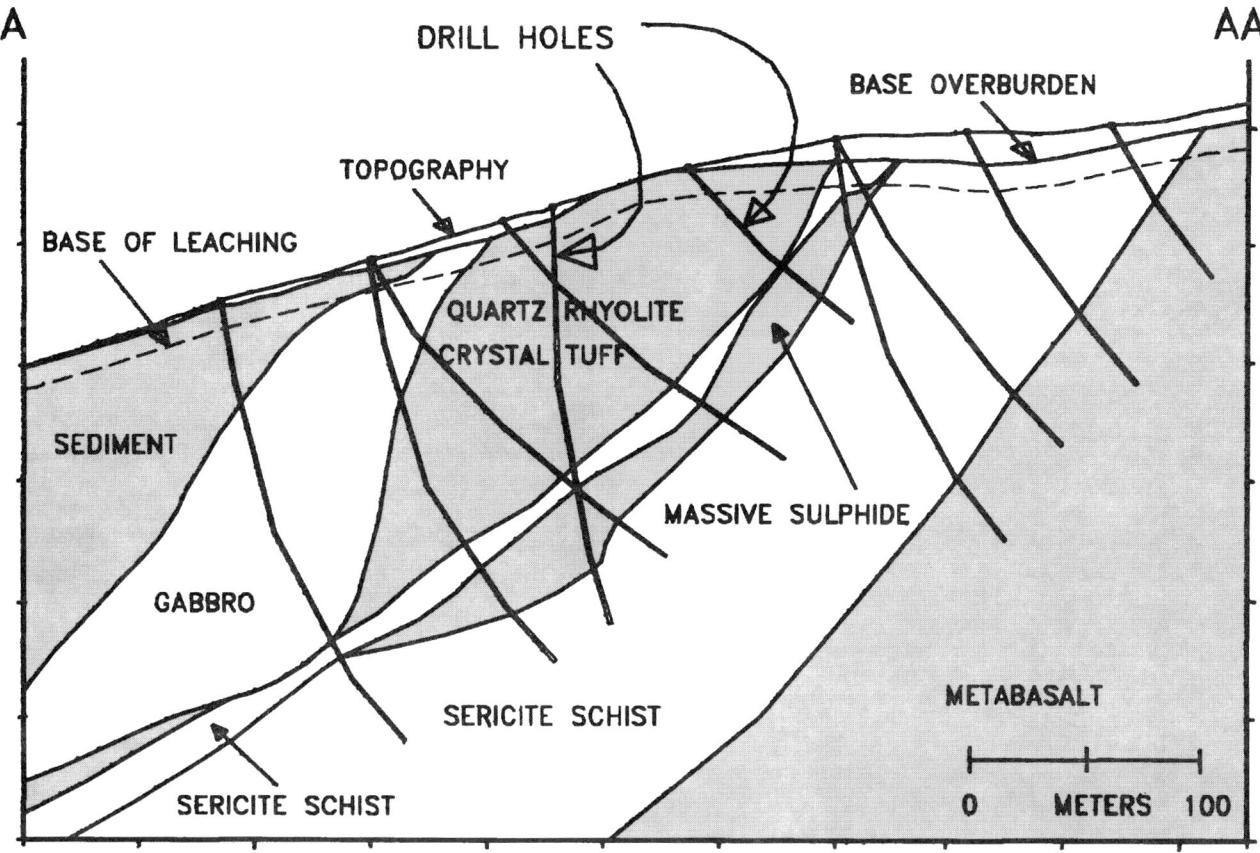

Figure 2. Typical cross section through Kutcho Creek deposit showing all major rock units. Drill holes are projected onto section. See Figure 1 for location of section.

decreasing with depth to 5 percent. Within the lower portion of the Sericite Schist is a sheet-like, continuous basic rock unit termed metabasalt which represents a metamorphosed basaltic tuff.

The Kutcho zone is approximately 1500 m long, 300 m wide, up to 30 m thick, and plunges 5 to 20 degrees westward. It is wedge-shaped in cross section with the thicker portion near the surface. The deposit averages 65% sulfides and consists of multiple sulfide lenses with significant metal values in both massive and disseminated sulfide zones. The massive sulfide portion is typically 90% sulfides and characterized by fine-grained pyrite with lesser chalcopyrite, sphalerite, and bornite. Minor amounts of chalcocite, covelite, galena, and tetrahedrite also occur.

A quartz-feldspar crystal tuff unit up to 200 m thick overlies the massive sulfides and laterally interfingers with the rhyolite lapilli tuff; these are collectively called the quartz rhyolite crystal tuff unit.

Basic rock units consisting of local tuffs and intrusives varying between amphibolite and feldspar porphyry form discontinuous piles above the quartz rhyolite crystal tuff and within the rocks above the Kutcho sequence. These basic rocks have been collectively termed gabbro for the modeling process.

Overlying the Kutcho sequence of volcanic and intrusive rocks are the sedimentary rocks which consist mainly of limestone and argillite formations.

Data

Data used for modeling came from a topographic survey, surface mapping, and drill-hole data. An aerial photo survey was flown and used to generate a topographic map with 10 m contours. Contours from this map were digitized and a grid generated. Surface mapping revealed many outcrops but no exposed contacts between rock units. The outline of each outcrop was digitized as a string of X-Y points.

Drilling was along north-south lines (sections oriented approximately normal to strike) with holes spaced 60 m apart. When this study began 140 diamond drill holes had been drilled, with up to 15 holes being found along some lines (Figure 2). Hole depths reached 250 to 750 m and were deviated to penetrate the rock units perpendicular to bedding. Several holes were drilled from one site by whip-stocking. Typically 5 to 10 directional surveys were run on each hole that was directionally drilled. Split core samples were assayed every meter through the ore zones and the entire core was logged by a geologist.

The elevation, X-Y location, and orientation of the

Table 1. Format and Representative Values for the Horizon

ID	X-Coord.	Y-Coord.	Topo.	Base Ovb	Top Gabbro	Top Sed.
DH51	38421	22534	1672.3	Null	Null	Null
DH51	38424	22525	Null	1663.0	Null	Null
DH51	38427	22481	Null	Null	1629.7	Null
DH51	38425	22469	Null	Null	Null	1607.5
DH06	38396	22623	1592.9	Null	Null	Null
DH06	38391	22577	Null	Null	1548.4	Null
DH06	38382	22558	Null	Null	Null	1526.3

Table 2. Format and Typical Values of Interval Data

Beginning (ft)	Ending (ft)	Cu%	Zn%	Au%	Ag%	SpGr
234.5	235.5	1.40	0.80	0.23	0.09	2.94
235.5	236.5	2.67	1.11	0.43	0.08	3.27
236.5	237.5	5.22	1.43	0.61	0.27	Null
237.5	238.5	2.19	1.23	0.31	0.37	3.09

beginning of each hole, plus azimuth, deviation, and distance to each survey point, were recorded and stored for each hole in the database. As each file was read into the modeling program, this survey information was used to convert automatically the depth to each pick and to the top and bottom of each data interval to elevation and X-Y location.

Top Picks Data

The following stratigraphic horizons were used to model this deposit. Drilling depths where each hole penetrated a horizon (i.e., picks) were recorded.

- Surface topography
- Base of overburden
- Top of gabbro
- Base of gabbro
- Top of quartz rhyolite crystal tuff
- Top of hanging wall sericite schist
- Top of massive sulfide
- Base of massive sulfide
- Top of metabasalt

The modeling program automatically extracted top picks from the drill-hole database and placed them in a format compatible with the surface-mapping program: one record (row) for each X-Y location with multiple fields (columns) per record. Each field represents elevation of a top, base, or other value recorded at that location. A pick file extracted from the drill-hole database had the form shown in Table 1.

Because the holes are deviated, there is a different X-Y coordinate for each pick. Thus, in a record the elevation of one pick was listed and all other picks were set to missing (null). When a pick was not recorded, as for base of overburden in DH06, no record was created.

Interval Data

Information was obtained from the boreholes in one-meter intervals. The distance to the top and base of each interval was recorded, along with assay values (percent of Cu, Zn, Au, and Ag), and specific gravity. Drill-hole interval data had the form shown in Table 2. If any entry was missing for an interval, then a special code (null) was stored.

GRID CONSTRUCTION

Several 3-D modeling programs use grids to define the major boundaries between rock units. In addition, as discussed by Jones (1992), there are many ways to correlate data between drill holes when assigning values to the 3-D model cells within each rock unit. Exxon's 3-D modeling program allows either grids or ellipsoids of varying axial lengths to control these correlations. Because the Kutcho Creek deposit is stratified, grids were used to define the major rock unit boundaries. Grids were also used to guide the correlation of data between those rock unit boundaries. Before 3-D geologic block models could be constructed, these grids needed to be built.

All grids used in this project had the same origin, X-Y limits, and grid increments. The grid increment was 10 m in both the X- and Y-direction. This allowed all data to be honored along drill hole sections and resulted in six increments (3-D model cells) between sections. The gridding method used was a spreading-from-data algorithm similar to that described by Walters (1969) and Jones et al. (1986). Following are descriptions of the major gridding procedures.

Topography

The digitized topography contours were gridded, contoured, and error-corrected until an acceptable map was produced. Elevation was then interpolated from the digitized topography grid at each collar location and compared to the surveyed collar elevations to determine error between the topography grid and the collars. Most errors were plus or minus a few meters. A grid of error was built and added to the initial topography grid, smoothly tying the grid to the drill-hole collars (Jones et al., 1986).

Base of Overburden

The digitized outcrop locations were used to define absence of overburden. Elevations were interpolated from the final topography grid at these locations. These outcrop elevations were then merged with base-of-overburden picks retrieved from the drill-hole database. This combined base-of-overburden data was gridded, contoured, and error-corrected until an acceptable base-of-overburden grid was produced.

At outcrops, the base-of-overburden grid intersected and often projected above the topography grid. These projections were expected and were corrected by creating a new base-of-overburden grid which was the deeper of the two. This new base of overburden grid correctly represented the base of overburden where overburden existed and was near or coincident with topography at outcrops.

Other Structural Surfaces

Surfaces from the top of gabbro down to the top of metabasalt were layered and roughly conformable (quasi-parallel) in most areas. Because of this, these units were modeled using the shape-assist technique (Fierstien and Brewster, 1992). Since the hanging wall sericite schist (HWSS) is present over the entire area, has a dip similar to most of the surfaces, and was penetrated by nearly all drill holes, we selected it as control to assist in building the other surfaces. The hanging-wall- sericite-schist grid was built directly from its top picks. To transfer the general trend and not a detailed shape of the HWSS grid to each surface, we calculated a second-order trend of the HWSS grid and used that as control.

Grids of thickness between the HWSS Trend grid and each of the other surfaces were used to build structure grids for those surfaces. At each location in the data file, a value was interpolated from the HWSS Trend grid and placed in a new field called HWSS TREND. The HWSS TREND field was subtracted from each surface's field and the resulting values placed in a new residual field for each surface. A grid was constructed for each residual field and then added to the HWSS Trend grid, creating the structure grid for each surface.

For most surfaces, this shape-assist technique produced acceptable results. However, the top of metabasalt extrapolated too deep on the down-dip edge of data and the top of gabbro had several problems inside the data area. The gabbro was not continuous everywhere and pinched out in the middle of the data area. Since there were no data values where the gabbro pinched out, dummy data points were used to prevent extrapolation through those regions. Where gabbro data were absent and gabbro was interpreted to have pinched out, dummy data for the top of gabbro were entered with values below the base of gabbro. The resulting top-of-gabbro grid projected below the base of gabbro in these areas, forcing the pinchout. Dummy points similarly were used to prevent the top of metabasalt from projecting too deeply on the down-dip edge.

Because the drilling program was ongoing, a procedure was developed to automatically remove these dummy values from the file when they conflicted with newly acquired data. This procedure set dummy values for a surface that were within 30 m of a real data value for that surface to missing.

Final Grid Construction

Three more steps were needed before the grids were ready to use for 3-D model construction. First, although the initial grids were built and adjusted until they matched the geologic interpretation, the surfaces had not been compared to one another to prevent them from crossing. Second, it was possible that extrapolation of one surface would cause it to project above another surface's data. This had to be checked for and corrected. Third, additional grids were needed to guide the 3-D modeling process.

Baselap/Truncation Operations Between Grids

Jones et al. (1986) and Jones and Hamilton (1992) discuss how simply constructed grids, as described above, may cross one another at pinchouts or as they project away from the data. They describe procedures which compare adjacent grids and the higher or lower elevations of the two are output as a new grid, replacing one of the original grids. For example, during construction, the base-of-overburden grid was forced to project above the topography grid at rock outcrops. These grids were compared and a new base-of-overburden grid was constructed which was the minimum of the two. This new base-of-overburden grid correctly represented the base of overburden where overburden existed and was coincident with topography at outcrops.

To prepare the other grids, the top of hanging wall sericite schist was compared to the surface below it and the lower elevations of the two were output, creating a new lower grid (top of massive sulfide). That new lower grid was then compared to the surface below it and a new grid for the lower surface created in the same way. This process continued down to and including the top of metabasalt. Similarly, the top of hanging wall sericite schist was compared to the surface above it and the higher elevations of the two were output creating a new higher grid. This process comparing and outputting higher values continued up to the base of overburden. The final step involved truncating all surface grids below the base of overburden by comparing each to the base of overburden and outputting the lower of the two as a new, truncated, lower surface grid.

Quality Control For Honoring Data

Because the drill holes were deviated, tops for one surface were not vertically above tops for the next lower surface. This meant that one grid could potentially project up or down between its data points and violate data for the surface above or below it (Hamil-

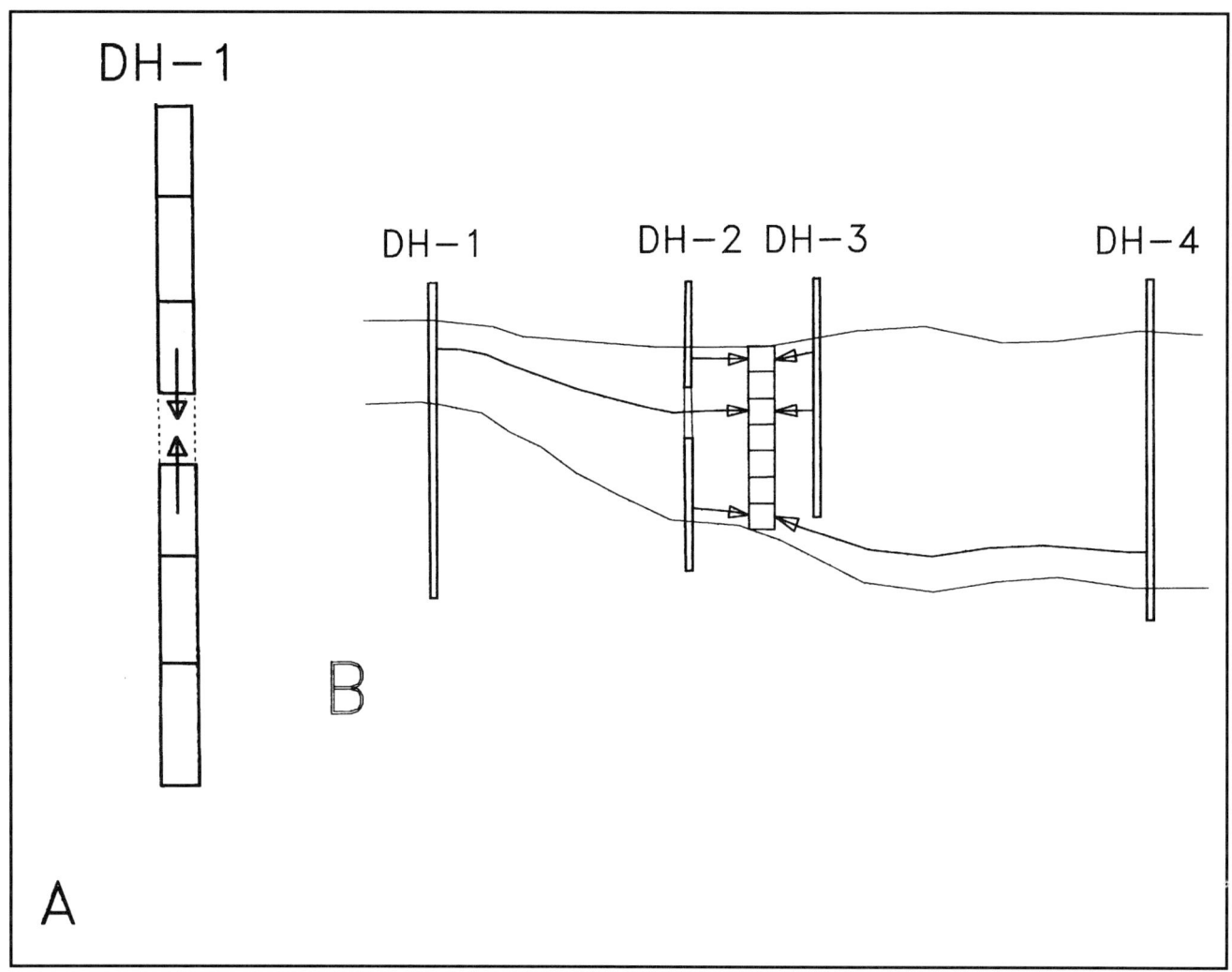

Figure 3. Schematic diagrams showing how missing data could be handled. (A) Drill hole with missing interval value assigned from interval above, below, or from both. (B) Cross section through two horizons with four drill holes displayed. Stack of 3-D model cells are displayed in middle of section. Arrows indicate which intervals from which holes are used for value calculation. Drill hole 2 is missing data through part of zone, and hole number 3 only partially penetrates zone.

ton and Henize, 1992). The baselap and truncation operations described above often forced the surface above or below a violating surface to be coincident with the violating surface where they cross. This adjusted surface no longer honored its original data. Interpolation was used to identify these problems. Elevations for the surfaces were interpolated from the final baselapped and truncated grid and stored in a new field. Those values were then subtracted from the original data for that surface, creating a field that represented the difference between the grid and the data. If the maximum error was less than 0.2 m the surface was considered acceptable. If the error was greater than this, cross sections were generated through the problem drill holes and the problem evaluated and corrected. Only a few of these problems occurred and they were associated with sharp changes in surface gradient.

Additional Grids For 3-D Modeling

As discussed above, Exxon's 3-D modeling program allows either grids or ellipsoids of varying axial lengths to control correlations between drill holes when assigning values to 3-D model cells. For this project, we controlled the correlations by adjusting grid geometries and by specifying which grids were used for correlation (Jones, 1988b).

The 3-D program was designed to default so that correlations would be proportionally distributed between grids (Figure 3). For example, consider a rock unit whose upper and lower surfaces were defined by grids. A cell located 1/5 of the distance from the upper grid to the lower grid would be correlated with drill-hole interval data that were 1/5 of the distance from the unit's upper pick to its lower pick (or grid, if

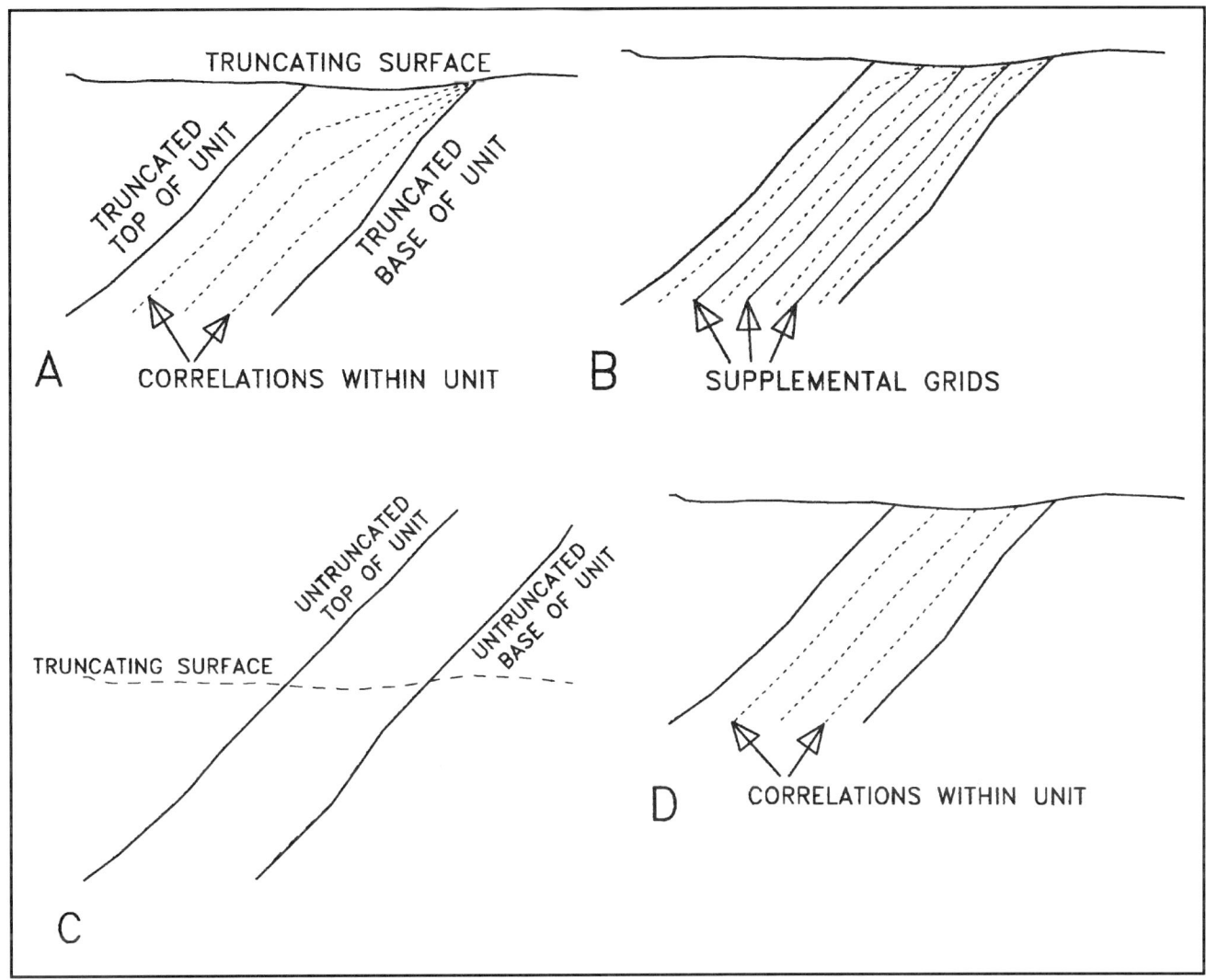

Figure 4. Cross sections showing the massive sulfide truncated by the base of overburden. The dotted lines represent the default (proportional) correlations. (A) Default (proportional) correlations during interpolation create an unreal pinchout. (B) Using supplemental grids reduces the unreal pinchout problem.
(C) Untruncated grids for the top and base of massive sulfide can be used to create a model that projects past the truncating surface. (D) Cell values that originally projected past a truncation were removed except in the area where the unit exists, avoiding unreal pinchout problem.

a pick was missing). When the drill hole ended before penetrating the base of a unit, the grid for the base of the unit was used just as if the drill hole had penetrated and the pick was missing. For deviated drill holes, the distance along the hole could not be used to determine reliably the correlative position of the interval data. In these situations each interval was tested, by interpolation on the grid above and below it, to determine its correlative position within the unit.

Mineralization occurred within the massive sulfide and slightly above and below it. Correlations within the massive sulfide unit were assumed to be proportionally distributed (conformable) between top and base. Correlations above and below the massive sulfide were assumed to parallel its upper and lower surfaces.

Within the massive sulfide unit, the program would automatically treat correlations proportionally. If this unit did not baselap or was not truncated, then this default approach would be acceptable. However, as described above, the dipping top and base of massive sulfide grids were truncated by the base of overburden, resulting in a final top-of-massive-sulfide grid which ran along the base of overburden until it encountered and became coincident with the truncated base of massive sulfide. Where these grids became coincident created a pinchout which did not really exist (Figure 4A).

Because correlations are proportionally distributed between grids, this incorrect pinchout caused the correlations to squeeze together as if a depositional pin-

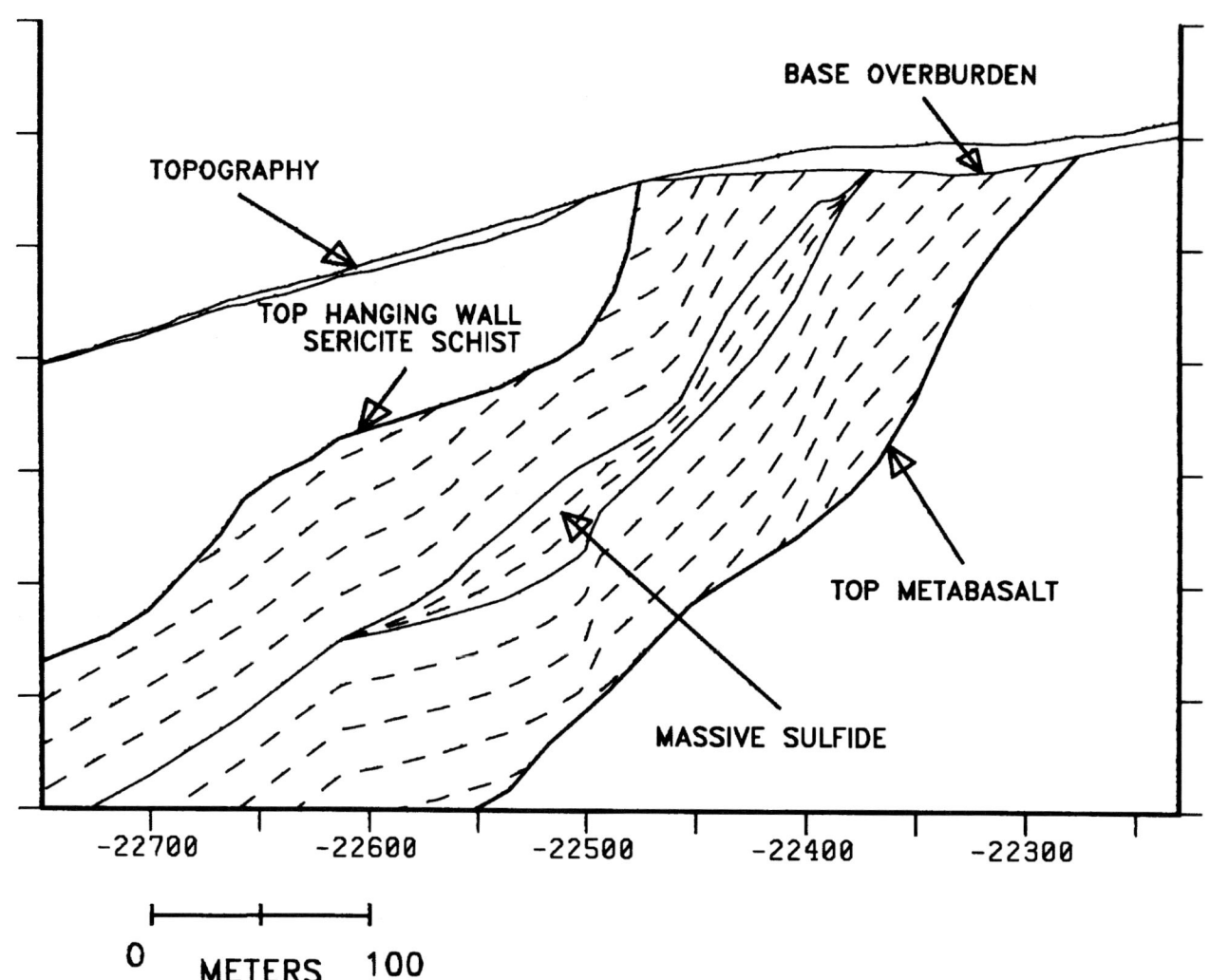

Figure 5. Cross section showing the primary rock unit boundary grids (solid lines) and the supplemental grids (dashed lines) that are used to control correlations during construction of the metal models.

chout had occurred rather than a truncation (Figure 4A). To prevent or minimize this pinchout effect, we created supplemental grids proportionally spaced between the untruncated top- and base-of-massive-sulfide grids. These grids were then truncated by the base of overburden and, when used for modeling, reduce the size of this depositional pinchout effect to the point of insignificance (Figure 4B).

Other approaches are available to handle this problem (C. Johnson, 1991, personal communication). For example, the untruncated grids for the top and base of massive sulfide could have been used to build a 3-D model that extended above the base of overburden and topography into the "air" (Figure 4C). Then, the correlations would have extended properly up to and past the truncation. After model construction, the fictitious extrapolation of the massive sulfide past the base of overburden is removed using a simple truncation operation on the model (Figure 4D). These extra steps were not necessary since the supplementary grids reduced the effect and the 3-D model was eventually leached to 10 m below the base of overburden (simulating the approach suggested by Johnson).

Supplemental grids were also used to force correlations above and below the massive sulfide unit to parallel the unit's boundaries. These supplemental grids were parallel to the top and base and generated by shifting the massive sulfide top and base grids (before they were truncated by base of overburden) away from the massive sulfide by 20 m increments a total of six times in each direction. Once created these supplemental grids were compared to other surfaces in the framework to ensure that they did not cross (Figure 5). Standard baselap and truncation operations were used.

Supplemental grids were built similarly for the other rock units. These were not used to control correlation of assay values but to control correlation of specific gravity values. All supplemental grids were properly baselapped and truncated by adjacent grids.

As with the truncated massive sulfide grids, these supplemental grids also became coincident with the base of overburden, creating a nonexistent pinchout and the potential for incorrect correlations (Figure 4A). The space between the supplemental grids was kept small to minimize this problem.

3-D MODEL CONSTRUCTION

Three-dimensional geologic block model construction involves (1) definition of the rock-unit boundaries, (2) definition of the correlation fabric within each rock unit, (3) definition of the 3-D block model geometry (cells) into which interpolated values are placed, (4) interpolating values for 3-D cell positions from neighboring interval data using the appropriate correlation fabric and interpolation procedure, and (5) construction of other models by performing operations on or between one or more models. These procedures are discussed by Badiozamani et al. (1992), Johnson and Jones (1988), Jones (1992), Yildirim (1985), among others.

Build 3-D Metal Models

A separate 3-D model of each metal (Ag, Au, Cu, and Zn) was constructed. Each model was built using the same set of controls.

Grids Used for Correlation

As described above, grids are used by this modeling program to control 3-D correlations when assigning values to the cells of the model. Twenty grids were used to guide construction of the 3-D metal models (Figure 5). They were from top to bottom:

- Topography
- Base of overburden
- Top of hanging wall sericite schist
- Six supplemental grids above the top of massive sulfide
- Top of massive sulfide
- Two supplemental grids within the massive sulfide
- Base of massive sulfide
- Six supplemental grids below the base of massive sulfide
- Top of metabasalt

No mineralization occurred above the hanging wall sericite schist, so no rock unit grids above that surface were required. Using only needed grids significantly reduced computing time.

Modeling Parameters

A vertical cell size of 2 m was used. This size was chosen primarily for speed and storage considerations, since many hundreds of reserve estimations were to be tested, although a finer cell may have provided more accurate results (Denver and Phillips, 1992). The horizontal cell size was the same as the grid cell size—10 m—and was selected to honor data and to allow about 6 3-D model cells between drill-hole sections.

The geologist interpreted a directional bias in the strength of correlations. This bias had a magnitude of 3:1 in the direction N75W. The magnitude and direction were used to define the axes of the weighting ellipse used to discount data values when calculating a cell's value (Jones, 1988a; Kushner and Yarus, 1992). The geometry of the ellipse was also used to define the maximum distance (search radius) allowed between a model cell and a data value (40 m in the direction N15E and 120 m in the direction N75W). Since the sections along which holes were drilled ran approximately north-south and were spaced 60 m apart, this guaranteed that a model cell close to one line of drilling could still acquire information from an adjacent line. This ellipse also restricted extrapolations in the down-dip direction to a maximum distance of 40 m.

The model extended to a depth of 480 m below topography. This is well below the proposed depth of the open-pit mine and significant computer time and disk space were saved by limiting the model depth. Again, cells were assigned values to that depth only if data were within the search ellipse for that cell.

Calculating a Cell's Value

Cell values were calculated at the center of an ellipse 40 m by 120 m with major axis oriented N75W. This ellipse was separated into quadrants whose boundaries were drawn NS and EW. One drill hole from each quadrant and within the search ellipse distance was selected. This was the drill hole that contained interval data at a position correlative with the cell and was closest to that cell. Top picks were used to define the correlative position of interval data (grids if picks were missing) and the 20 grids described above were used to define the correlative position of cells.

Once the appropriate drill holes and interval data were selected, the value from each hole was calculated. The correlative position of the top and base of the cell was determined and similar correlative positions in the selected drill hole intervals were found. Since the correlative position could span several intervals and include parts of intervals, the thickness (length)-weighted average of those intervals was used as the value contribution from that hole. The four quadrant values were then combined using their distances from the center of the ellipse and inverse-distance-squared weighting. However, before being used, the distances were discounted based on the cell-hole orientation relative to the weighting ellipse. In this way, the directional bias noted by the geologist was incorporated into the modeling process.

Handling Missing Data

Some intervals were missing from the drill hole data base. When these were encountered, three methods were available to handle the missing values: assign a value or calculated an average value from intervals above or below in the same hole (Figure 3A) and use that value for cell interpolation, look through the missing interval to the next nearest drill hole, or assign a missing value to the cell. Since the deposit is highly stratified and the drill holes typically penetrated the strata at right angles to bedding, replacing missing data using values from intervals above and below was not reasonable. We had the program look through the missing drill hole interval to get a replacement value. That replacement interval(s) had to lie within the same quadrant and search ellipse boundary as did the missing interval and be correlative to the missing interval (Figure 3B). Only a few intervals were missing from the mineralized zone. However, if a drill hole did not fully penetrate a unit (usually because the bottom of the hole was reached), then the program looked past that hole just as if it contained missing data (Figure 3).

Many cells in the resulting 3-D metal models were not assigned values because they were too far from drill hole data. This was true of all cells in rock units above the hanging wall sericite schist, since there was no assay data for those units. Model operations allowed each cell in a model to be either modified (set to null, clipped, etc.) to create a new model or operated (add, subtract, etc.) against the cell in a second model having the same position to create a new model. A model operation was performed on each metal model and missing cell values were replaced with a value of zero, indicating no mineralization.

Build 3-D Specific-Gravity Model

The specific-gravity model was used to calculate tonnage of material. Specific gravities of the massive sulfide (ore), gabbro (waste), and mineralized areas of the Sericite Schist (ore) were very high relative to other units (waste); therefore, the estimated ore and waste tonnage were very different. Understanding the amount and distribution of these tonnage variations was important to mine planning.

The specific-gravity model was constructed in the same manner as were the metal models. However, whereas the metal models had data only in certain zones, specific-gravity data were available in all zones. Twenty nine grids were used to control correlation of these data. They were from top to bottom:

- Topography
- Base of overburden
- Two supplemental grids above the gabbro
- Top of gabbro
- Two supplemental grids within the gabbro
- Base of gabbro
- One supplemental grid above the quartz rhyolite crystal tuff
- Quartz rhyolite crystal tuff
- One supplemental grid within the quartz rhyolite crystal tuff
- Hanging wall sericite schist
- Six supplemental above the top of massive sulfide
- Top of massive sulfide
- Two supplemental grids within the massive sulfide
- Base of massive sulfide
- Six supplemental grids below the base of massive sulfide
- Top of metabasalt

Because specific-gravity measurements were not made at all intervals along the drill holes, considerable missing data were encountered. Since only correlative data within the search ellipse could be used to calculate a cell's value, many of the model's cells were not assigned a value (i.e., were set to null). Unlike the metal models, where these missing values were replaced with zeros, these cells had to be supplied representative values of specific gravity. The project geologist reviewed the specific gravity data for each rock unit and determined a replacement value for the unit. Then, after construction of the specific-gravity model, all cells which were missing values were assigned a specific gravity value appropriate for the rock unit that contained that cell.

Build 3-D Copper-Equivalence Model

Evaluation of ore reserves for the deposit required that the four metal models be combined into one model. This model represented the amount of metal if all metals were converted to their equivalent assays in copper dollar-value and then summed. Copper was selected because it was the dominant metal and the milling process would be designed to optimize its recovery. Because the final copper-equivalence 3-D Model was in terms of the milling process and market value, this model could be updated easily to reflect changes in either or to test the sensitivity of the deposit to shifts in either.

Convert Metal Models to a Copper-Equivalence Model

Calculating the equivalence of a given metal in terms of copper depends on two factors: (1) The ratio of percent mill recovery of the metal to that of copper, and (2) The ratio of market price (net smelter return) of the metal to that of copper. The following equation was used to convert zinc (Zn) to copper (Cu).

$$\% Cu \text{Eq for } Zn = \frac{(\% Zn) \times (\% \text{Recovery } Zn) \times (\$ \text{ per kg} Zn)}{(\% \text{Recovery } Cu) \times (\$ \text{ per kg} Cu)} \quad (1)$$

A factor for the copper equivalence of zinc is produced by this equation. The assay value in every cell of the zinc metal model was multiplied by that factor, converting it to its equivalent value in copper. The modeled assay values in the silver and gold models were also multiplied by the appropriate copper-equivalence factors. Table 3 shows representative values of the recoveries and market values when the first models were built.

After the three metals were converted to their equivalence in copper, their values were added to the copper model, creating a combined copper-equivalence model.

Leach the Copper-Equivalence Model

The upper 8 to 12 m of rock was leached of most significant mineralization, leaving a gossan visible in some outcrops and in core. Assays for this region were not used during the modeling process because their affect could have extended too deep into the deposit due to data distribution. However, during modeling, high-grade mineralization was projected into this leached zone. This mineralization was removed from the copper-equivalence model by leaching it to a depth of 10 m. To do this, all cells down to 10 m below the base-of-overburden grid were set to a zero assay value, creating the final copper-equivalence model (Figure 6).

USING THE COPPER-EQUIVALENCE MODEL

Once constructed, the model was used to estimate volumes of ore and waste, construct maps of ore distribution and waste/ore ratios, construct cross sections, aid in the design of an open-pit mine, and evaluate the economic potential of that mine design.

Geologic Reserves

Geologic reserves are estimated for a deposit and include all mineralized material within the area of the deposit down to a specified depth. These reserves are usually estimated at several grade cutoffs to determine volume and grade sensitivities to these cutoffs. For the Kutcho Creek deposit, geologic reserves were calculated for the entire deposit and by lease (the IOL portion and the SML portion). Each of the following calculations and displays were generated at several cutoffs for the entire deposit and by lease.

Calculating Reserve Estimates

The specific-gravity model and copper-equivalence model were used to determine tonnage, volume, and waste/ore ratio for a specified grade cutoff; a typical copper-grade cutoff might be 1.0 percent. Operations between the copper-equivalence model and the specific-gravity model were used to create an ore model and a waste model, depending upon whether grade was above or below cutoff. All cells of both models were initially set to zero. If a cell's grade was below cutoff, its specific gravity was output to the waste model; if above or equal to cutoff, then it was output to the ore model.

Since a cell's dimension was $10 \times 10 \times 2$ m, its volume was 200 m^3. By counting the number of cells greater than zero in each model and multiplying the results by 200, the total volume of ore and waste in the model could be determined quickly. In addition, the waste/ore ratio by volume could be generated. Multiplying each cell's volume by its specific gravity and an appropriate units-conversion factor allowed us to calculate the tonnage of each cell. By summing tonnage values in all cells of each model, the tonnage of ore and waste in the model could be determined. In addition, the waste/ore ratio by tonnes could be generated. Specific gravities averaged 2.87 for waste and 3.98 for ore; therefore, waste/ore ratio by tonnes differed significantly from that based on volume.

To determine average ore grade above cutoff, a cutoff was selected. Then, the grades of all cells above the cutoff were added together and divided by the total number of cells above cutoff.

Maps Generated from the Model

An ore-thickness map represents the total thickness (ore and waste) of those cells above cutoff. Before a map of this type can be generated, a grid of ore thickness has to be built. This was done by counting the number of cells in each column of the copper-equivalence model that were above cutoff. The result was multiplied by the cell thickness (2 m) and that value stored at that column's location in a grid. The resulting grid was contoured to produce a map of ore thickness. Because values in the grid were even multiples of two, the contours were blocky. For aesthetics, a filter or other smoothing process could be applied to the grid.

Multiplying the grade (as a proportion) of a cell by its thickness gives a representation of the copper content in that cell. Doing this for each cell above cutoff, summing the results for each column, and then assigning the result to a grid node at the column location produced a grid of grade × thickness. Contouring this grid generated a grade × thickness map (Figure 7). Planning engineers utilize this type of map to focus attention on high metal concentration areas which may be amenable to early mine recovery.

Grids that define the top and base of ore are useful for mine planning and can be used to more precisely define correlation controls during model construction (Lindsay, 1992).

A rarely generated map, useful for mine planning on a bench-by-bench basis, is the horizontal slice. This shows the modeled property as it lies along a horizontal plane, such as the surface on a mine bench.

Cross Sections

The most valuable tool for evaluating the correctness of the 3-D model is the cross section. Grids, drill holes with assay values, and shaded or colored representations of the 3-D model's contents were drawn on the sections. Model contents were displayed using different shading patterns or colors to show different ore grades. The sections were normally dip sections (Figure 6), although strike sections and sections zig-zagging through the model were created.

Pit Design and Optimization

For this project, pit optimization meant determining the economic limit of the pit. This is the point at which the cost of extracting and processing a unit of ore exceeds the net value of that ore. When performing manual optimization, mining engineers often work one cross section at a time. The ore data is posted along drill holes on a cross section and contoured in the plane of section. Several pits are then drawn on the section. Labor-intensive grade averaging, projections, and planimetering are used to determine, for each pit, average ore grade, waste/ore ratio, and other

Table 3. Mill Recovery Rate and Market Price for Each Metal (January, 1982)

Metal	% Recovery	Price (Canadian Dollars)
Copper	88	$2.25/kg
Zinc	70	$0.60/kg
Silver	75	$0.42/g
Gold	75	$15.76/g

information pertinent to making pit design decisions.

A similar approach was applied to the copper-equivalence model. On a series of computer-generated cross sections through the model, the most likely pit was drawn, with about four other pits drawn above or below it at 10 to 20 m vertical intervals (Figure 8). Pit geometry (wall angles, etc.) was determined from preliminary geotechnical studies and experience with other mines. Each mini-pit was made into a grid that intersected topography, extended 15 m to either side of the section and had vertical walls on sides paralleling the section.

To build a mini-pit grid, the engineer specified the elevation at the pit base, points around the base where the walls began, and the angle of each wall. The pit grids were generated automatically from this data. Changes in angle of the wall could be incorporated by adding inflection points and new angles. Twenty-four sections, each 60 m apart, were analyzed. A total of 103 mini-pit grids were generated and evaluated in a period of two days.

The average grade of ore and stripping ratio (waste/ore) were automatically calculated for each pit. A plot of stripping ratio to ore grade was then produced (Figure 9): a break-even line defined by the equation

$$\text{Recoverable Ore Value} = \text{Mining Cost} + \text{Milling and Surface Cost} + \text{Administrative Cost} \quad (2)$$

separated the graph into economic and uneconomic areas. The optimal mini-pit (recover most ore economically) selected for each section was the deepest pit that plotted in the economic area of the graph. Some of these plots indicated that a pit intermediate to the ones tested was appropriate; it was then generated and tested.

After the optimal mini-pit for each section was selected, the X-Y location, elevation, and pit-wall angles for the bottom edges of each selected section pit were merged into one polygon. This polygon, together with additional points and pit-wall angles to round out the pit ends, defined the bottom of the total pit. The pit-generation program was then used to build a pit grid covering the entire deposit. All geo-

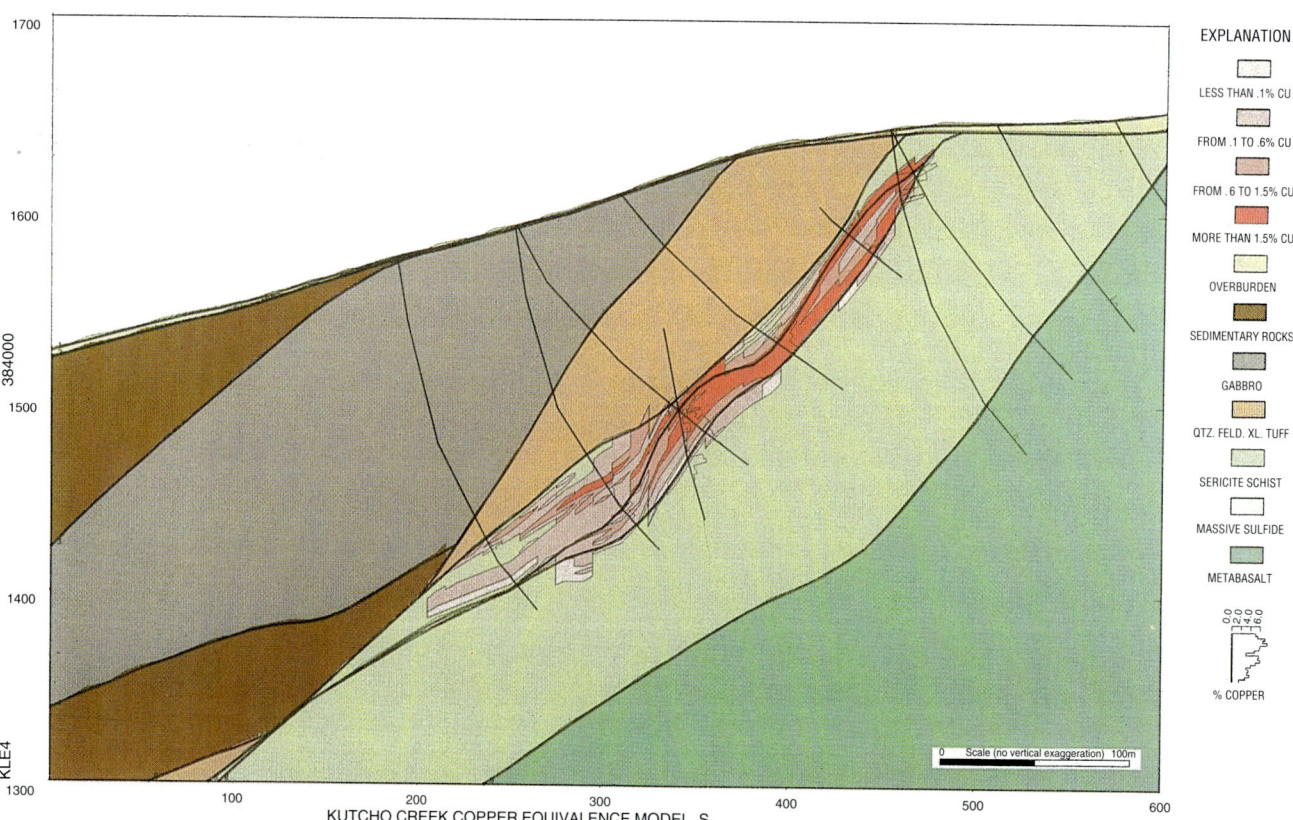

Figure 6. Cross section through the final copper-equivalence model. High-grade mineralization is represented by varying shades of red. See Figure 1 for location of section.

logic reserve calculations were performed on portions of the model that lay both above (within) and below the pit. The grade-thickness maps for material below the optimal pit were very useful. These allowed the engineer to identify material left behind and adjust the pit, where appropriate, to incorporate that material and create the optimal pit (Figure 10).

The final pit was used to produce tonnages, volumes, average grades, and stripping ratios for the entire pit and on a bench-by-bench basis. Each bench represented a ten-meter-thick horizontal slab through the pit. Pit evaluations for a number of cut-off grades were generated allowing the definition of tonnage/grade/cutoff relationships (Figure 11). This set of curves permitted extrapolation of total pit production figures at any cut-off.

Three intermediate pits were developed and analyzed as part of a production schedule to help identify production bottlenecks. The program identified ore and waste cells, and included or excluded them based on how they combined with cells above or below. This allowed internal waste lenses of significant size to be excluded from the model and smaller lenses that could not be segregated during mining to be included. Dilution factors were also included in the analysis. This produced changes in the curves of Figure 11 that had significant impact on mine planning. Similar procedures were used to test varying bench height and its influence on tonnage, grade, and metal content (grade × tonnage).

CONCLUDING REMARKS

Three-dimensional modeling of the Kutcho Creek massive sulfide deposit produced geologic block models that, when checked with cross sections along each line of drill holes, were found to honor all data, honor the geologic interpretation, and to incorporate detail from data off the line of section. Rapid updating (three days for the entire model) and the ability to generate displays in any direction went well beyond manual capabilities. Modeling each metal independently allowed its distribution to be understood and provided a more realistic copper-equivalent model than that obtained by combining the assays and then modeling. Maps, sections, and other displays of the final copper-equivalent model allowed the engineer to quickly sketch several mine designs and quickly build mine-pit grids. Rapid interrogation of the model to produce tonnage, volume, and waste/ore ratio values within and below these mine-pit grids and on a bench-by-

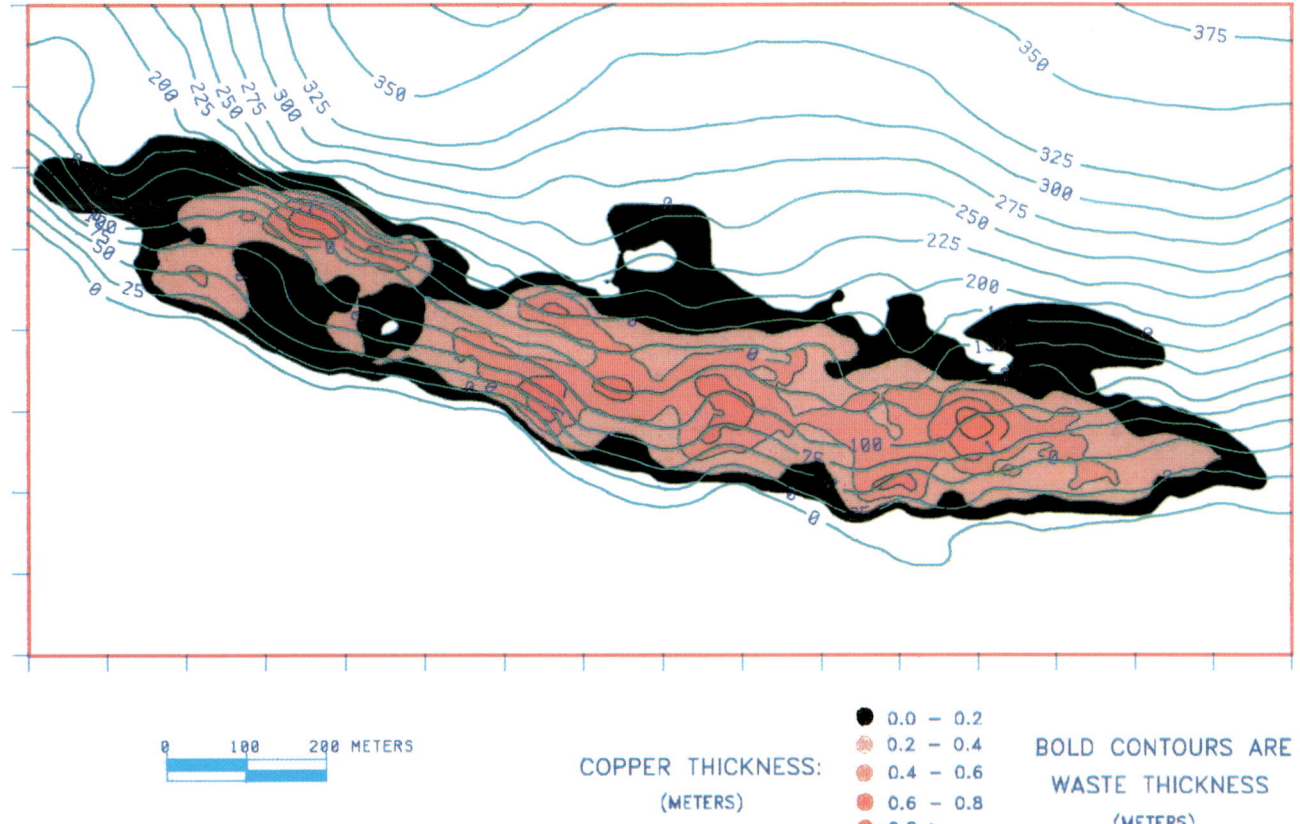

Figure 7. Map showing contours of grade times thickness for copper (solid lines) with contours of waste thickness (dashed lines) superimposed.

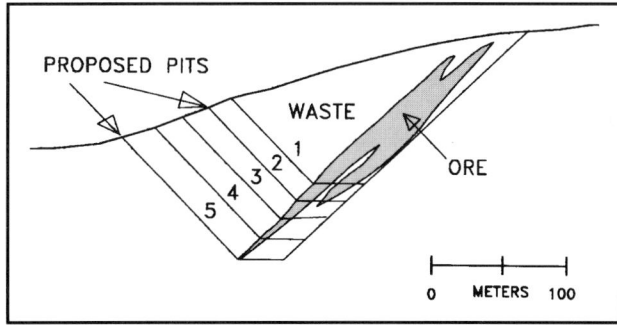

Figure 8. Diagrammatic cross section showing high grade mineralization in terms of copper equivalence and five proposed pit designs.

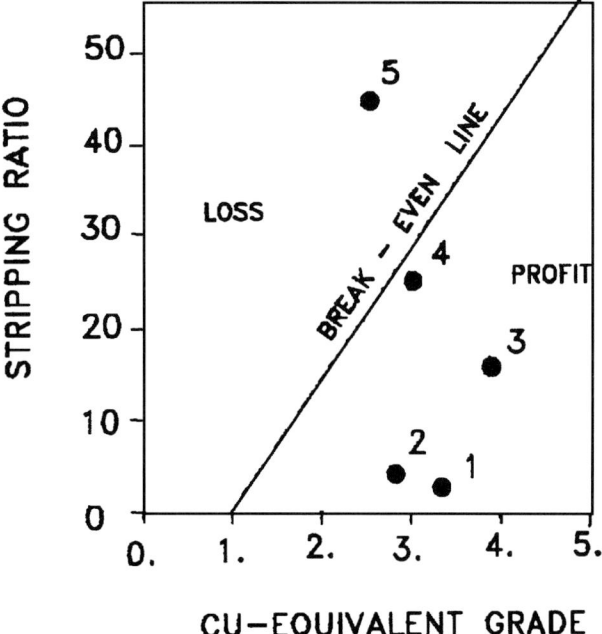

Figure 9. Plot of stripping ratio (waste/ore) versus copper-equivalent ore grade. Each dot represents a proposed mini-pit design for one cross section. Line represents the break-even point where operating costs equal recoverable ore value.

bench basis, allowed more than 100 mine configurations to be tested in two days. Plots of relationships between grade, tonnage, and stripping ratio allowed testing of the deposit's sensitivity to varying ore-grade cutoffs and varying bench height. Use of recovery rate and current market value when building the copper-equivalent model allowed rapid reconstruction of the model to test the effects that market and milling fluctuations had on economics of the deposit.

Although this is clearly a minerals application, the programs used were written for the oil industry. Because this deposit is highly stratified, techniques learned through modeling layered reservoirs transferred smoothly to this project. The Kutcho Creek model was developed and used in the period 1979-1982, so computer hardware to display contents of the model three-dimensionally were not available. Since that time, the programs have experienced major enhancements and concepts for incorporating heterogeneities, interfingering facies, and forms characteristic of geologic processes have improved dramatically. Development and distribution of similar programs on modern hardware are providing economic and fast 3-D model construction and visualization capabilities to all sectors of the petroleum and minerals industries.

The major modeling hurdle in 1979, just as today, is the difficulty of getting geologic interpretation into rock-unit geometries and into attribute distributions within those units. However, even more significant is the lack of geologic understanding as to what the shape and size of geologic features are. Programs now allow these geometries to be entered through stochastic modeling or interactive editing but we often find we know too little about the features (e.g. mineralized lenses, point bars, reefs, lava flows) to properly represent them (C. Johnson, 1991, personal communication).

The baselap, truncation, and conformable techniques for 2-D modeling and the use of supplemental surfaces to define correlation fabrics for 3-D modeling are standard tools that have been used since the early 1970s. New editing techniques, ways to handle fault separation, normal and reverse faulting, and correlation of attributes across faults when 3-D modeling are slowly being developed. The speed of these developments has increased dramatically in the past few years due to the display and processing speed of workstations. Even with these improvements, geologic and engineering demands quickly exceed program functionality.

ACKNOWLEDGMENTS

We would like to thank Carlton Johnson and his associates at Exxon Production Research Company for helping us with this study and for giving us the opportunity to describe their 3-D geologic modeling program and how it was applied to this project.

Special thanks goes to those who contributed significantly to the success of this project. Dan Hayba handled all ore reserve estimations, open-pit mine displays, and mine design and redesign iterations with the engineer. Sterling Helwick developed the program that automatically created the mine-pit grids. Raj Paul was the mining engineer who evaluated output from the model, designed the mini pits, and designed and fine tuned the final open-mine pit design.

REFERENCES CITED

Badiozamani, K., F. Roghani, and G. Hawes, 1992, Applications of variable zone modeling and map-

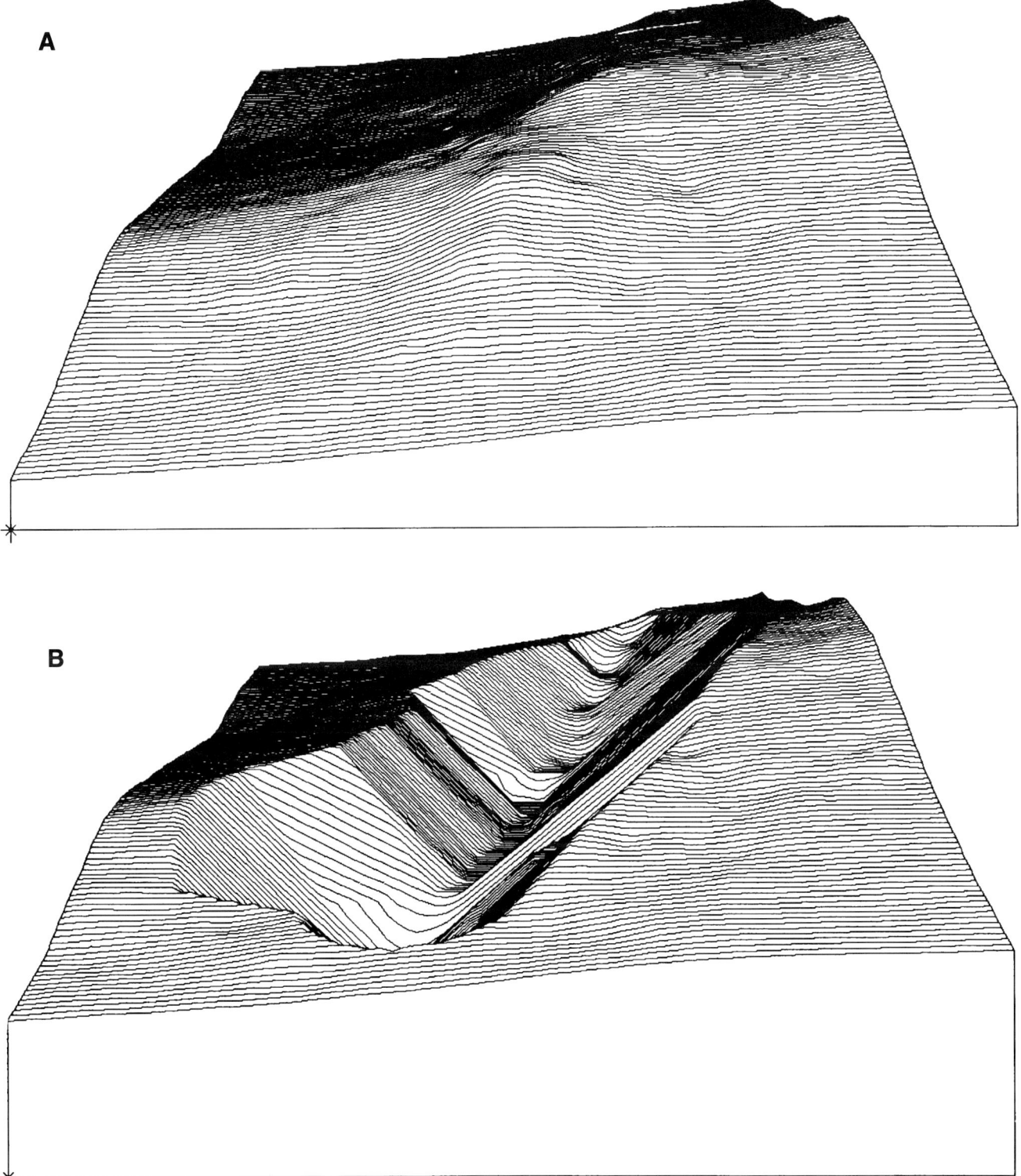

Figure 10. Perspective views looking from the east. (A) Topography before development. (B) The final pit superimposed on topography.

ping of Canadian oil sands, (this volume).
Bridge, D. A., J. M. Marr, K. Hashimoto, M. Obara, and R. Suzuki, 1983; Geology of the Kutcho Creek volcanogenic massive sulfide deposits, Northern British Columbia, in Mineral Deposits of Northern Cordillera, C.I.M. Special Volume 37, p. 115–128.
Denver, L. E., and D. C. Phillips, 1992, The impact of vertical averaging on hydrocarbon volumetric cal-

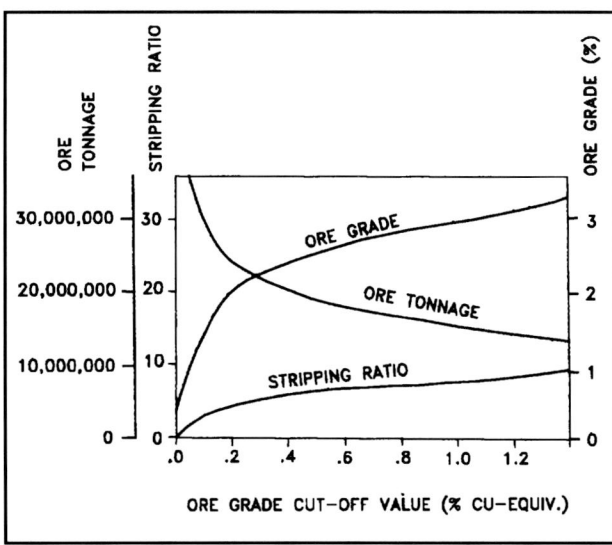

Figure 11. Relationship between Y-axis: stripping ratio, ore tonnage, and ore grade (% Cu-Equivalent) as ore-grade cutoff (X-axis) in terms of percent copper-equivalent increases.

culations: a case study, (this volume).

Hamilton, D. E., J. L. Marie, G. M. Moon, F. J. Moretti, and W. P. Ryman, 1985, Application of three-dimensional computer modeling for reservoir and ore-body analysis (abs.): AAPG Bulletin, v. 69, no. 2, p. 262.

Hamilton, D. E., and S. K. Henize, 1992, Computer mapping of Pinnacle reefs, evaporites, and carbonates: Northern trend Michigan basin, (this volume).

Johnson, C. R., and T. A. Jones, 1988, Putting geology in reservoir simulations: a three-dimensional approach: Society of Petroleum Engineers Preprint 18321, 63rd Annual Technical Conference, p. 585–594.

Jones, T. A., 1988a, Modeling geology in three dimensions: Geobyte, v. 3, no. 1, p. 14–20.

Jones, T. A., 1988b, Geostatistical models with stratigraphic control: Computers & Geosciences, v. 14, p. 135–138.

Jones, T. A., 1992, Extensions to three dimensions: introduction to the section on 3-D geologic block modeling, (this volume).

Jones, T. A., and D. E. Hamilton, 1992, A philosophy of contour mapping with the computer, (this volume).

Jones, T. A., D. E. Hamilton, and C. R. Johnson, 1986, Contouring geologic surfaces with the computer: New York, Van Nostrand Reinhold, 314 p.

Kushnir, G., and J. M. Yarus, 1992, Modeling anisotropy in computer mapping of geologic data, (this volume).

Lindsay, J. F., 1992, Three-dimensional geologic block model of a polar fan-delta complex, Canning River, North Slope, Alaska, (this volume).

Walters, R. F., 1969, Contouring by machine: A user's guide, AAPG Bulletin, v. 64, p. 916–926.

Yildirim, E., 1985, A block value modeling application in tar sands, AOSTRA Seminar on The Advances in Petroleum Recovery and Upgrading Technology: Edmonton, AOSTRA, 57 p.

Chapter 16

The Impact of Vertical Averaging on Hydrocarbon Volumetric Calculations—A Case Study

Larry E. Denver
Stratamodel, Inc.
Houston, Texas, U.S.A.

Danny C. Phillips
Stratamodel, Inc.
Houston, Texas, U.S.A.

ABSTRACT

Accurate hydrocarbon volumetrics can only result from detailed reservoir description. Two-dimensional techniques have traditionally served to quantify reservoir description. However, two-dimensional characterization of three-dimensional heterogeneities may not yield acceptable volumetric analysis in heterogeneous reservoirs. The primary pitfall in the traditional methodology is the inherent vertical averaging. In the case studies that follow, volume discrepancies as high as 21% are documented between three-dimensional modeling and two-dimensional mapping. This paper focuses on the deleterious impact of vertical averaging on reservoir characterization and hydrocarbon volumetric calculations and demonstrates that three-dimensional geologic block modeling is a more accurate alternative to traditional volumetric techniques.

INTRODUCTION

Hand mapping and computer mapping are well suited for defining structural surfaces or the distribution of attributes that vary significantly in two dimensions. Computer mapping systems especially are well suited for performing mathematical operations between attribute maps or surfaces, or for yielding trend analyses of surfaces. In general, computer mapping serves to eliminate the need to manually cross-contour surfaces and provides a realistic means of mathematically analyzing surfaces. For example, computer mapping has been employed for computing volumes because it automates the tedious and time-consuming tasks of manual cross-contouring and planimetering. Although computer mapping is a powerful tool, it does little more than automate traditional hand-mapping techniques. Both approaches fail to characterize the three-dimensional distribution of geological attributes necessary to yield accurate hydrocarbon volumetrics.

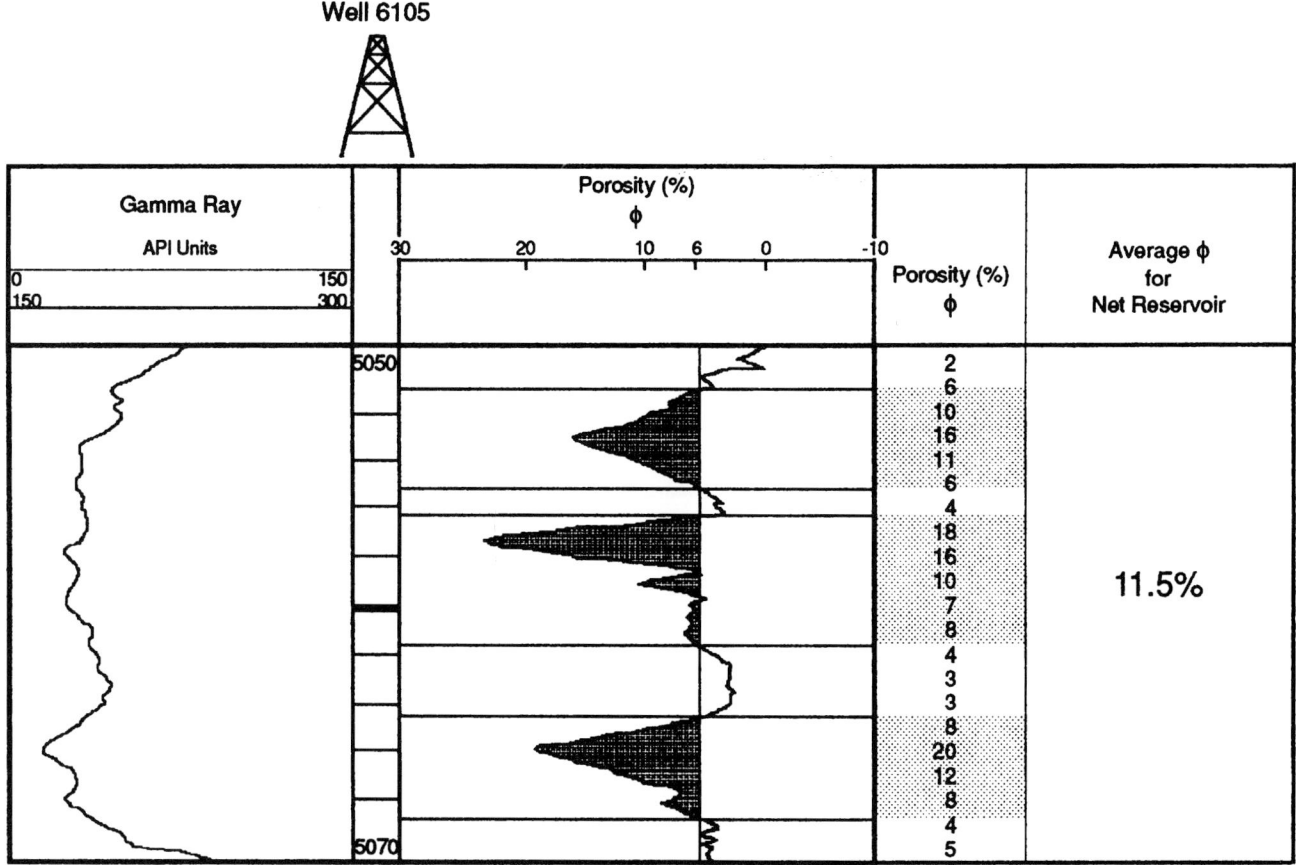

Figure 1. A 6% porosity cutoff is used in this well to delineate net reservoir. Once net reservoir is defined, average porosity is calculated based only on the net reservoir intervals.

ANALYSIS OF THE AVERAGING PROBLEM

Volumetric mapping methodology requires that detailed porosity and saturation values measured in the reservoir interval be reduced to average values for each zone at each well (Jones et al., 1986) (Figure 1). Typically, once porosity and hydrocarbon saturation are gridded using the appropriate controls, net reservoir thickness is determined using net-to-gross ratios or other similar techniques. These three attributes are interpolated laterally before being multiplied since each may vary independently of the others (or in relation to other reservoir parameters, such as height above the oil-water contact (OWC)) and each may define a unique trend (Figure 2).

Cutoffs and Loss of Resolution

Integration of the net-feet-hydrocarbon (NFH) grid produces volume. In relatively homogeneous reservoirs, vertically averaged data yield reasonable volumetrics, especially if neither porosity nor saturation cutoffs are employed at the wells. In heterogeneous reservoirs, however, cutoffs and over-averaging yield poor results because (1) cutoffs used to define net reservoir at the wells bias the interpolation of porosity, saturation, and net reservoir thickness, (2) averaged data make it difficult to correlate and spatially define a detailed porosity and saturation distribution within the reservoir, and (3) multiplying grossly averaged porosity and saturation values does not yield accurate volumetrics because extreme values are not maintained.

The technique used to define net reservoir affects the distribution of well attributes. Cutoffs are imposed in the well data to prevent gross reservoir from biasing the net reservoir averages. This data manipulation is often accomplished by defining gross reservoir based on gamma-ray or SP cutoffs and then delimiting the net reservoir data based on porosity and saturation cutoffs (Figure 1). Once the net reservoir rock has been defined for each well, average porosity and saturation values are based solely on the net reservoir sub-intervals. Subsequent interpolation of the averaged values may be biased, however, by the cutoffs defined at the wells. Thus, although cutoffs help define reservoir production limits, they may change the characterization of the reservoir attributes and make interpretation of geologic trends difficult.

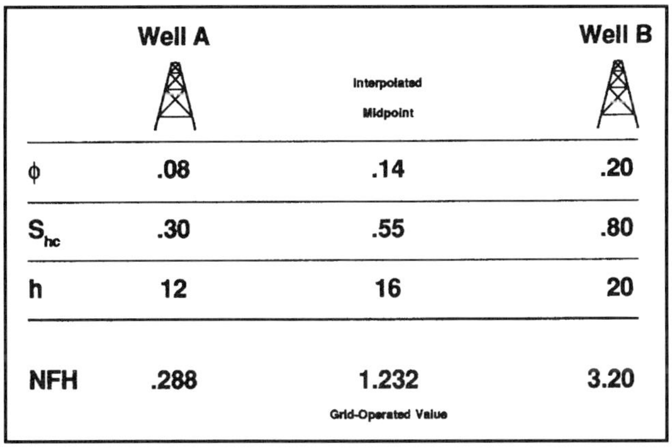

	Well A	Interpolated Midpoint	Well B
φ	.08	.14	.20
S_{hc}	.30	.55	.80
h	12	16	20
NFH	.288	1.232	3.20
		Grid-Operated Value	

Figure 2. Porosity, saturation, and net reservoir thickness are independent variables. Each variable should be interpolated into a grid prior to any operations between the variables. The value represented at the Grid-Operated Midpoint (1.232) is a product of the correct methodology. Porosity, saturation, and thickness are interpolated to the midpoint and then multiplied to yield the net feet hydrocarbon value (After Swanson and Giovinco, 1988.) (Copyright 1988, Society of Petroleum Engineers. Swanson, Donald C., and Sam Giovinco, 1988, Improved volumetrics with the computer: Hugoton Gas field, Kansas, Paper no. 18313, presented at the 1988 SPE Annual Technical Conference and Exhibition, Houston, Texas, October 2-5.)

The vertical averaging required to map three dimensions onto two dimensions severely limits the geologist's ability to correlate data within key intervals. The reservoir must be subdivided to maintain the heterogeneities (Jones et al., 1986, pages 218–219) that delimit the extreme values since the extremes significantly affect volume calculations. Figure 3 shows detailed porosity and gamma ray curves for two wells. Each well has approximately 40 ft (12 m) of gross and 10 ft (3 m) of net reservoir rock. In this example, regardless of whether the net thickness is directly interpolated or a net-to-gross technique is used, the grid nodes between these wells would indicate that the net reservoir is stratigraphically connected. In reality, the reservoir is compartmentalized and should be modeled as such. In general, the distribution bias imposed by economic cutoffs and the lack of stratigraphic correlation stems from the loss of vertical resolution that occurs when mapping in only two dimensions. This loss of resolution results in inaccurate hydrocarbon volumetrics because multiplying averaged heterogeneous values is not appropriate. Multiplication of highly averaged reservoir values does not yield the same result as multiplication and summation of the population that yields the averages because increased averaging dampens extreme values (Clark, 1979, Chapter 3). The operation between the averaged attributes in a heterogeneous volumetric data set may not yield the same answer as the summation of foot-by-foot operations between porosity and saturation values recorded by the log analysis. Therefore, even at the well, calculation of NFH is incorrect before beginning the interpolation step. It may appear, then, that the solution lies in multiplying the detailed porosity and saturation provided by the well log analysis and interpolating the product. This product grid would then be multiplied by the net reservoir isochore to yield the NFH. This technique, however, does not maintain individual trends that may exist in porosity or saturation. They will instead be masked by the trend of the product of the independent variables. The volume difference obtained by operating on the extremes, versus operating on the averages, varies according to the following factors:

- the degree of reservoir heterogeneity as defined by the extreme values, and
- the relationship (positive/negative) between porosity and saturation.

Figure 4 uses a simple one-dimensional model to illustrate the principles stated above. In this example, porosity and saturation have been calculated for every foot of the 12-ft interval at the well. NFH has been calculated in each well using two different techniques. The NFH calculated in Example 1–Method 1 results from the multiplication of average porosity, average saturation, and net reservoir thickness. This product is approximately 10% greater than the result of summing the individual products as done in Example 1–Method 2.

Example 2 assumes the same data as Example 1, except the relationship between porosity and saturation has been adjusted so that increasing saturation values are mirrored by increasing porosity values (positive correlation). Consequently, Example 2–Method 2 shows a 10% increase in NFH when compared with Example 2–Method 1 because the greater porosity values are now multiplied by the greater hydrocarbon saturation values. Example 2–Method 1 shows the same NFH calculated as Example 1–Method 1 because inverting the porosities has no effect on the average porosity value calculated.

The effects of vertical averaging are not new to geoscientists working reservoir characterization projects. Slice mapping is sometimes used to provide more detailed reservoir description. This technique reduces vertical averaging by simply increasing substantially the number of grids used to define a reservoir (Casavant, 1988). However, as shown in Figure 4, for slice mapping to be effective, the geoscientist must painstakingly create grids that honor the vertical heterogeneity of the well data. Grids may need to be generated for every foot of reservoir in order to capture the heterogeneity of the reservoir system. In addition, if new reservoir cutoffs are chosen as economic conditions change, the data must be completely recalculated and all grids must be regenerated.

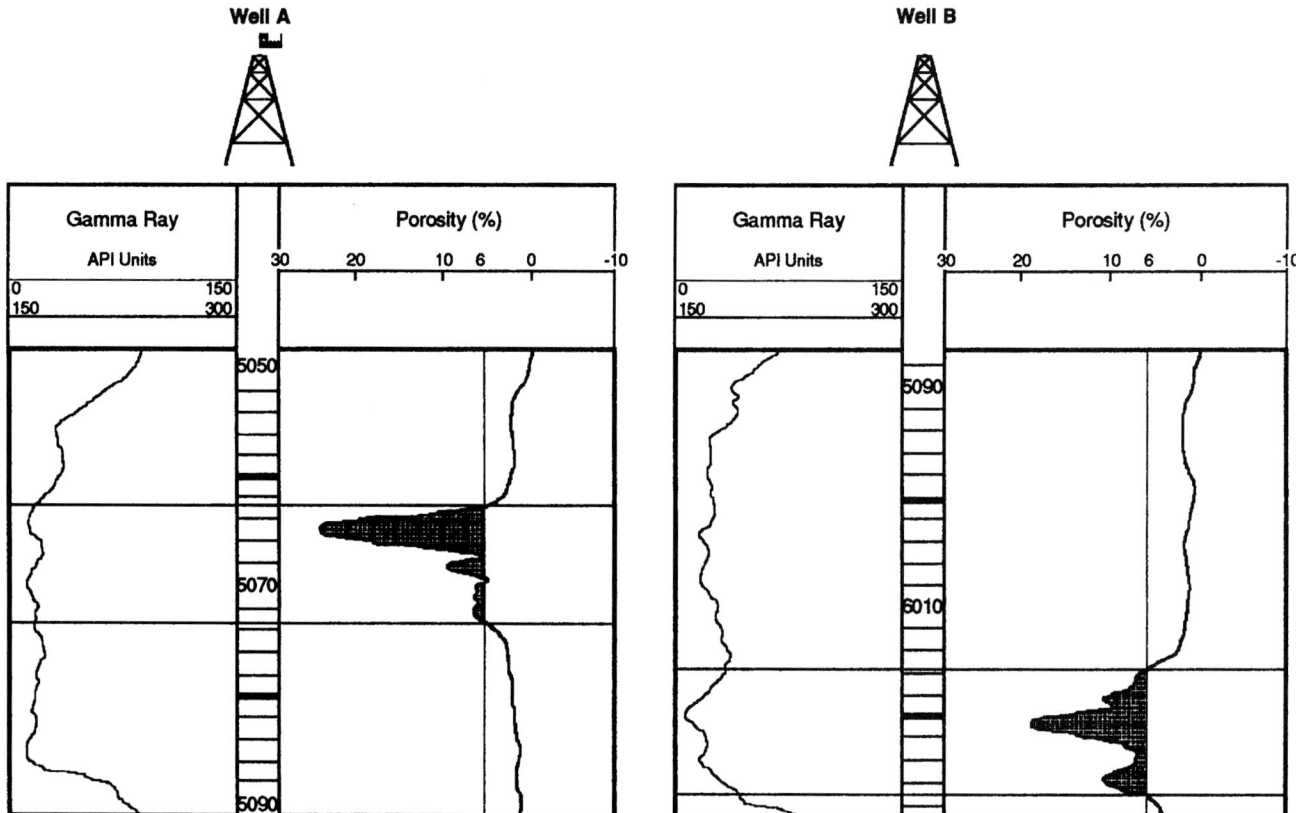

Figure 3. The net reservoir defined in these two wells has approximately the same thickness but exists in different zones of the gross reservoir. The interpolation of net thickness in this example would imply that the net reservoir is connected between these wells, when actually it is not.

Effects Of Averaging On Interpolated Values

It is important to first analyze the effects of averaging at the well to establish the pitfalls of the averaging technique versus the impact that lateral interpolation may have on the volumetric calculation. Since the NFH calculation shown in Figure 4 includes no lateral interpolation as part of the process, the averaging technique is established as the cause of the NFH discrepancy. The examples in Figure 5 illustrate the impact that averaging has on volumetric calculations between wells. Each well in the example records porosity and hydrocarbon saturation based on different degrees of averaging. The left-most column of each well assumes an average value for the entire interval while the right-most column records porosity or saturation at a 1-ft resolution. The tables between the wells show the interpolated midpoint values for the respective intervals. In this example, it is assumed that the 1-ft intervals correlate stratigraphically. The hydrocarbon-pore-thickness values near the bottom of the example illustrate the difference at the midpoint depending on the level of averaging. Because of the positive relationship between porosity and hydrocarbon saturation, the far-right column shows an increase in hydrocarbon pore thickness if compared with the same calculation in the left-hand column. These differences are relatively small since the interval is very thin (eight units thick) and the attributes are limited in range. The geologic and hydrologic parameters that control porosity and hydrocarbon saturation ultimately determine where hydrocarbon volumes may increase or decrease when compared with a two-dimensional approach.

THREE-DIMENSIONAL GEOLOGIC BLOCK MODELING METHODOLOGY

The data sets in the case study discussed below were modeled using Stratamodel's Stratigraphic Geocellular Modeling (SGM) system. The system provides a means of generating a stratigraphic framework which allows faults, unconformities, and stratal units to be defined (Swanson, 1988; Denver and Phillips, 1990). The framework consists of cell layers positioned to match the stratigraphy of the reservoirs and is based on surface grids imported into the system from seismic interpretation systems or computer-mapping packages. Generally, the grids mark the tops of sequences

Example 1

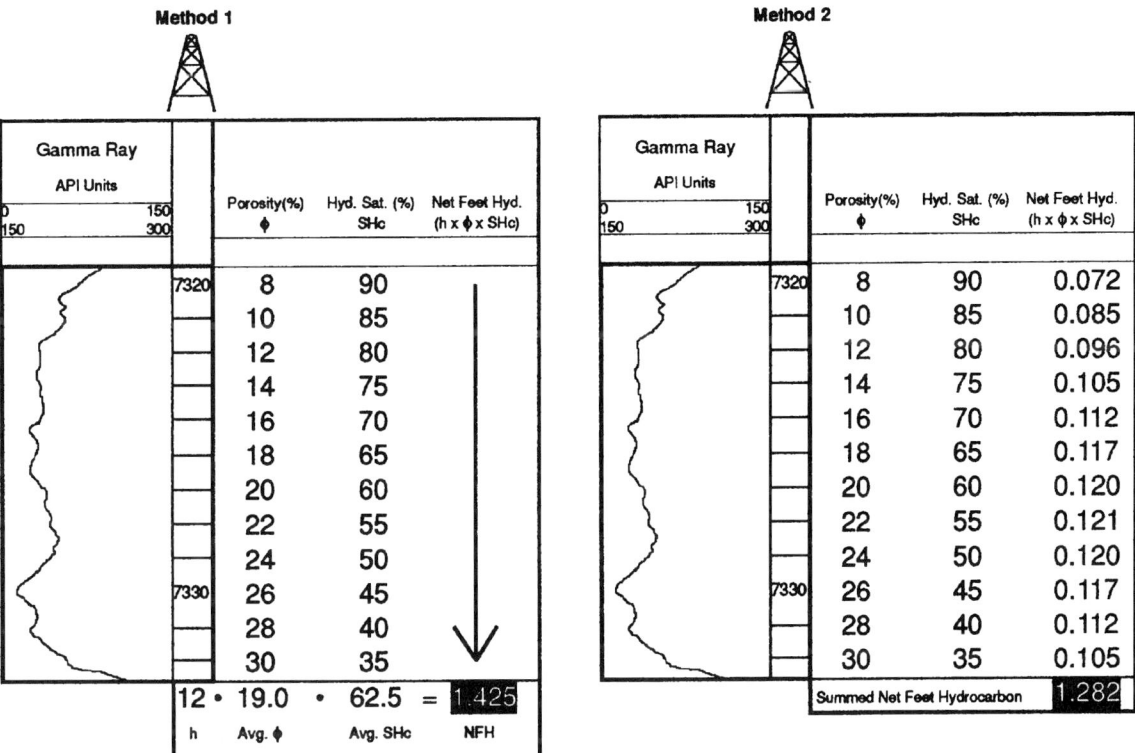

Percentage **decrease** between Method 1 and Method 2 = 10.035%

Example 2

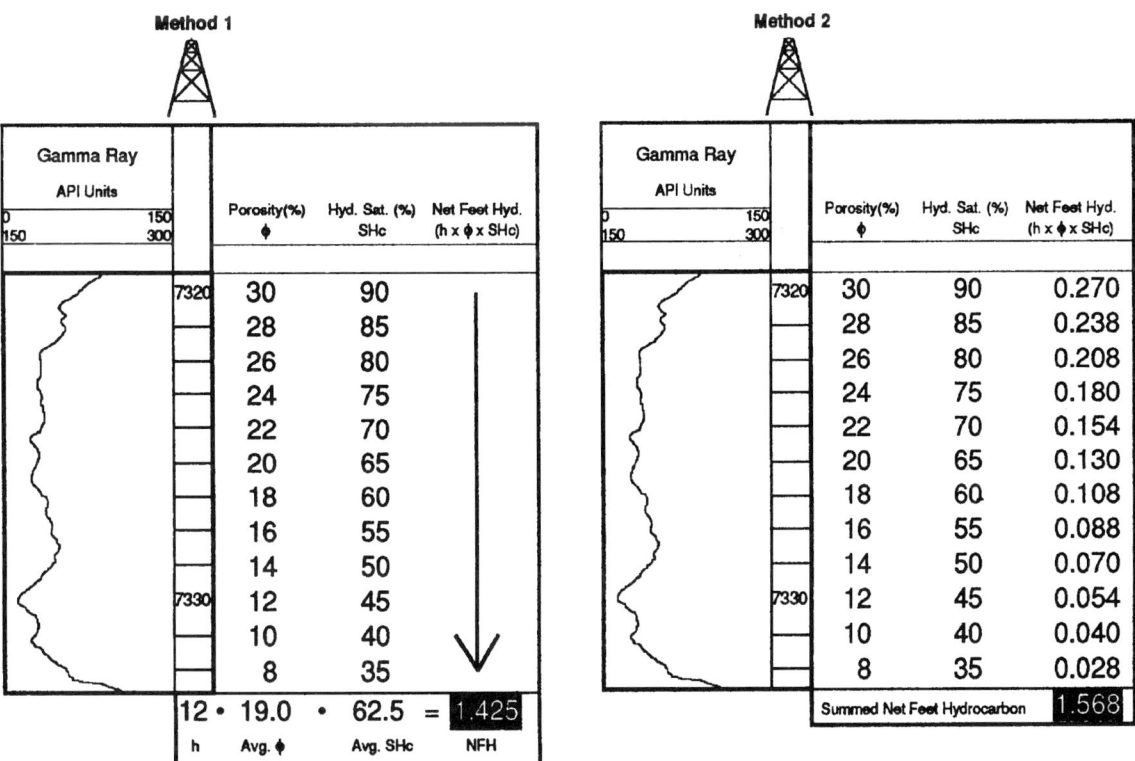

Percentage **increase** between Method 1 and Method 2 = 10.035%

Figure 4. Examples that illustrate the impact of vertical averaging when calculating Net Feet Hydrocarbon (NFH) at a well. NFH is calculated in Method 1 by multiplying average porosity × average saturation × net reservoir thickness. In Method 2, the porosity and saturation from each foot of the log analysis is multiplied. The product from each foot is then summed, yielding NFH.

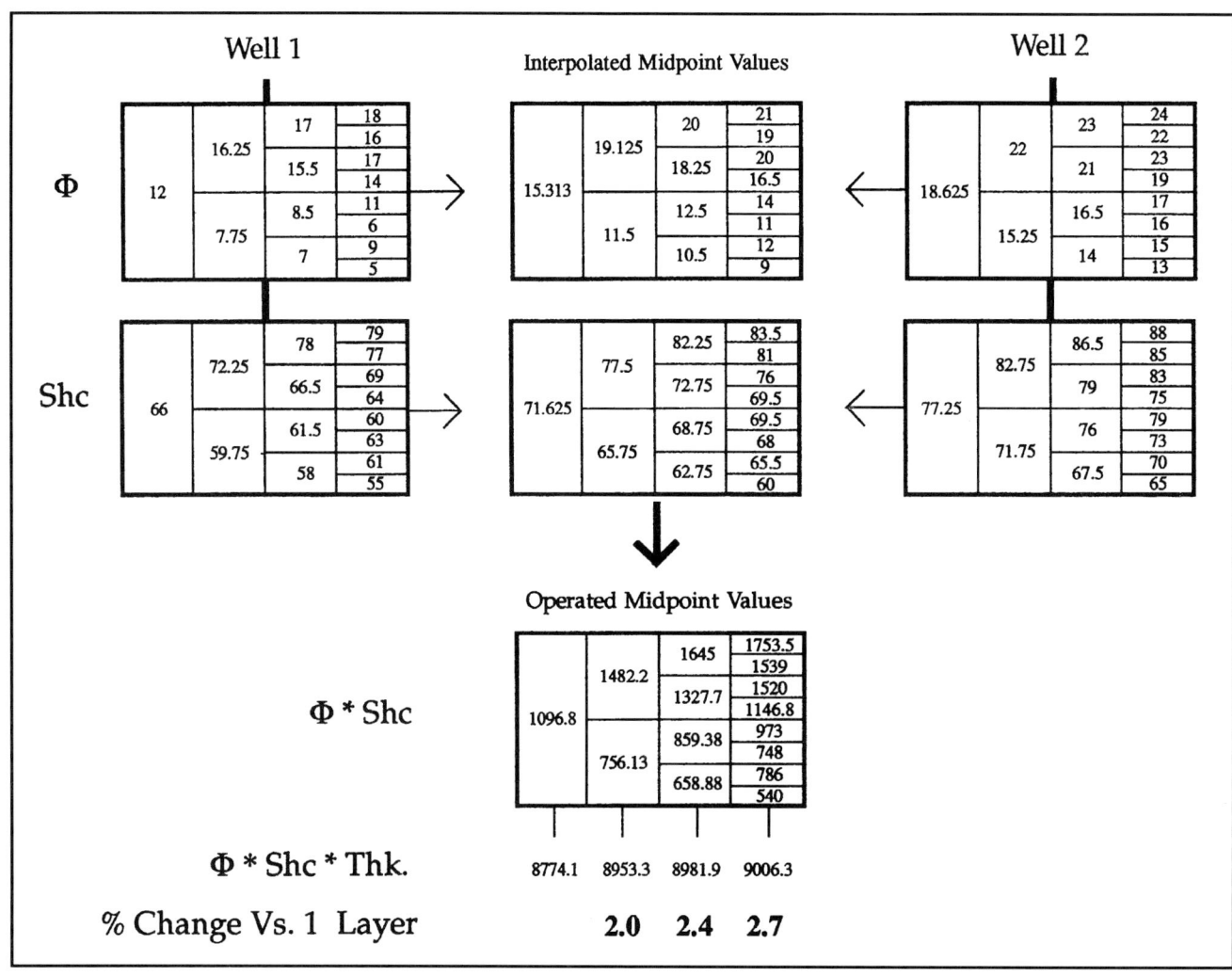

Figure 5. Effects of varying thicknesses on data averaging and interpolation. Wells 1 and 2 represent 8-ft (2.4-m) intervals of data for porosity and hydrocarbon saturation with values for each interval shown on the right and progressive averaging of adjacent intervals shown in columns to the left. The center column represents interpolated (averaged) values for a column of 3-D model cells halfway between the wells. Cell height increases from right to left. The bottom box represents 3-D model cells resulting from multiplying the porosity cells by the hydrocarbon saturation cells.

(modified from Sloss, 1963) or define fault boundaries. Each sequence is assigned a depositional pattern and layers are generated dependent on the depositional geometry specified (Figure 6). The user controls the cell-layer thickness for each sequence based on stratigraphic complexity and reservoir heterogeneity. The system creates a three-dimensional framework of cell layers, which allows correlation of stratigraphically equivalent data intervals and control of cell resolution to preserve vertical heterogeneity of the reservoir attributes (Figure 7). This framework is used to guide interpolation of well attributes in building the final, three-dimensional attribute model (Figure 8). If certain attributes do not correlate as defined by the initial framework, a second framework can be built with geometries appropriate for that attribute. After construction, the second model can be resampled to match the first model, combining all attributes into the same framework and allowing further operations between attributes.

CASE STUDY RESULTS

Two data sets, one from the Hugoton embayment in southwestern Kansas (Figure 9) and the other from the Delaware basin in southeastern New Mexico (Figure 10), were chosen for this study. Selection of these data sets was based on their moderate level of heterogeneity and lack of structural complexity. No faults or fluid contacts are present, ensuring that the volume discrepancies are due to three-dimensional versus two-dimensional methods and not to variation in the way the methods handle these discontinuities. Table 1 summarizes both data sets.

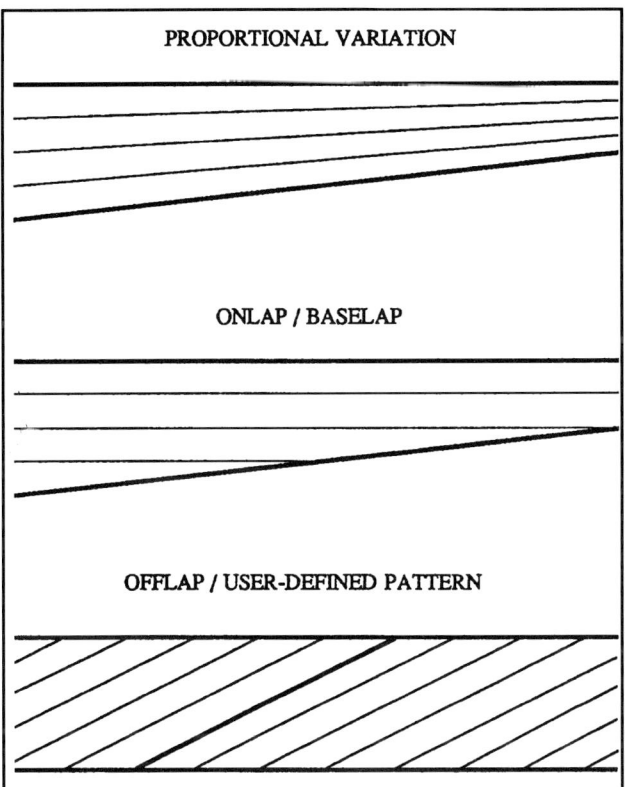

Figure 6. Various patterns are used in Stratigraphic Geocellular Modeling (SGM) to subdivide geological sequences into layers that provide correlations between wells. Proportional variation, onlap/baselap, and offlap are some of the default patterns available. In addition to these, the user may generate a separate grid to define a correlation pattern not related to grids at the top or base of the sequence.

Hugoton Data Set

The Hugoton log-analysis data were prepared by Mesa Petroleum with assistance from Swanson Geological Services as part of a 1200-well reservoir characterization study that targeted infill drilling locations and calculated volumes (Swanson and Giovinco, 1989). Net reservoir thickness, average porosity, and average hydrocarbon saturation were calculated at each well based on a 6% porosity cutoff and a 10% hydrocarbon saturation cutoff. These parameters were field-wide cutoffs but had little impact on the 28-well subset used for this study because most data fell within the net reservoir range. Average porosity and hydrocarbon saturation were modeled using only the net reservoir interval. First, however, the formation was vertically subdivided into homogeneous subintervals based on porosity in the wells (Figure 11). Average porosity and hydrocarbon saturation were calculated for the entire net reservoir but each sub-interval average was also maintained for subsequent modeling in SGM. A commercial mapping package was used to generate porosity, saturation, and net thickness grids. These grids were multiplied together to create a net-feet-hydrocarbons grid that was integrated to obtain volumes. The same data were modeled by SGM.

The formation top and base grids (the same as those used in the mapping step) were used to define the sequence. Correlation within the sequence was designated as proportional to match the interpretation of cyclothem deposition. Proportional variation results in cells that thicken and thin as the formation thickness changes; the number of cell layers in the formation remains constant (see Figure 6). A moving-average algorithm was used to interpolate porosity, saturation, and net thickness in these models, as well as in the mapping phase, because it does not project above the maximum or below the minimum well data values.

Six models were generated that contained 1, 2, 4, 8, 16, and 32-cell layers (Figure 12). If only one layer is defined, the system automatically calculates the weighted-average porosity and saturation within the sequence for each well, and interpolates using the same well values as used in the mapping method. This model yielded volumes that differed by less than 1% from the mapping calculation. Without changing parameters, the sequence was proportionally divided into two layers. The system vertically subdivided the well data and calculated two porosity and saturation values for each well sub-interval. These values were then interpolated using stratigraphically equivalent values in adjacent wells. The volume calculated in this model increased more than 7% from the one-layer model. Progressive increase in the resolution of the 3-D model (more layers) continued to increase volumes, as shown in Table 2.

The volume increases level off at about 21.5% and indicate that 8 cell layers is probably an acceptable resolution to characterize the heterogeneities within the reservoir.

It is important to recognize that volume differences between these models result from changes in the vertical modeling resolution, not from the interpolation scheme. In fact, when difference grids were calculated between the one-layer model and the 32-layer model, the greatest differences were consistently at the well locations (Figure 13). Because a moving-average interpolation scheme was used in this study, the distributed values within each cell are always between the maximum or minimum of the data set. Therefore, the grid operated differences calculated between the wells are consistently less extreme than those calculated near the wells. Thus, the volume discrepancies between models can not be attributed to extrapolation of values greater than those seen in the surrounding wells. The increase of resolution from 8 layers to 16 produced only a 1% change in volume, and the difference between 16 and 32 was even less. This doubling of the layer resolution produces little change in hydrocarbon volume because the well-data heterogeneity are nearly captured with the eight-layer model. This is

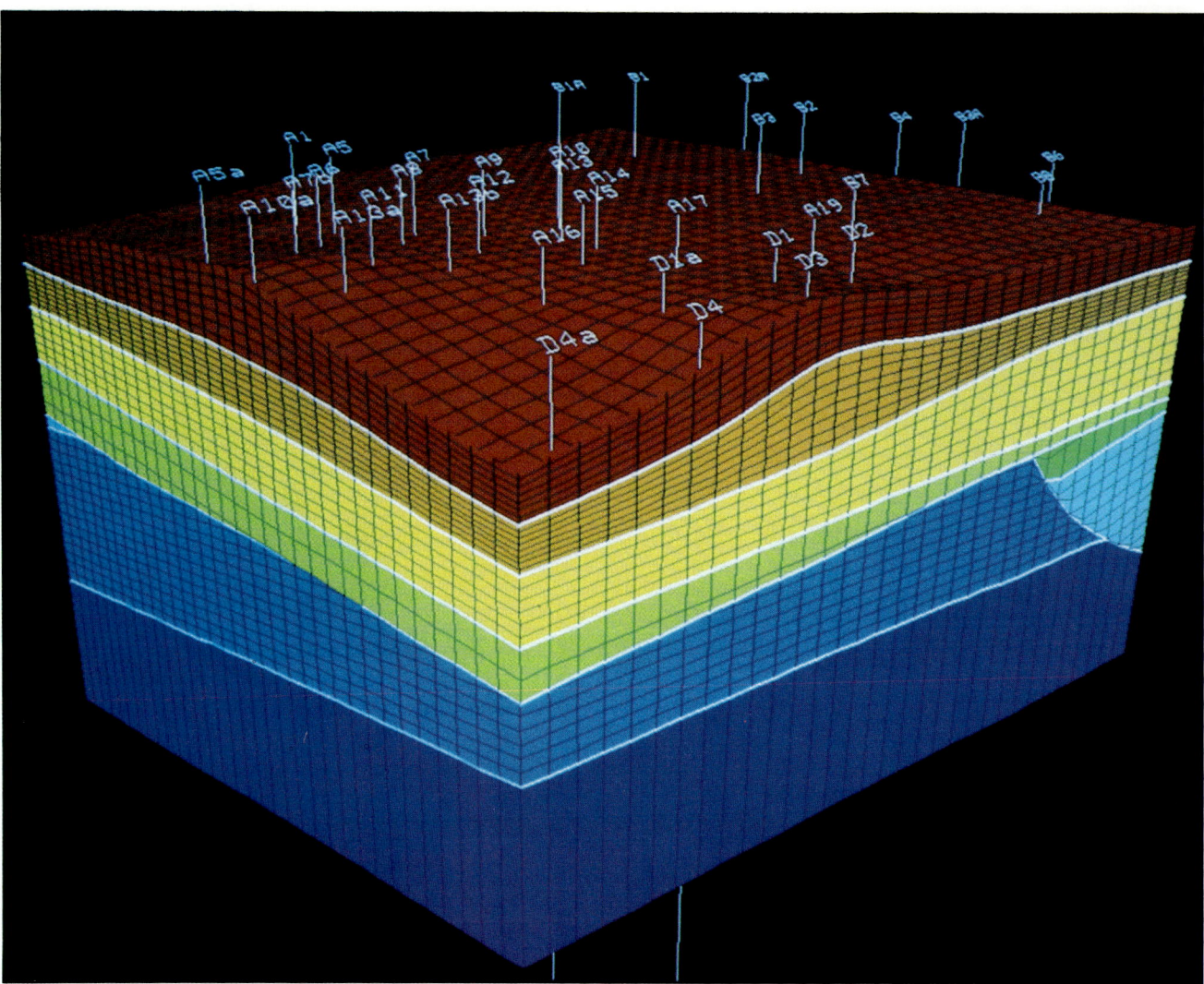

Figure 7. This stratigraphic framework consists of 70 layers subdivided into 175,000 cells. The layers help determine the wells and well-interval values that will be used in the interpolation of attributes into the cells.

not surprising since the wells were pre-processed to define relatively homogeneous subintervals before being imported into SGM (see Figure 11). Thus, once the layer resolution exceeded that of the original data, the extreme values defined at the wells were maintained in the system and the ensuing operations generated negligible changes. Had the raw well data been passed directly into the system at the original one foot sampling, the model may have required more than 8 layers to capture the well heterogeneity.

Delaware Basin Data Set

The Delaware basin data are the property of Consolidated Natural Gas (CNG) and was prepared by Swanson Geological Services. The data were loaned to the authors for use in this study. Data consisted of 23 wells, covering approximately 30 sections (see Table 1 for data summary). Log analysis was performed at a resolution of one foot and no cutoffs were imposed.

The fluvio-deltaic setting of these data showed much greater vertical variability than the carbonates in the Hugoton field. In part, the difference is somewhat manufactured due to processing of the Hugoton well data. On the other hand, the Delaware basin data are characterized by interbedded sand and shale sequences unlike the lithologically consistent Hugoton data. In addition to increased variability, average reservoir thickness was approximately 150 ft (45 m), nearly three times that of the Hugoton data set. Based only on these facts, more layers would be required to accurately calculate hydrocarbon volumes. Nine models were generated in this study with volumetric results as seen in Table 3.

The vertical resolution was doubled for each new model using the proportional technique to subdivide the reservoir. Although establishing lateral correlations was difficult, thin shales (time lines) seen in the

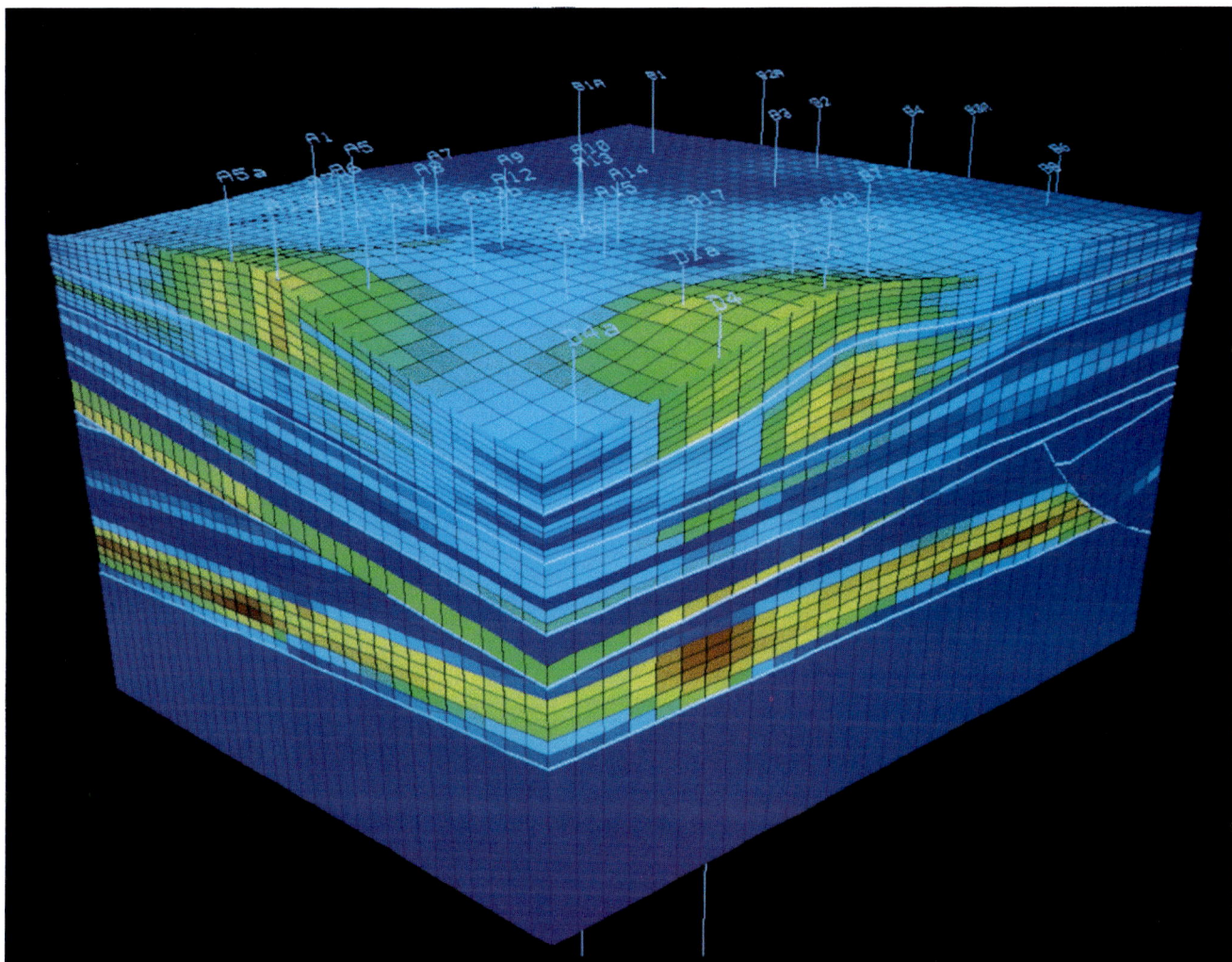

Figure 8. Porosity is the attribute displayed in this model.

high-resolution 3-D model support the correlation method chosen (Figure 14b). The volume difference calculated between the two-layer model and single-layer model was a 3.2% decrease. Thus, the estimated hydrocarbon volume decreased with increased resolution. The four-layer model produced a volume that showed a 0.6% decrease when compared with the single-layer volume. Each additional increase in resolution, however, produced an increase in the calculated volume (Table 3). The final 256-layer model generated a volume increase of 7.7% when compared with the original one-layer model. Like the Hugoton, the initial increases in layering generated significant volume discrepancies. Unlike the Hugoton data set, however, volumes continued to change slightly even up to 256 layers. As mentioned earlier, the difference in the well data preparation is largely responsible for this phenomenon. The fact that layers defined by the model do not necessarily match the one-foot intervals defined by the regularly sampled log analysis also contributes to continued volume increases. SGM is averaging log analysis values at the modeling resolution, so if a one-foot model layer happens to sample from depth 5000.5 to 5001.5, as opposed to 5000.0 to 5001.0, the lower half of that interval will be averaged with the adjacent lower interval, blurring slightly the log analysis data. In general, more heterogeneity requires increased model resolution.

The decrease in volume in the two- and three-layer models is due to an inverse relationship that exists between saturation and porosity in only some of the sub-intervals. In fact, in this case certain non-reservoir intervals of the model (as defined by porosity in the model) actually showed high-hydrocarbon saturation values in the same cells. This phenomenon may occur due to log analysis that records a low porosity value while reporting saturation as indeterminate in the same one-foot zone. Under normal circumstances, a cell penetrated by a well would record a porosity and saturation value based almost entirely on the values of

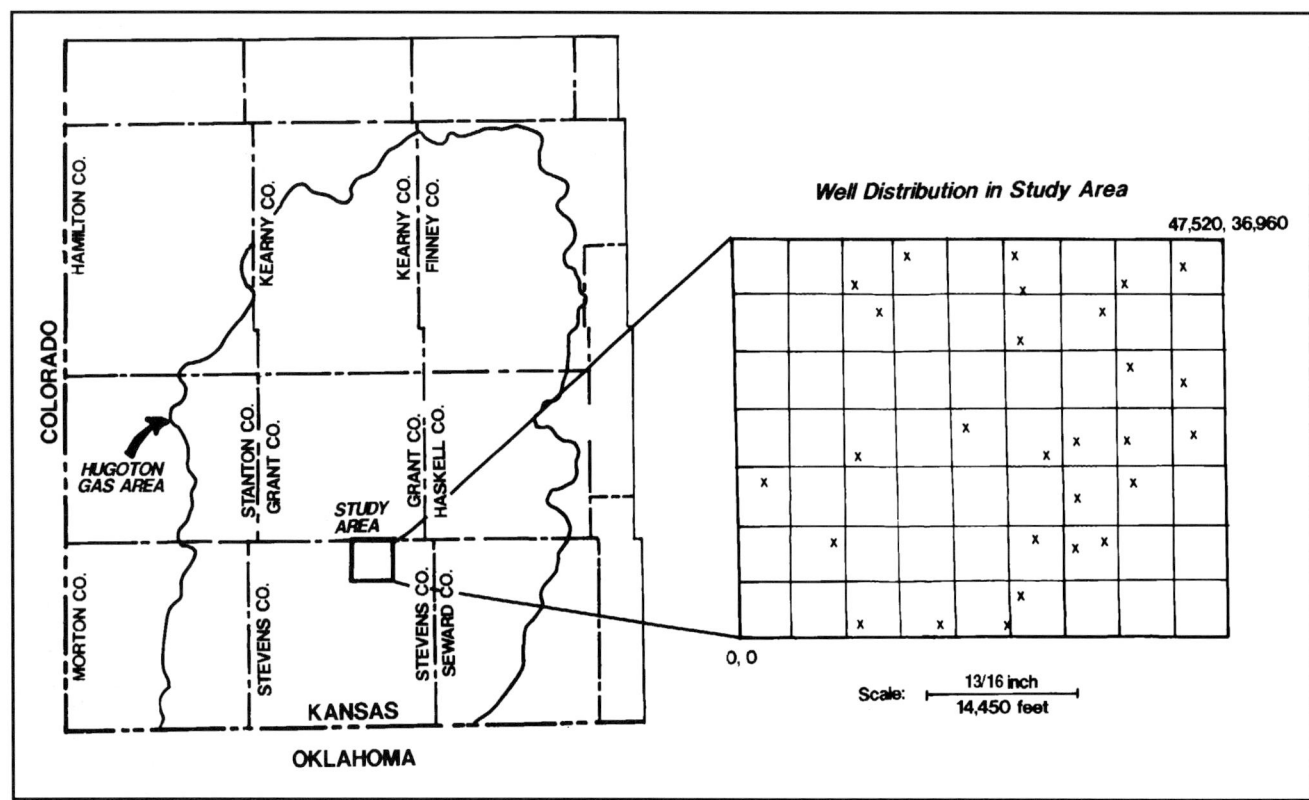

Figure 9. Base map of the Hugoton embayment study area.

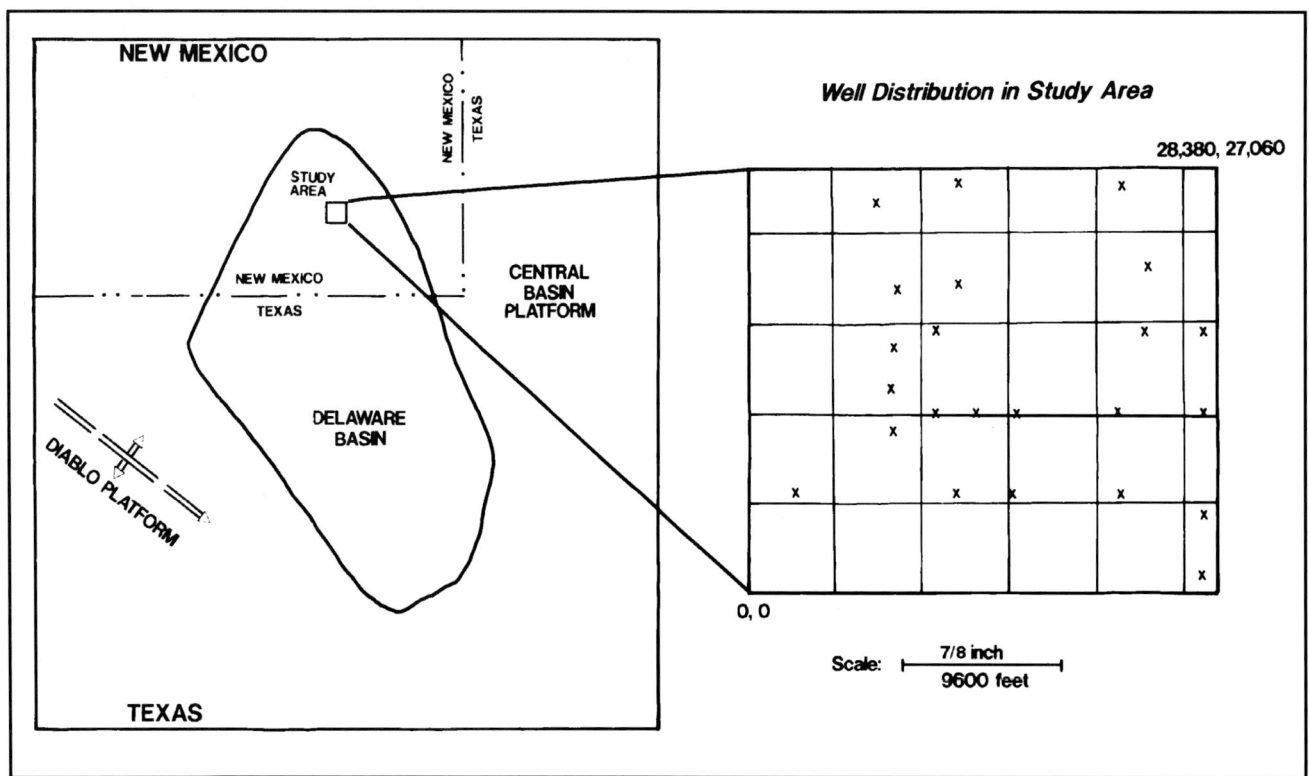

Figure 10. Base map of the Delaware basin study area.

Table 1. Summary of Datasets

Field Parameters	Hugoton Embayment	Delaware Basin
Geologic Setting	Cyclothems	Fluvial Deltaic
Number of Wells	28	23
Area of Interest	63 mi^2	29+ mi^2
Maximum Reservoir Thickness	54 ft	161 ft
Minimum Reservoir Thickness	38 ft	135 ft
Porosity Range	0.00–0.36	0.00–0.22
Mean Porosity	0.17	0.09
Porosity Std. Dev.	0.05	0.05
Hydrocarbon Saturation Range	0.00–0.88	0.00–0.86
Mean Hydrocarbon Saturation	0.60	0.49
Hydrocarbon Sat. Std. Dev.	0.18	0.24

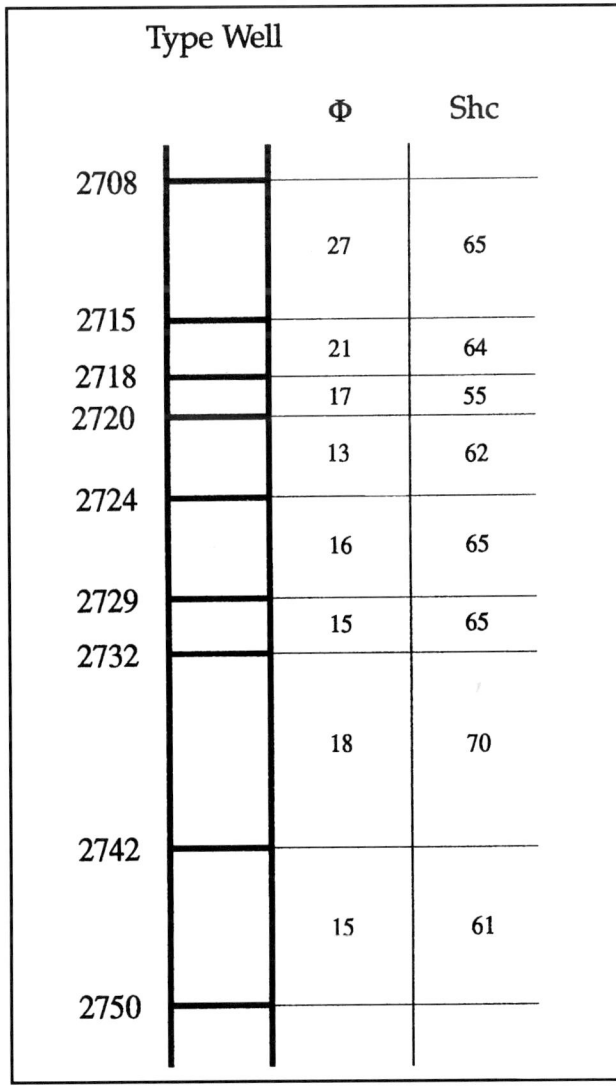

the well that passed through the cell. However, if saturation is indeterminate in this same well, the saturation value recorded in the cell would be calculated based on other wells in the search radius other than the well penetrating the cell. Under this scenario, a high hydrocarbon-saturation value could be calculated for a cell that recorded a very low porosity value. Thus, high saturation values can be interpolated into porosity-defined non-reservoir. In a typical volumetrics study using SGM, cutoffs on porosity, V_{shale}, saturation, etc.. would probably be employed to ensure that only reservoir quality rock was being included in the volume calculation. These cutoffs are typically imposed after interpolation and at the cell resolution. Because all volumes in this study were being compared with a one-layer model with no cutoffs, cutoffs were not used in the high-resolution models. In summary, as model resolution increased, the general positive trend between porosity and saturation produced increased volumes.

Although this data set showed increased volumes in the high-resolution models, the abundant indeterminate values found in the log analysis have an impact on the overall results. The indeterminates make it difficult to ascertain whether all observed porosity and saturation trends are real or artificial.

Figure 11. This illustration typifies the processing technique used in the Hugoton data set. The 1-ft log analysis was summarized into relatively homogeneous zones and maintained for subsequent modeling.

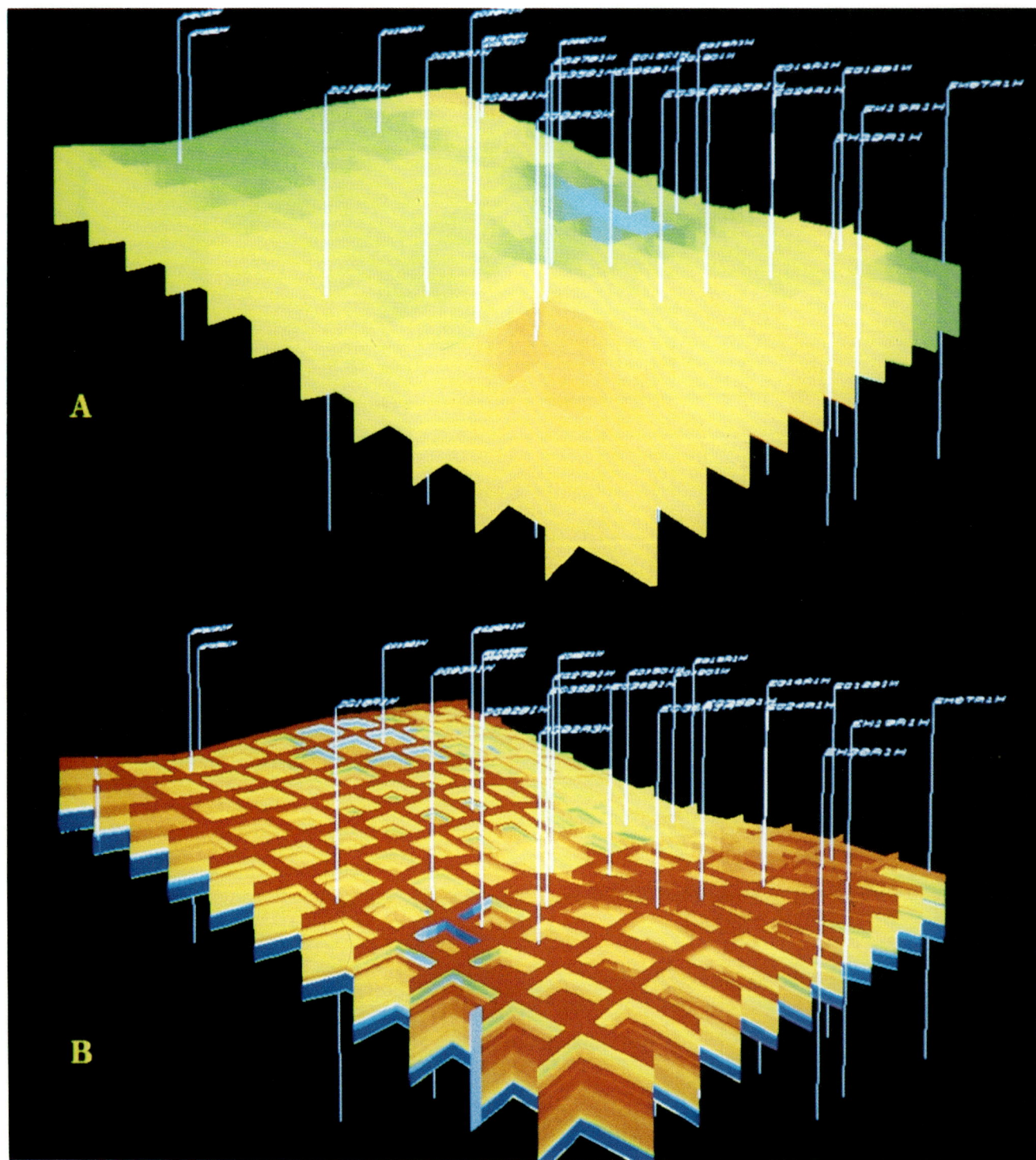

Figure 12. (A) One-layer model of porosity. (B) 32-layer model of porosity showing detailed characterization.

Since it is the relationship between porosity and saturation that largely determines whether the calculated volumes increase or decrease with increased resolution, the 7.7% increase found in this data set probably has a more significant margin of error than the Hugoton data set. The general increase in volumes, however, is probably quite real since the percentage of log indeterminates to real log values is quite small.

Table 2. Summary of Volume Results for Hugoton Dataset

Number of Layers	Volume Increase Relative to 1 Layer
2	7.0%
4	16.0%
8	20.0%
16	21.0%
32	21.5%

CONCLUSIONS

Both data sets showed increased volumes in these case studies. As mentioned above, the change in volume is controlled largely by the positive or negative correlation between porosity and saturation and the level of heterogeneity. Clearly, in heterogeneous rocks showing layered correlations, it is important to bias the distribution according to the geology while maintaining a vertical resolution that matches the level of heterogeneity suggested by the log analysis. Assuming accurate correlations, a positive correlation between hydrocarbon saturation and porosity will yield increased volumes when compared with 2-D techniques that require significant averaging. A negative correlation between hydrocarbon saturation and porosity will show a comparative decrease in volumes. One must be careful, however, that during interpolation the data-interval correlations are relatively accurate so false trends are not introduced. Poor data correlations not only impact basic interpretation, but may also yield inaccurate volumetrics.

Although these data sets showed increased volumes from 3-D geologic block modeling, 3-D modeling may not always generate volumes that differ significantly from 2-D methods. In examples where saturation is constant (homogeneous), even if porosity shows variability, greater resolution will not yield different volumes (Figure 15). It is also important to recognize that the final volume difference between two- and three-dimensional techniques is not the only concern. Characterizing reservoir attributes spatially is important even if the final volume calculation is unchanged. It is the spatial definition that enables geoscientists to more effectively exploit reservoirs by understanding the reservoir parameters better. As we have seen in the Delaware basin data set, it is quite possible that some models may show both positive and negative correlations between porosity and hydrocarbon saturation in different portions of the reservoir. It then becomes important to further investigate these relationships to better understand whether interpretations are incorrect, data quality is poor, or whether the physical parameters of the reservoir support the observed trends.

This study focused only on the impact that vertical averaging has on volumetrics. Lateral distribution of reservoir parameters was based solely on deterministic methods. Although deterministic techniques can provide accurate distributions in densely drilled fields, the technique is of limited use in fields with sparse well control if no additional tools are available to impose further geologic bias. Stochastic approaches should be used when modeling data sets with limited well control, especially in complex data sets dominated by fluvial deltaics or other geologically heterogeneous reservoirs. Similarly, tools that allow specific, detailed trends to be incorporated into the model in order to bias subsequent attribute interpolations are useful in sparse-data areas. Current development of SGM is focused on tools that allow input of these trends interactively or from outside data. Future work needs to address the effectiveness of these approaches and their impact on volume calculations.

Depending on the technique, it is clear that 3-D geologic block modeling can potentially improve our ability to characterize reservoirs and yield improved volumetrics. Although this study does not comprehensively discuss all the variables that impact volumetrics, we hope it raises some questions about the current methods used for calculating volumes. At the very least, it behooves us to evaluate volume accuracy by comparing model resolution with the data's precision.

ACKNOWLEDGMENTS

Thanks are given to CNG and Southwestern Energy for their contribution of data and information. Thanks also to John Curtin, who drafted all figures and edited the initial drafts. We also greatly appreciate the efforts of Dave Hamilton and Tom Jones, who edited this document with great patience and skill.

REFERENCES CITED

Casavant, R. R., 1988, Slice mapping: Reservoir characterization technique—West Yucca Butte field, Pecos County, Texas (abs.): AAPG Bulletin, v. 72, no. 3, p. 169.

Clark, I., 1979, Practical geostatistics: Essex, Applied Science Publishers, 129 p.

Denver, L. E., and D. C. Phillips, 1990, Stratigraphic geocellular modeling: Geobyte, v. 5, no. 1, p. 45–47.

Jones, T. A., D. E. Hamilton, and C. R. Johnson, 1986, Contouring geologic surfaces with the computer: New York, Van Nostrand Reinhold, 314 p.

Sloss, L. L., 1963, Sequences in the cratonic interior of North America: GSA Bulletin, v. 74, p. 93–114.

Swanson, D. C., 1988, A new geological volume computer modeling system for reservoir description, Paper no. 17579, presented at the 1988 SPE International Meeting on Petroleum Engineering, Tianjin, China: Richardson, Texas, Society of Petroleum Engineers.

EXPLANATION

X – 100% Water Saturated

+ – Gas Producer

– Contour Interval: .15 ft.

Scale: 1 : 68000

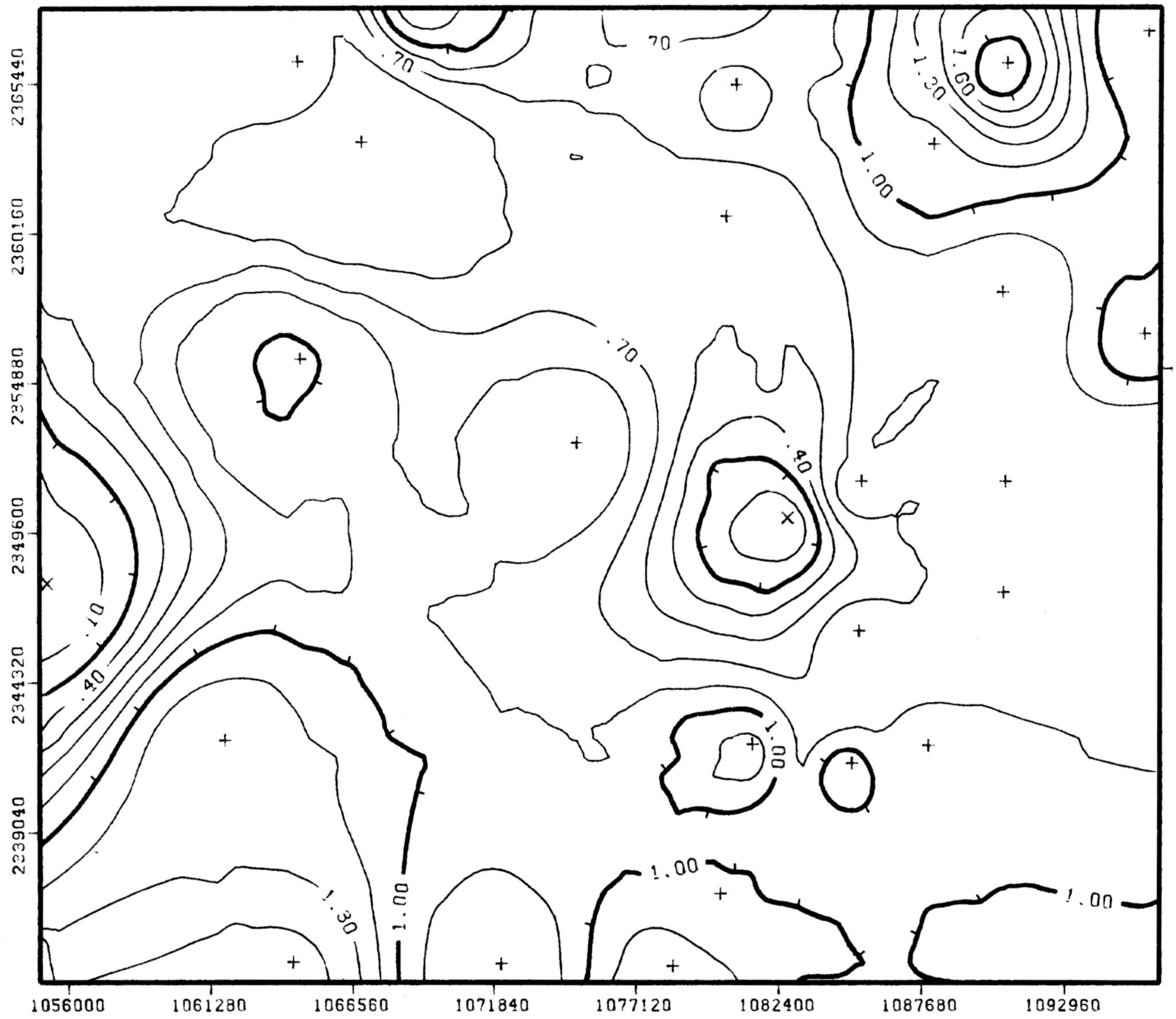

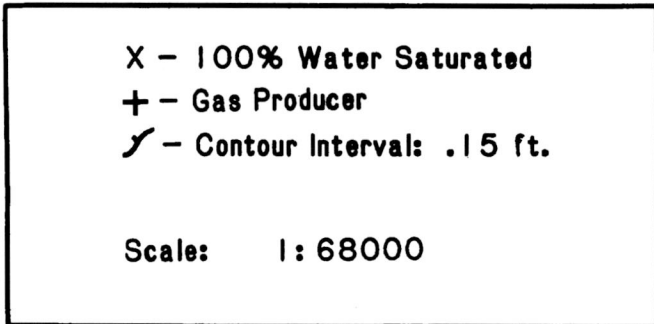

Figure 13. Contour map representing the difference (in feet) between Net Feet Hydrocarbon (NFH) calculated in the one-layer model and the 32-layer model. The greatest differences are consistently at the well locations, except for two of the wells, which have 100% water saturation. Since those wells have a homogeneous saturation, increased resolution has no impact on NFH calculations.

Table 3. Summary of Volume Results for Delaware Basin Dataset

Number of Layers	Volume Increase Relative to 1 Layer
2	3.2%
4	0.6%
8	4.8%
16	5.4%
32	6.3%
64	6.5%
128	6.9%
256	7.7%

Swanson, D. C., and S. Giovinco, 1988, Improved volumetrics with the computer: Hugoton gas field, Kansas, Paper no. 18313, presented at the 1988 SPE Annual Technical Conference and Exhibition, Houston, Texas: Richardson, Texas, Society of Petroleum Engineers.

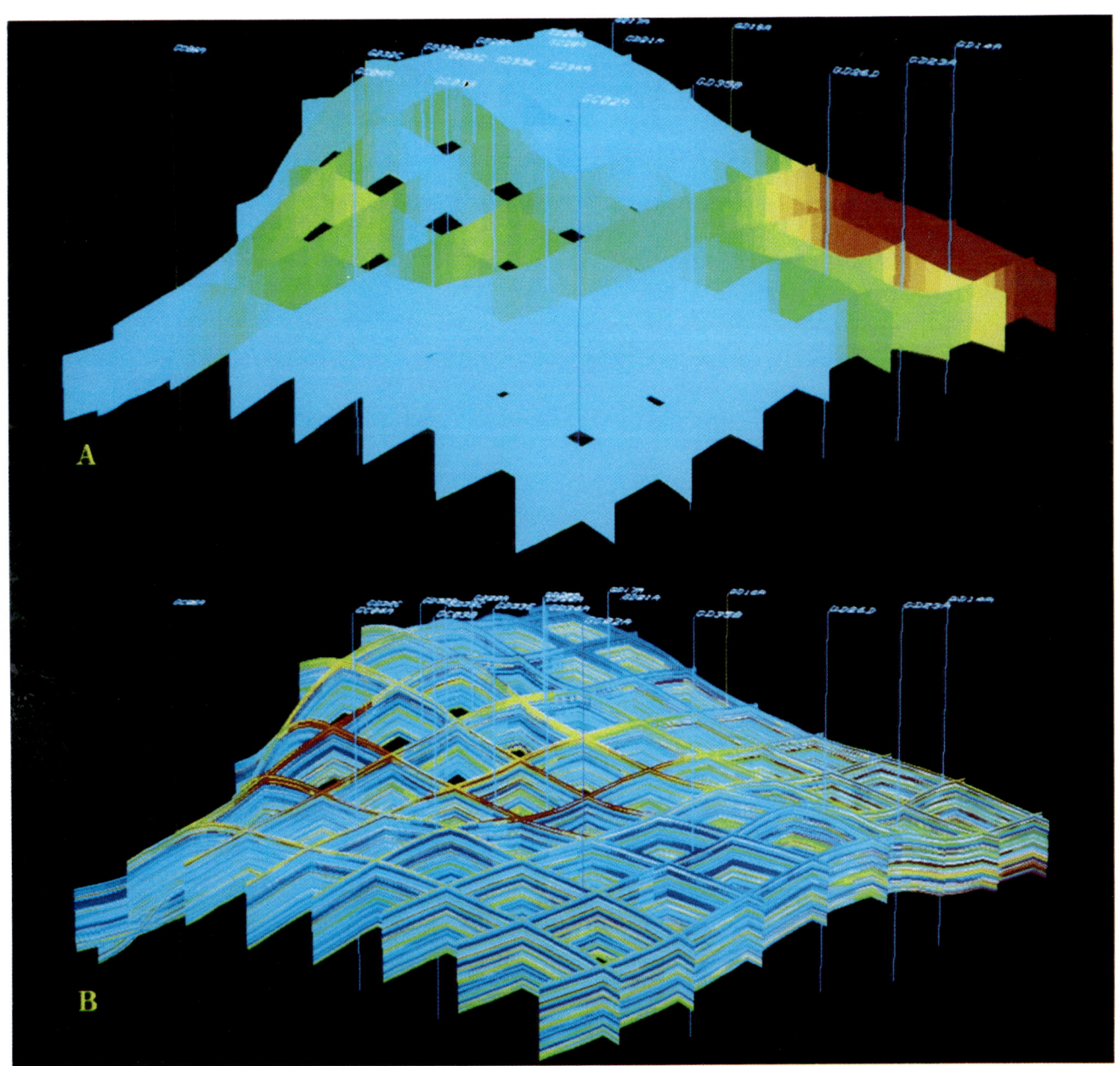

Figure 14. Nine total models were generated of the Delaware basin data set. The following examples illustrate the porosity distributions in the one-layer and 256-layer models. (A) One-layer model of porosity. (B) 256-layer model of porosity showing detailed characterization.

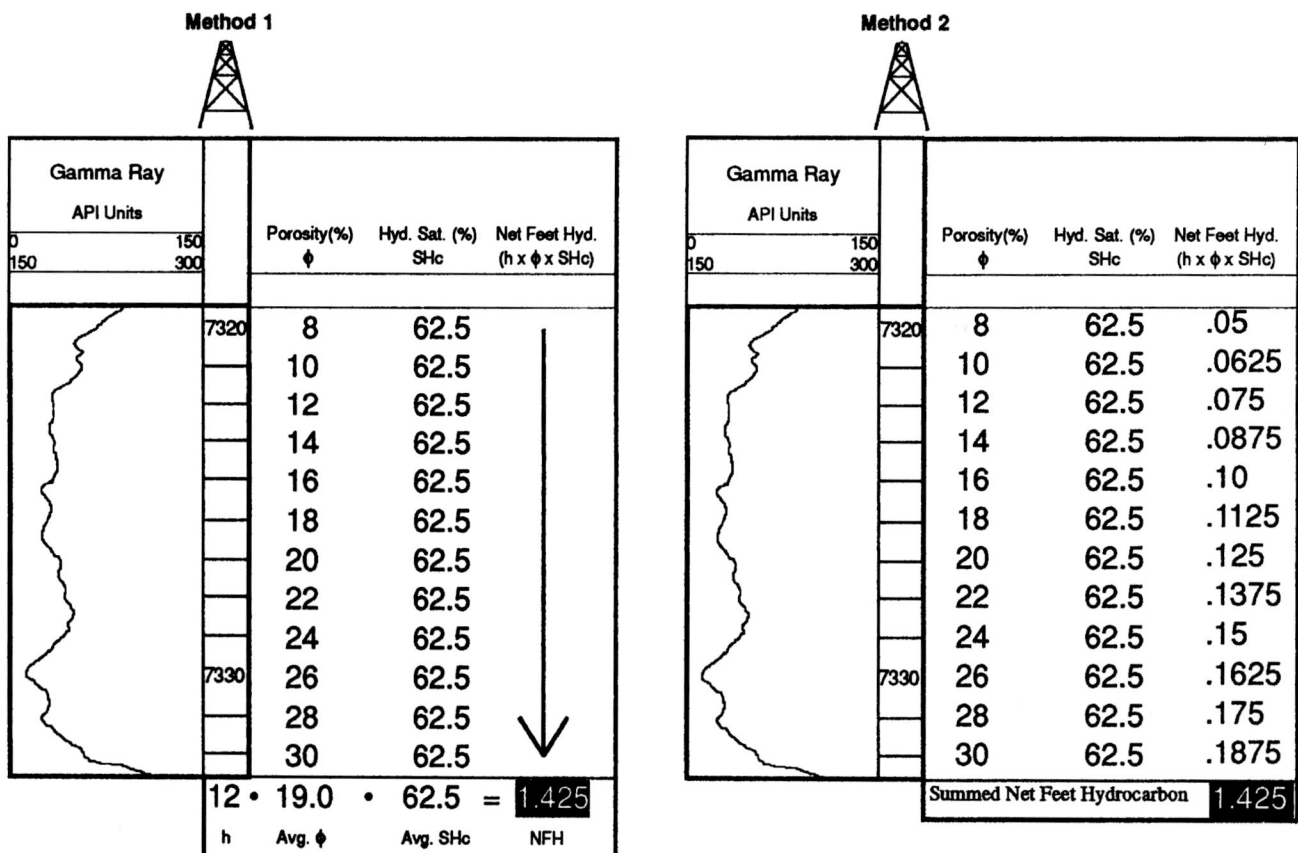

Figure 15. If porosity or saturation is homogeneous, detailed operations yield the same results as the operation of average values.

Chapter 17

Application of Variable Zone Modeling to Modeling and Mapping of Canadian Oil Sands

Khosrow Badiozamani
Morrison Knudsen Corporation,
Boise, Idaho, U.S.A.

Foad Roghani
Morrison Knudsen Corporation
Boise, Idaho, U.S.A.

George Hawes
Esso Resources, Canada
Calgary, Alberta
Canada

ABSTRACT

The Variable Zone Modeling (VZM) concept was introduced to resolve the problem of accurate estimation of mineral concentrations in an epigenetic deposit where concentration of an attribute is a function of local and small-scale geologic variations. VZM not only allows block height variation from one stack to the next, but it also allows block height changes within each stack. This approach provides the ability to fit the blocks exactly to geologic units (zones and subzones). A computer program was designed to automatically generate the subzone definitions through the use of predefined criteria.

VZM was applied to the Athabasca oil sand deposits of Alberta, Canada, which contain many depositional facies. Use of VZM in modeling these sand units produced excellent results in estimating bitumen content where previous models have failed. The application of VZM to these deposits in the vicinity of a surface mine at Fort McMurray, Alberta, is discussed in detail in this paper.

INTRODUCTION

Consideration and understanding of the geological conditions and depositional environment prior to 3-D geologic block modeling is essential. Numerous examples can be given where lack of understanding of the geologic conditions has resulted in disappointing models. Users commonly have attributed the disappointing results to the lack of proper mathematical models rather than to a need for examination of the true causes of failure. The full understanding of geology and the relationship between units of a reservoir or an ore body can provide the answer to the proper interpolation technique.

Many users of computerized modeling and mapping systems expect the programs to automatically model any type of deposit without the need on their part to understand the basic fundamental concepts involved. In some instances, the application of standard modeling algorithms to simple and straightforward deposits may provide satisfactory results; however, blind usage of a computerized modeling and mapping system usually will not result in an accurate representation of a deposit. With increased complexity, a thorough understanding of the geology, as well as the algorithms used, is required. Knowledge of the depositional environment and appropriate modeling techniques are essential for accurate modeling and representation of a deposit.

The value of an attribute, such as porosity in a reservoir or grade in a mineral deposit, is not only dependent upon its position in the mineralized zone, but also on the material surrounding it. As a result, the *area of influence* concept is commonly applied during modeling and has been incorporated into most programs to account for the gradational changes from ore to waste. Numerous techniques such as polygonal, triangulation, and inverse-distance weighting have been developed to accommodate this intuitive weighting process (Popoff, 1966), along with the use of traditional statistical approaches (Hazen, Jr., 1967; Agterberg, 1974; Barnes, 1980, p. 36–40).

The major difficulty with the previous estimation techniques is their inability to fully account for regional variations. The occurrences of high- and low-grade regions within a deposit and lower-grade fringe areas make ore grade a *regionalized variable* (Royle, 1980). A regionalized variable is one whose value is dependent upon its spatial position.

Using empirical techniques developed by Krige, Matheron (1963) developed mathematical concepts to account for the regionalized variables that have been applied successfully in modeling various mineral deposits (David, 1977; Huijbregts, 1973; Journel and Huijbregts, 1981). To quote Ramani and Stanley (1979, p. 245): "The selection of a method to develop an ore body depends on geological considerations; exploration and/or sampling methods; availability, reliability, and volume of data; specific purpose of the estimation; and the requirements of accuracy." We would like to add to this list: the complexity of the deposit and mineralization stage.

During evaluation of bitumen content in oil sand deposits in northern Alberta, we found standard modeling methods useful but somewhat limited in their ability to incorporate geologic interpretation. As a result, a new modeling technique, Variable Zone Model (VZM), was developed. VZM combines the algorithmic techniques of standard modeling methods with detailed interpretation of geologic changes in both lateral and vertical directions plus changes in geometry and bitumen distribution (Badiozamani and Roghani, 1988; Badiozamani, 1988). VZM is used extensively in modeling oil sand deposits of the McMurray Formation with successful results. VZM is equally applicable to deposits where variation in geologic conditions plays a significant role in mineral content.

THE VARIABLE ZONE MODELING METHOD

VZM tries to model, as accurately as possible, the interrelationships between geologic components (zones and subzones) and the attribute being modeled. To accomplish this, all important geologic and depositional zones are identified, and then each zone is further divided into subzones. These subzones form the lowest common denominator during model building, where the geologic information associated with each subzone is used, therefore ensuring interpolation between related materials.

After geologic specification and subzone definition, VZM follows the four steps of conventional block modeling: input of geologic and assay data; compositing of assay data into predefined subzones; designation of block size (x,y); and estimation of attribute content for each block. However, in VZM the compositing, block size specification, and attribute estimation are all controlled by geological factors such as lithology, facies, and attribute variation.

In the following sections, the steps involved in the variable zone modeling process are discussed in detail. All of the steps prior to interpolation apply to raw data (e.g., drill hole information).

Zone Definition

The first step in developing a variable zone model is division of the rock body to be modeled into correlatable or related zones. The criterion for division into zones usually is gross depositional environment, dominant rock type, or ore/hydrocarbon genesis. For example, a formation could be divided into its correlatable components, such as marine, transitional, and non-marine or continental zones. Such division reduces the possibility of interpolation between unrelated zones. If additional information is available, each of these zones could be further subdivided to improve geologic control during modeling. If information is scant or smaller zones cannot be correlated throughout an area, a single zone may represent the entire deposit.

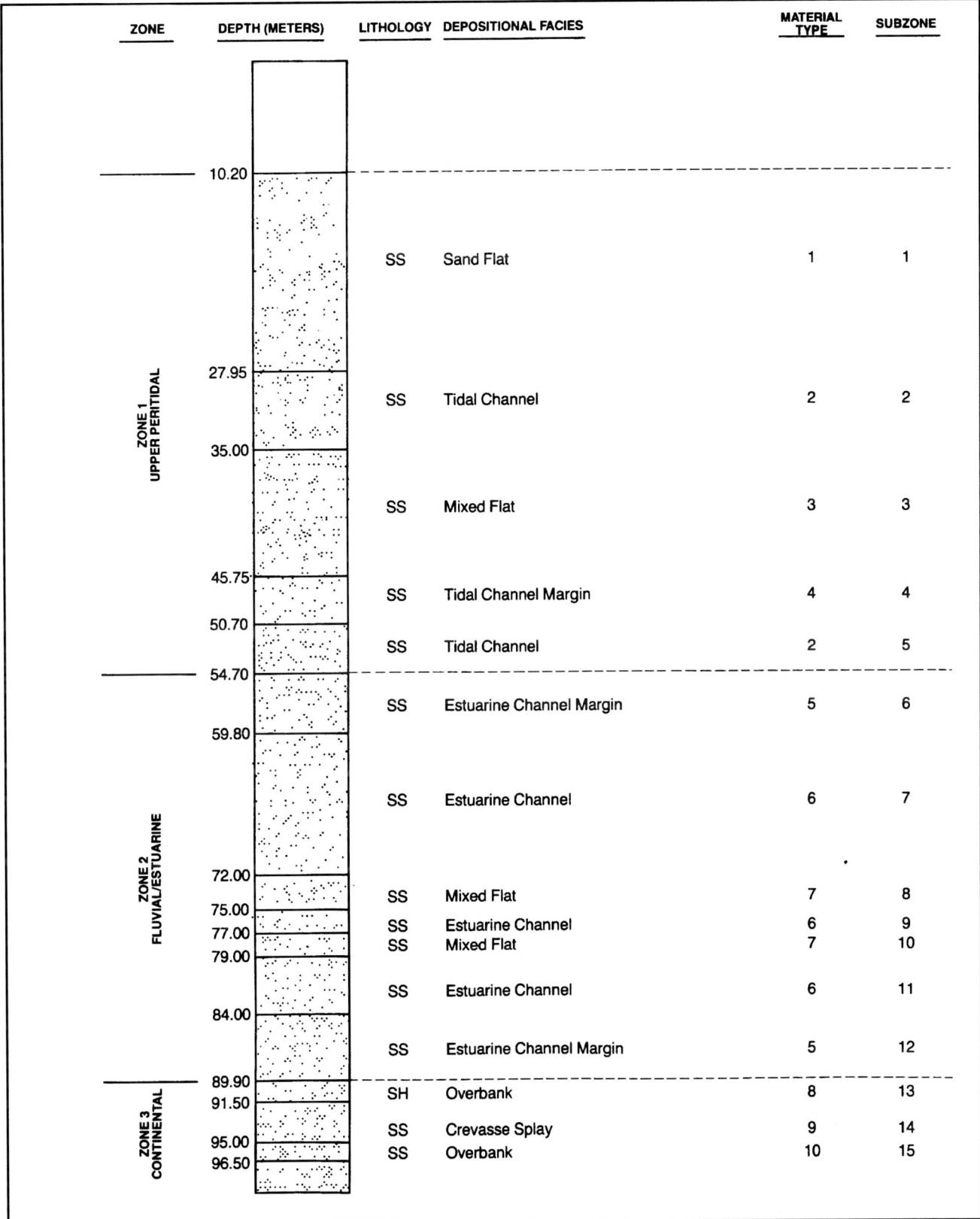

Figure 1. Division of geologic data to material types, zones and subzones. The diagram shows three major zones with 10 material type which contain 15 subzones. Subzones are sequentially numbered from top to bottom.

Material Type Definition

Once the zones have been identified, the unique geologic features of significance within each zone are defined. Each unique feature is assigned a specific *material type* designation. Material type is defined as a combination of geologic properties that together uniquely distinguish a rock unit from other rock units in the same zone. Material type may be based on such properties as lithology, depositional facies, or degree of weathering. For example, channel and point-bar sands constitute two different material types in a fluvial environment. Similarly, two sand bodies belonging to different channels could be considered separate material types. During modeling, regardless of the algorithm used, only data from similar material types are used to interpolate values for blocks found within that material type.

Subzone Definition

Where appropriate, zones are further subdivided into subzones. Material type forms the basis for creating these subzones. For example, in a fluvial channel with coarse-grained sand at the base, silty sand in the middle, and silt on top, three material types could be defined, consisting of fluvial-channel coarse sand, fluvial-channel silty sand, and fluvial-channel silt. Therefore, the zone would be separated into three subzones based on distribution of these material types. If this fining-upward cycle is repeated again within the channel, the channel would be divided vertically into six subzones, based on the three material types. These subzones occupy the same depositional facies (fluvial channel) and the boundaries between them are based on a change in lithology that the geologist felt was significant for the purpose of modeling.

Figure 1 illustrates the relationship between material type, zone, and subzone selection. As shown in this geologic log, zone two contains three material types; however, it was divided into seven subzones based on multiple occurrences of the three material types.

Compositing Data

After zone and subzone designation and prior to interpolation between drill hole data to estimate the value of each block, various analytic values (e.g., porosity, bitumen content, ore grade) must be determined for each subzone. Sample intervals generally do not coincide with the top and bottom of each subzone. Multiple samples may be available for each subzone. If the subzone and sample intervals coincide, the value of the resultant sample is assigned to the subzone. If there is more than one sample per subzone, the samples at each drill hole location are composited and the composite value assigned to the subzone. The composite value is weight averaged by thickness of the samples, or by the product of thickness and density if there is substantial density difference between samples. Weight averaging by thickness and density is performed only for elements whose value is reported as weight percent (e.g., grade), and not for elements that are based on volume percent (e.g., porosity).

Defining Block Dimensions

Block dimensions are a function of different variables, such as the distance between sample locations (drill holes), magnitude of local variation of the variable under investigation, and required degree of accuracy. Generally, closer drill hole spacing and greater local variation requires smaller block dimensions. The block size for mining and excavation purposes may differ from the geologic block size, since it is dictated by such factors as mining method and size of equipment.

The block height used in most 3-D geologic modeling systems is either fixed at a constant value (Figure 2), or variable, with a specified number of blocks proportionally spaced between the top and base of the zone (Figure 3). As shown in Figures 2 and 3, the conventional and variable block modeling systems lack granularity, resulting in different material types to be mixed in each block. Decreasing block height alone may not necessarily fit all geologic variations to meet the geologic boundary limits. VZM, on the other hand, takes into account all material type variations and automatically adjusts the block height to coincide with each material type boundary (Figure 4). Because the block size is a function of geologic units in VZM and is not an arbitrary user-defined limit, it minimizes the number of blocks needed to accurately represent a deposit.

Block height in fixed-height block models commonly is determined by factors such as type of mining operation and equipment used. When this is done, attribute samples are composited at fixed intervals without regard to geologic boundaries. Results of modeling with these blocks and composited values make it impossible to separate pay and nonpay (ore and waste) material. More correctly, the fixed block height value should be set at a resolution commensurate with sample thickness. However, this approach may increase the number of blocks in a model substantially if the sample intervals are small. If interpolations of attribute values are then guided by correlation surfaces (Hamilton and Didur, 1992), there is a chance that values from like material will be used to determine a block's value. There is also a strong possibility that samples from one material type will be correlated with samples from another material type during interpolation, resulting in a poor estimate at that block's position.

Block height in proportionally spaced variable block models varies throughout the model. The number of blocks for each zone in the model is set and is usually based upon resolution of the data, although occasionally it is based upon mine operation and equipment. The thickness of the zone at a location is divided by the specified number of vertical blocks for that zone to determine block height at that position. Thus, block height for each stack of blocks is fixed, but varies from one stack to the next. The surfaces defining the top

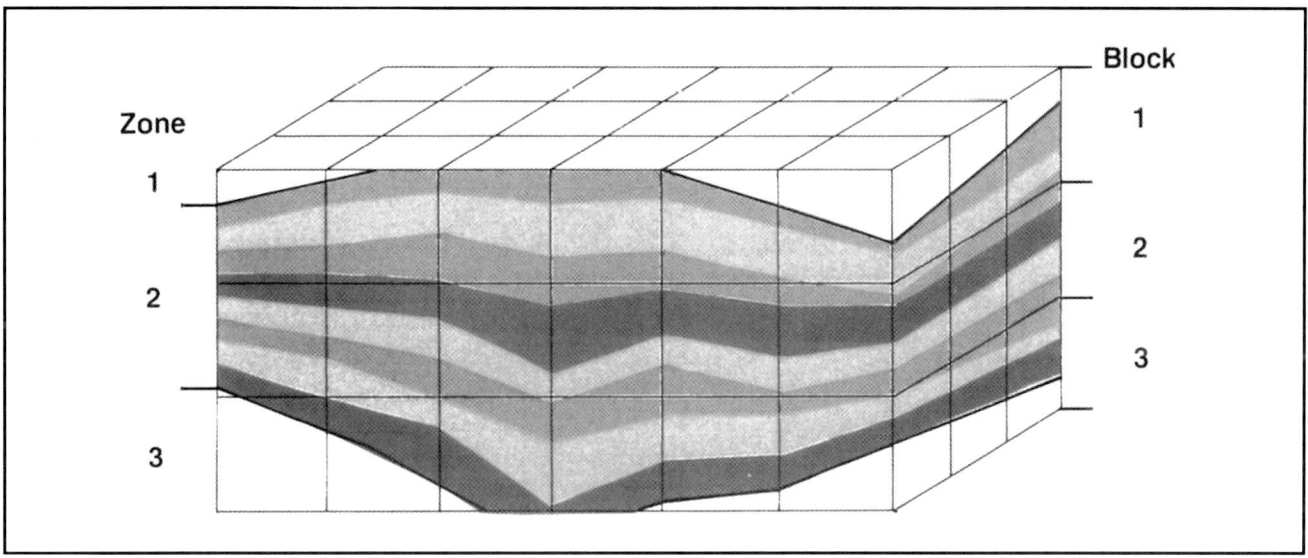

Figure 2. Conventional block model with fixed, equal thickness blocks. The blocks are superimposed on the geologic zones which are shown as oblique lines passing through the blocks. Different shades within zone 2 represent various lithology.

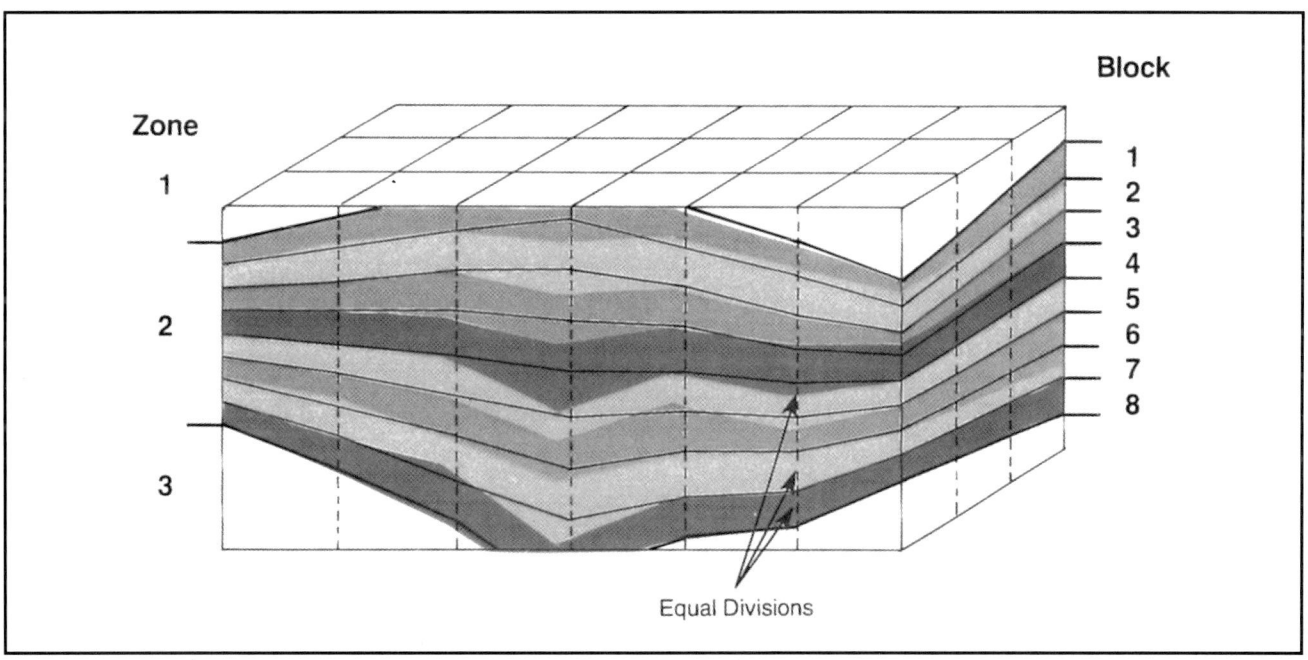

Figure 3. Variable block model with proportional thickness blocks. The blocks match the top and the bottom of the geologic zones, but do not necessarily coincide with lithologic boundaries shown with different shades for zone 2.

and base of a zone typically are geologic boundaries. If the deposit is highly stratified, then interpolations along these block layers will produce reasonable estimates of the attribute's value at each block. However, if significant lateral and vertical variations in material types occur, then samples from unrelated material types are used during interpolation, resulting in a poor value estimate.

Block height in VZM is defined by the geologic boundaries and varies from block to block both laterally and vertically in the model. Block boundaries are defined to coincide with boundaries between different material types. Because interpolations are performed only using samples having material type similar to the block being estimated (i.e., within that subzone), incorrect interpolations do not occur. If mining bench

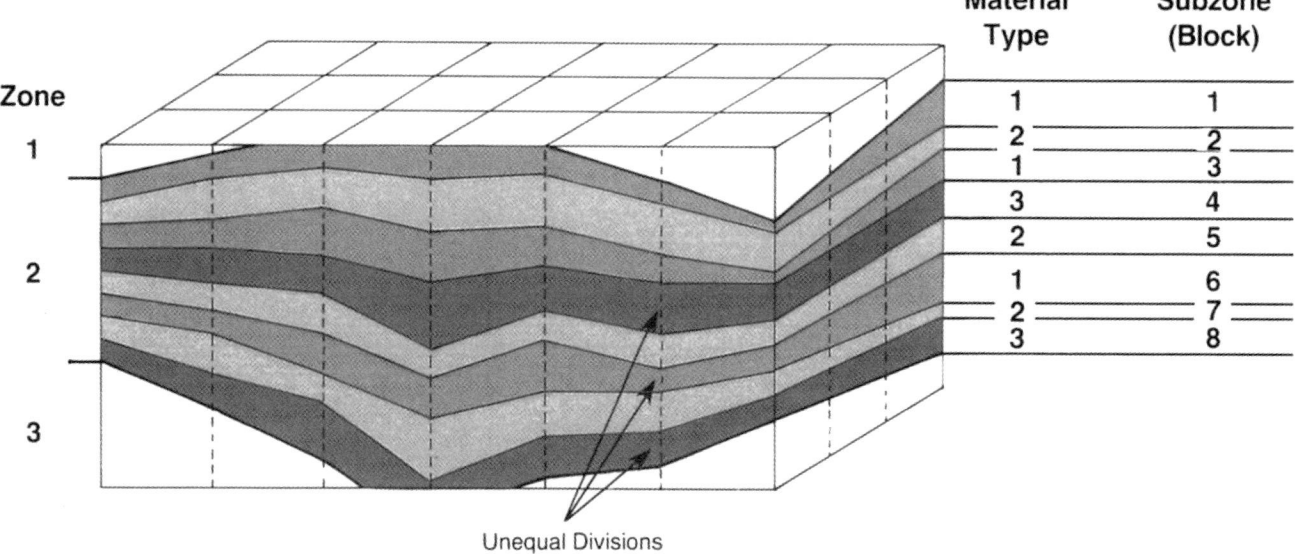

Figure 4. Variable Zone model (VZM) with variable thickness subzone blocks. The blocks not only match the top and bottom of the geologic zones, but they also match the limits of each subzone. Constructed based on variation in the lithology. Various lithologies are shown with different shades.

height is greater than the block height, multiple block values may need to be averaged to arrive at a bench value, which is the case with other 3-D modeling systems. However, unlike those methods, since interpolations are performed only between like material types, pay and nonpay material have not been averaged by interpolation, resulting in more accurate estimated values.

Interpolation

The final step in model construction is the interpolation of value estimates for each block. Any interpolation algorithm can be used, and different algorithms may be appropriate for different attributes. For example, bitumen content may be modeled using geostatistics and porosity may be modeled using inverse-distance squared.

Interpolation is done on a subzone basis. That is, only composited data for a subzone are used to calculate value estimates for blocks in that subzone. Since a subzone is defined by material type, this means that interpolation is done only with material of the same type. All of the search limits, sectors, and distance weighting parameters common to interpolation algorithms are still appropriate.

APPLICATION OF VZM TO AN OIL SAND DEPOSIT

VZM was used to build a 3-D geologic block model of bitumen content distribution for an Athabasca oil sand deposit in the McMurray Formation.

Geologic Setting

The Athabasca oil sand deposits in northeastern Alberta contain one of the largest oil deposits in the world (Demaison, 1977). The majority of this oil is contained in the Lower Cretaceous McMurray Formation. In general, the McMurray Formation is divided into three informal members—lower, middle, and upper (Flach and Mossop, 1985) shown in Figure 5. According to Flach and Mossop, this formation was deposited in a north-south-trending depression created by the removal of Middle Devonian evaporites.

The study area, known as the OSLO oil sand project, is located about 40 mi (64 km) north of Fort McMurray, Alberta, Canada, in and around Township 95 of Range 8 West (Figure 6). The McMurray Formation in the study area unconformably overlies the Devonian Waterways Formation. The Waterways Formation developed in a relatively shallow, warm-water marine environment. The sediments are composed of skeletal and micritic carbonates and light greenish-gray calcareous clay. A major Devonian low trends northward through the area. Salt removal from the underlying Middle Devonian Elk Point Group and subsequent collapse of the overlying Beaverhill Lake Group is believed to have played a major role in controlling the initial geometry of the drainage basins developed on the Paleozoic erosional surface. Normal faults possibly extending into the Precambrian basement complex, which may have been active as late as early McMurray time, are also thought to have played a role in the development of these early drainage basins.

The McMurray Formation represents sediment infilling of a paleo-drainage system that developed on

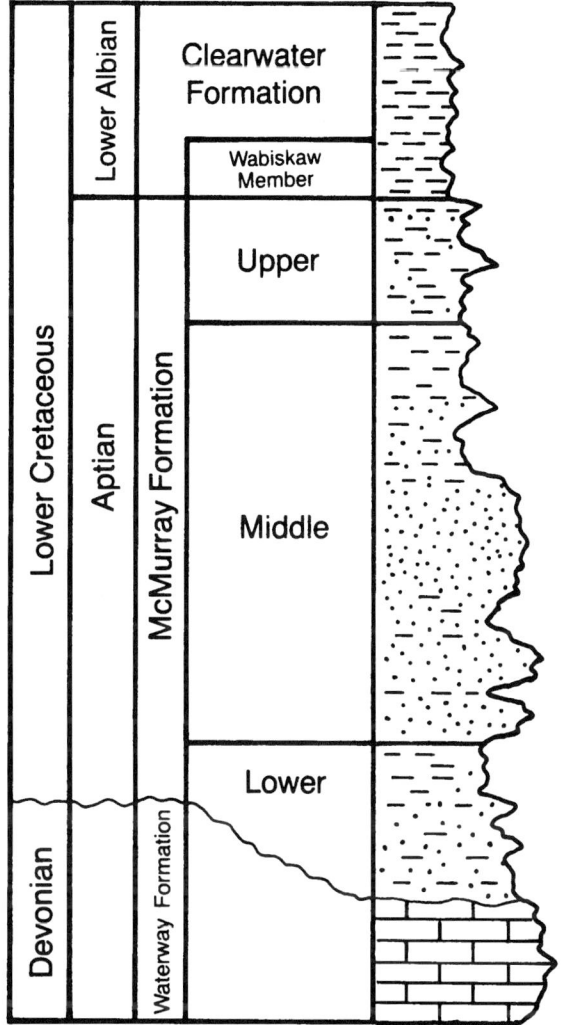

Figure 5. General stratigraphy of the McMurray and adjacent formations (Modified after Flach and Mossop, 1985).

the Devonian erosional surface. The formation is a complex mixture of continental, fluvial-estuarine, and marine environments (Figure 7). These environments are discussed in detail below.

McMurray sedimentation is terminated by a major transgression of a boreal sea during Early Cretaceous (Albian) time. Above this is the basal Wabiskaw member of the Clearwater Formation. Where preserved, the Clearwater consists of the Wabiskaw member and previously undifferentiated Clearwater silty-clays and clays.

Pleistocene glacial and glacio-fluvial geologic units lie unconformably above the Cretaceous sediments. In the study area the Pleistocene consists predominantly of tills, sourced mainly from the McMurray oil sand and Grand Rapids Formation, with a small percentage derived from the Clearwater Formation. Pleistocene meltwater channels, although prominent subsurface features, do not have surficial expression and can be more accurately defined as burial valleys. They are typically filled with glacio-fluvial sand and gravels and glacial tills. Glacial channels with depths exceeding 60 m and widths of more than 1 km are not uncommon.

Holocene and Recent sediments consisting of water-saturated organic-mineral soil, muskeg, peat, minor occurrences of Holocene lake bottom sediments, as well as fluvial and eolian silts and sands, complete the stratigraphic sequence.

Depositional Environment of the McMurray Formation

Continental

The Continental environment (Figure 8A) is dominated by coastal plain sediments, with muddy flood plain, pond mud, overbank facies, sandy crevasse splay, and levee and fluvial channel facies. The continental fluvial sands were deposited in the deepest parts of the basin (Devonian lows) and were subsequently capped by marsh or back beach deposits that forms a good time line. The overall continental sediment package has the effect of flattening the irregular Devonian topography. The continental sands are water saturated with minor isolated oil saturation.

These continental deposits onlap onto Devonian highs. There is an increase in the occurrence of muddy marsh deposits toward the top of the continental succession indicating a decrease in channel activity and an overall upward-fining trend. This suggests a stagnation of the basin's drainage system in response to an overall rise in sea level.

Mixed Fluvial–Estuarine

Interplay of transgressive and minor regressive cycles by flooding from the north subsequently formed an estuary in the drainage basin. Fluvial or continental marsh sedimentation prevailed in the upper reaches of the partially drowned valleys, while estuarine processes influenced by tidal currents deposited the bulk of the sediments within the lower part of the estuary. Lateral migration of tidal channels into marsh, tidal flat, and older tidal channel deposits account for the dynamic juxtaposition of depositional environments and the complex stratigraphy exhibited in the middle McMurray member.

The fluvial estuarine environment (Figure 8B) created a thick sequence of stacked sands ranging from more fluvial-dominated channel sands and channel breccia at the base to more estuarine-dominated channels, channel tops and channel margins at the top. The sands range from coarse to fine-grained, are poorly sorted, and have the highest overall bitumen saturation of the oil-bearing sand facies. This zone may contain occasional tidal flat zones. Towards the end of McMurray time a strong transgressive pulse flooded the estuary and sedimentation became more coastal and open marine in nature.

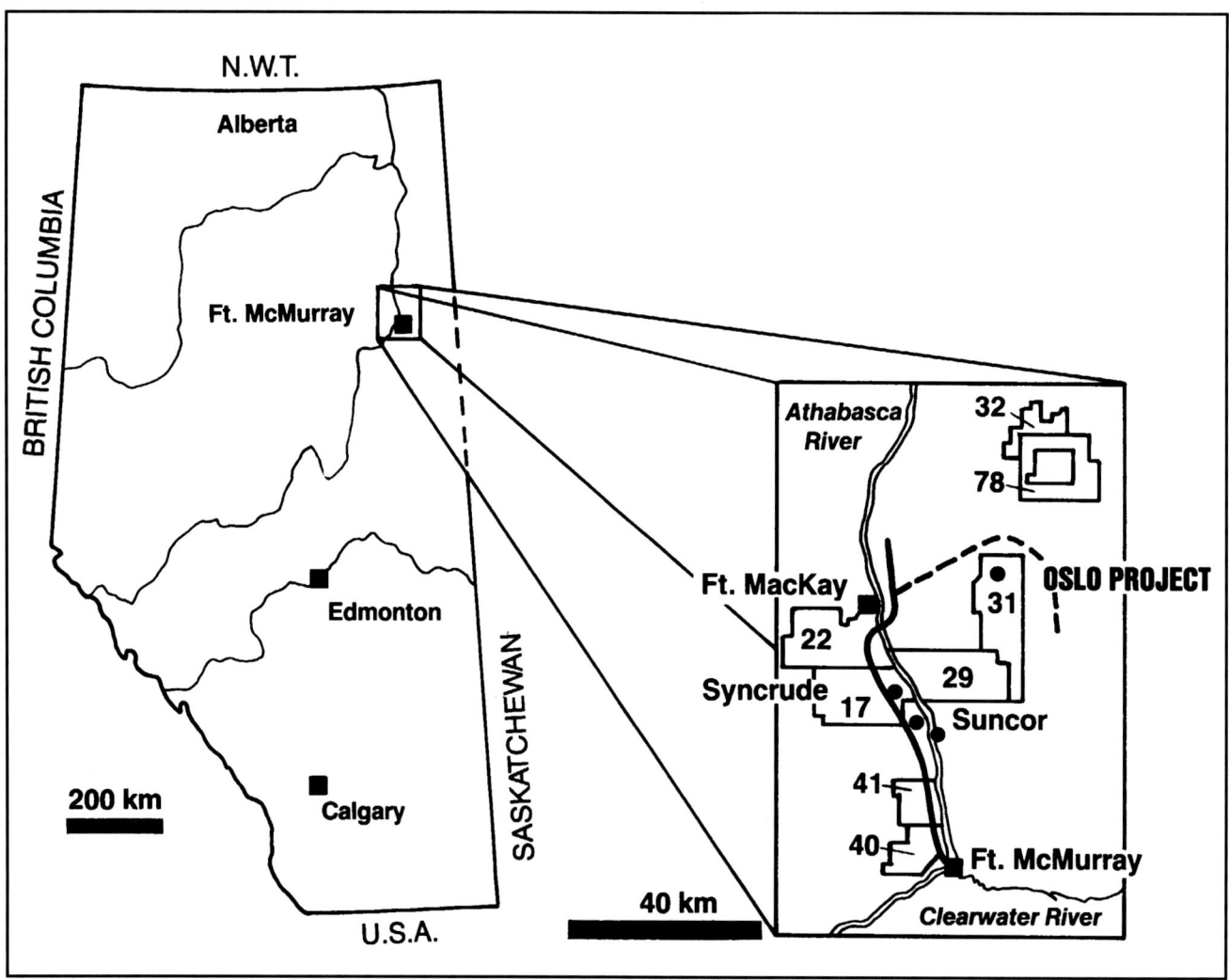

Figure 6. Index map showing location of the study area within the Province of Alberta, Canada.

Tidal/Estuarine

Rocks of the tidal/estuarine or "upper tidal" environment (Figure 8C) generally consist of tide-dominated sediments comprising sand flats, mixed flats and mud flats, which are characteristically lower in bitumen saturation. Occasionally, fine-grained, well-sorted, and highly bitumen-saturated tidal channels are interspersed throughout the area.

Marine

The upper portion of the McMurray Formation was deposited in a near-shore marine setting with facies ranging from offshore mud to shoreface sand (Figure 8D). In general, the upper shoreface facies consists of fine-grained, well-sorted sands with medium oil saturation. The lower shoreface/offshore transition facies has a higher percentage of very fine sands and clays, and hence is lower in oil saturation. Post-depositional erosion has limited the extent of marine sediments in the study area.

Facies within the above key environmental groups have been identified on the basis of grain size, texture, and sedimentary structures, as well as through lateral and vertical facies relationship. The facies types and their inherent characteristics have a direct bearing on oil saturation, fines content, geometry, and quality of the deposit.

Data

The data consist of geologic information and geophysical logs for 1000 drill holes, 200 km of reflection seismic data for delineation of Pleistocene erosional channel, and field work information. Using the above data, geologic information was interpreted to define the detailed stratigraphic and depositional environment. Most of the drill holes were sampled at 30-cm intervals. These samples were analyzed for porosity, bitumen, water, and solid content. EAGLES (a proprietary software developed by Morrison Knudsen Corporation) was used for geologic evaluation, analytic

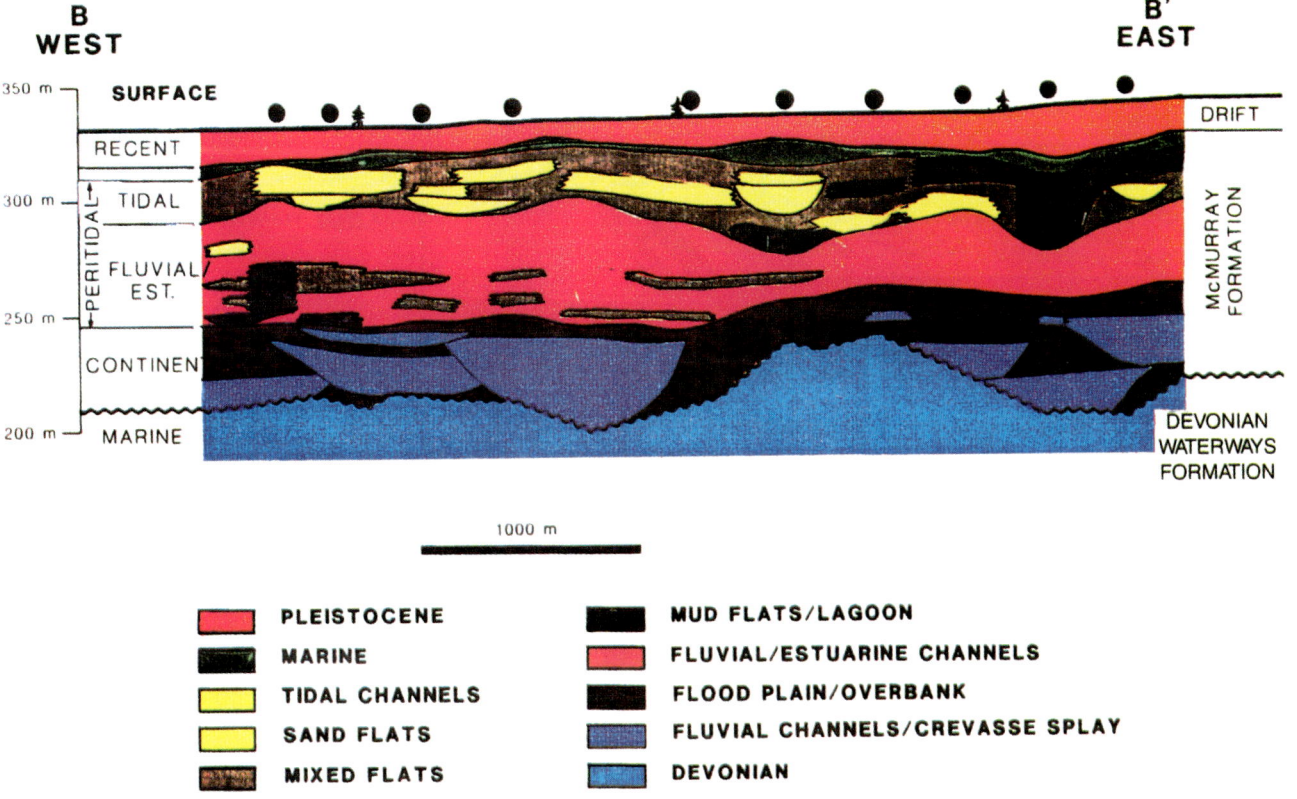

Figure 7. Schematic cross section of the McMurray Formation showing depositional facies in the study area.

data analysis, modeling, and mine planning of the project. EAGLES stands for Engineering, Analytic, Geologic, Land Management and Evaluation System.

Building the Model

Grids were built for the top structure of the McMurray Formation and thickness of each subzone. The top structure grid was used as a reference elevation for hanging different zones and subzones. A 3-D geologic model was then built for porosity, bitumen, water and solid contents, using these zone and subzone definitions.

The key to modeling this deposit successfully was to understand the deposit's detailed geology and the influence of its depositional environment on hydrocarbon accumulation. As discussed above, the stratigraphy of the McMurray Formation is complex. Sand bodies are the major hydrocarbon-bearing zones. Each depositional facies in the McMurray Formation may contain multiple sand units with different hydrocarbon-bearing characteristics. The inherent physical characteristics of these units, such as porosity, permeability, and grain size (fines content), control the oil saturation level. It was essential, therefore, to identify these sand units within each depositional facies.

An important characteristic of the McMurray Formation, that had to considered during model construction, is the lack of lateral and vertical facies continuity. Adjacent boreholes, only a few hundred meters apart, usually contain no beds or markers that can be correlated. Physical features, such as grain size and texture, also vary substantially in both directions within a short distance. Horizontal variations are a function of magnitude and extent of the local depositional features, whereas vertical variations are a function of energy level of the water at time of deposition. These physical features affect porosity, permeability, and grain size, and therefore, oil saturation. As a result, not only must each facies be identified during modeling, but the sand bodies belonging to each depositional system must also be identified.

Zone Definition

Prior to modeling, the McMurray Formation was divided into five zones based on depositional envi-

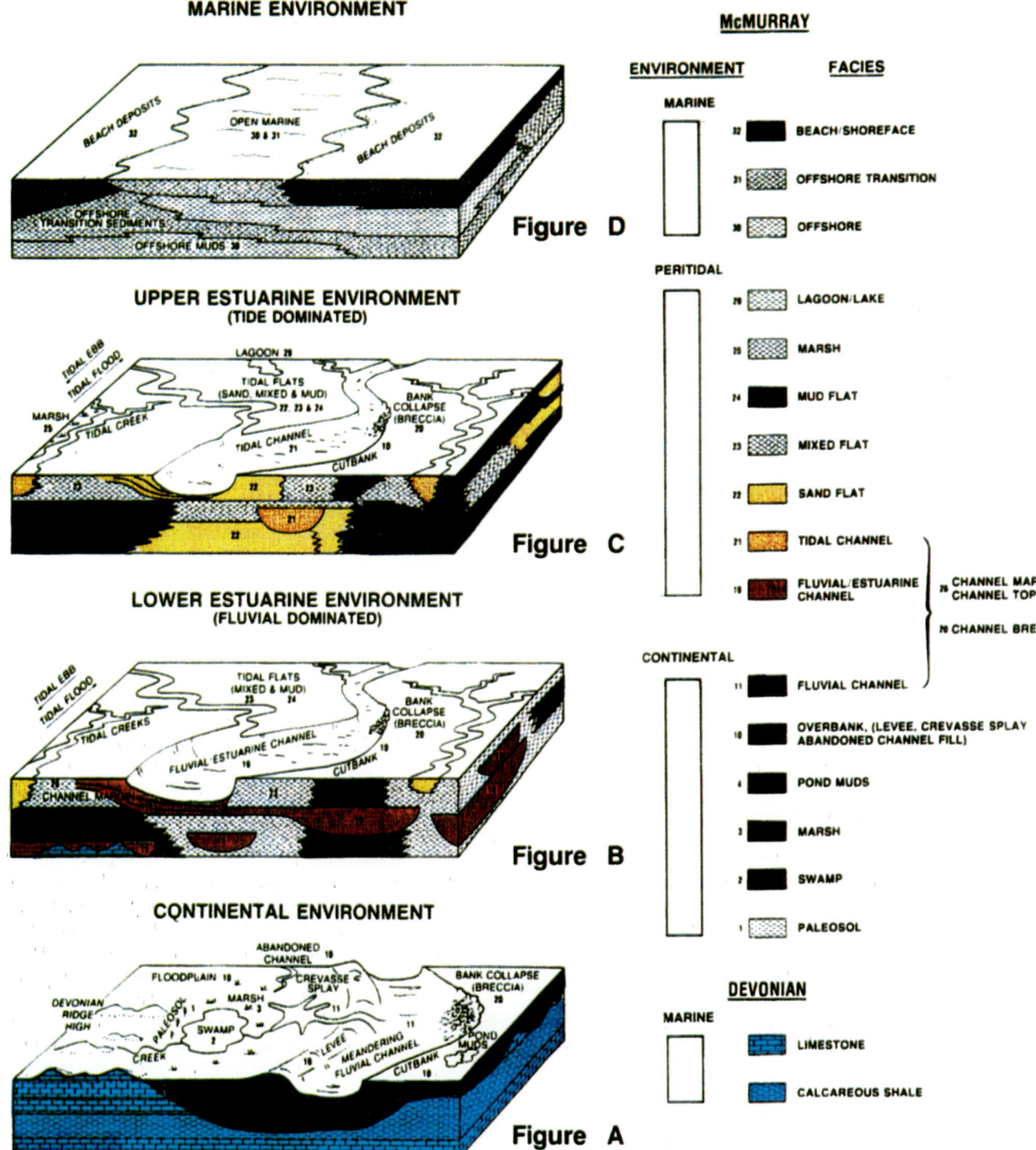

Figure 8. Depositional model of various zones of the McMurray Formation.

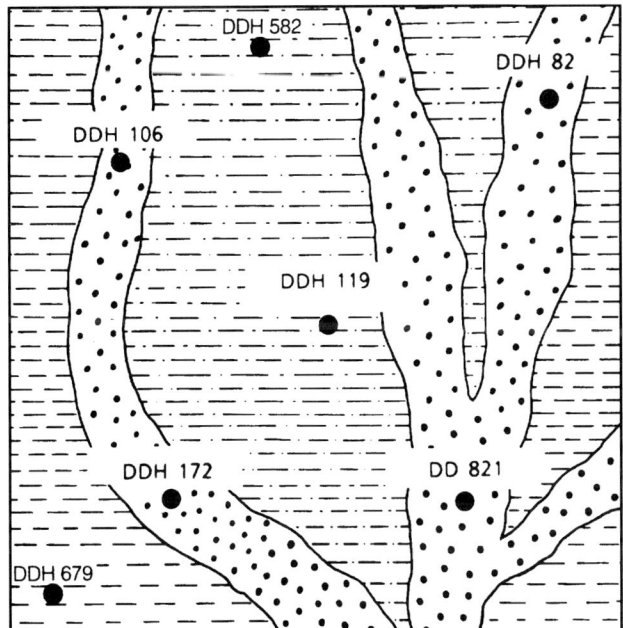

Figure 9. Schematic map showing drill holes intersecting two different depositional facies such as tidal channel and tidal flat.

ronments. These zones, from bottom to top, are continental, lower pretidal, fluvial-estuarine, upper pretidal, and marine. Four of these zones have been discussed in detail. The lower pretidal zone does not exist in most of the project area and can be observed only in a few locations. It consists mostly of fine-grained material of mud flat and mixed flat composition.

Material Type Definition

A combination of lithology and depositional facies was used to define 40 material types. Using this many material types would result in too many subzones, so some of these material types were grouped together. Grouping was done where there was a close relationship between the physical and geological properties of different material types as related to its hydrocarbon distribution. For example, shoreface and lower shoreface sand units have many similar geologic properties and could be combined into one group.

Use of these material types clearly demonstrates the validity of the VZM methodology. One may expect these material types to have different porosities and permeabilities and, therefore, to respond differently to hydrocarbon migration. Figure 9 is a schematic diagram showing seven drill holes penetrating two material types, i.e., tidal channel and tidal flat. If the material types are ignored during interpolation, data from these unrelated environments will be mixed, resulting in erroneous value estimates for blocks in those environments. However, if material type is considered when estimating values for blocks within the range of influence of each material type, then results are more representative of the geologic interpretation.

Subzones, Block Dimensions, and Compositing

Each zone was divided into subzones based on material types. This resulted in subdivision of the five zones into 125 subzones. The top and base of each subzone defined the top and base of each block. The X- and Y-dimensions of the blocks were defined by the X- and Y-grid increments. All sample data between the top and base of each subzone in a drill hole were composited to determine the value of each element.

Figure 10 is a cross section showing zones 1, 2, and 3 and their subzone boundaries. In zone 3 (bottom zone) the subzones are fairly continuous. In zones 1 and 2, the subzones are more lenticular and discontinuous with zone 1 missing in the center of the section. Both zones 1 and 2 are cut by a Pleistocene channel on the left side of the section.

Interpolation

Grids were built for the top of McMurray Formation and thickness of each subzone within the formation. A trend existed along the top structure, so a trending algorithm (regression plane method) was used to model this surface. Seismic information was also used as control for grid construction. The seismic data provided detail needed to accurately define the Pleistocene erosional channel that cut into this formation.

Because of lack of continuity between drill holes, algorithms that build trends, such as regression or polynomials, were not appropriate for modeling thickness. Instead, a linear inverse-distance weighting algorithm was used to build the thickness grid for each subzone. These thickness grids were hung from the top of the McMurray Formation to define the 3-D model of the deposit.

A 3-D geologic block model containing four attributes (porosity, bitumen, solid, and water content) estimates for each block of each subzone was constructed using a quadratic inverse-distance weighting algorithm. Again, this algorithm was used because lack of continuity between drill holes prevented the use of algorithms that build trends. The final 3-D model contained 5 zones and 125 subzones.

Figure 11 is a cross section through the 3-D VZM model showing contours of bitumen content. Changes in value across adjacent subzone boundaries are apparent. Lack of continuity is seen in horizontal and vertical directions. Such extreme variations can be observed along the high walls of operating mines in the nearby area. Clearly the transition from an area dominated by one environment to an area dominated by another environment is seen in the modeled bitumen content values.

Analysis

For comparison purposes, a block model without respect to the subzone boundaries of each zone was also constructed. At each drill hole, all data for each

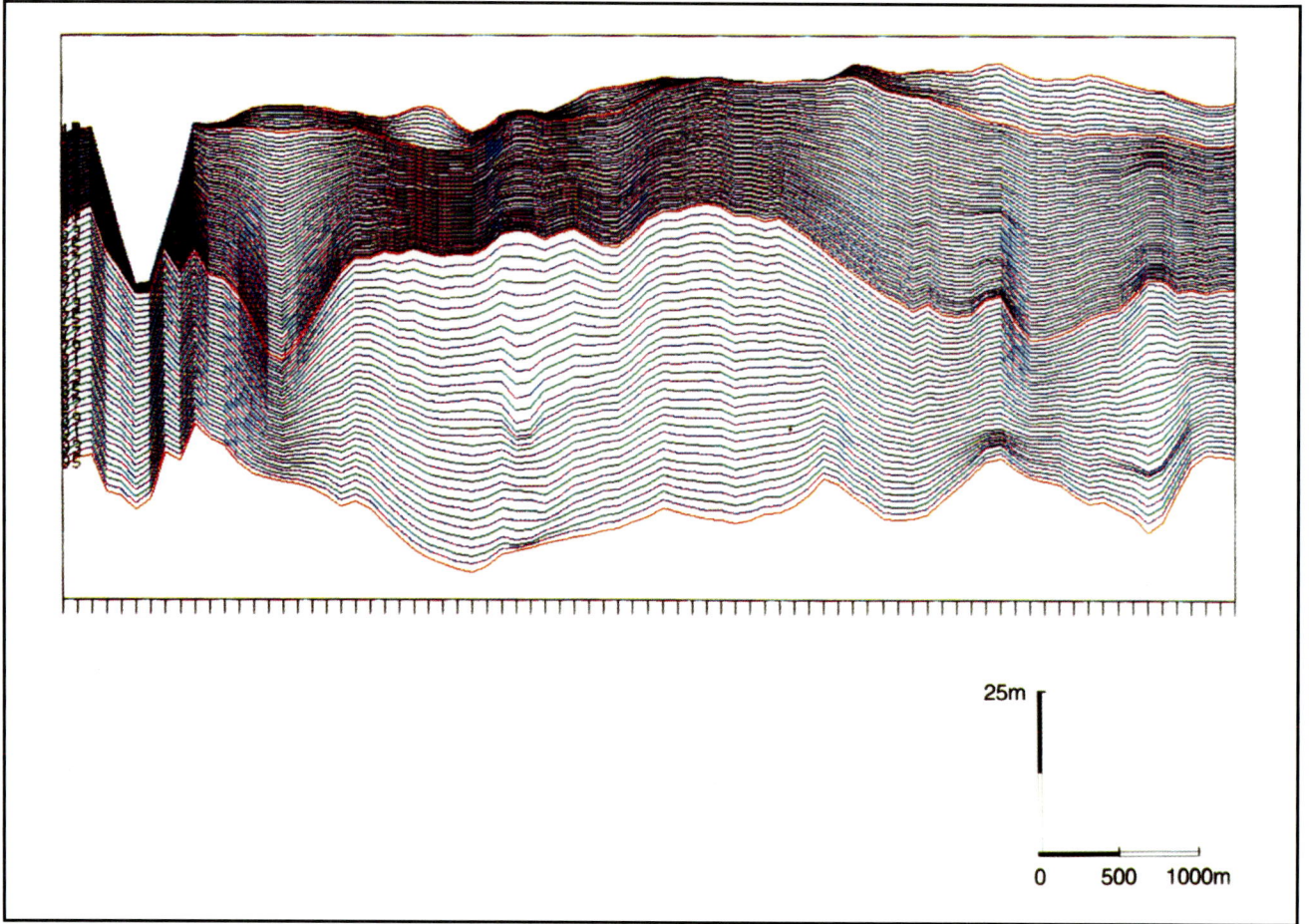

Figure 10. Cross section through the VZM model showing subzone geometries for zones 1, 2 and 3. Zones are identified by red lines. Note discontinuous and variable nature of zone 1 at the top of the section. Also, note the major Pleistocene glacial channel on the left side of section which has eroded zones 1 and 2.

zone were composited (averaged) into one value for that zone. These data were then used to interpolate value estimates for each block within that zone. The heights of these blocks were equal to the thickness of the zone.

Contour maps for bitumen content in Zone 3 are used to compare results of 3-D models built with VZM and with the fixed-height block method. Estimated bitumen content values for VZM blocks in all subzones of Zone 3 were thickness weighted averaged to produce a grid and contour map (Figure 12). Figure 13 shows a contour map of estimated values for blocks of the fixed-height in Zone 3. In areas where few material types exist vertically or laterally in the drill holes (northwest or upper left corner), VZM and block modeling techniques produce similar results. However, where many material types are present, VZM more accurately portrays the variation in bitumen content. Sharp gradients are produced by VZM at boundaries between significantly different material types. Even though Figure 13 is more pleasing to the eye, it is far from reality, as seen in the operating oil-sand mines, where bitumen variation is closer to Figure 12.

The conventional approach to three-dimensional geologic modeling of a mineral deposit is to generate a block model with blocks of fixed X, Y, and height. Each block is then assigned a percentage value for specific attributes of interest using various estimation techniques. This approach is generally satisfactory when it is applied to disseminated deposits. However, for most of the epigenetic mineral deposits, where distribution of an attribute is a function of geologic conditions such as porosity or permeability, the block modeling approach falls short of expectation. Therefore, the estimated block concentrations may vary substantially from actual.

As discussed by Denver and Phillips (1992), reduction in vertical block height may increase the accuracy of the model. However, unless block boundaries coincide with the geologic boundaries, the results may not be very representative. Because of vertical and lateral variations, averaging across material types will still occur in the fixed-height block modeling method; therefore, the estimated block values will not contain the geologic accuracy of the VZM model.

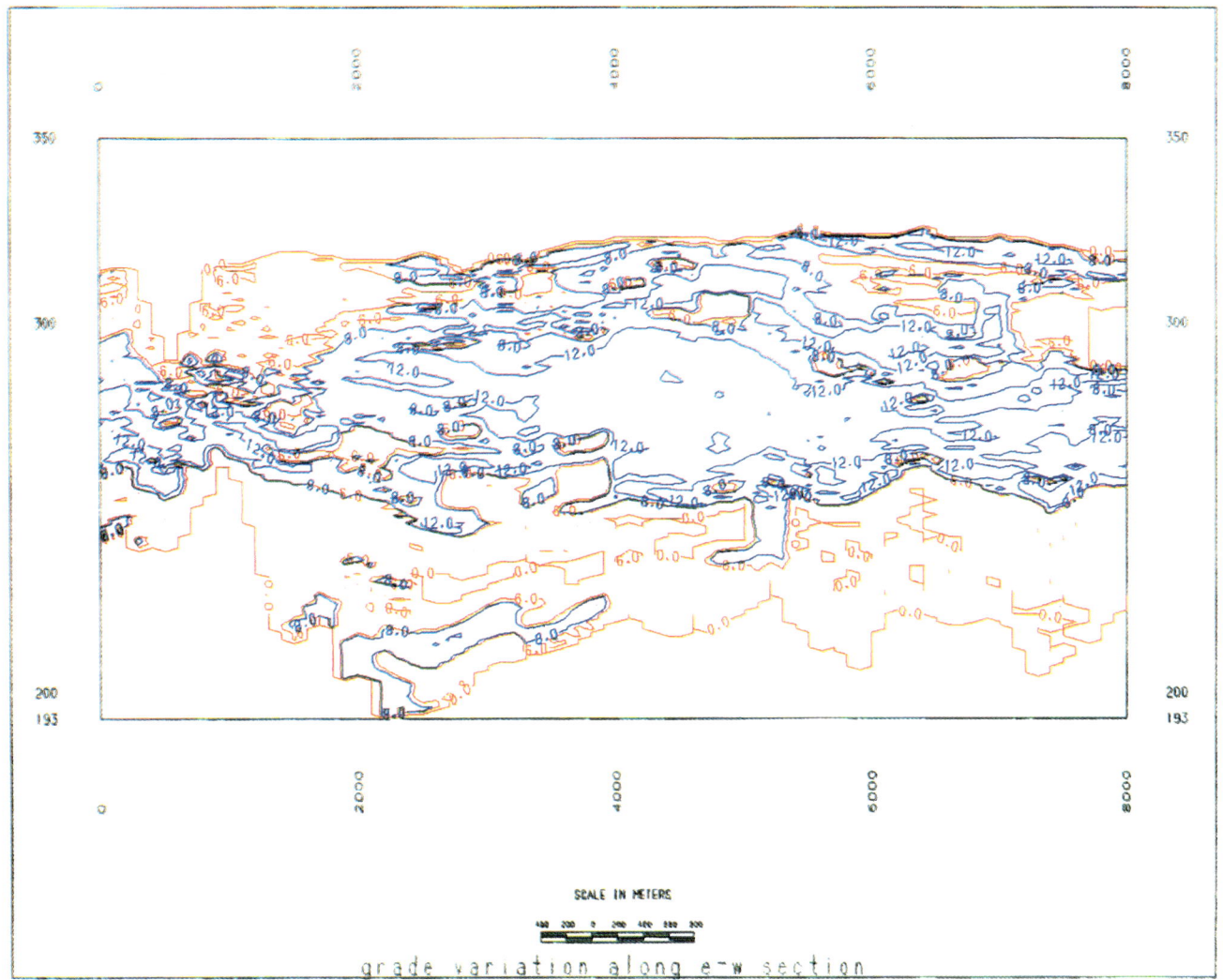

Figure 11. Cross section showing vertical and lateral bitumen grade variations within the oil-sand unit. Note the degree of variation throughout the region which is indicative of the facies changes within a short distance.

CONCLUSION

Blind application of various modeling techniques and algorithms to complex geologic deposits may have undesirable results. In modeling bitumen content of the McMurray Formation of Alberta, Canada, results from using existing techniques were deemed unacceptable. Therefore, a new Variable Zone Modeling technique was developed to accommodate the need to model the highly changing depositional environments in this formation. Results from using this technique accurately reflected the variation in material and depositional environment in the estimated bitumen values of the model. These results confirm the need to (1) thoroughly understand the geologic conditions and the interplay between the depositional environment and epigenetic deposits, (2) have a 3-D computer modeling program that allows the incorporation of these geologic interpretations, and (3) properly blend the interpretation with the data inside the program to produce accurate 3-D geologic block models.

REFERENCES CITED

Agterberg, F. P., 1974, Trend analysis: Geomathematics, mathematical background and geoscience applications: New York, Elsevier, p. 277–309.

Badiozamani, K., 1988, A computer mine planning approach for oil sand operations, in R. K. Singhal, ed., Proc. Mine Planning and Equipment Selection, Calgary, Rotterdam, Balkema, p. 49–56.

Badiozamani, K., and F. Roghani, 1988, Modified block modeling—a different approach to ore deposit modeling, in K. Fytas, J. Collins, and R. K. Singhal, eds., Proceedings 1st Canadian Conference on Computer Applications in the Mineral Industry, Quebec: Rotterdam, Balkema, p. 197–205.

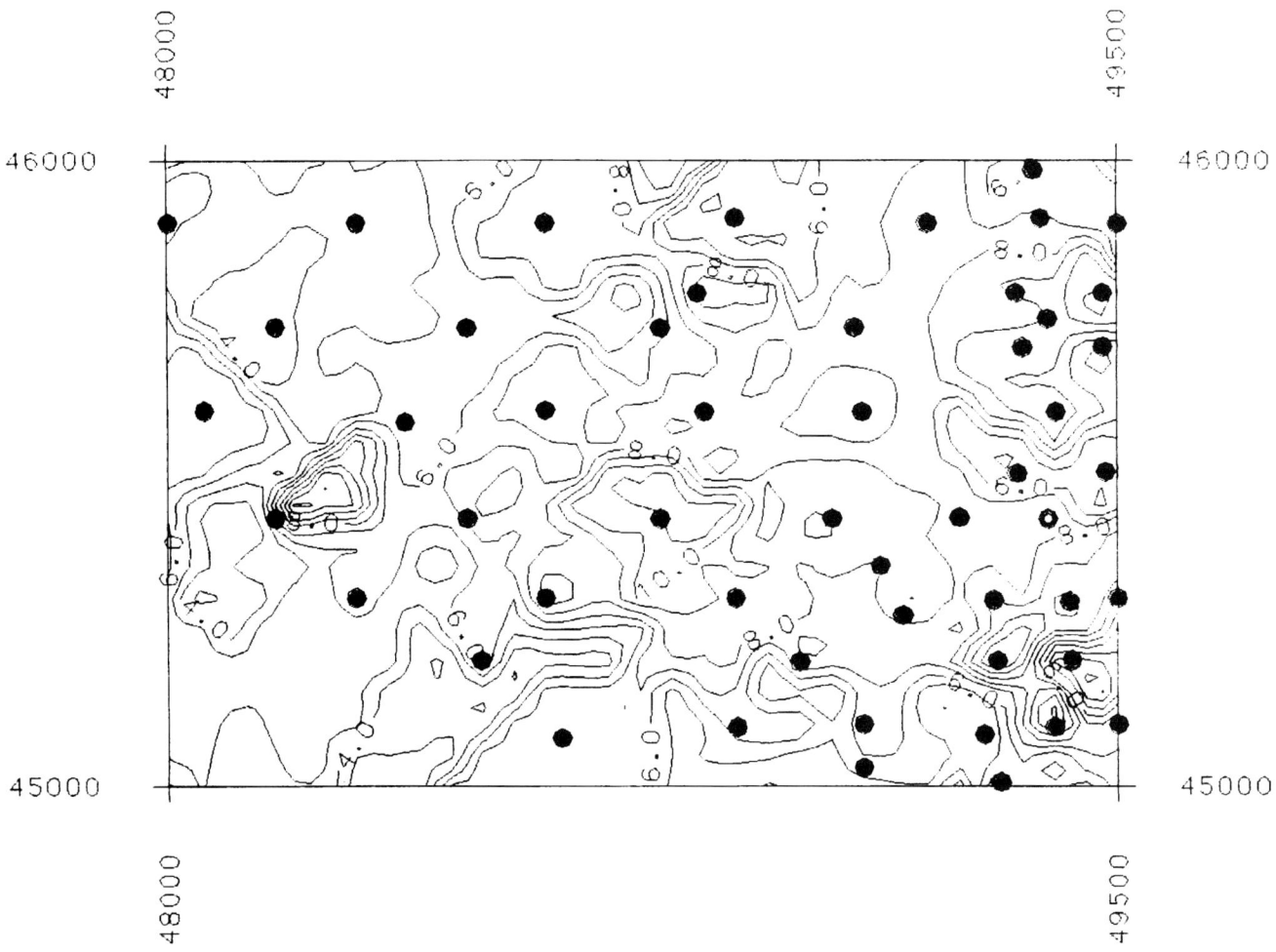

Figure 12. Contour map of bitumen content variation within Zone 3 using VZM approach. Comparison of this map with the one on Figure 13 shows the amount of variability in the bitumen content which cannot be depicted by conventional block modeling approach.

Barnes, M. P., 1980, Compositing drill hole sample values, computer-assisted mineral appraisal and feasibility: New York, Society of Mining Engineers of the AIME, 167 p.

David, M., 1977, Geostatistical ore reserve estimation: Amsterdam, Elsevier, 364 p.

Demaison, O. J. 1977, Tar sands and supergiant oil fields: AAPG Bulletin, v. 61, p. 1950–1961.

Denver, L. E., and D. C. Phillips, 1992, The impact of vertical averaging on hydrocarbon volumetric calculations—a case study, (this volume).

Flach, P. D., and G. D. Mossop, 1985, Depositional environment of lower cretaceous McMurray Formation, Athabasca oil sands, Alberta: AAPG Bulletin, v. 69, no. 3, p. 1195–1207.

Hamilton, D. E., and R. S. Didur, 1992, Three-dimensional geologic block modeling of the Kutcho Creek Massive Sulfide deposit, British Columbia, (this volume).

Hazen, S. W., 1967, Some statistical techniques for analyzing mine and mineral deposit sample and assay data: U.S. Bureau of Mines Bulletin, 621 p.

Huijbregts, C., 1973, Regionalized variables and applications to quantitative analysis of spatial data, Proceedings from NATA Advanced Study Institute, Display and Analysis of Spatial Data: London, John Wiley and Sons, p. 38–53.

Journel, A. G., and C. J. Huijbregts, 1981, Mining geostatistics: London, Academic Press, 600 p.

Matheron, G., 1963, Principles of geostatistics: Economic Geology, v. 58: p. 1246–1266.

Popoff, C. C., 1966, Computing reserves of mineral deposits: Principles and conventional methods. U.S. Bureau of Mines Information Circular 8283.

Ramani, R. V., and B. T. Stanley, 1979, Ore-body modeling, in A. Weiss, ed., Computer methods for the 80's: New York: Society of Mine Engineers Port City Press, p. 245–252.

Royle, A. G., 1980, Why geostatistics?, Geostatistics: New York, McGraw-Hill, p. 17–40.

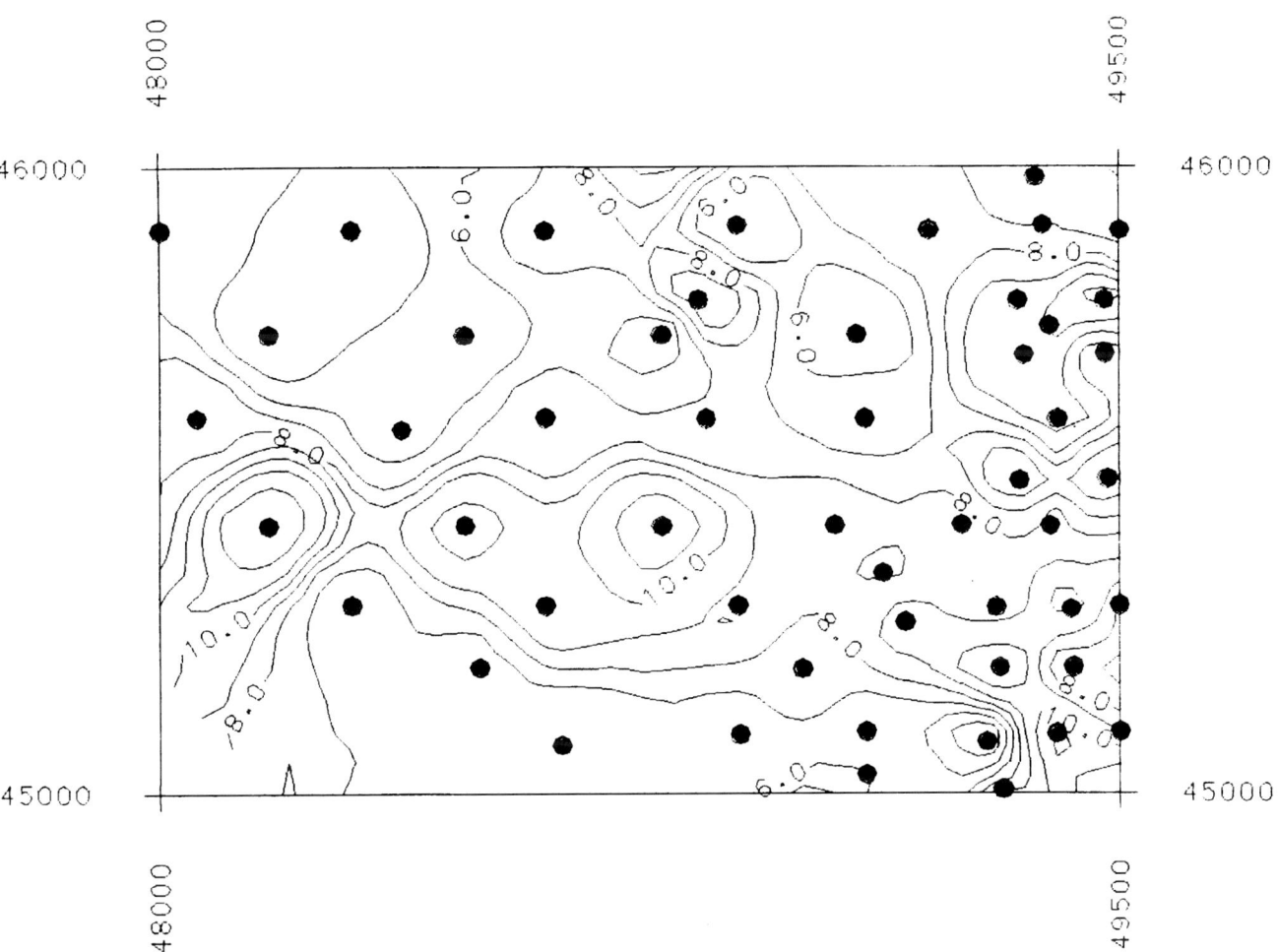

Figure 13. Contour map of bitumen content variation within Zone 3 using conventional fixed height block model approach. Note smooth variation of bitumen content as compared with Figure 12, however smooth variation does not match the actual observation of the mine.

Chapter 18

Three-Dimensional Modeling Techniques in the Analysis of a Mature Steam Drive

Ramsay A. Barrett
Mobil New Exploration Ventures Company
Dallas, Texas, U.S.A.

Jeffery Bailey
Mobil Exploration & Producing U.S. Inc.
Denver, Colorado, U.S.A.

ABSTRACT

The analysis of a mature steamflood in the South Belridge field was made possible due to an infill drilling program that provided detailed data from the reservoir. Conventional well log and temperature log data were used to build 3-D models of the sand-shale geometrics and temperature structure over the entire reservoir thickness, and oil saturation maps for 14 sands within the reservoir. Graphic visualization of these models allowed insight as to the sand-shale geometry, thermal architecture, and steam-chest development in this steamflood. Volume analysis matched historical production data. Changes in operating policy based on the results of this analysis improved ultimate recovery along with operating efficiency.

INTRODUCTION

The South Belridge field is located in the southern San Joaquin Valley, Kern County, California, approximately 40 mi (65 km) west of Bakersfield (Figure 1). The area of interest is a 40-acre (16 hectare, or ha) steamflood developed on 10-acre (4-ha) inverted 9-spots designed to displace heavy oil (12–13°API). Steam is injected into multiple layered sands that are separated by impermeable clay layers. The sand-shale geometries are controlled by a prograding fluvio-deltaic depositional system. The oil is produced by open-hole gravel pack wells, open to all five zones between 500–1100 ft (150–330 m). This commingled production presents tremendous problems in determining which zones are being efficiently exploited and which zones are being bypassed. Several wells drilled in 1987 indicated that the profile, or sweep efficiency, may be significantly worse than previously suspected. On the strength of these data, it was decided to convert four 10-acre (4-ha) patterns to 5-acre (2-ha) inverted 9-spots.

The complexity of the sand-shale geometry cannot be defined using conventional two-dimensional mapping techniques. Two-dimensional maps use average values vertically, which blurs individual sands together and often makes complex reservoir heterogeneities seem deceptively simple. In this reservoir, complex

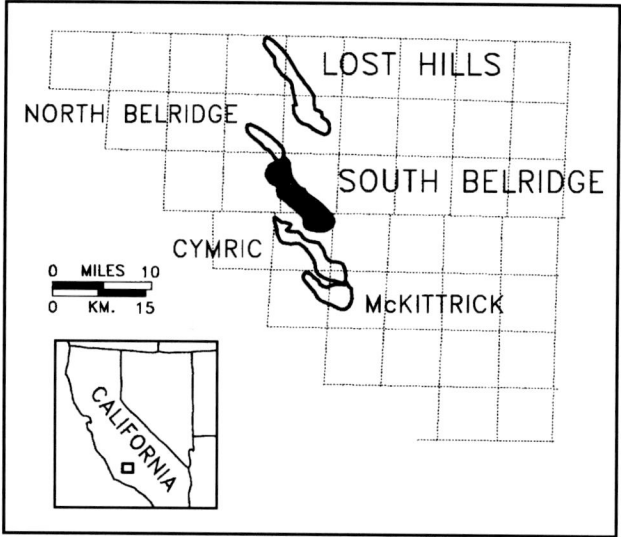

Figure 1. Location map showing South Belridge field and nearby fields.

sand-shale geometries controlled the dispersal of steam and caused the temperature and oil saturations to exhibit extreme variations vertically and laterally. A method to view and analyze these three-dimensional variations was needed to efficiently produce the large quantities of oil bypassed by the steam. Three-dimensional modeling allowed data to be honored both laterally and vertically. It also allowed details of the reservoir's thermal architecture and oil saturations to be seen, understood, and used to plan further field development.

BACKGROUND

Geology

The South Belridge field is a northwest–southeast-trending anticline (Figure 2). The study area is on the southern edge of the structure, with gentle dips of 3–4°. The productive zone is the Pleistocene Tulare Formation, which consists of unconsolidated oil-saturated sands separated by impermeable clay layers. Table 1 summarizes the reservoir properties of the Tulare.

The sand-shale architecture in the reservoir is controlled by a prograding fluvio-deltaic depositional system (Miller et al., 1990). For modeling purposes, the productive interval has been divided into five zones: the lowermost E zone through the uppermost A zone. Figure 3 is a schematic cross section through the reservoir showing these zones and their sand geometries. These zones loosely correspond to the following depositional environments:

- The lowermost E zone consists of delta-front sands that are thin bedded (2–8 ft [0.6–2.5 m]) well-sorted, fine-grained, and areally extensive. The E zone's permeability is 4-10 times less than that of the upper zones.
- The D zone typically consists of 10–15 ft (3–4.5 m) thick channel and delta-mouth bar sands that are fine to medium grained and areally extensive.
- The C zone is characterized by 5–15 ft (1.5–4.5 m) thick meandering upper delta-plain channels that are encased in clay and rarely amalgamated. These sands, due to their sinuous nature, have limited areal extent.
- The A and B zones are stacked braid-plain channels that have numerous thin shale baffles intercalated within. The stacked and amalgamated nature of these sands make them areally extensive.

Although the depositional framework is well understood, the degree of interconnectivity of each zone's sands and shales is not. We must characterize and visualize the geometries of these features in order to gain a clear understanding of how they impact steam dispersal.

Production and Operating History

The South Belridge field was discovered in 1911, but it experienced only sporadic development until the 1970s, when large-scale steamfloods were installed. Due to low-gravity oil (12–15°API) and high viscosity at original reservoir temperature (1500 cp at 100°F), steam must be injected to reduce viscosity. The study area was initially developed on a 10-acre inverted 9-spot pattern configuration (Figure 4A). In 1987, several cored wells were drilled in the interior of a mature pattern. The cores revealed areas where the steam chest was fully developed with oil saturation values approaching 0%. Conversely, some sands exhibited near original oil saturations and reservoir temperatures. This indicated that the steam profile, or sweep efficiency, was much poorer than originally assumed. It was hypothesized that increasing the well density to a 5-acre (2-ha) pattern size (Figure 4B) would increase the probability of completing the injector and all surrounding producers in the same sand.

In the fall of 1988, four 10-acre inverted 9-spot patterns were converted to 5-acre inverted 9-spots. The program consisted of 24 producers, 5 injectors, and 4 temperature observation wells. Figure 5 shows the original wells and the new wells. Data from these wells afforded a snap shot of a mature steamflood in this field. Dual Induction and Density Neutron logs were run in the open hole, and mud temperature logs were run several days after completion to allow the borehole cooling effect of drilling to abate and reflect true reservoir temperature.

PROGRAM DESCRIPTION

The Interactive Volume Modeling program (IVM), developed by Dynamic Graphics Incorporated (DGI), was used to build the 3-D geologic models (Paradis

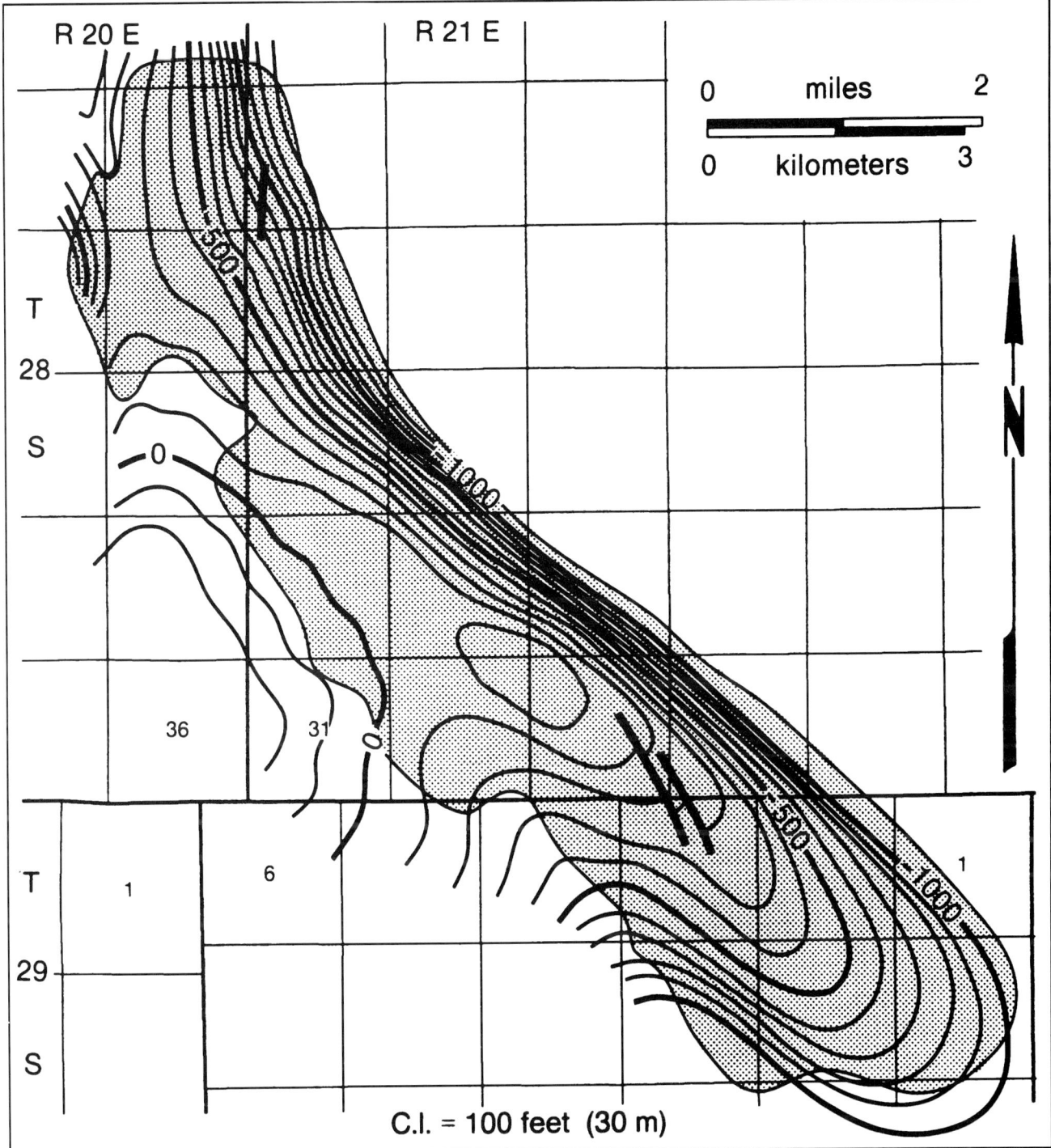

Figure 2. Structure map showing the South Belridge field as a NW–SE-trending, gently folded anticline. (From Miller et al., 1990, used with permission of Springer-Verlag.)

and Belcher, 1990). IVM takes as input, data consisting of one or more property values and their X-Y-Z coordinates (Table 2). Usually these data are a series of measurements along a borehole, section, or rock face.

Values are interpolated from these data and assigned to nodes in a 3-D matrix covering the same X-Y-Z space. These interpolations can be constrained by upper and lower bounding surfaces (grids), typically built with DGI's Interactive Surface Modeling program (ISM). Parameter controls for interpolation restrict the X-Y distance to search for data, number of wells used per quadrant, and distance to extrapolate,

Table 1. Properties of the Tulare Formation in the Area of the Reservoir

Property	Average	Range
Depth		400–1400 ft
Porosity	35%	30–42%
Permeability	3000 md	100–5000 md
Water Saturation	24%	17–60%
Oil Gravity		12–15°API

among others. The interpolation process is three-dimensional gridding and results in a model that represents a continuously changing property between points in the 3-D matrix. This differs from 3-D block models where the block's value is assumed to be homogeneous throughout the block (Jones, 1992), although both are stored similarly.

Once constructed, the model is passed to a 3-D contouring algorithm just as a grid is passed to a 2-D contouring algorithm in a mapping program. However, rather than generating contour lines of equal value (c.f. Figure 2), the 3-D contouring algorithm produces contour surfaces of equal value, referred to as isosurfaces (c.f. Figure 12). The user specifies which isosurface values are desired. The processing time is dependent on the size of the 3-D matrix and the number of isosurfaces being generated. The figures shown in this paper each took 10–15 minutes to generate on a Silicon Graphics POWER SERIES machine, model 4D/80GT (400K vectors, 55K polygons).

Various display techniques are used to visualize the distribution of the isosurfaces in 3-D. Controls on these displays include color, size, viewing direction, and which isosurfaces are visible, among others. The display program also lets you step through the volume with cutting planes orthogonal to the axes. A *chair* option permits removal of a quarter cube from any corner of the model and viewing of the remainder (c.f. Figure 7).

3-D Model Construction

Three types of models were generated: sand-shale, oil saturation, and temperature. The model construction process involved three steps: (1) build grids for the top and base of areally extensive sands to use as vertical constraints when modeling oil saturation within those sands, (2) build the 3-D models, and (3) generate and view isosurfaces from the models.

2-D Grid Construction

Structure grids on the top and base of 14 areally extensive sand units were generated. Each structure-top grid was built using the minimum-tension algorithm in ISM. This algorithm is similar to least squares and results in a surface with a smooth, pleasing appearance. Values were interpolated from this grid at original data locations and compared with the input

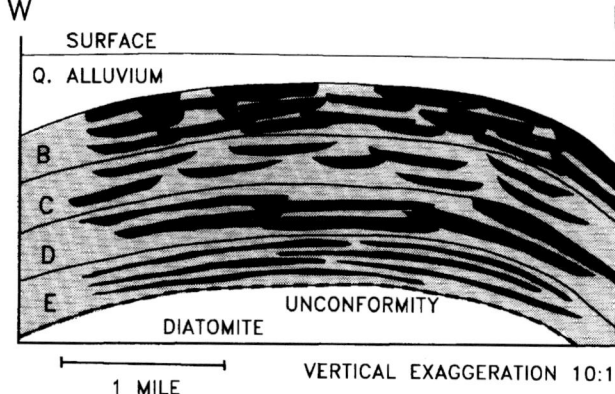

Figure 3. Schematic cross section through the study area depicting sand-shale geometries: Zone B (and A), stacked braid-plain channel sands; zone C, meandering upper delta plain channel sands; zone D, areally extensive distributary channel and delta mouth bar sands; and zone E, areally extensive delta front sands.

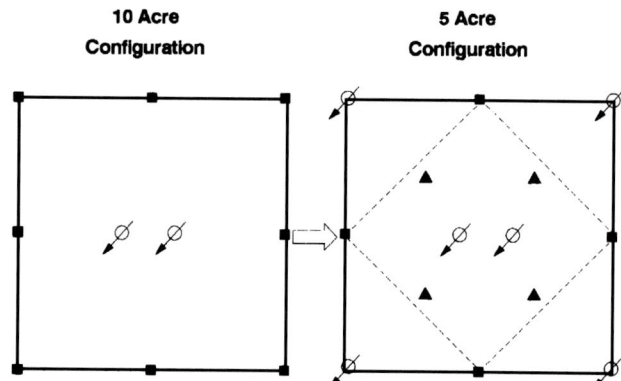

Figure 4. Initial and current steamflood injection and production well patterns. (A) 10-acre inverted 9-spot with 8 producers surrounding 2 injectors. (B) Original 10-acre (4-ha) pattern converted to a 5-acre (2-ha) inverted 9-spot pattern. Note that corner producers are replaced by injectors.

data values to check the grid's accuracy.

Grids of each sand's thickness were built and subtracted from that sand's structure-top grid, creating the structure-base grid. ISM's isopach algorithm was used to build these thickness grids. This algorithm forced the thickness grid to extrapolate negative at zero-thickness locations, causing a smooth zero line to be created that ran between zero (non-thickness) values and real thickness values. Each thickness grid was clipped to a minimum value of zero before being subtracted from the sand's structure-top grid. Each resulting structure-base grid was compared with the original data values to ensure its accuracy.

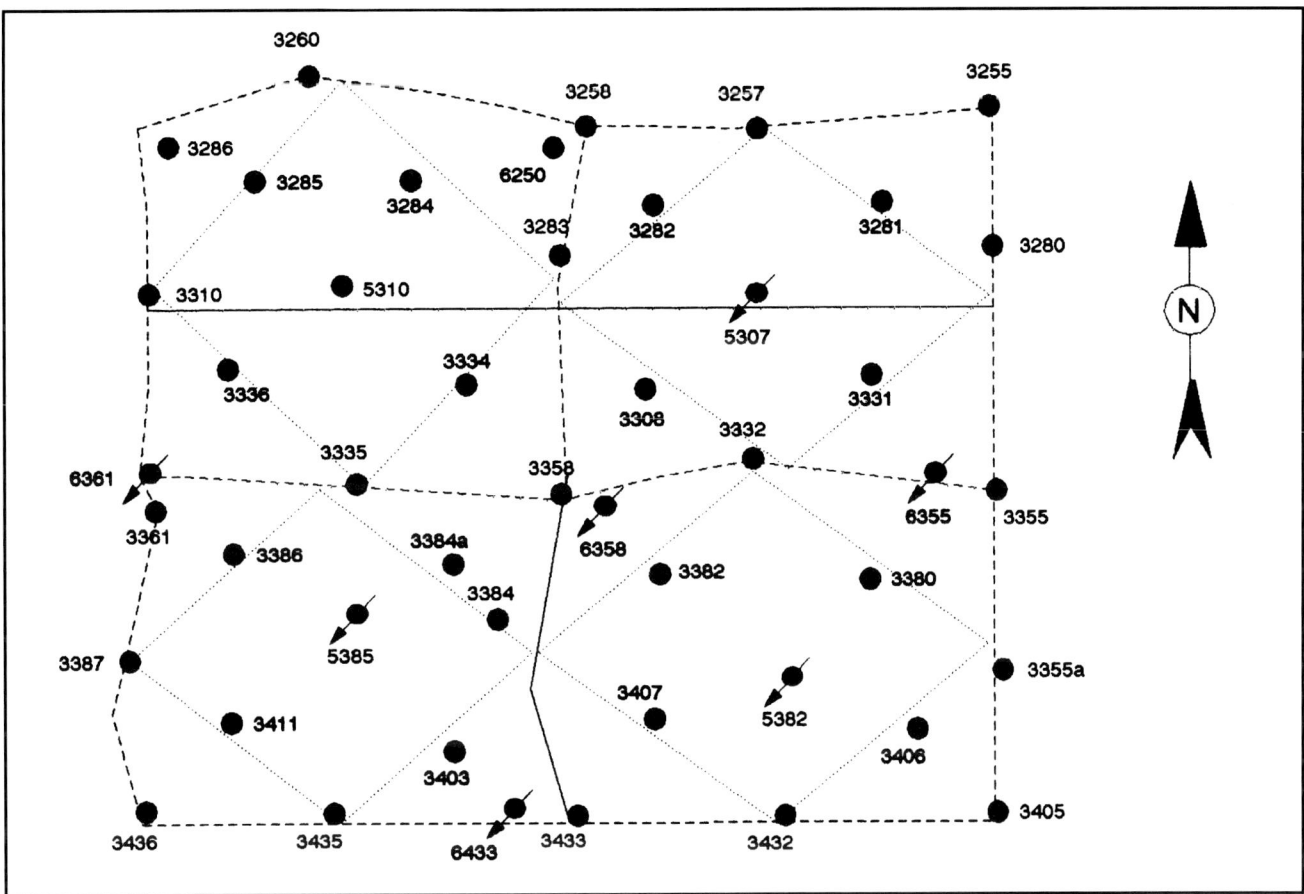

Figure 5. Map of the study area showing well distribution in the four 10-acre (4-ha) patterns that were converted to 5-acre (2-ha) patterns.

Modeling Parameters

The X-, Y-, and Z-increments used for all models in this study were the same. Wells were spaced 200 ft (60 m) apart in the X- and Y-direction, with samples every 10 ft (3 m) in the Z-direction. We used X- and Y-increments of 100 ft (30 m) and a Z-increment of 10 ft (3 m) (the Z-resolution of the data). These increments were small enough to portray shale pinchouts between wells in the sand-shale model, the model most sensitive to variations in these values.

A different search radius was used for each dataset. The process of selecting a search radius for the sand-shale model was typical of the decision-making process. In this case, we wanted the shales to be as continuous as possible. To achieve this, we needed 2 wells per quadrant within the search circle. The wells were 200 ft (60 m) apart, so a search radius of 600 ft (182 m) allowed this to happen. In the Z-direction, a search radius of 10 ft (3 m) was used. Larger Z search radii caused pinching of continuous shales away from wells because data values from sand units above and below the thin shale body were used during interpolation.

Lateral extrapolation was allowed only 100 ft (30 m) outside the data area. This was achieved by constraining the X- and Y-dimensions of the model.

Table 2. Hypothetical Data from One Well Representing the Form of the Data Input to the IVM Program

X	Y	Z-Depth (ft)	Temperature (°F)	Porosity (%)
100	100	50	85	20
100	100	60	90	18
100	100	70	95	24
—	—	—	—	—
—	—	—	—	—

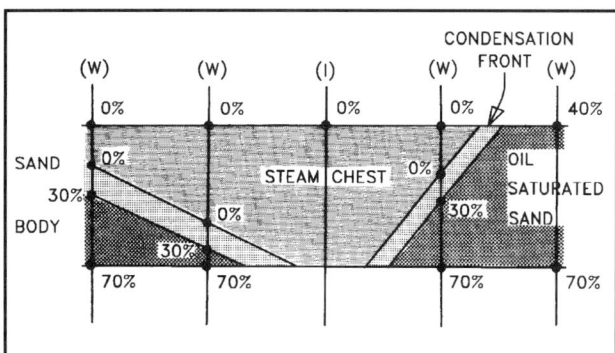

Figure 6. Schematic cross section showing the oil-saturation to steam-chest relationship. The values shown are inferred from well-log data and are supported by whole-core data taken nearby. These assumptions were used in building the oil-saturation maps.

Modeling Sand-Shale (Discrete Data)

The sand-shale data were discrete, that is, at a data location the value was either sand or shale with no gradation between them. The sand and shale layers were determined from Dual Induction logs for the 41 wells. Due to the effect of temperature on resistivity, an ohm cutoff could not be assumed. An arbitrary value of 10 was used for shale, 20 for sand, and 15 for their contact. A subset of the data for one well is shown in Table 3. X- and Y-values are Lambert coordinates describing the surface location of the well, and depth is in feet.

The algorithm used in IVM to build 3-D models is a continuous interpolator. Therefore, for 3-D model nodes that lay between a well with shale (10) and a well with sand (20), it assigned values that change smoothly from 10 to 20. The isosurface value 15 represents the lithologic boundary between the two material types in this 3-D model. Because of this interpolation method, a code of 15, bounded closely above and below by 20 and 10 (Table 3) were input as data in each well at the sand–shale boundaries. This ensured that the boundary was correctly positioned in areas of the model that were near wells.

Modeling Oil Saturation And Temperature (Continuous Data)

Oil Saturation Data

Figure 6 is a diagrammatic cross section through a typical sand unit showing the distribution of oil saturation relative to the steam chest. Oil saturation values for each sand unit were derived from several sources. The steam chest (live steam) is defined by the density-neutron cross-over and is assumed to indicate 0% oil saturation. This is supported by analysis of large volumes of core data. From core data we learned that the transition from the steam chest to condensate (condensation front) is about 3 ft (1 m) and that the oil saturation at that 3-ft position is about 30%. Zones where temperatures were high relative to the initial reservoir temperature were assumed to have 30% oil saturation. Zones where temperatures were at or near original reservoir temperature were assumed to indicate original oil saturation, 70%.

Maps of the 14 areally continuous sands were made showing temperature distribution and live steam position. Using these maps, well logs, and the assumptions described above, oil saturation data were recorded for each of the 14 sands in each well (c.f. Figure 6). Four typical conditions were seen within a sand unit:

(1) Where crossover was seen from top to bottom, oil saturations of 0% were assigned to the entire sand. (This crossover condition occurred only at injectors.)
(2) Where partial base of crossover was seen within the sand (typically wells within about 100 ft [30 m] of the injector) the crossover extended from the top to the middle of the sand; oil saturations of 0% were assigned from the top down to the base of the crossover; 0% to 30% at a position from 3 ft (1 m) below the crossover; and 70% at the bottom (if the temperature was low, 30% if it was high; c.f. Figure 6).

Table 3. Data from One Well for the Material Types: Sand (20), Shale (10), and Sand-Shale Boundaries (15)

Well Name	X	Y	Z-Depth	Material Type (Sd-Sh)
3355A	—	—	500	20
3355A	—	—	510	20 Sand
3355A	—	—	513	20
3355A	—	—	514	15 ———
3355A	—	—	515	10
3355A	—	—	520	10 Shale
3355A	—	—	530	10
3355A	—	—	538	10
3355A	—	—	539	15 ———
3355A	—	—	540	20
3355A	—	—	540	20 Sand
3355A	—	—	540	20

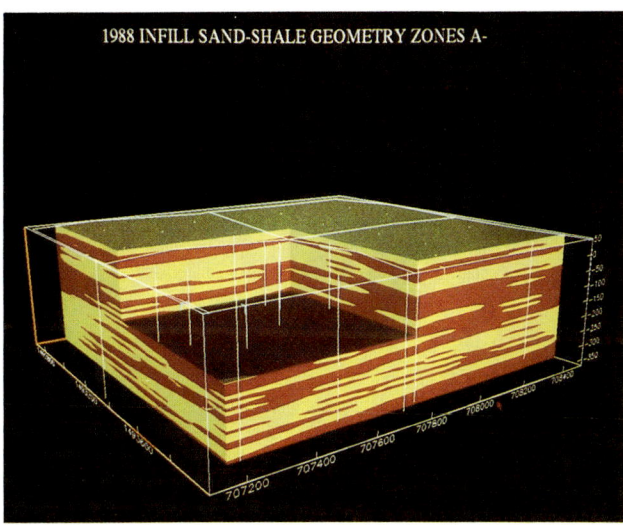

Figure 7. Display of the sand-shale model using the *chair* option. The vertical lines depict wells used for control. The continuous sands (yellow) in the upper and lower thirds contrast sharply with the less continuous and more shale-rich (red) middle third.

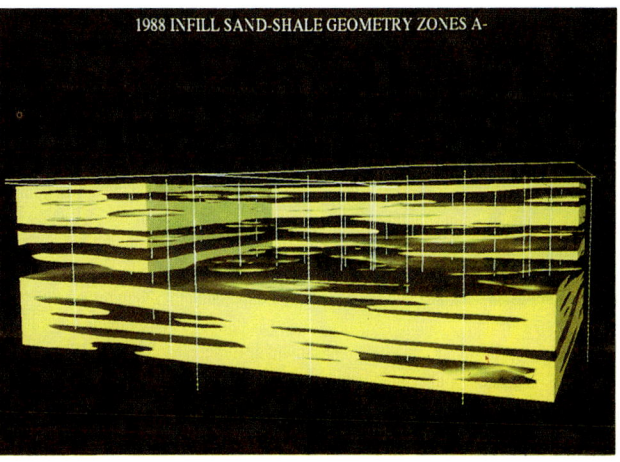

Figure 8. Display of the sand-shale model. The sands are shown with the shale lenses turned off. Note that sands in the middle are relatively discontinuous and surrounded by shales of complex geometries that easily compartmentalize the steam chest.

(3) If no crossover was seen in the sand but high temperatures were seen at the top and low temperatures at the base (wells from 100–300 ft [30–90 m] from the injector), oil saturations of 30% were assigned to the top and middle sand and 70% to the base.

(4) If no crossover was seen and original temperatures existed throughout (sand missed or beyond effective limit of steamflood), oil saturations of 70% were assigned to the entire sand.

The above was empirically derived and a unique set of assumptions must be generated for any individual steam flood.

Temperature Data

Temperature and depth were measured directly in the borehole. Data quality problems occurred where convection cells were established within the borehole fluids across hot to cold zones. Temperature values in this area are poorly resolved. Use of contact temperature tools in the open hole environment improved resolution, but the absolute reservoir temperature is not known.

Model Construction

Both oil saturation and temperature were continuous and smoothly varying parameters, although steep gradients did occur. IVM's 3-D gridding algorithm was designed to handle this type of data and was applied with no modifications to the data or special processing of the resulting models. Fourteen 3-D models were built for oil saturation, one for each areally continuous sand unit. One model of temperature was built for the entire reservoir.

MODEL EVALUATION

Sand-Shale

Figures 7 and 8 show the sand-shale model, with yellow representing sand (values greater than 15) and red representing shale (values of 15 or less). The rock bodies in the sand-shale model match the geologic interpretation in gross form and spatial distribution. However, it is obvious that the shapes of the rock bodies are mathematically defined and do not incorporate familiar geologic shapes (channels and bars) that geologists associate with these depositional features. Models built strictly on a mathematical basis (i.e., no interpretation) usually have problems with rock-body form and continuity or heterogeneity between data. Subtleties of form were not critical for this study and the high well density (one well per acre) allowed the heterogeneity of the reservoir to be acceptably modeled.

The value of this model was not found in the credible reproduction of a complex fluvio-deltaic system, but rather in the power it gave for visualizing how complex the sand-shale geometries are and the obvious impact those geometries have on propagation of the steam chest through the reservoir. Our geologic understanding of this reservoir was considerably enhanced by viewing this model from many angles and by slicing through it. These images also proved helpful in communicating the sand-shale spatial relationships to engineers and others not used to thinking about geological bodies in 3-D.

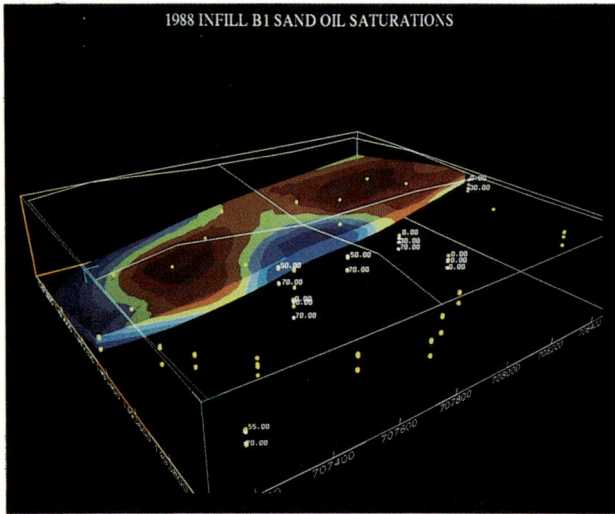

Figure 9. Display of the oil saturation model for the B-1 sand unit. Red indicates zero oil saturation, grading to blue at 70% oil saturation. The image has been cut in half to show the effect of steam override.

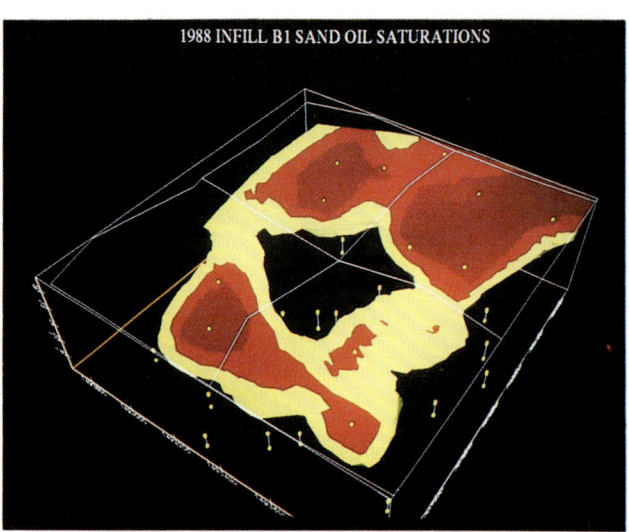

Figure 10. Display of the B-1 sand unit oil saturation model. Only sands with 0 to 20% oil saturation are displayed.

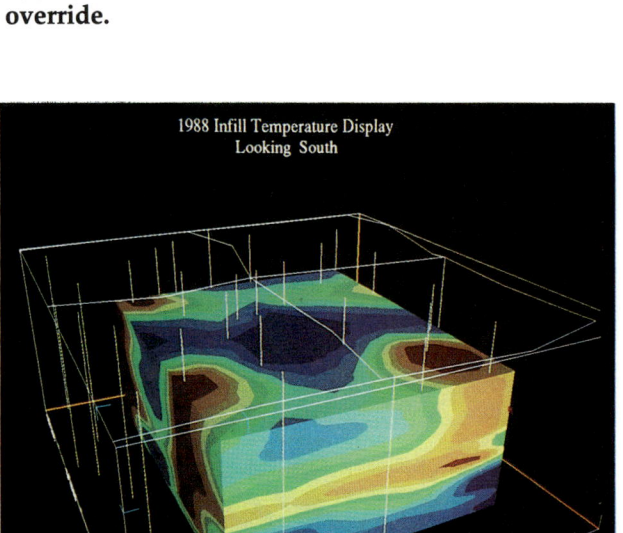

Figure 11. Display of the temperature model. The thermal architecture of the reservoir is depicted by live steam at about 300° F (148.9° C) (red) grading to 100° F (37.8° C) (blue). The model is sliced along each axis, close to the injection wells, giving an internal view of the reservoir. Note the excellent thermal communication in the lower third, compared with the poor thermal profile in the middle third. This correlates well with continuity of the sands depicted in the sand-shale model (Figure 7).

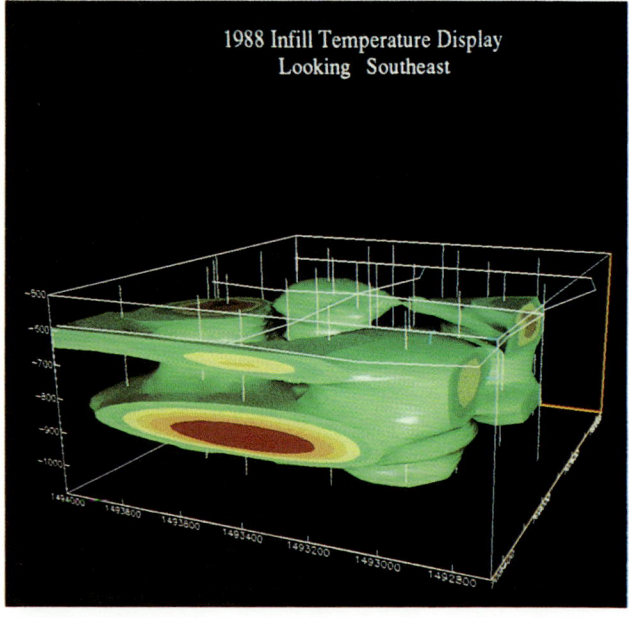

Figure 12. Display of the temperature model. Only temperatures greater than 180° F (82.2° C) are displayed.

Oil Saturation

Figure 9 is a cut-away view through the oil saturation model of the B-1 sand unit. It clearly shows the effects of steam override and implies a credible thermal propagation. The information contained in this model gives a better understanding of recovery from a pattern than is possible using production history, which is often difficult to interpret due to the allocation factors used in assigning production to a borehole.

Figure 10 is a view of the same model from above, with only sands having less than 20% oil saturation displayed. The nonradial development of the steam chest seen here was also seen in most of the other sands. This development differed dramatically from the common assumption that sweep was radial and clearly showed the need for infill drilling.

Volumes calculated from these oil saturation models varied from production histories by 3% to 35%. Comparison of these numbers with the volume of injected steam, produced data on steamflood efficiency. This efficiency information, together with estimated volumes of remaining oil, permitted the optimization of steam injection rates and led ultimately to a better oil-steam ratio (OSR), the barometer of steamflood efficiency.

Temperature

Figure 11 displays the temperature model sliced along each axis close to the injection wells. Figure 12 shows the entire model but only the high temperature isosurfaces are visible. Because individual sands were not considered during the construction of this model, the level of detail seen in the sand-shale and the oil saturation models is not seen. However, because of the volume of temperature data available along each borehole, the model provides a useful definition of the gross zones of good and poor heat transfer. The poor thermal profile in zone C ($^1/_3$ down from the top) is attributed to the isolated sand channels and low sand-shale ratio. Increasing well density should improve the chances of injection-producer communication in this zone.

Images from this model, in conjunction with images from the sand-shale model, showed that zone E sands (bottom) were cold and had taken only a small amount of steam. Most of the steam went into zone D, as indicated by its excellent thermal profile. Further analysis of zone E indicated that it should have a dedicated injector, rather than sharing one with zone D. Excellent production response in zone E has occurred since this operating policy changed.

SUMMARY

Three-dimensional modeling proved to be an excellent tool in the analysis of the complex relationships found in a mature steamflood. Specifically, the oil saturation models clearly demonstrated that the areal distribution of the steam chest, within a discrete sand body, was asymmetric (non-radial). The 3-D visual representation of the model clearly showed that, although certain portions of a discrete sand body were swept well by steam, other portions of the same sand in the same pattern had been completely bypassed. This bypass left significant quantities of hydrocarbons unproduced, and justified continuing the infill drilling program, which ultimately proved exceedingly profitable.

The temperature model showed gross inequities in the vertical distribution of heat over the entire reservoir interval. Specifically, the heat was concentrated in zones B and D at the expense of zones C and E. This indicated that steam injection efficiency was not equal at each perforation within an injector and led to the drilling of 3, rather than 2, injectors per pattern with an associated increase in pattern oil-steam ratio.

The sand-shale model, although not a duplicate of the geologist's interpretation, provided insight as to how the geometric relationship between sand and shale bodies impacted division and isolation of the steam chest both vertically and horizontally. This partly explained the asymmetric sweep observed in the oil-saturation maps. In addition, this model supported the concept that decreasing the distance between injector and producer (i.e., infill drilling) would reduce the influence of shales on areal and vertical steam migration, promoting more efficient radial steam chest development.

Finally, in terms of communication the 3-D images proved invaluable in relating the complex and dynamic interplay of heat, steam, and changing fluid saturations within a layered sand-shale reservoir.

REFERENCES CITED

Jones, T. A., 1992, Extensions to three dimensions: Introduction to the selection on 3-D geologic block modeling, (this volume).

Miller, D. M., T. E. Covington, and J. G. McPherson, 1990, Fluvio-deltaic reservoir, South Belridge field, San Joaquin Valley, California: in J. W. Barwis, J. G. McPherson, and J. D. Studlik, eds., Sandstone Petroleum Reservoirs: New York, Springer-Verlag, 332 p.

Paradis, A., and B. Belcher, 1990, Interactive volume modeling: Geobyte, v. 5, no. 1, p. 42–44.

Chapter 19

Three-Dimensional Modeling Of Complex Geological Structures: New Development Tools For Creating 3-D Volumes

Raphael Mayoraz
Laboratory of Geology
Swiss Federal Institute of Technology (EPFL)
Lausanne, Switzerland

Carol E. Mann
Dynamic Graphics, Inc.
Alameda, California, U.S.A.

Aurele Parriaux
Laboratory of Geology
Swiss Federal Institute of Technology (EPFL)
Lausanne, Switzerland

ABSTRACT

The interpretation of complex geological data has always been an obstacle in geological model building. These data, frequently well constrained vertically by boreholes and/or at the surface by geological outcrops, tend to be sparse laterally, thereby increasing the difficulty in interpretation. While geological modeling has been performed successfully in less complex areas, until recently, the lack of calculation and graphical computer power has prevented complex problems from being addressed. With the advent of new hardware technology, software developers, in conjunction with researchers, have been developing powerful new application software specific to the requirements of the geologist and engineer that can address these geologic modeling obstacles.

The Laboratory of Geology at the Swiss Federal Institute of Technology in Lausanne (EPFL), in collaboration with Dynamic Graphics, Inc., is studying, developing, and testing new software tools for more detailed and sophisticated modeling and visualization of any type of geological structure. The development of such tools is important not only for the geologist and geophysicist attempting to understand geologic history and structural relationships, but also for the civil engineer who must construct man-made structures, such as tunnels, within the confines of the geologic world. Scientists and engineers alike require these tools in order to achieve a better description of the geological structures at hand, to better understand and analyze their geometry, and to optimize the implementation of their works, including boreholes and tunnels. For example, several large-scale tunnel projects for future railway systems have been proposed through the Swiss Alps. The investment of time and money in modeling prior to construction of such projects is minimal compared with the savings obtained by optimizing the path and the boring system.

INTRODUCTION

The project described here, undertaken by the Laboratory of Geology at the Swiss Federal Institute of Technology (EPFL), in cooperation with Dynamic Graphics, Inc., was designed to test current software and hardware capabilities and to develop new tools for geological model building. This test case, a tunnel through geological structures similar to the Swiss Alps, has three primary objectives:

- to model and visualize the complex structures below the surface topography, as well as reconstructed structures above;
- to demonstrate how this type of modeling can be useful in terms of civil engineering works by evaluating the type of geology that will be encountered prior to construction (e.g., unstable areas, faults); and
- to determine and overcome the difficulties and pitfalls in performing such research.

Previous studies using existing software have been successful in visualizing simple geological structures and in addressing the second of these three objectives (Mayoraz and Parriaux, 1991). Due to the complexity of this test case, however, these existing tools alone have not been sufficient. To this end, our study has been instrumental in developing methods and corresponding software programs that allow the end user to model any geological or man-made surface or volume. These 3-D structure models can then be intersected with other such structures to create a complex geological model.

BACKGROUND

The needs and goals of geologists in terms of 3-D modeling are highly variable, but can be separated into two parts: (1) visualization of geologic structures, and (2) analysis of property models (attributes) within those structures.

First, geologists need full 3-D geometric descriptions of geologic structures to better understand the local geology, to evaluate mechanisms of folding and faulting, to determine the intersection locations of the geological structures with civil engineering works, and so on. This need is particularly apparent with structural geologists and civil engineers involved in building tunnels, bridges, or buildings. This need is the basis for our study, justified in part by the future development of hydropower utilities and railway systems, including three major tunnels (each exceeding 30 km) that have been proposed through the Swiss Alps. These projects can greatly benefit from a complete 3-D description of the complex geologic structures that these tunnels will intersect. By visualizing the geologic and man-made structures, the boring paths can be optimized and practical problems involving general stability, rock and landslides, water wells, hydrostatic pressure, intersection angles with unstable rock zones, etc., can be analyzed and predicted. The cost in terms of time and money is much lower if these obstacles are discovered, evaluated, and taken into consideration prior to construction, rather than suddenly being encountered in the field.

Second, many geologists, especially in the petroleum and environmental industries, are not only interested in the 3-D geometric descriptions of geologic structures, but also in introducing property data (porosity, velocity, temperature, pressure, density of pollutant, etc.) within those structures. The goal is to see how those properties vary within some region.

Available Conventional Tools

Two-dimensional and 3-dimensional gridding techniques are the main tools used in geological modeling. Using 2-D gridding techniques with 3-D (X, Y, Z) data alone produces 2-D single-valued surface information (i.e., a grid) for defining geological layers. This grid is a two-dimensional matrix or array of Z-values that usually represent depth, elevation, or time. These 2-D surfaces can be used as geological boundary limits and simple normal faults, but not as complex 3-D geological surfaces, such as recumbent or reverse-faulted surfaces (i.e., a single (X,Y) location cannot have multiple Z-values, thereby limiting the types of structures that can be modeled).

3-D gridding interpolation techniques are typically used with four-dimensional data (X, Y, Z, and some property, P) to produce a 3-D property model of, for example, porosity, velocity, or temperature. A 3-D grid

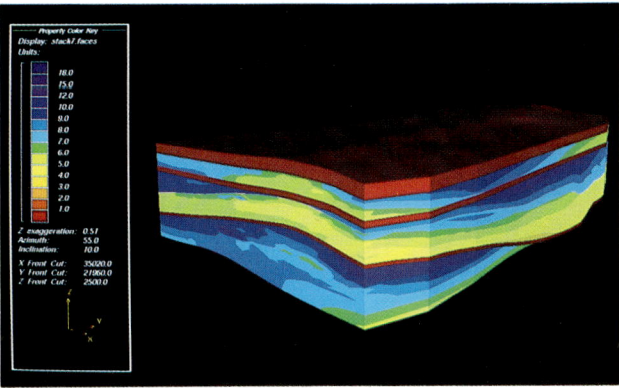

Figure 1: A 3-D display file showing a slice through a seven-layer property model that includes relatively simple 2-D surfaces. Structurally, layer 4 pinches out on either side of the anticline of layer 5, and layer 5 is truncated by layer 3 (layer 1 being the topmost zone; layer 7, the bottom-most zone); however, none of the surfaces represents complex 3-D structures. The appropriate property models, in this case porosity, are represented between the structural boundaries of all seven layers, with layers 1, 3, and 6 representing shale units.

is a three-dimensional matrix or array of these property values (P-values). Each 3-D grid represents only the data within a single user-specified zone. This zone could be defined simply on the basis of depth or elevation, or by one or several geologic boundaries, such as between a sandstone and a shale. Generally, these geological boundaries are the available single-valued 2-D surfaces.

To visualize the 3-D grid, a 3-D display file or model is produced as essentially a 3-D contour map, where the contours are 3-D isosurfaces (i.e., surface boundaries that join points having the same P-values). For example, an isosurface may represent the location where temperature equals 0°F. Several 3-D display files, each bounded by a structure (thereby creating a zone), can be merged together. Hence, this merged file is more than just a tool for visualizing a single 3-D grid: it allows us to look at property variations within several zones (usually lithological layers). These merged 3-D display files contain both structural and property information.

Application of Conventional Techniques

These 2-D and 3-D gridding techniques have been used in geologic modeling to produce a 3-D distribution of properties within simple structural boundaries. In environmental research and groundwater management, these properties frequently represent the distribution of atmospheric and underground pollutants. The 2-D boundaries usually represent topography or bathymetry. These studies primarily address continually varying properties, and, therefore, have been successful using these techniques. Such models are easily produced using currently available versions of Dynamic Graphics' Geological Modeling Program (GMP) (Mahoney, 1991; Manley and Tallet, 1990).

Most oil and gas companies also use 2-D and 3-D gridding techniques to visualize the vast amounts of data coming from seismic surveys and boreholes. Borehole data, accurate and dense vertically but sparse laterally, can be gridded in 3-D to produce property models. Seismic surveys and well picks are generally used to define single-valued geological boundaries. These data, when used with the techniques described above, are extremely useful in creating 2-D surface and 3-D property models, but not complex 3-D structures (Figure 1). In spite of this limitation, reservoirs can be modeled and analyzed using this information.

Reaching the Limits of the Conventional Techniques

As explained above, the limits of the 2-D and 3-D conventional techniques are reached as soon as geological boundaries are more complex: recumbent and reverse-faulted surfaces.

We can attempt to use 3-D gridding techniques to model geological boundaries by assigning each lithologic boundary a numerical value based on its depth. This value is used as the property to be gridded. The assumption is that the interpolated values represent the lithologic boundaries. For example, if lithologic boundaries are sampled at 1 m, 2.5 m, and 3 m along one well path, then these same values are given to those same boundaries throughout the data set regardless of depth. In the resulting grid, therefore, 1 will always represent the first boundary, 2.5 the second, and 3 the third.

It turns out that this method works relatively well and has been used successfully where the layering is flat-lying and parallel. For example, EPFL has modeled groundwater reservoirs in a region of flat intersecting local moraine and lacustrine deposits. Final locations for water-producing boreholes were determined by analyzing the 3-D model of the aquifer. In another EPFL study, geological structures and a proposed hydropower pipeline were modeled to analyze the relationship between the proposed path and a particularly unstable formation. Once the model was created, it became apparent that the path should be changed: the proposed tunnel ran near and parallel to the unstable zone (which is only 20 to 40 m thick) for approximately 0.5 km. The suggestion was made, based on the model, to alter the path to cross perpendicular to the layer.

These simple cases have shown the utility of this approach; however, they have also defined limitations of the methods: problems involved with intersecting surfaces, pinchouts, etc., and with artifacts caused by the 3-D gridding. In fact, this type of interpolation does not give usable results when treating several geologic layers together that are non-parallel or discontinuous. The mathematics of the property gridding tech-

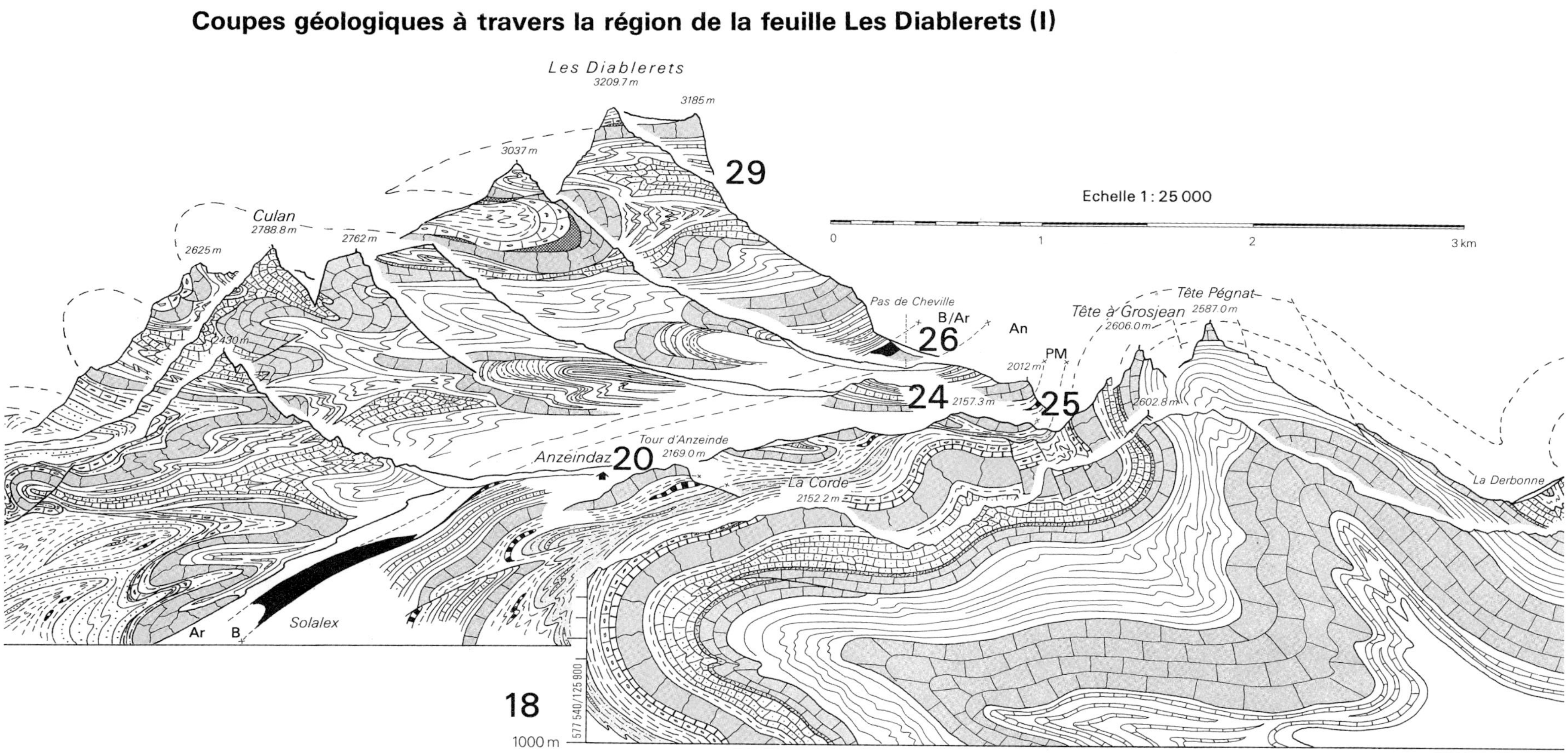

Figure 2: Serial geological cross sections through the Diablerets country showing complex geological structures in the Swiss Alps. (After Badoux and Gabus, 1991.)

nique break down under these situations. To avoid these problems and to model geological boundaries or layers of any complexity, a new approach abutting on new software tools is required. This is what we call 3-D structure modeling.

A NEW APPROACH OF 3-D STRUCTURE MODELING: THE BASIC PRINCIPLE

Geological structures have classically been defined by using geologic maps and cross sections. This type of information is particularly abundant and precise in mountain chains where topography produces natural cross sections through the structures; however, interpretation of this information is difficult. The complex geometry of Alpine-like geological structures is predominantly represented by recumbent folds and reverse faults, which, as discussed above, means that standard 2-D surface mapping and 3-D property modeling cannot easily describe them (Figure 2).

The best way to model these geological structures is to create a 3-D surface representing each boundary. These surfaces represent the stratigraphic and tectonic limits (e.g., geological layer boundaries, normal faults, thrust faults, etc.) as well as the civil works (e.g., bridges, tunnels, houses, etc.). Merely viewing the surfaces is generally insufficient for one to visualize the full nature of the problems: with a large number of surfaces, it is difficult to appreciate the true depth and interaction of the structures. Viewing a cross section through the 3-D model of a single geological layer produces only two thin surfaces (the lower and upper limits of the layer), but does not give an indication of its volume and lateral variations in thickness. A more useful way to visualize and understand the immensity of the geological structures is to create a full 3-D volume model.

The approach needed then is to create the 3-D surfaces that define the geological and man-made objects, and to fill the volume between the surfaces. In this study, we merely fill the layers with a single value or property, although the volume could also be filled with a varying property. The methods described next illustrate how 3-D volumes for geologic structures and civil engineering works are generated. These 3-D models can then be displayed and analyzed in the visualization portion of GMP, GeoDraw.

METHODS FOR BUILDING 3-D GEOLOGICAL STRUCTURE MODELS

Initial Data Entry

The first step is to define surfaces that honor all the points supplied by seismic data, geologic maps, cross sections, and well core data. For this study, several types of data were available, but to simplify our study,

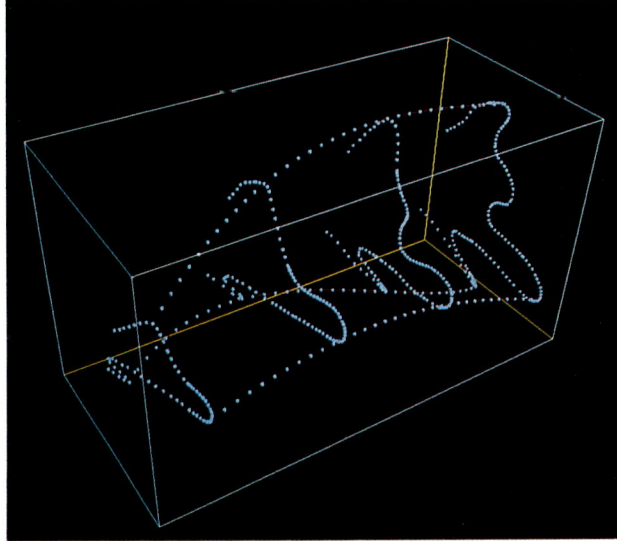

Figure 3: Example of 3 connection lines with regularly spaced connection points between the digitized cross sections. A nonlinear interpolation was used between cross sections. For clarity, most of the connection lines that would be used to create a structure have been removed from this example.

only cross sections were used. Eventually, however, information from the other data will be incorporated when creating those sections. The integration of all these data types into a single series of cross sections is critical to creating the 3-D structures, as this interpretation is the basis for all subsequent models. These structural models can then be used to control property modeling by limiting the input data and adding shape control.

The interpreted cross sections may be vertical, horizontal, or oblique. Each stratigraphic boundary on the profiles is digitized using software that places each point of the sections in the proper (X,Y,Z) location. For a given surface, the program merges all the corresponding cross section and base map data into one file.

Constructing a Single 3-D Structure

The next step is to interpolate connection points between the cross sections. A special program was developed to do this: first, control points are selected, between which regularly spaced points are created along each digitized section. Second, those points are connected from one cross section to the next along tie lines. Third, new points are interpolated along those tie lines (Figure 3). A nonlinear interpolation is typically used to generate smooth geological surfaces. For civil engineering structures, however, such as a hydropower underground pipeline, a linear interpolation is desired, since the orientation changes are abrupt.

The density of points along the profiles and connection lines is user-chosen and should depend on the

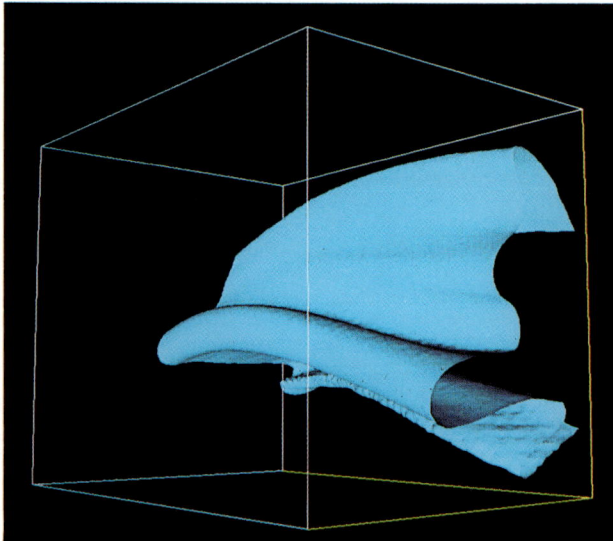

Figure 4: Example of a 3-D geological boundary between 2 geologic layers. This surface is produced by interpolating a 3-D mesh of points from the connection lines between the cross sections (Figure 3). A 3-D grid and then the isosurface are calculated from the mesh.

complexity of the structure: a complex structure requires a higher density of points than a simple one. For example, within the tight hinge of a fold, many points are necessary to accurately define the structure, whereas the limbs on either side may require fewer points to be modeled successfully. The density of connection lines and interpolated points along those lines is specified interactively using a 3-D cursor.

The result of this interpolation is a mesh of points that defines the digitized surface. Each point in the mesh is assigned the same arbitrary property value. This same process is repeated to create two additional meshes of control points (one on each side of the structure), which are then added to the model. Each mesh has a different assigned value (e.g., the surface may have a value of 0, while the meshes of points on either side of it have values of –1 and 1). Usually the two control meshes are parallel to the surface. This set of bounding control meshes allows an isosurface to be generated.

After the control points are determined, GMP is used to interpolate a 3-D grid using both the original surface points and control points. From the 3-D grid, a 3-D isosurface is interpolated by extracting, in this example, the 0-value surface. This isosurface represents the geological boundary (Figure 4). All the problems described above regarding 3-D gridding techniques are avoided in this situation, since we treat only one surface at a time and we add control points on both sides of the surface, increasing the precision of our model. As development continues, users will eventually be able to go directly from the mesh of points to the 3-D surface.

This process works nicely for a single folded surface, but it becomes more complex if discontinuities are added (e.g., faults). Using development versions of GMP, it is possible to incorporate most types of discontinuities.

Merging 3-D Structures

After creating 3-D surfaces for each geological boundary, geological layers are built by merging one 3-D surface with another. This set of merged or superposed surfaces can then be cut by other surfaces: faults, topography, tunnels, bridges, excavations, etc. The user specifies, during the creation of each surface, which side of the structure is kept during these merging operations. Each cut, therefore, creates a zone limited by the surfaces; these zones can then be treated as independent objects during display. Figure 5 illustrates, using a simple example, how these relationships are built. This technique can also be used with faults and man-made structures, such as tunnels, as illustrated in our study in Figure 6.

After the boundaries of the zones are defined, the volume between the surfaces can be filled with either an arbitrary value representing the rock type or a 3-D property model. Once the volumes are filled, the zones or objects are merged into one file to create the final geological model. This 3-D model can then be viewed and manipulated using GeoDraw.

APPLICATION OF THE METHOD

To test and illustrate the power and quality of 3-D volume modeling and visualization of complex geological structures, we chose a fictitious example, mostly inspired from alpine structures. This example consists of three folded geologic layers cut by two faults and the topography. The two faults also intersect one another. An imaginary tunnel that could represent a hydropower pipeline extends through this geological model.

The geological structures were initially defined by four cross sections, while the tunnel was defined by a single section duplicated eight times along its projected path (Figure 7). From these cross sections, using the methods described above, a complete model was derived. The final model is a 74 Mbyte file that contains the complete 3-D geometric description of the geological objects as well as the tunnel.

Once constructed, the model can be viewed in many different forms from any point. The intersection of the geological structures with topography produces essentially a 3-D geological map (partially shown in Figure 8). The complete or partial model can also be sliced interactively perpendicular to the X, Y, or Z axes, and a rectangular subsection of the model can be removed (known as *chair* mode; Figure 8). Models that are clipped obliquely can also be calculated and then displayed (Figure 9). In addition, the display colors can be altered to emphasize certain structures or

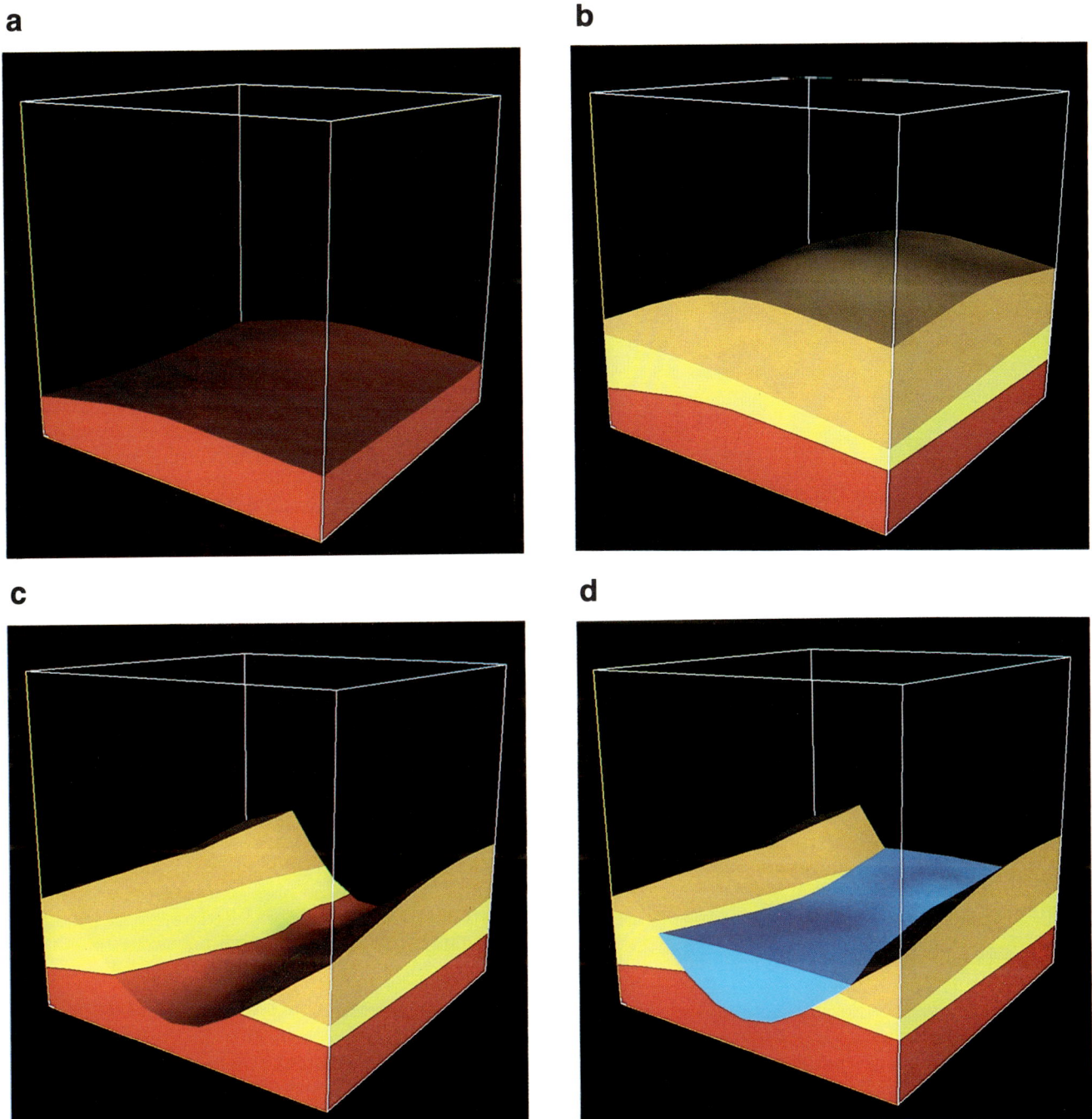

Figure 5: This series of 4 pictures illustrates the concept of intersecting several surfaces. For this example, the volume between the surfaces has been filled with a single property value for each layer. The views in Figures 5a and 5b show 3 layers, each being "deposited" on top of the one below. In Figure 5c, an "erosional" surface is introduced, which removes portions of the first 3 layers. The last view (Figure 5d) includes another depositional surface, which fills in the gaps left from the previous step.

properties, or to conform to any standards. Using an object display menu, each of the model's components can be displayed independently or in combination. For example, a single geological layer can be viewed extending above the topography and between the faults (Figure 10).

This model is an excellent example of the geometric description of the geological structures. Primarily due to the nonlinear interpolation, the smoothed geological surfaces and volumes can be easily visualized. The model allows us to observe, for example, the curved axis of the front fold. Similar deformation is visible on

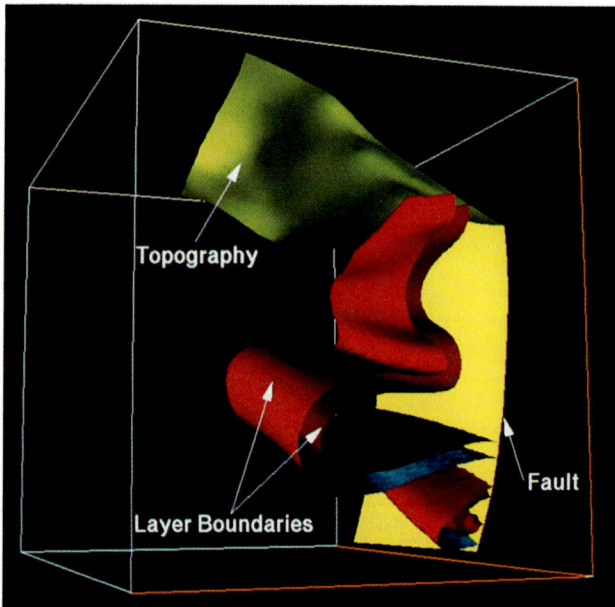

Figure 6: Example of sectors defined by four 3-D surfaces: two geological boundaries (corresponding to three geological layers), one fault, and topography. The boundary surfaces are first cut by the fault and then by the topography (evidenced by the topography truncating the fault, as opposed to vice versa).

the top fold where curvature of the axis is more accentuated. Increased folding from left to right of the model is also clearly visible (Figure 11). These kinds of observations are instrumental in understanding the different stages of folding that have produced these structures, and, ultimately, in understanding the geologic history and relationships of the area.

For civil engineers, the main interest of the model is to see precisely where and at what angles the tunnel crosses the modeled geologic structures. For example, the location where the two faults intersect would be a particularly hazardous region during boring of the tunnel, because of instability and water problems. Viewing the model, we see that the tunnel intersects the first fault just below this location. To reduce risk, it would be better to alter the tunnel's path to avoid the unstable zone. The exact location of the intersection can be determined interactively (Figure 12).

Similarly, the intersection of the geological layers by the tunnel is equally important (for example, to determine the type, quality, and hardness of the rock that will be drilled). Slicing the model along the tunnel path allows the viewer to observe the dip of the layers on the sides of the tunnel (Figure 9). A more realistic view is achieved by going inside the tunnel and seeing exactly what intersections occur (Figure 13). The general geology that the tunnel crosses can also be observed along the entire extent of the tunnel (Figure 14). From such a view, it is possible to calculate the volume of material that must be extracted in order to build the tunnel. An on-screen instant volumetrics calculation can be set to display for each view represented on the screen. For example, the extracted volume of rock from layer 1 within the tunnel is 5.7 cubic data scale units, based on the model.

CONCLUSIONS

Although currently available software has been used successfully in modeling simple structures, more powerful interpretation, modeling, and visualization tools are needed and are being developed to better understand and visualize complex geological structures, such as those described in this chapter. While property modeling can be used in limited circumstances to model geological structures, our study illustrates the usefulness and necessity for the 3-D structure modeling techniques used here. In point of fact, a full-scale study using these same techniques and software tools is currently underway at EPFL.

From the tools developed here, complex 3-D surfaces and volumes can be created and analyzed to increase our understanding of the geologic history and relationships that exist in study areas. As a decision-making tool, 3-D modeling of geological structures can be instrumental in determining projected paths for man-made facilities, thereby avoiding hazardous areas and decreasing the potential for financial overruns. Ultimately, fundamental research can take advantage of this type of modeling by finally being able to visualize and interactively analyze data and interpreted structures in three dimensions.

ACKNOWLEDGMENTS

We give personal thanks to Art Paradis and Agnis Kaugars of Dynamic Graphics, Inc., for accepting us in their development team, showing such a particular interest, and spending the time and energy to solve our problems. Thanks are also extended to EPFL, which has provided the material and funding for this research, and to Peter Irwin and Dynamic Graphics, Ltd., for all of their help and software support. Special thanks go to Francis Lapique (SIC-EPFL) and to all the Silicon Graphics S.A. members (Lausanne) for technical support. The authors would also like to thank David Hamilton, Tom Jones, and Skip Pack for their invaluable editorial comments. All modeling and dis-

Figure 7: The geological data used to create the 3-D model: four geological cross sections, their base map locations, and the base map location of the tunnel.

Section 1
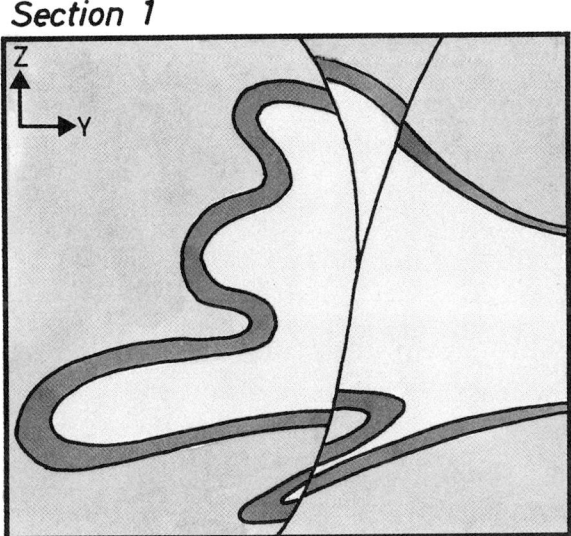

Section 2
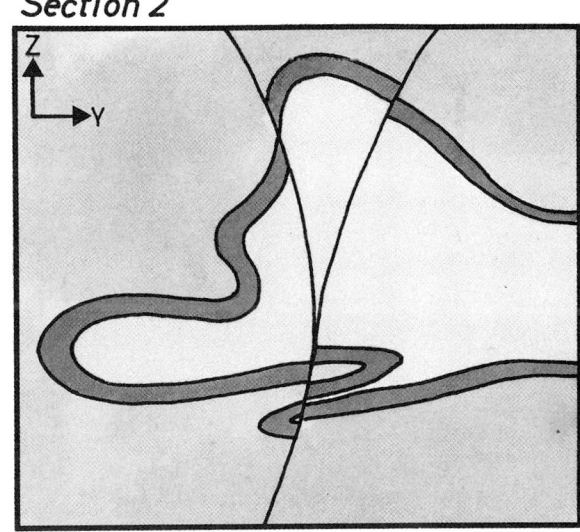

Section 3
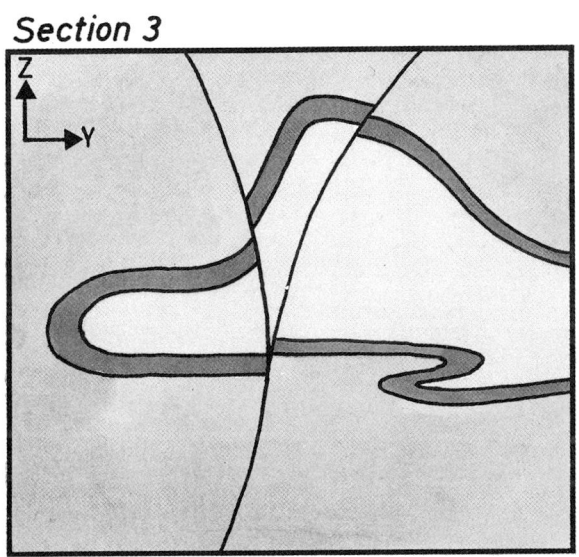

Section 4
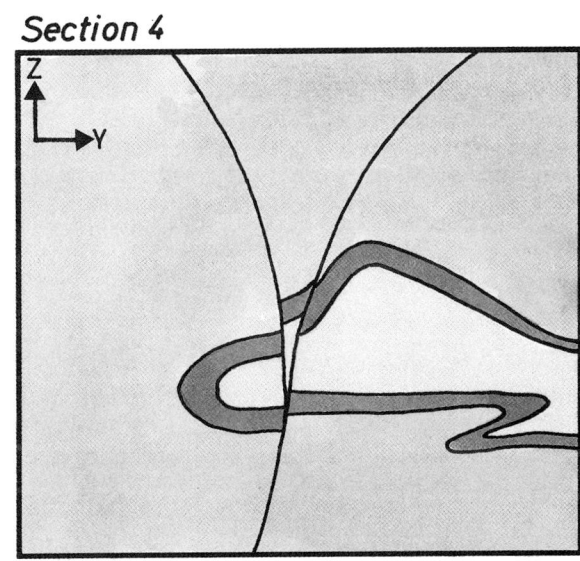

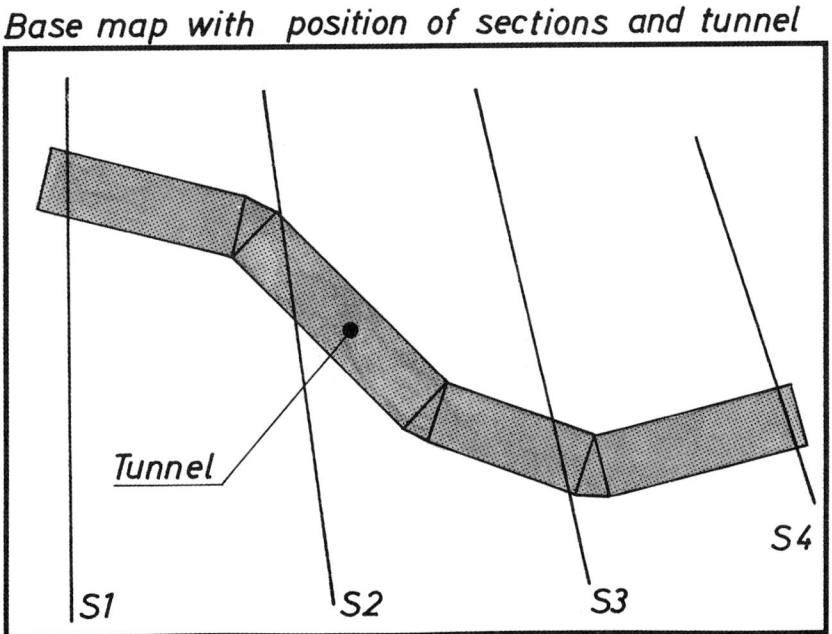

Base map with position of sections and tunnel

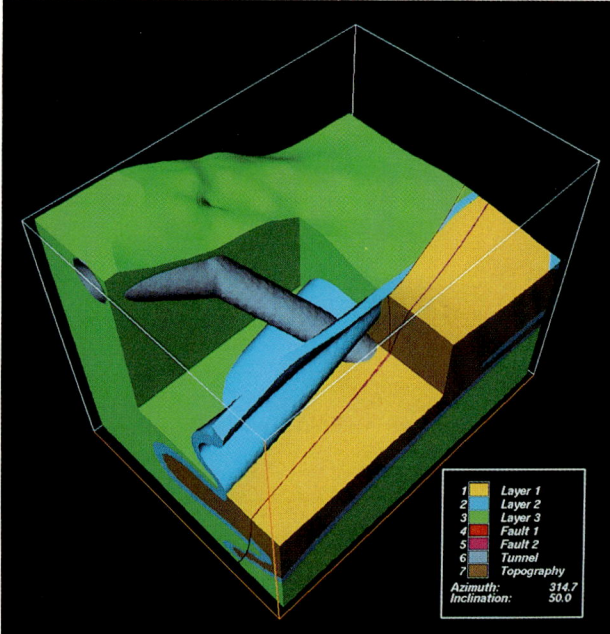

Figure 8: View of the complete model (using chair mode) showing the tunnel, faults, and geologic layers. The chair subsection has been removed to show the geometry of layer 2 and the position of the tunnel relative to the geologic layers. The surface geology is represented on the topography to create a 3-D geological map.

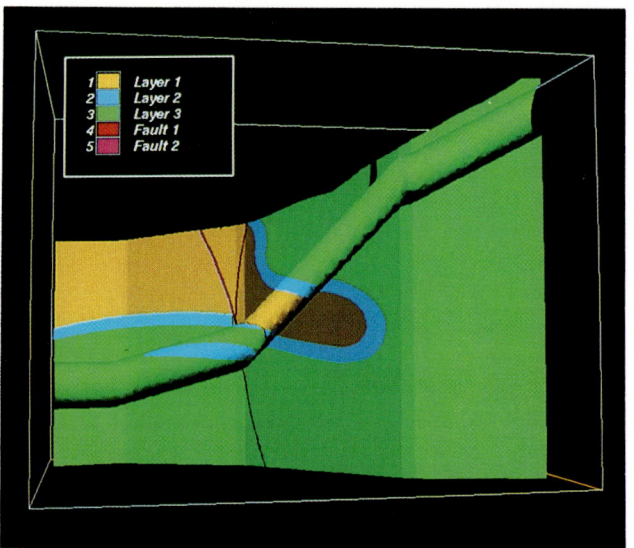

Figure 9: The complete model below the topography is shown clipped along the tunnel path, making visible the intersection angles between the geologic layers and the tunnel. This view also shows the rock-type distribution along the tunnel path.

plays of the data were performed on a Silicon Graphics Iris 4D/35 using released and development versions of Dynamic Graphics' Geological Modeling Program.

REFERENCES CITED

Badoux, H., and J. H. Gabus, 1991, Atlas géological de la Suisse: feuille 1285 Les Diablerets, Notice explicative: Service Hydrologique et Géologique National, 1:25,000, no. 88.

Mahoney, D. P., 1991, Mapping toxic spills: Computer Graphics World, v. 14, no. 1, p. 89–90.

Manley, T. O., and J. A. Tallet, 1990, Volumetric visualization: an effective use of GIS technology in the field of oceanography: Oceanography, v. 3, no. 1, p. 23–29.

Mayoraz, R., and A. Parriaux, 1990. Infographic tools to visualize the ground works in the geological context: example of the Bornels gallery, western Switzerland, in A. Parriaux, Memoirs of the 22nd Congress of IAH, v. 22: Lausanne, International Association of Hydrologists, p. 1293–1299.

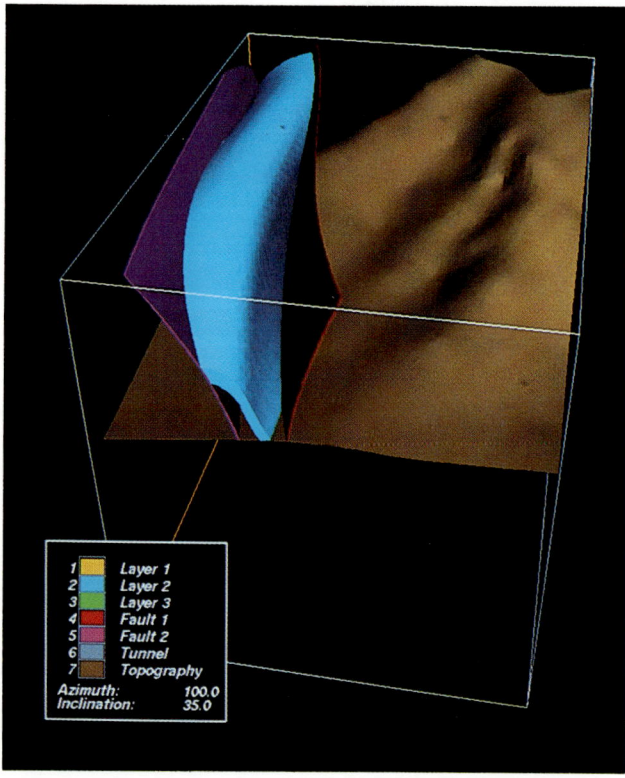

Figure 10: Example of selective display, showing only the portion of layer 2 that is above the topography and between the two faults.

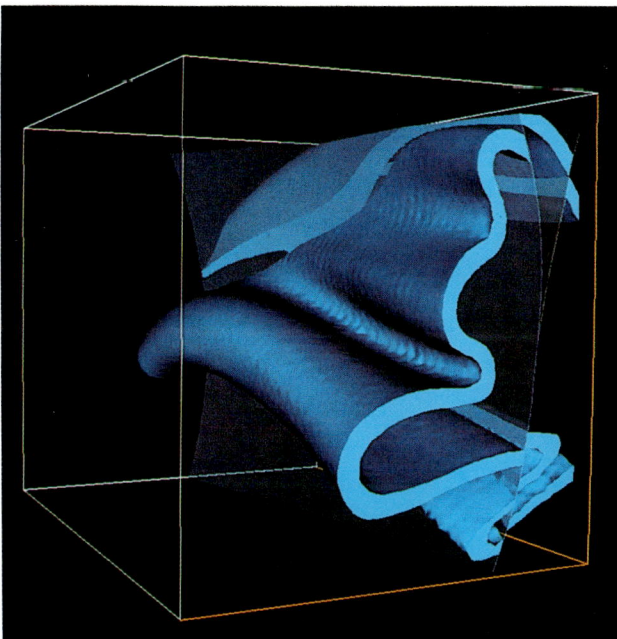

Figure 11: View of layer 2: the two faults are displayed using a transparency technique, allowing a complete view of the layer volume. In terms of structural geology, this view shows the curvature of the main fold axis and that folding increases from left to right within the layer.

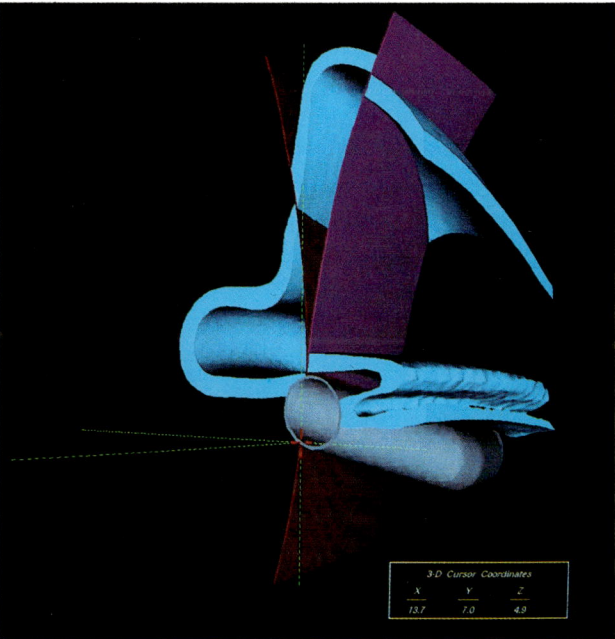

Figure 12: Slicing the model reveals the intersection of the tunnel and the geological structures. (Layers 1 and 3 have been removed so that more of layer 2 is visible.) The 3-D cursor, shown in red and green, is used to determine the exact location of the lower intersection of the tunnel and fault 1.

Figure 13: An inside view of the tunnel, at the place where its orientation changes. The intersections of the tunnel with fault 1 and the three geologic layers, along with their relative dips, are clearly visible. Knowing the location and position of such intersections is very helpful in civil engineering for determining the best tunnel path. (Note: the azimuth and inclination of the model are shown in the key in the upper left hand corner; this information allows users to orient themselves.)

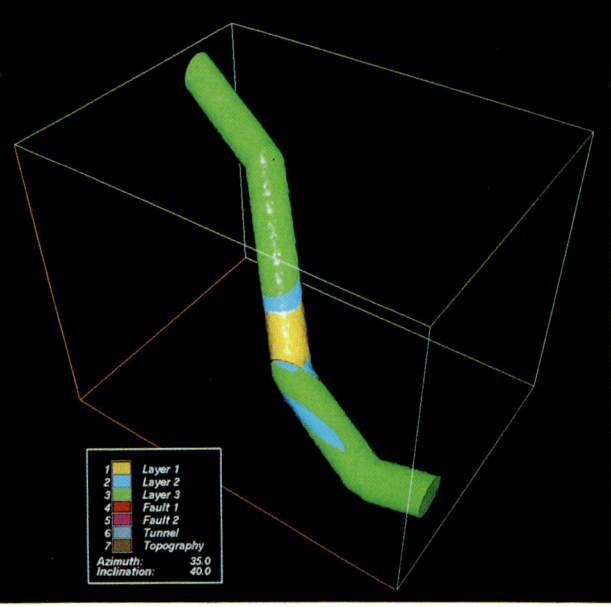

Figure 14: General view of the tunnel, showing its predicted geology.

Appendix A

Algorithm Comparison With Cross Sections

INTRODUCTION

Contouring with the computer has expanded rapidly during the past decade. This expansion has led to a variety of algorithms, methods, and programs to create contour maps. The number of choices is now so great that the user, particularly the beginner, may have difficulty in selecting which method to use. Summaries of gridding and contouring algorithms have been published (c.f. Walters, 1969; Waters, 1981; Jones et al., 1986; Heine, 1986; Banks, 1991), and comparisons typically have involved the generation of several contour maps with a variety of algorithms or controls. However, it is often difficult to see even substantial variations in surface form in a contour map.

Cross sections provide an excellent mechanism for studying the detailed form resulting from an algorithm. Applying an algorithm to a dataset and viewing the results as a cross section allows both significant and subtle variations between surfaces to be seen easily. Using real datasets to compare and evaluate algorithms may not be the best approach, however, because (1) features that cause problems for algorithms may not be found in the data, (2) it is difficult to find the best cross sections to show significant variations in surface form for algorithm evaluation, and (3) data points off the line of section may influence the form of the surface in the plane of section, making evaluation difficult. Therefore, a specially designed dataset with an understood form and with no off-line data influence is recommended.

DATA FOR ALGORITHM COMPARISON IN CROSS SECTION

A synthetic dataset (Table 1) has been built specifically for cross section analysis of algorithms. Figure 1 is a map showing the distribution and values of the synthetic data with contours drawn using a least-squares gridding algorithm and biharmonic filter. The dataset consists of points evenly spaced along north–south lines. Points along a line have the same value but the value changes from line to line. The spacing between the north–south lines (east–west direction) varies. On Figure 1, the line A–AA defines where cross sections used for evaluating the algorithms are generated. Because data values are the same in the north–south direction, there is little or no influence from data off the line of section and all surface forms can be attributed to points seen in the line of section.

We recommend that this or a similar dataset be used for part of the evaluation of surface modeling algorithms. To demonstrate the utility of this approach, we show the evaluation of five algorithms. This is not a thorough analysis of these algorithms but demonstrates the approach and provides some insight into how these algorithms compare.

ALGORITHMS TO BE COMPARED

Five algorithms or methods of creating contour maps are compared here. The algorithms are weighted-average and least-squares gridding, with and without application of a smoothing filter. An added variant involves isopach projection. All of these algorithms are grid based because such a mapping system was easily available. Triangular networks (e.g., Banks, 1991) are the other commonly used method for generating contours and are equally appropriate for applying the cross section analysis method, although not demonstrated here.

Gridding is the process of using data points to estimate values at a set of grid nodes. The process involves (1) defining the grid lattice by selecting a grid spacing and origin; (2) processing each grid node independently by (a) selecting nearby data points to use in calculating a value at the grid node, and (b) estimating a value for that node; and optionally (3) applying a filter to the just-calculated intermediate node values to smooth the resulting contours and tie the grid closely to the data values.

How data distribution affects the definition of grid lattice and the data selection process is best analyzed in plan view rather than cross section. Algorithms are best compared using the cross section approach described here. For this reason, the algorithms and filters (steps 2b and 3) used here are discussed in more detail.

Weighted-Average Algorithm

This method of calculating a value at each grid node is as simple as calculating an average. In this

Table 1. Synthetic Data

X	Y	Z
5.0	0.0	20.0
5.0	10.0	20.0
5.0	20.0	20.0
5.0	30.0	20.0
5.0	40.0	20.0
5.0	50.0	20.0
20.0	0.0	50.0
20.0	10.0	50.0
20.0	20.0	50.0
20.0	30.0	50.0
20.0	40.0	50.0
20.0	50.0	50.0
30.0	0.0	40.0
30.0	10.0	40.0
30.0	20.0	40.0
30.0	30.0	40.0
30.0	40.0	40.0
30.0	50.0	40.0
35.0	0.0	45.0
35.0	10.0	45.0
35.0	20.0	45.0
35.0	30.0	45.0
35.0	40.0	45.0
35.0	50.0	45.0
50.0	0.0	0.0
50.0	10.0	0.0
50.0	20.0	0.0
50.0	30.0	0.0
50.0	40.0	0.0
50.0	50.0	0.0
72.5	0.0	0.0
72.5	10.0	0.0
72.5	20.0	0.0
72.5	30.0	0.0
72.5	40.0	0.0
72.5	50.0	0.0
75.0	0.0	25.0
75.0	10.0	25.0
75.0	20.0	25.0
75.0	30.0	25.0
75.0	40.0	25.0
75.0	50.0	25.0
85.0	0.0	25.0
85.0	10.0	25.0
85.0	20.0	25.0
85.0	30.0	25.0
85.0	40.0	25.0
85.0	50.0	25.0
95.0	0.0	25.0
95.0	10.0	25.0
95.0	20.0	25.0
95.0	30.0	25.0
95.0	40.0	25.0
95.0	50.0	25.0
105.0	0.0	25.0
105.0	10.0	25.0
105.0	20.0	25.0
105.0	30.0	25.0
105.0	40.0	25.0
105.0	50.0	25.0
115.0	0.0	15.0
115.0	10.0	15.0
115.0	20.0	15.0
115.0	30.0	15.0
115.0	40.0	15.0
115.0	50.0	15.0
125.0	0.0	20.0
125.0	10.0	20.0
125.0	20.0	20.0
125.0	30.0	20.0
125.0	40.0	20.0
125.0	50.0	20.0
140.0	0.0	40.0
140.0	10.0	40.0
140.0	20.0	40.0
140.0	30.0	40.0
140.0	40.0	40.0
140.0	50.0	40.0
155.0	0.0	25.0
155.0	10.0	25.0
155.0	20.0	25.0
155.0	30.0	25.0
155.0	40.0	25.0
155.0	50.0	25.0
165.0	0.0	20.0
165.0	10.0	20.0
165.0	20.0	20.0
165.0	30.0	20.0
165.0	40.0	20.0
165.0	50.0	20.0
175.0	0.0	30.0
175.0	10.0	30.0
175.0	20.0	30.0
175.0	30.0	30.0
175.0	40.0	30.0
175.0	50.0	30.0
182.5	0.0	25.0
182.5	10.0	25.0
182.5	20.0	25.0
182.5	30.0	25.0
182.5	40.0	25.0
182.5	50.0	25.0
185.0	0.0	20.0
185.0	10.0	20.0
185.0	20.0	20.0
185.0	30.0	20.0
185.0	40.0	20.0
185.0	50.0	20.0

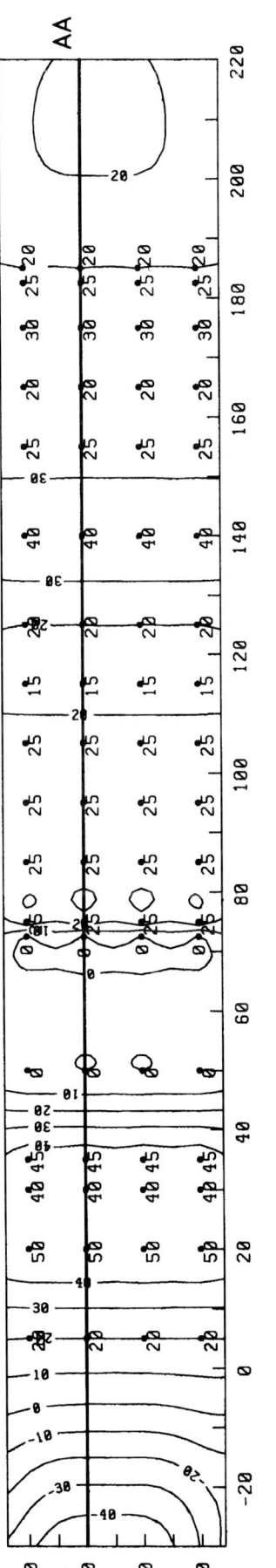

Figure 1. Map of synthetic data with contours and cross section baseline (A-AA).

Algorithm Comparison 275

case, the observed values at the selected data points are averaged to create a value for the node. This could be a simple average, but nearby points are generally considered to be more representative of the node than are distant ones, so the average typically is weighted by distance from the node. The weights may be inverse-distance, but inverse-distance raised to a power (commonly 2) or other functions of distance are also used.

It is rare that a weighted-average grid is used without further processing. The grid normally is filtered (smoothed) and tied to the data. Filters are applied because only a relatively few points are used to calculate a node's value. When moving from one node to the next, different data points may be selected, making significant differences between adjacent node values and irregularities in the resulting surface contours. We show the raw (unfiltered) grid below for illustrative purposes.

Least-Squares Algorithm

This method of calculating a nodal value consists of fitting a simple surface through the nearby selected data points, and interpolating a value from the surface at the location of the node; this interpolated value is assigned to the node. The form of the surface typically is planar or quadratic, and it is fit to the data points by least squares.

If many data points are used when calculating each node's value, the grid will project the global trend of the data beyond the edge of the data area. If the search is restricted, the grid will project a local trend (which can be unpredictable if the data are noisy) past the edge of the data. As with a weighted-average grid, the grid normally is filtered (smoothed) and tied to the data before being used.

Filters

The simple grid-generating algorithms (e.g., weighted-average, least-squares) can create grids that make irregular, noisy maps or that do not honor input data values. Filters and algorithms have been derived to smooth and adjust the grid values so as to better honor the data. Accordingly, the smoothing filter normally is used as a standard part of the gridding process.

Briggs (1974) presents methods that modify a grid in such a way that total curvature is minimized. This has the effect of removing small irregularities while leaving the overall form of the contours, since the filter uses 13 nodes for operating on each original node. The algorithm is based on the iterative solution of a biharmonic equation and is equivalent to fitting bicubic splines (Terzopoulos, 1988). The biharmonic filter is sometimes compared to a stiff sheet of metal (hand saw). When opposite ends of the metal sheet are held and pressure is applied to the center, the metal deforms in a smooth fashion with no sharp bends.

Smith and Wessel (1990) present methods that similarly modify a grid, but with the property of splines in

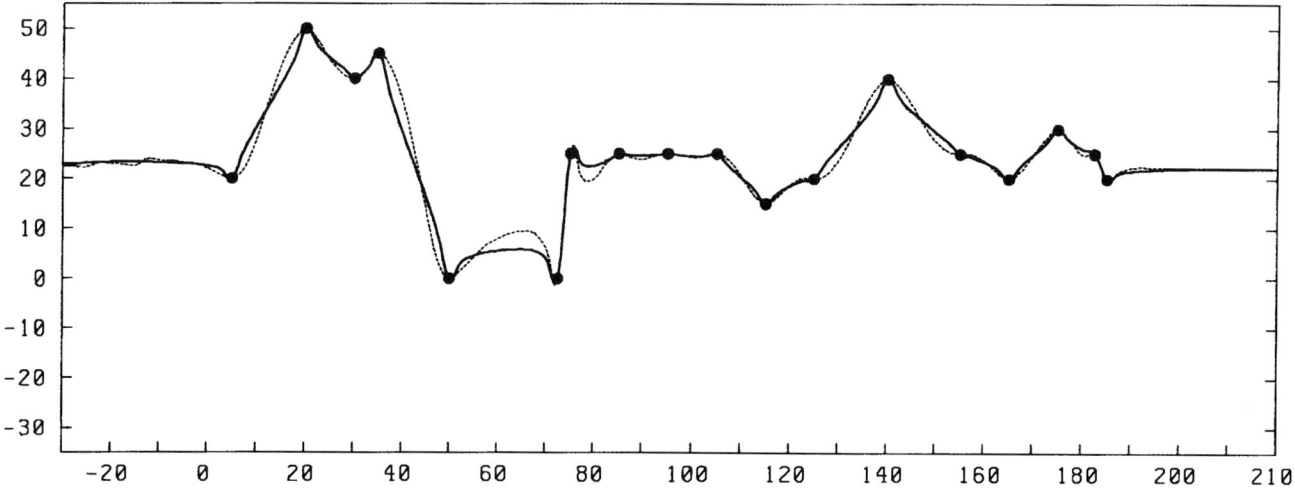

Figure 2. Cross section through synthetic data showing weighted-average algorithm with no filter (dashed line) and weighted-average algorithm with Laplacian filter (solid line).

tension. This filter, called by some the Laplacian method, uses five nodes and gives a peaked or sharp appearance to the resulting grid in the vicinity of data points. The Laplacian filter is sometimes compared to a soap bubble with a finger pushed into it to symbolize a data point.

Some users apply the Laplacian filter to a grid generated by weighted averages because their properties lead to similar forms. The biharmonic filter similarly is used with least-squares generated grids. Other mappers use both filters alternately during iteration to get a combined, intermediate result.

Most sophisticated implementations of these filters also allow local, direct adjustment of the grid nodes to honor the data. The exact algorithm used to tie the smoothed grid to the data is usually considered proprietary and not described by vendors of mapping software. A data-constrained filter implementation is discussed by Smith and Wessel (1990).

The grid spacing commonly is reduced as part of the filtering process. Grid initialization is performed on a relatively coarse grid and the first use of a filter removes low frequency noise from the gridded surface. As the grid spacing is reduced, subsequent passes remove higher frequency noise from the refined grid.

Zero-Line Projection of Thickness

A difficulty when gridding thickness data is dealing with the zero contour, or zero-line. As discussed by Jones et al. (1986) and Jones and Hamilton (1992), the observed zero values should not be used as ordinary data points; these zeros do not indicate thickness, they indicate that the unit is absent at that location. If the points are not used, however, the grid (and notably the zero-line) will be uncontrolled.

Jones et al. (1986) describe several methods for controlling the zero contour; these include insertion of negative values to replace zero observations. A simple approach is to select a single negative value that is consistent with trends near the pinchout, and just replace all zeros by that value. A second option uses available algorithms to individually calculate the negative values. These values are chosen to be consistent with nearby (local) trends and data to cause the map to project through the zero contour in a reasonable way. In either case, the resultant map will have a zero contour between the positive and zero data points, rather than forcing the zero contour incorrectly to honor all the zero values.

CROSS SECTION COMPARISON OF THE ALGORITHMS

The synthetic data were input to the weighted-average, least-squares, and isopach gridding algorithms. The weighted-average and least-squares algorithms were applied with and without filters; the isopach algorithm was applied with a filter.

Parameter settings for selection of nearby points to be used for calculating nodes included the following: the search distance was unlimited, potentially allowing any point to be used in calculations; eight search sectors were used; up to four data points were used per sector; and calculations were made if as few as one sector contained data. Nodes were calculated as described above for the weighted-average and least-squares methods.

A peaked (Laplacian) filter was applied to the weighted-average grid and a smooth (biharmonic) filter was applied to the least-squares grid. The biharmonic filter was also applied during construction of the isopach grid. The filter controls allowed up to 10 smoothing iterations over the grid, although all of the grids stopped smoothing before the tenth pass. Data

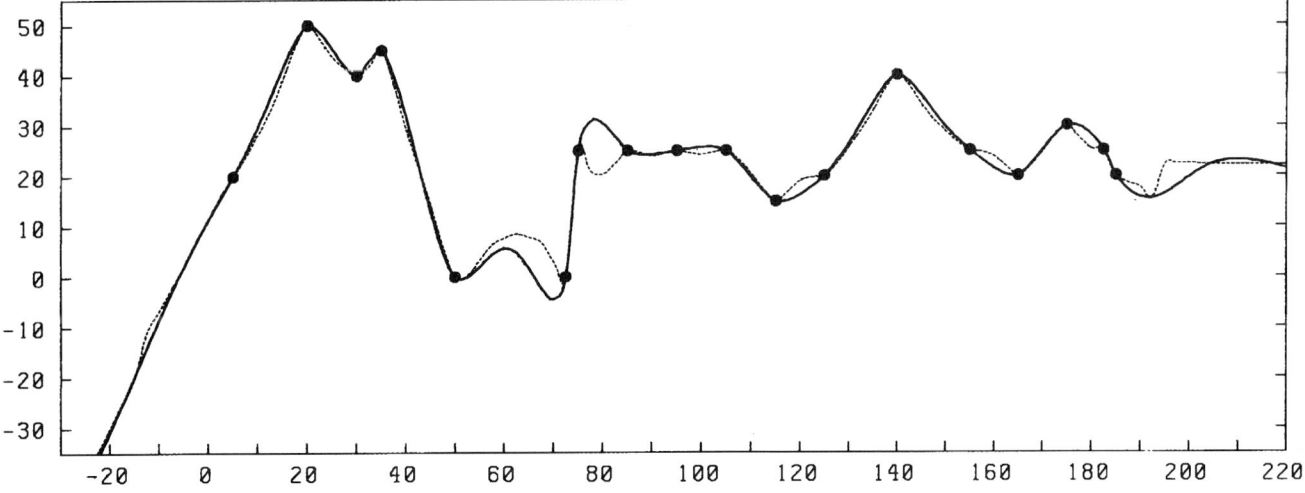

Figure 3. Cross section through synthetic data showing least-squares algorithm with no filter (dashed line) and least-squares algorithm with biharmonic filter (solid line).

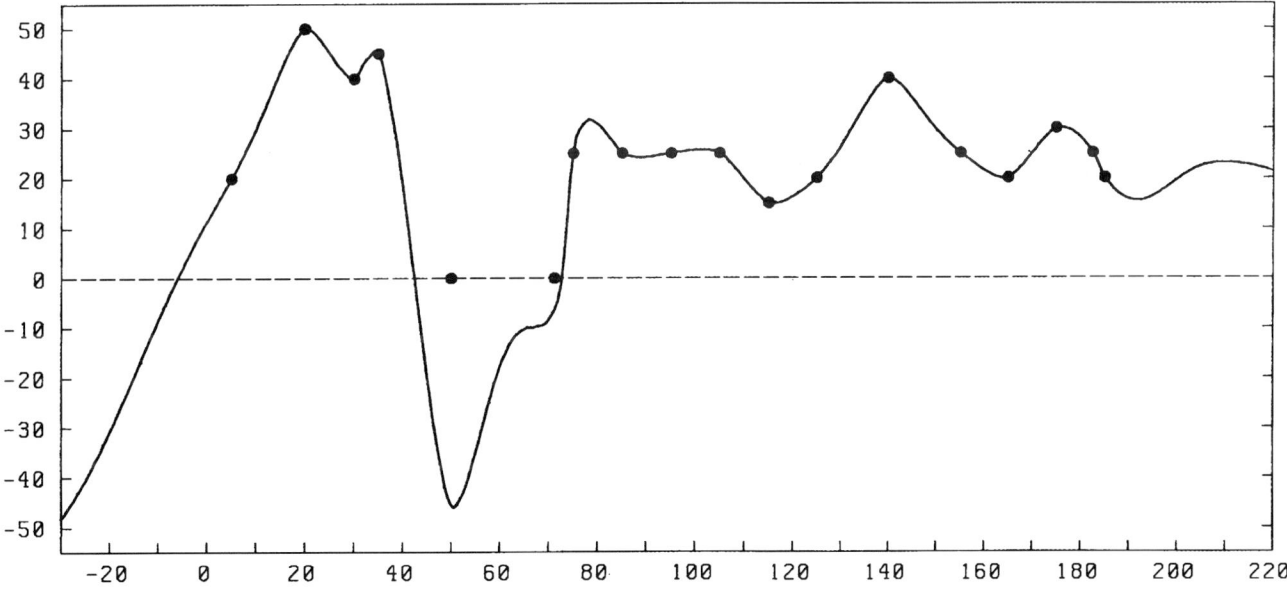

Figure 4. Cross section through synthetic data showing the effect of preprocessing the thickness data (replacing zero values with negative values) and gridding with least-squares algorithm and biharmonic filter.

were used during filtering to ensure that the filtered grids continued to honor the data.

Figure 2 shows profiles through grids built with the unfiltered weighted-average (dashed line) and filtered weighted-average (solid line) algorithms. Unfiltered grid nodes that are distant from data (e.g., near the edge of the grid or in empty areas) tend to flatten to an average. If virtually all data points are used, the grid tends to a global average; if the search is restricted, the grid tends to a local average. In any event, the user should be aware that this algorithm tends to project a flat, horizontal surface away from data, even if a uniform dip exists in the area containing data. High and low peaks in the surface only occur at data points with no over projection of surface form.

A Laplacian filter was applied to the grid built using the weighted-average algorithm without refinement (reduction) of the grid increment. As seen in cross section (Figure 2), the already peaked nature of the weighted-average algorithm is accentuated. The filter smoothens rough spots in the initial grid and honors the data.

Figure 3 shows profiles through grids built with the unfiltered least-squares (dashed line) and filtered least-squares (solid line) algorithms. The least-squares algorithm projects differently than does the weighted-average algorithm. Any trend found in the selected points is incorporated in the least-squares fit surface and causes the surface to project above or below the data used to create it. This projection creates broader, more rounded highs and lows than those created by the weighted-average algorithm. At the edge of the data or in large spaces between data, these projections can be highly variable, as seen on the left and right sides of the cross section (Figure 3). This variability is accentuated here by use of only 4 points per sector and the even data distribution, making the least-squares calculation a localized fit.

A biharmonic filter was applied to the grid built using the least-squares algorithm without refinement (reduction) of the grid increment. As seen in cross section (Figure 3), the relatively smooth nature of the least-squares algorithm is accentuated. The filter smoothens rough spots in the initial grid, broadens closures, increases projections, and honors the data.

Figure 4 shows a profile through a grid built with a thickness gridding algorithm. As discussed above, the algorithm automatically replaces zero thickness values with negative values. The negative value depends upon the closeness and values of surrounding positive thickness data points. After zero values were replaced, a least-squares algorithm with a biharmonic filter was used. Therefore, the surface in the areas of positive thickness is similar in form to corresponding areas of the filtered least-squares surface (Figure 3). The zero line is seen to fall between the zero and positive data values and its closeness to zero data values varies from one zero point to the next.

SUMMARY

Many geoscientists have several contouring programs available to them and many programs contain several contouring (gridding) algorithms. Experience is the best teacher of which algorithm to use for a particular geologic interpretation. However, application of available algorithms to the same datasets, and viewing the results in cross section, can quickly show how algorithms respond to data and where they are appropriate.

REFERENCES CITED

Banks, R., 1991, Contouring algorithms: Geobyte, v. 6, p. 15–23.

Briggs, I. C., 1974, Machine contouring using minimum curvature: Geophysics, v. 39, p. 39–48.

Heine, G. W., 1986, A controlled study of some two-dimensional interpolation methods: Computer Oriented Geological Society Computer Contributions, v. 2, no. 2, p. 60–72.

Jones, T. A., and D. E. Hamilton, 1992, A philosophy of contour mapping with the computer, (this volume).

Jones, T. A., D. E. Hamilton, and C. R. Johnson, 1986, Contouring geologic surfaces with the computer: New York, Van Nostrand Reinhold, 314 p.

Smith, W. H. F., and P. Wessel, 1990, Gridding with continuous curvature splines in tension: Geophysics, v. 55, p. 293–305.

Terzopoulos, D., 1988, The computation of visible surface representations: *in* IEEE Transactions on Pattern Analysis and Machine Intelligence, v. 10, no. 4, p. 417–438.

Walters, R. F., 1969, Contouring by machine: A user's guide: AAPG Bulletin, v. 53, p. 2324–2340.

Waters, N. M., 1981, Computer mapping: A review of what is available and what is useful for exploration purposes: Canadian Society of Petroleum Geologists Bulletin, v. 29, p. 182–196.

Appendix B

Five Geological Datasets

The most common use of a computer contouring system is to generate maps for economic and scientific applications. In addition, however, some users must also develop and test new mapping methods, compare algorithms, and teach others how to map. One common problem when doing these tasks involves finding realistic, adequate datasets. Appendix B contains 5 geological datasets. Three of the datasets represent non-faulted structural data. Two datasets contain sediment thickness, including pinchouts or zero-lines.

The data are generally well distributed, so most mapping algorithms should be able to produce maps that honor the data and look geologically reasonable. Two of the structure datasets (top of coal and bathymetry) and one of the thickness datasets (delta) were selected because of their simple geologic forms. The other two datasets (structure on pinnacle reefs and thickness of stream channels) are more complex and may require special effort to give good maps. In any event, the datasets represent geological data typically encountered in mapping projects.

STRUCTURAL TOP OF PINNACLE REEFS

These data represent Niagaran Brown tops from the Michigan basin and include points on and between pinnacle reefs. The surface can be thought of as a tilted, planar surface upon which are situated flat- to round-topped reefs with steep sides. These reefs are 100 to 800 ft (30 to 240 m) high, 500 to 4000 ft (150 to 1220 m) across, and are circular in form. Most wells penetrate the reefs, but many wells instead penetrate the sides of reefs and the flat surface between them. The geologic setting is described in Mesollela et al. (1974) and Droste and Shaver (1985). Improved but complex methods for mapping these data are presented by Hamilton and Henize (1992).

The 145 data points are listed in Table 1. We thank Petroleum Information Corporation for permission to publish this dataset. Figure 1 shows a map of the data, with contours at an interval of 100 ft (30 m). The grid interval used was 1000 units, and the least-squares gridding algorithm with smoothing filter was used (see Appendix A, this volume).

STRUCTURAL TOP OF A COAL UNIT

These control points represent elevations on top of a coal unit in West Virginia. These points are situated over a large anticlinal structure. The coal data have a relatively uniform distribution although clustering is present.

The 233 data points are listed in Table 2. This dataset was provided by Morrison-Knudsen Corporation of Boise, Idaho. Figure 2 shows a map of the data, with contours plotted at an interval of 25 ft (7.6 m). The grid interval used was 500 units, and the least-squares gridding algorithm with smoothing filter was used (see Appendix A, this volume).

UNDULATING BATHYMETRIC SURFACE

These data originated from bathymetry in an area with a dipping sea bottom. To make the data more challenging to mapping algorithms, residuals between the data points and a dipping plane were calculated. These residuals make up this dataset. Regional dip is thus missing from the data, but now several closed highs and lows are present. The data points may be thought of as representing topography-bathymetry, with the zero contour corresponding to the shoreline. On the other hand, we may think of this as structural data in which several closures, rather than a single high, must be mapped.

The 176 data points in this dataset are listed in Table 3. Figure 3 shows a map of the data, with contours plotted at an interval of 100 units. The grid interval is 2 units, and the least-squares gridding algorithm with smoothing filter was used (see Appendix A, this volume).

SEDIMENT THICKNESS IN A DELTA

The dataset consists of points indicating thickness of sedimentation in the Holocene Guadalupe River delta on the central Texas Gulf Coast. Geologic details are discussed by Donaldson et al. (1970). This dataset was chosen to show effects of trends and elongation on algorithms, as well as to test zero-line control.

Table 1. Dataset of Elevation on the Niagaran Brown Top, Michigan Basin (Data courtesy of Petroleum Information Corporation.)

X–coordinate	Y–coordinate	Elevation
2472596	3754	−5538
2472827	−17341	−5942
2472994	−16753	−5888
2473358	1660	−5265
2473702	5998	−5233
2474025	1011	−5270
2475006	−8923	−5707
2475910	2238	−5413
2475967	3020	−5405
2476059	7099	−5407
2476205	8836	−5467
2477194	10292	−5210
2477401	−1725	−5627
2477446	−22470	−5591
2477459	10750	−5234
2477526	6141	−5534
2477620	−10383	−5796
2477639	−12378	−5600
2477669	10080	−5491
2477996	7523	−5532
2478174	−12793	−5420
2478283	2412	−5260
2478719	3342	−5258
2478794	5817	−5543
2478923	8636	−5517
2478992	13346	−5213
2479144	839	−5612
2479172	−18351	−6050
2479993	13570	−5231
2480554	19231	−5649
2481200	7040	−5419
2481481	−18975	−6067
2481951	−4206	−5364
2482193	−9646	−5836
2482529	8406	−5479
2482819	−3092	−5404
2483064	9867	−5210
2483138	−21122	−5873
2484071	−2025	−5414
2484519	11116	−5287
2485045	−11669	−5876
2485308	10263	−5344
2485830	−2812	−5810
2486093	−8482	−5833
2486362	12881	−5561
2486663	7683	−5644
2486800	5304	−5686
2487332	4431	−5554
2487485	3697	−5417
2487620	−6290	−5760
2488095	4040	−5373
2488422	15154	−5496
2488840	−4484	−5833
2489085	9075	−5369
2489692	10410	−5299
2489814	−16922	−6004
2489940	−10211	−5886
2490126	−5410	−5858
2490144	15696	−5232
2490591	11336	−5279
2490630	−12496	−5477
2490751	−3177	−5803

Table 1. Continued.

X–coordinate	Y–coordinate	Elevation
2490832	10438	−5350
2490875	−19622	−6094
2490936	−4034	−5794
2491017	16556	−5252
2491755	12568	−5329
2492393	−20699	−6163
2493374	14278	−5322
2493381	13153	−5428
2493826	7499	−5732
2494200	15285	−5375
2494383	−6869	−5913
2494959	8169	−5739
2495720	12699	−5291
2495732	11569	−5397
2495984	3634	−5784
2496116	8931	−5745
2497321	−14387	−5957
2497452	1294	−5475
2498291	−5345	−5905
2498356	18231	−5341
2498376	16952	−5579
2499716	−3688	−5911
2499870	−1418	−5887
2500087	10858	−5416
2500250	2187	−5506
2500257	−9522	−6005
2500463	−5373	−5951
2501231	15627	−5693
2501300	−2981	−5914
2501339	−4031	−5930
2501827	−9151	−5933
2502206	12849	−5386
2502391	7214	−5593
2502496	−2777	−5929
2503707	14339	−5758
2503907	13237	−5666
2504073	7833	−5639
2504635	564	−5927
2504667	−8896	−5993
2505061	13835	−5745
2505135	8952	−5464
2505201	17154	−5364
2505306	−2532	−5966
2505372	−7788	−5566
2506499	18295	−5452
2506550	−8200	−6029
2506629	13013	−5453
2506752	12195	−5527
2507352	13180	−5446
2507902	14645	−5756
2508133	19404	−5439
2508199	−9368	−6115
2510121	−11593	−6157
2510606	7818	−5628
2510894	−137	−5989
2511214	12196	−5510
2511265	10960	−5797
2512666	1183	−5989
2513435	−3017	−5818
2513616	−419	−5990
2514229	−7999	−6166
2514973	−15350	−6255
2515027	−2937	−5510
2515083	−1857	−5676

Table 1. Continued.

X–coordinate	Y–coordinate	Elevation
2515174	16308	–5731
2515185	–6085	–6089
2515679	14249	–5630
2516010	–3789	–6120
2516811	17702	–5628
2516883	13807	–5862
2517224	9873	–5929
2517991	–13074	–6249
2518230	2866	–6010
2518504	4984	–5974
2518779	12655	–5878
2519933	45	–5722
2519938	–16567	–6338
2520350	11948	–5813
2521246	591	–5584
2521360	9310	–5937
2521411	17176	–5582
2521607	–6702	–6189
2522233	17496	–5509

The 154 data points are listed in Table 4. The dataset was obtained by visual interpolation from maps presented by Donaldson et al. (1970). Figure 4 shows a map of the data, with contours at an interval of 1 ft (0.3 m). The grid interval used was 100 units, and the least-squares gridding algorithm with smoothing filter was used (see Appendix A, this volume).

A significant difficulty when gridding thickness data is dealing with the zero contour, or zero-line. As shown by Jones et al. (1986) and Jones and Hamilton (1992), the observed zero values should not be used as ordinary data points; these zeros do not indicate thickness, they indicate that the unit is not present at that location. In order to give a reasonable zero-line in the map, the zero-data values were changed to –2 ft prior to generating the grid. The choice of –2 was somewhat arbitrary, but was based on "eyeball" estimates from the observed trends in thickness.

Special programs are available to calculate negative values for the data points, based on local gradients

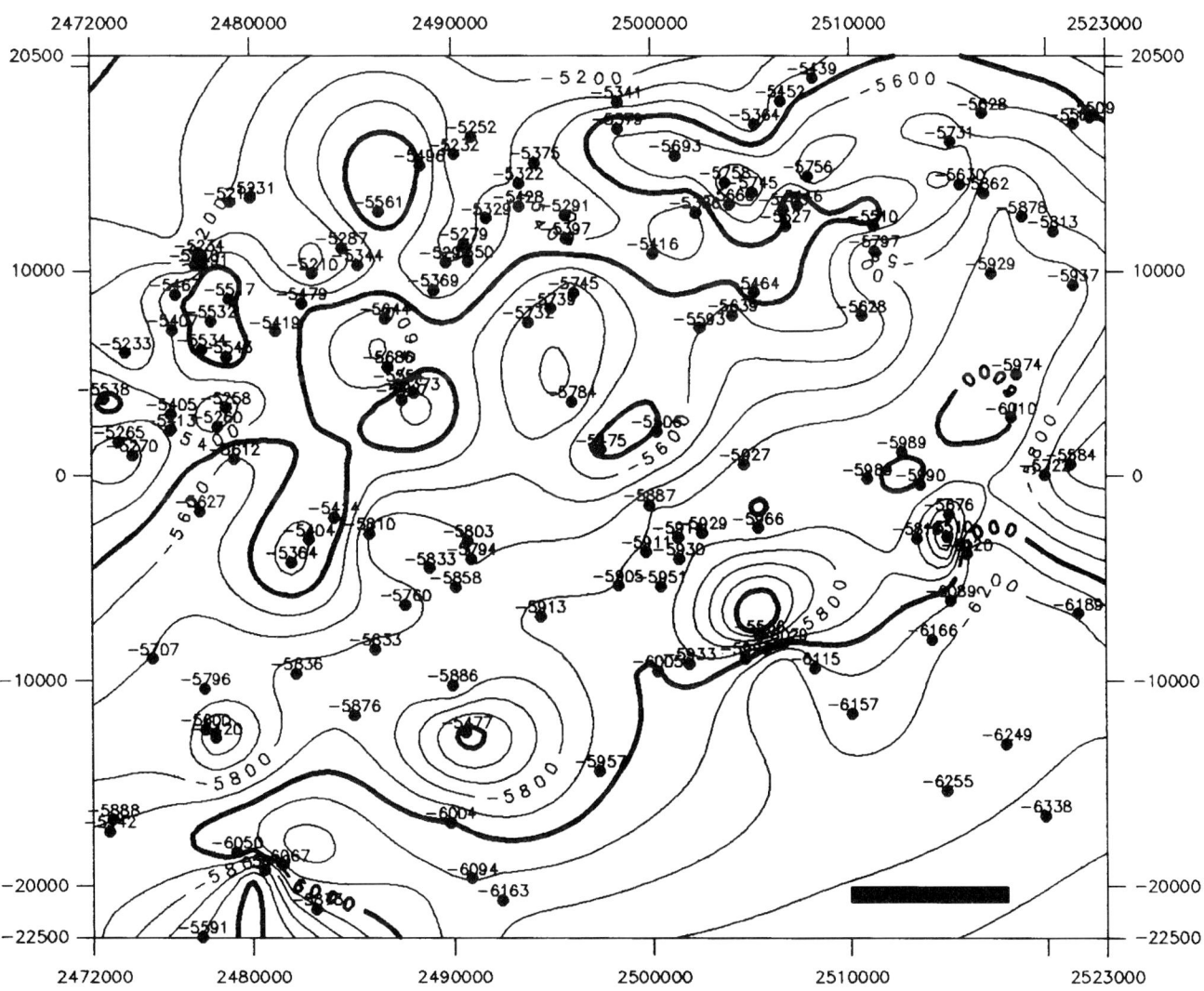

Figure 1. Map showing data distribution and contours of structure on pinnacle reefs (Table 1). Contour interval is 100 ft (30 m). Scale bar represents 8000 units.

Table 2. Dataset of Structural Elevation on a Coal Unit (Data courtesy of Morrison–Knudsen Corporation.)

X–coordinate	Y–coordinate	Elevation
−66421	117753	1263
−66741	131672	1282
−67447	128820	1268
−67614	130275	1275
−67795	116848	1266
−68484	124773	1304
−68732	116066	1274
−69210	117198	1299
−69352	116587	1277
−69393	115016	1264
−69399	123938	1327
−69528	123587	1332
−69714	114372	1255
−69950	127300	1322
−69995	115476	1274
−70450	131000	1280
−71028	117937	1355
−71035	122711	1331
−71195	115057	1293
−71210	115996	1308
−71300	111100	1278
−71300	113650	1267
−71727	121644	1307
−71887	115350	1332
−72002	122254	1322
−72005	120527	1331
−72093	119915	1349
−72177	117096	1360
−72235	125541	1340
−72270	109460	1268
−72275	127950	1329
−72856	118509	1367
−72908	113879	1342
−72975	114853	1351
−73034	121366	1340
−73282	117525	1379
−73438	115916	1383
−73485	113763	1353
−73560	114915	1343
−73653	115503	1375
−73849	126579	1311
−74022	131804	1220
−74049	131411	1223
−74264	131885	1225
−74274	115753	1351
−74287	131736	1224
−74317	114218	1345
−74400	133750	1217
−74459	118582	1352
−74518	132259	1221
−74536	117290	1366
−74650	125000	1324
−74855	120040	1331
−74943	116497	1365
−74971	119287	1353
−75400	108500	1296
−75550	110300	1352
−75700	104200	1436
−75935	130638	1232
−75935	131064	1219
−76676	125356	1296
−76713	127737	1256
−76794	110932	1328

Table 2. Continued.

X–coordinate	Y–coordinate	Elevation
−76803	110406	1334
−76950	115300	1352
−77255	112284	1354
−77302	132107	1212
−77645	111468	1347
−77846	109308	1317
−77906	109567	1325
−78005	113499	1372
−78140	110695	1333
−78382	123966	1289
−78475	112027	1376
−78495	109790	1352
−78795	113854	1393
−78955	111295	1380
−79005	128885	1204
−79024	110507	1376
−79075	114675	1391
−79174	121573	1315
−79205	113922	1401
−79250	110330	1384
−79305	112681	1377
−79401	109566	1375
−79482	125877	1256
−79545	110692	1381
−79647	108996	1378
−79718	106485	1321
−79817	108762	1367
−79905	112277	1383
−80095	115195	1375
−80155	111537	1387
−80185	112025	1384
−80205	113397	1390
−80205	114307	1375
−80263	108992	1379
−80455	111288	1378
−80642	119311	1328
−80705	109559	1386
−80705	112675	1394
−80705	114404	1368
−80855	111535	1379
−80858	110514	1377
−80902	122535	1288
−80919	108163	1373
−81059	108768	1377
−81150	112130	1397
−81305	112022	1393
−81395	111286	1387
−81461	108985	1389
−81560	109937	1382
−81578	104054	1330
−81605	111533	1384
−81705	113797	1377
−81705	115818	1358
−81835	110681	1384
−81905	112876	1380
−81959	108757	1391
−82005	112124	1395
−82105	111282	1385
−82105	115620	1352
−82200	106725	1366
−82325	114500	1380
−82357	110319	1389
−82525	112575	1386
−82525	113587	1387

Table 2. Continued.

X–coordinate	Y–coordinate	Elevation
−82545	111279	1390
−82751	109418	1407
−82755	110689	1390
−82755	111601	1400
−82816	128939	1165
−82823	115462	1347
−82825	112475	1387
−82916	109316	1402
−82925	114799	1358
−83095	113188	1382
−83144	114654	1367
−83146	110317	1400
−83185	115593	1327
−83255	110903	1402
−83255	111084	1402
−83255	111901	1400
−83255	112370	1389
−83260	109617	1403
−83265	116211	1320
−83292	118511	1281
−83357	108970	1407
−83405	113584	1369
−83455	111701	1406
−83495	113188	1373
−83537	104967	1349
−83637	110142	1405
−83655	110896	1401
−83655	112000	1403
−83660	109335	1410
−83910	107344	1385
−83985	112375	1382
−83990	108889	1409
−84075	110001	1412
−84075	112075	1393
−84085	110998	1400
−84085	113798	1355
−84191	114928	1329
−84275	110508	1407
−84385	112930	1365
−84416	109340	1407
−84472	108879	1403
−84475	110002	1406
−84475	110705	1401
−84475	111293	1396
−84485	112150	1387
−84485	115712	1311
−84747	125141	1153
−84800	123690	1158
−84805	114498	1326
−84805	116626	1285
−84828	133111	1019
−84837	113991	1341
−84895	112184	1379
−84895	112779	1357
−84905	113606	1345
−84905	114011	1340
−84935	110840	1393
−84940	116135	1285
−84995	112587	1362
−85000	119180	1250
−85005	115403	1304
−85095	111620	1381
−85095	112487	1365
−85095	113298	1345

Table 2. Continued.

X–coordinate	Y–coordinate	Elevation
−85099	118572	1253
−85201	119879	1224
−85205	115606	1295
−85275	117806	1260
−85295	111457	1387
−85295	112088	1385
−85295	113502	1346
−85715	114110	1327
−85815	112586	1374
−85825	112905	1359
−85825	114699	1305
−85825	115400	1285
−86125	112890	1347
−86185	119101	1203
−86215	113800	1315
−86215	114504	1302
−86215	116887	1240
−86225	113500	1326
−86225	114999	1287
−86284	118384	1212
−86650	104550	1392
−87497	127128	1115
−87498	119179	1163
−87550	109900	1335
−87800	113650	1284
−87852	121375	1118
−88107	116585	1179
−88130	118380	1144
−88541	123467	1054
−88625	115811	1189
−88835	116928	1147
−88886	117827	1137
−88908	119180	1096
−88950	108400	1354
−89571	118249	1111
−90674	126522	941
−91105	120864	1006
−91804	113692	1127
−92300	117175	1031
−93134	130318	826
−93600	109450	1098
−94253	126885	813

and data control; these give individual negative values for each zero data point and generally provide better maps.

SAND THICKNESS IN STREAM CHANNELS

This dataset represents sand thickness in fluvial-deltaic stream channels in the Red Fork Sandstone (Middle Pennsylvanian) of Oklahoma. An interpretive channel-mapping method applied to these data was described by Fierstien and Brewster (1992).

The 309 data points are listed in Table 5. The dataset was provided by MASERA Corporation of Tulsa, Oklahoma. Figure 5 shows a map of the data,

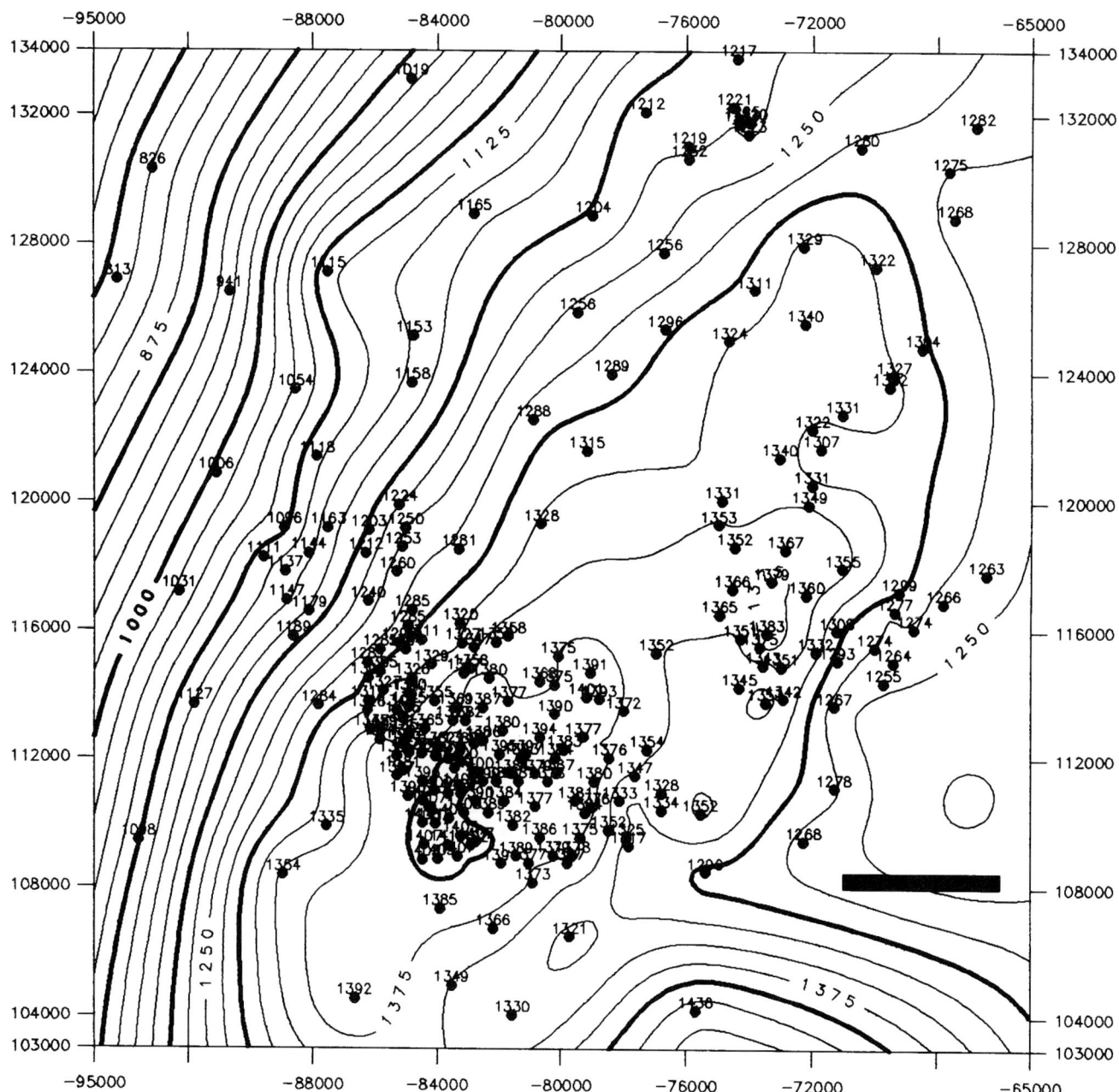

Figure 2. Map showing data distribution and contours of structure on a coal unit (Table 2). Contour interval is 25 ft (7.7 m). Scale bar represents 5000 units.

Table 3. Dataset of Depth to an Undulating Bathymetric Surface

X–coordinate	Y–coordinate	Depth
613	635	−204
592	627	−365
566	615	−265
550	612	−88
570	619	−234
586	625	−411
600	630	−304
605	632	−265

Table 3. Continued

X–coordinate	Y–coordinate	Depth
596	649	−37
583	646	12
569	640	−245
553	632	−217
543	624	−143
535	616	245
543	616	157
549	626	−247
556	636	−195

Table 3. Continued.

X–coordinate	Y–coordinate	Depth
565	637	−275
577	643	−33
609	658	−238
614	678	247
606	671	36
590	665	−86
567	655	−112
562	651	−300
552	646	−377
543	641	−296
530	637	53
534	643	−365
552	650	−377
557	657	−339
585	675	24
607	683	343
617	685	570
598	686	423
584	680	15
571	672	−83
559	667	−325
547	663	−567
535	658	−360
529	653	−255
535	668	−511
547	675	−719
564	682	−439
578	686	−181
555	678	−508
545	673	−734
537	671	−646
618	645	−167
606	642	−259
583	632	−285
563	631	−290
548	620	−104
535	684	−813
540	633	−168
545	612	23
540	619	−16
536	625	−48
530	652	−397
531	646	−239
537	636	−191
541	629	−160
546	625	−120
551	616	−81
558	618	−327
555	623	−201
554	626	−209
548	638	−257
547	645	−415
545	655	−582
541	665	−614
541	678	−766
537	681	−797
546	685	−728
555	673	−507
563	664	−294
585	641	28
589	638	−90
597	627	−327
603	615	−129
580	618	−307

Table 3. Continued.

X–coordinate	Y–coordinate	Depth
574	627	−204
571	634	−228
567	644	−111
561	648	−307
560	659	−317
556	671	−349
552	677	−531
551	682	−689
569	681	−400
577	655	264
582	651	153
597	628	−327
595	635	−193
589	651	58
584	657	18
582	664	2
578	677	−180
578	681	−30
587	684	−111
587	679	189
599	668	−16
604	655	−276
611	650	−221
609	646	−236
617	634	−23
544	646	−289
546	645	−423
563	650	−142
578	655	122
594	664	−204
605	667	−120
611	669	−74
615	671	106
603	666	−135
600	664	−158
590	656	−84
575	641	−48
571	640	−229
556	633	−194
546	628	−271
533	624	78
532	632	−229
552	641	−226
560	644	−315
589	653	57
601	654	−449
610	658	−380
607	646	−252
604	644	−275
594	640	−201
579	633	−316
574	631	−355
556	623	−193
544	619	14
539	618	125
551	619	−231
560	623	−312
575	632	−347
595	637	−343
605	638	−266
606	640	−259
614	627	104
614	620	105
530	628	55

Table 3. Continued.

X-coordinate	Y-coordinate	Depth
537	631	−191
550	632	−240
561	640	−306
564	638	−283
582	638	−144
597	638	−328
603	641	−282
611	647	−221
610	644	−228
607	640	−251
599	630	−312
591	620	−372
583	613	−133
546	616	30
560	617	−311
561	624	−304

Table 3. Continued.

X-coordinate	Y-coordinate	Depth
582	624	−292
589	628	−389
597	630	−327
612	630	−211
618	639	−16
617	681	270
611	677	224
599	669	−16
593	666	−62
585	661	25
561	654	−158
551	653	−385
544	654	−440
537	651	−493
532	644	−231

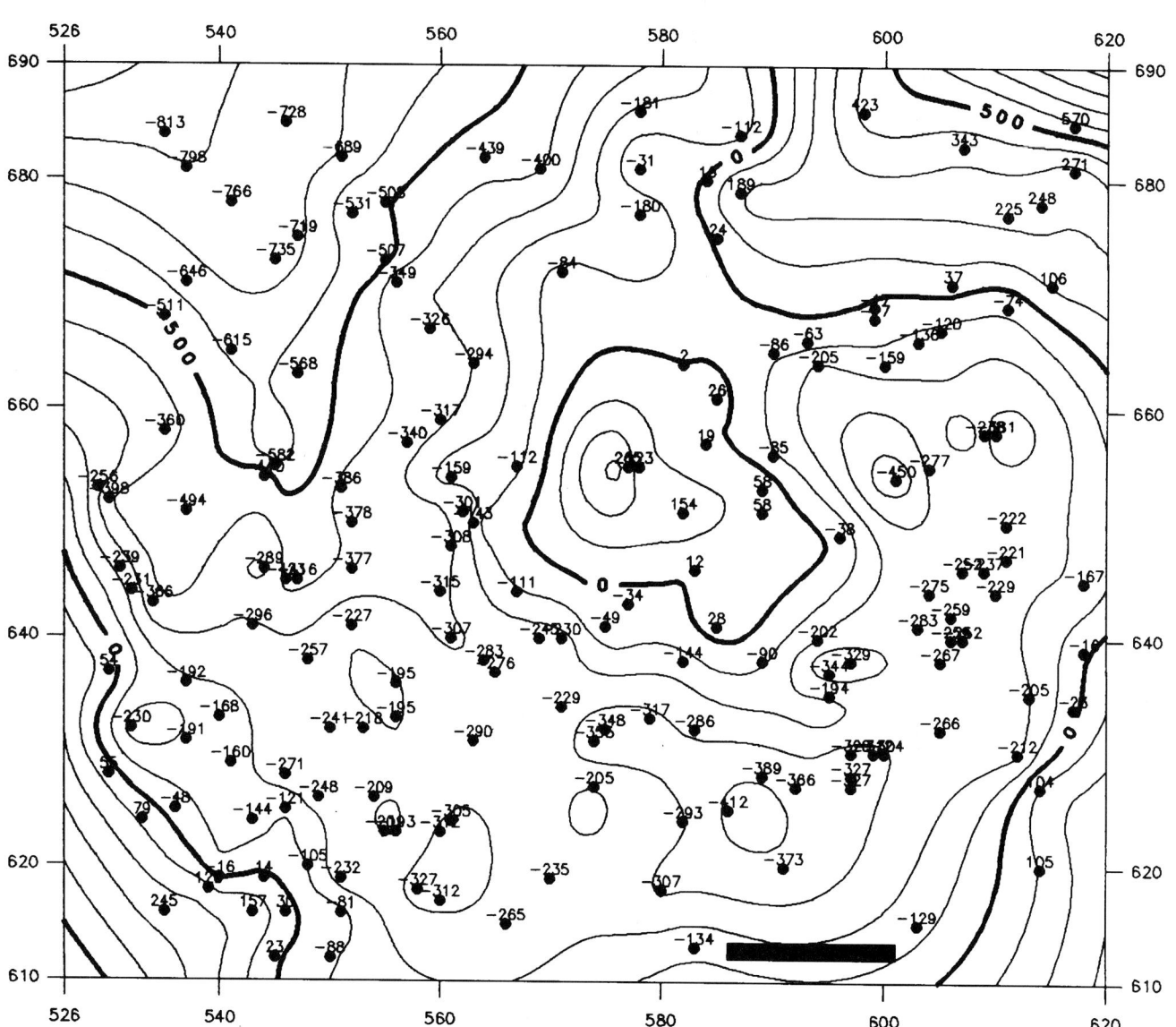

Figure 3. Map showing data distribution and contours on undulating bathymetric surface (Table 3). Contour interval is 100 ft (30 m). Scale bar represents 15 units.

Table 4. Dataset of Thickness of Sediments in Guadalupe River Delta (Data interpolated from maps in Donaldson et al., 1970.)

X–coordinate	Y–coordinate	Thickness
10698	21730	1.5
11072	22295	3.0
10351	21471	0.0
10705	21474	0.0
10876	21311	0.0
11058	21282	0.4
11619	21178	1.0
11554	20833	0.0
12061	20692	0.0
12346	20699	0.0
11418	22729	0.0
10993	22777	3.0
11397	22615	3.0
11324	22533	6.2
11566	22290	3.0
11719	22300	0.0
11734	22158	3.3
11265	21775	4.7
11803	21358	4.9
11935	21642	4.2
12005	21749	2.2
11965	21804	3.0
12293	21121	0.4
12362	21836	7.8
12541	21902	3.0
12334	22149	0.0
12558	22153	0.0
12409	22690	0.0
12763	23228	0.0
13268	23397	0.0
12449	21641	6.0
12643	21322	2.5
12818	21515	3.0
13000	21668	3.0
13108	21661	1.7
12979	21400	3.0
13540	23042	0.0
13394	23013	0.9
13166	23085	1.5
13059	23096	4.0
12820	22811	2.5
13163	22792	0.0
12970	22619	1.5
12566	22539	1.5
12803	22374	2.4
13014	22465	0.4
13109	22351	0.0
13042	22133	0.0
12855	22024	1.9
13123	22065	0.0
13237	22310	0.0
13184	21870	2.0
13316	21757	1.4
13463	22713	1.1
13490	22806	0.0
14076	22857	0.0
14002	22806	1.5
13742	22741	1.5
13737	22619	0.0
13447	22210	0.0
13368	22288	0.9
13434	22209	0.0
13438	22201	0.0
13714	21760	0.0
13637	21322	0.0
14002	21287	0.0
14642	21703	0.0
15188	21948	0.0
12819	21181	5.7
13140	21130	5.5
12793	20881	5.6
12953	20447	0.0
13356	21128	3.0
13615	21073	3.0
13571	20888	5.0
13369	20725	4.9
13592	20677	5.9
13474	20616	6.2
13471	20454	1.3
13460	20369	0.0
13719	20717	6.7
13660	20580	5.9
13804	20495	0.0
13884	20610	1.1
13985	20658	0.0
14233	20885	0.0
14415	21228	3.5
14237	21030	5.0
13947	20898	4.5
14374	20568	0.0
14509	20572	1.9
14536	20320	0.0
14663	20231	0.0
14947	20028	0.0
15130	19909	0.0
15301	19845	0.0
15356	19957	0.0
15148	20018	1.0
15210	20356	0.0
15099	20368	1.5
14957	20345	1.5
14874	20486	1.9
14595	20591	2.3
14623	20384	1.3
14801	20734	1.7
15069	20760	2.4
15080	21031	2.5
14613	20671	3.0
14693	20883	3.0
14593	21175	4.9
14837	21181	6.4
15082	21321	4.8
14508	21047	5.4
14663	20963	4.2
15418	20637	0.0
15515	20796	0.0
15380	20742	1.2
15426	21051	0.0
15374	21278	2.0
15252	21131	1.5
15531	21347	2.0
15567	21242	2.7
15749	21250	2.2
15194	21553	2.0
15238	21768	1.0
15441	21532	1.4

Table 4. Continued.

X–coordinate	Y–coordinate	Thickness
15407	21902	1.4
15582	21900	0.9
15659	22127	0.0
15780	21900	0.0
15768	21683	0.0
16069	21423	0.0
16038	21197	0.0
15689	21103	0.0
14678	21504	1.5
14019	21162	3.0
14874	21291	4.5
11288	22029	4.5
11337	22321	4.5
11053	22631	6.0

Table 4. Continued.

X–coordinate	Y–coordinate	Thickness
10525	22105	3.0
11693	21772	4.5
11467	21771	4.5
11575	21645	4.2
11502	21474	3.0
11105	21600	3.0
10889	21881	1.5
11440	21358	1.5
11925	21263	3.0
11936	21095	1.5
12109	21276	1.5
12253	21332	3.0
13442	21515	1.5
13256	20963	4.5

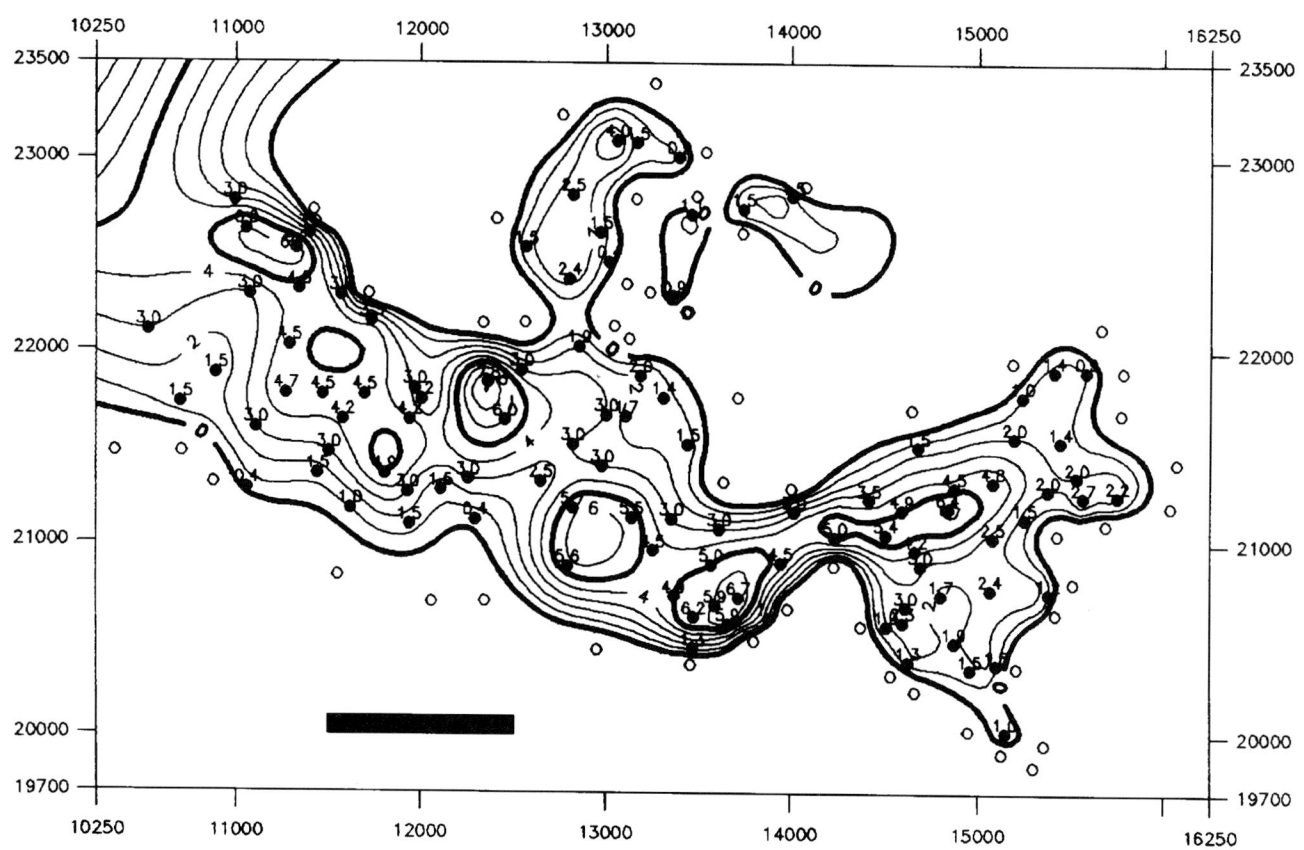

Figure 4. Map showing data distribution and contours of sediment thickness in Guadalupe delta (Table 4). Solid dots represent positive thickness, and open dots represent zero thickness. The zero values were replaced in the calculations by –2 values to induce correct projections. Contour interval is 1 ft (0.3 m). Scale bar represents 1000 units.

with overlain contours at an interval of 10 units. The grid interval used was 1500 units, and the least-squares gridding algorithm with smoothing filter was used (see Appendix A, this volume).

As with the delta dataset, to create this map, zeros were changed to negative values prior to gridding. We assigned values of –20 at the zero-data points.

REFERENCES CITED

Donaldson, A. C., R. H. Martin, and W. H. Kanes, 1970, Holocene Guadalupe delta of Texas Gulf Coast, in J. P. Morgan, ed., Deltaic sedimentation: Modern and ancient: SEPM Special Publication 15, p. 107–137.

Table 5. Dataset of Thickness of Sand in Stream Channels (Data courtesy of MASERA Corporation.)

X–coordinate	Y–coordinate	Thickness
1795867	13191332	0
1795870	13190027	24
1795897	13179475	0
1796101	13163764	0
1796138	13161123	6
1797056	13201787	0
1797191	13187398	0
1797824	13194561	0
1798387	13209670	0
1798404	13207036	0
1798609	13176836	0
1798832	13190004	30
1799693	13210985	0
1799725	13203123	0
1799818	13187390	0
1799963	13172919	0
1799982	13192848	42
1800687	13163153	3
1801075	13197877	0
1801440	13193604	49
1802394	13199201	0
1802421	13193944	21
1802439	13191316	26
1802457	13186074	0
1802539	13178198	0
1802824	13181161	0
1803128	13184105	0
1803680	13205791	0
1803890	13175550	0
1804088	13187738	0
1804257	13172586	0
1804949	13166430	0
1805100	13194031	47
1805131	13183477	0
1806294	13209723	0
1806376	13192709	22
1806386	13191392	19
1806394	13190071	10
1807648	13200568	0
1807709	13191408	22
1807880	13172994	0
1808017	13193689	29
1808027	13192727	23
1808898	13212383	0
1808921	13209761	0
1809143	13178258	0
1809318	13195015	0
1809335	13193695	25
1809993	13193702	22
1810045	13187794	0
1810078	13183184	0
1810526	13169054	0
1810547	13162519	6
1811111	13160953	0
1811204	13162539	0
1811547	13204540	0
1811731	13180884	0
1812817	13191271	0
1813136	13171702	0
1813159	13163816	4
1813693	13182220	0
1814099	13222983	0
1814153	13217710	0
1814203	13199290	0
1814236	13201918	0
1814236	13203236	0
1814333	13184864	58
1815271	13187842	27
1815305	13183231	24
1815474	13213768	0
1816139	13160307	9
1816694	13225628	0
1816780	13212464	0
1816797	13209838	0
1816959	13182256	16
1816981	13179625	0
1817445	13162847	10
1818122	13213798	0
1818262	13181709	39
1818887	13184258	14
1819375	13223049	0
1819423	13217775	0
1819487	13194091	0
1819512	13187550	0
1819691	13171739	0
1819746	13162680	62
1820537	13182616	22
1820643	13224388	0
1820765	13203310	0
1820790	13196724	0
1820960	13175704	0
1820988	13166496	0
1821024	13161330	0
1822037	13212550	0
1822039	13209919	0
1823385	13203343	0
1823555	13176374	0
1824268	13160657	0
1824563	13224405	0
1824576	13219170	0
1824679	13207319	0
1824723	13196777	0
1824732	13194149	0
1825085	13187914	0
1825228	13162957	42
1825882	13168839	0
1826037	13195479	0
1826108	13181673	0
1826553	13162944	42
1827227	13215255	0
1827488	13174437	0
1828522	13221839	0
1828779	13176427	0
1829828	13223145	0
1829847	13215256	0
1829913	13188889	16
1829986	13192890	0
1831104	13227072	0
1831197	13211302	0
1831476	13161326	0
1832562	13204875	0
1832586	13199475	0
1832622	13191594	4
1832633	13189621	31
1832751	13171854	0
1833097	13168239	0
1833867	13206052	0

Table 5. Continued.

X-coordinate	Y-coordinate	Thickness
1833881	13203424	0
1833936	13192925	20
1835113	13215278	0
1835124	13210016	0
1835138	13212651	0
1835149	13194577	14
1835186	13204751	0
1835200	13202116	0
1835212	13199495	0
1835231	13187674	30
1835321	13173191	0
1835668	13164314	32
1836243	13178140	0
1836499	13193930	56
1836964	13169571	0
1837147	13181767	0
1837233	13232215	0
1837653	13223223	0
1837666	13219263	0
1837695	13215323	0
1837737	13210067	0
1837751	13194169	28
1837754	13207433	0
1837768	13204785	37
1837785	13202134	20
1837814	13191654	25
1838293	13160506	0
1838996	13211396	28
1839008	13208762	0
1839058	13189038	0
1839069	13198211	0
1839268	13203188	0
1839586	13161780	0
1840045	13192180	0
1840245	13223245	0
1840262	13217965	0
1840262	13220604	0
1840276	13215335	0
1840284	13212705	22
1840340	13204810	0
1840368	13202158	0
1840420	13182469	0
1840845	13169603	0
1841578	13216652	40
1841591	13213359	16
1841596	13211386	0
1841620	13208775	0
1841637	13206132	0
1841652	13203496	0
1842097	13178209	0
1842870	13220615	0
1842891	13215336	32
1842978	13202182	0
1843022	13217974	28
1843459	13167048	16
1843467	13164404	24
1844205	13211404	0
1844794	13163147	0
1844802	13165734	18
1845370	13228457	0
1845483	13223263	0
1845503	13220632	30
1845509	13217987	23
1845518	13215352	20

Table 5. Continued.

X-coordinate	Y-coordinate	Thickness
1845609	13199588	0
1845696	13177259	0
1846103	13167044	0
1846821	13214049	0
1846830	13211419	0
1847115	13162823	0
1847134	13221956	24
1848120	13220647	44
1848124	13218003	21
1848179	13202254	0
1848238	13186517	0
1848250	13194391	0
1849536	13190481	0
1850049	13167077	21
1850661	13215415	0
1850719	13218056	0
1850737	13220681	39
1850865	13194411	0
1851299	13171017	0
1851325	13167080	28
1851370	13160546	0
1852010	13222020	10
1852119	13200978	0
1853161	13185575	0
1853311	13223329	0
1853329	13220725	32
1853342	13218123	0
1853398	13207575	0
1853414	13202311	0
1853926	13164474	0
1853947	13161834	0
1855240	13164483	0
1855790	13218145	0
1855903	13220748	0
1855967	13215499	0
1856037	13202347	0
1856053	13199702	0
1856496	13173712	0
1857338	13189222	0
1857403	13195464	0
1857840	13168447	20
1857854	13165818	8
1858471	13223369	0
1858516	13220765	14
1858551	13215523	0
1858610	13210264	0
1858646	13202385	0
1858779	13176044	0
1858802	13228280	0
1859537	13214548	0
1859817	13219479	21
1860027	13193189	0
1860073	13177367	0
1860440	13171107	0
1860486	13164515	0
1861091	13226007	0
1861113	13220770	37
1861147	13218166	12
1861244	13205022	0
1861281	13200084	0
1861695	13179779	0
1861789	13165841	33
1862366	13180353	0
1862400	13229957	0

Table 5. Continued.

X–coordinate	Y–coordinate	Thickness
1862404	13222100	14
1863102	13165848	22
1863717	13223397	22
1863739	13220788	28
1863778	13218177	14
1863795	13215548	0
1863845	13207662	0
1863993	13189270	0
1863999	13181356	0
1864051	13170799	0
1864954	13216868	28
1865071	13219498	30
1865099	13214234	0
1865548	13196176	0
1866414	13215550	0
1866416	13214236	0
1866480	13202400	0
1866582	13181041	0
1866638	13184005	0
1866686	13177414	0
1866779	13160288	0
1866879	13164248	6
1867332	13218841	22
1867640	13222116	22
1867646	13230002	0
1867647	13224749	16
1867659	13216867	21

Table 5. Continued.

X–coordinate	Y–coordinate	Thickness
1867683	13214233	0
1867735	13207654	0
1867912	13190604	0
1868011	13174788	0
1868087	13198577	0
1869400	13166939	0
1869943	13216871	0
1870268	13223452	26
1870545	13187988	0
1870587	13183707	0
1870935	13226740	0
1871425	13228709	12
1871566	13230028	8
1871569	13215567	0
1871594	13224775	0
1871627	13210306	0
1871740	13199811	0
1871802	13194562	0
1872038	13165637	0
1872269	13180420	0
1872858	13220835	0
1873220	13182743	0
1873236	13181096	0
1873586	13179114	0
1873686	13160683	22
1874202	13227413	0
1874430	13194586	0

Droste, J. B., and R. H. Shaver, 1985, Comparative stratigraphic framework for Silurian reefs—Michigan basin to surrounding platforms, in R. R. Cercone, and J. M. Budai, eds., Ordovician and Silurian Rocks of the Michigan basin and its margins, Michigan Basin Geological Society Special Paper 4, p. 73–94.

Fierstien, J. F., and A. V. Brewster, 1992, The shape assist technique: Incorporating stream channel interpretations into computer-generated surface models: Clark County, Kansas, (this volume).

Hamilton, D. E., and S. K. Henize, 1992, Computer mapping of pinnacle reefs, evaporates, and carbonates: Northern trend, Michigan basin, (this volume).

Jones, T. A., and D. E. Hamilton, 1992, A philosophy of contour mapping with the computer, (this volume).

Jones, T. A., D. E. Hamilton, and C. R. Johnson, 1986, Contouring geologic surfaces with the computer: New York, Van Nostrand Reinhold, 314 p.

Mesollela, K. J., K. D. Robinson, L. M. McCormick, and A. R. Ormiston, 1974, Cyclic deposition of Silurian carbonates and evaporites in Michigan Basin: AAPG Bulletin, v. 58, p. 34–62.

292   Appendix B

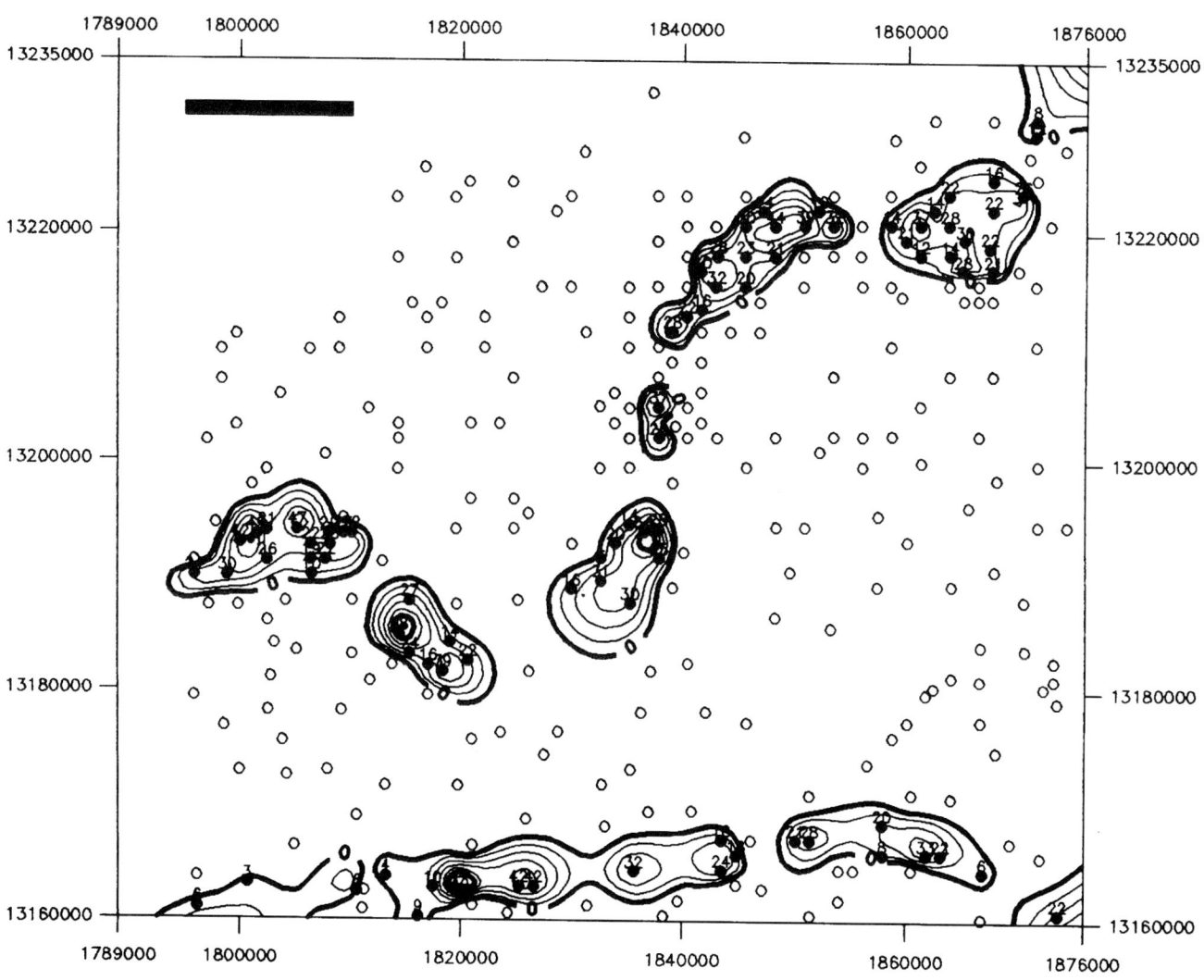

Figure 5. Map showing data distribution and contours of sand thickness in stream channels (Table 5). Solid dots represent positive thickness, and open dots represent zero thickness. The zeros were replaced by −20 values to induce correct projections. Contour interval is 10 ft (3 m). Scale bar represents 15,000 units.

Index

237 economic zone, Long Beach Unit, 142
2-D digital array, 100
2-D gridding techniques, geologic structures, 262
2-D mapping, 175, 176
3-D digital array, 100
3-D geologic block modeling, 175
3-D gridding techniques, geologic structures, 262
3-D model construction, 211
3-D modeling, 178, 181
    complex geological structures, 261
    development tools, 261
    mature steam drive, 251
3-D modeling process, 179
3-D structure models, 265
3-D subsurface modeling, 123
3-D subsurface modeling program, GOCAD, 138
A-1 Carbonate, 49, 56, 57, 58, 59, 62, 67, 68, 69, 70, 71
    Salina Formation, 47
A-1 Evaporite, 49, 56, 57, 58, 59, 62, 67, 68
    Salina Formation, 47
A-2 Carbonate, 49, 56, 58, 67, 68, 69, 70
    Salina Formation, 47
A-2 Evaporite, 49, 56, 58, 62, 67, 68, 69, 70, 71
    Salina Formation, 47
Absolute permeability value, 109
Aeromagnetic data, contour display, 13
Aeromagnetic survey, Vancouver Island, 10
Affine transformation, 79
Air-mercury capillary pressure curves, 109
Alps, Swiss, 181
Anastomosing fault, 150
Anisotropic Data Mapping (ADM), 86
Anisotropic direction, determination, 77
Anisotropic magnitude, determination, 77
Anisotropy, 76
    localized, 78
    modeling, 75
    regional, 78
    types, 77
Anomalies, graphical techniques for locating, 9
Antrim County, Michigan basin, 47, 50, 64
Arctic Circle, 184
Area of influence concept, 236
Athabasca oil sand deposits, Alberta, 235
Attribute grids, construction, 118
Automated modeling
    fault representation, 123
    geologic structures, 123
Average oil saturation, model, 105
Average oil saturation model, 115
Average porosity, model, 112
Baselap, 180
Basin, sedimentary, simulation, 94
Basin simulation, 93, 94, 99
    example, 98
Basin simulation model, 103
Bathymetric surface, computer contouring, 279
Beaufort Sea, Alaska, 183, 185, 200
Beaverhill Lake Group, Alberta, 240
Biased gridding technique, 79
Big Horn basin, Wyoming, 106
    stratigraphic section, 108

Biharmonic filter, 53, 118, 128, 278
    grid modeling, 52
Biharmonic filtering, 55
Block diagram, example, 60
Boundary-element modeling, 94
Brooks Range, Alaska, 184, 186, 188, 199
Brownlow Point, Alaska, 186, 196, 197
Bullen Point, Alaska, 195
Bull's eye contour patterns, 147
C-608 fault, Wilmington oil field, 146
Canadian oil sands, variable zone modeling, 235
Canning River, Alaska
    climate, 184
    drainage basin, 184
    fan-delta, 190
    hydrologic cycle, 184
    polar fan-delta complex, 183
Canning River delta, 193, 194, 197, 199, 200
    Alaska, 181
    block diagram, 187
    borehole location, 191
    correlation framework, 191
    environmental zones, 185
    Holocene sands isopach, 198
    sea-ice thickness, 189
    topographic map, 185
Capillary pressure curves, 105
    porosity, 108
Carbonates
    computer mapping, 47
    Michigan basin, 47
Carbonate surfaces, modeling, 55
Cassiar Mountains, British Columbia, 204
Chair option, 254
Challenge Channel, Alaska, 198
Chester Formation, Kansas, 28, 31, 32
Clearwater Formation, Alberta, 241
Colville River
    Alaska, 184
    delta, 186
Compaction
    estimating, 65
    modeling techniques, 61
    reefs, 64
Compaction procedure, 65
    cumulative lost thickness, 72
    precompaction top, 72
    restored top, 72
Complex geological structures, 3-D modeling, 261
Computer algorithms, contour mapping, 3
Computer contouring, 273
    bathymetric surface, 279
    pinnacle reefs, 279
    sediment thickness, 279
    structure mapping, 279
Computer contour mapping, philosophy of, 1
Computer-generated surfaces, 37
Computer gridding algorithms, 76
Computer mapping, anisotropy, 75
    direction, 78
    magnitude, 78
    carbonates, 47

evaporites, 47
examples, 1
faulted surfaces, 162
pinnacle reefs, 47
Computer mapping methods, 10
Computer modeling
multiple surfaces, 159
techniques, 124
Continuous structural surfaces, 163
Contour display, aeromagnetic data, 13
Contour map, erroneous projections, 5
Contour mapping
computer algorithms, 3
data clusters, 76
data considerations, 2
eyeglassing, 76
funneling, 76
hourglassing, 76
interpretation, 2
philosophy of computer, 1
pinocchio effects, 76
reefs, pinnacle, 3
Copper, Kutcho Creek deposit, 204
Copper concentrations, Vancouver Island, 19
Copper-equivalence model, 213
3-D, 212
optimization, 213
pit design, 213
CPS-3 Advanced Mapping Software, Radian Inc., 161
Cumulative lost thickness, compaction procedure, 72
Data clusters, contour mapping, 76
Data distribution
rectangular grid, 4
triangular network, 4
Degradation potential, 94
Degree of nesting, 100
Delaunay triangulation, 80, 81, 82, 83, 86, 95, 96, 97, 98, 99, 102, 103
Delaware basin, 181, 226
data set, 233
field parameters, 229
location map, 228
New Mexico, 224
Diablerets County, Swiss Alps, 264
Digital array
2-D, 100
3-D, 100
Digital arrays, 93, 99
Digital Elevation Models, 99
Dinwoody Formation, Wyoming, 106
Displays, combined data, 22
Domain triangulation, 98
Dry-oil production contact, 114
Dynamic Graphics Inc.
Geologic Modeling Program (GMP), 263
Interactive Surface Modeling program (ISM), 253
Interactive Volume Modeling program (IVM), 252
single-horizon-method fault model, 128
EAGLES software program,
Morrison Knudsen Corporation, 242
Elk Point Group, Alberta, 240
Error evaluation, modeling, 58
Evaporites
computer mapping, 47
Michigan basin, 47
Evaporite surfaces, modeling, 55
Exxon cellular modeling system, 181
Eyeglassing, contour mapping, 76
False color image, LANDSAT Thematic Data, 17

Fault
anastomosing, 150
growth, 150
opaque barrier, 124
Fault activity, timing, 156
Fault cut, 127, 144
Fault gap, 144
Fault groupings, 146
Faulting, scissors effect, 125
Fault modeling
innovations, 127
methods, 146, 163
problems, 162
single horizon methods, 128
Fault models, nonvertical, 162
Fault penetration, 144
Fault plane attitude, 164
Fault polygons, regrid within, 163
Fault representation, automated modeling, 123
Faults
as separate surfaces, 163
vertical separation, 144
Fault surface
gridded, 141
maps, 143
multiple, 154
rectangular gridded, 127
Fault terminology, 124
Fault-throw gridding, 163
Fault-to-fault intercepts, 145
Fault type, 127
Fault zones, isochore thinning, 162
Filters, 275
Fraser, 12
Finite-difference modeling, 94
Finite-element modeling, 94
Flaxman Formation, Alaska, 188, 194, 195, 196, 199, 200
isopach, 197
Flaxman Island, Alaska, 190, 195, 196, 197, 198
Ford economic zone
Long Beach Unit, 142
Wilmington oil field, production map, 150
Form grid map, 29
Fort McMurray, Alberta, 235
Fraser filter, 12
Free-energy surface, 108, 114, 115, 119
Full Fault Modeling System (FFMS), 161, 164
methodology, 164, 165
Radian Inc., 127
Funneling, contour mapping, 76
Geochemical data, display of, 15
Geochemistry, lead in soil, 20
Geologic block modeling, 175, 183
Kutcho Creek, 203
simulation, 181
Geologic Modeling Program (GMP),
Dynamic Graphics Inc., 263
Geologic structures
2-D gridding techniques, 262
3-D gridding techniques, 262
automated modeling, 123
Geophysical maps, methods for plotting, 11
Geostatistical contouring, 77
GOCAD, 3-D subsurface modeling program, 138
Gold
concentration, Vancouver Island, 25
Kutcho Creek deposit, 204
Graphical techniques, locating anomalies, 9
Gray unit, Niagaran Formation, 52

Grid
    least-squares, 6
    rectangular, data distribution, 4
    weighted-average, 6
Gridded fault surface, 141
Gridding, 273
Gridding methods, problems, 55
Grid modeling
    biharmonic filter, 52
    least-squares algorithm, 52
Gross pay, model, 112
Growth fault, 150
Growth faults, 156
Guadalupe River delta
    sediment thickness, 288
    Texas, 279
Gubik Formation, Alaska, 195
Hourglassing, contour mapping, 76
Hugoton embayment
    field parameters, 229
    Kansas, 224
    location map, 228
Hugoton gas field, 181, 226
    data set, 227
Hydrocarbon migration, 106
Hydrocarbon-pore-thickness grid, 115
    building, 119
Hydrocarbon-pore-thickness grids, model, 112
Hydrocarbon-pore-thickness model (HPT), 106
Hydrocarbon saturation models, testing, 105
Hydrocarbon volumetric calculations,
    vertical averaging, 219
Ice scouring, 188
Imperial Oil Limited, 204
Indonesia, lead concentration in soil, 20
Interactive Surface Modeling program (ISM)
    Dynamic Graphics Inc., 253
    software, 40
Interactive Volume Modeling program (IVM),
    Dynamic Graphics Inc., 252
International Atomic Energy Agency, 39
Isochore thinning, fault zones, 162
Ivanhoe field
    cross section, 167
    gross pay volumetrics, 173
    Kimmeridge Clay Formation, 167, 168
    location map, 167
    Main Piper Sandstone, 170, 171
    Outer Moray Firth basin, 161
    Piper Sandstone Formation, 166, 168
    Supra Piper Sandstone, 171
Junipero fault, Wilmington oil field, 157
Kalkaska County, Michigan basin, 47, 50, 64
Kern County, California, 251
Kimmeridge Clay Formation, Ivanhoe field, 167, 168
Klucewicz interpolation scheme, 96
Kriging, 39, 40, 41, 77, 93
    universal, 39
Kuparuk River, Alaska, 184
Kutcho Creek
    cross section, 205
    geologic block modeling, 203
    geology, 204
    grid construction, 206
    location map, 204
    Sericite Schist unit, 204, 205
Kutcho Creek deposit
    copper, 204
    gold, 204
    massive sulfide, 203, 204, 215
    silver, 204
    sulfide mineral, 181
    zinc, 204
Lag intervals, 77
Landmark/Zycor, single-horizon-method fault model, 128
LANDSAT Thematic Data, false color image, 17
Laramide orogeny, 106
Layer-cake model, 162
Lead, in soil, 20
Least-squares algorithm, 53, 118, 275, 277
    filtered, 278
    grid, 6
    grid modeling, 52
    unfiltered, 278
Limits data, adjusting, 58
Linear variogram, 39
Line of equal influence, 76
Localized Anisotropic Data Mapping (LADM), 84
Long Beach Unit, 142, 143
    237 economic zone, 142
    Ford economic zone, 142
    Ranger economic zone, 142
    Tar economic zone, 142
    Terminal economic zone, 142
    Union Pacific economic zone, 142
    Wilmington oil field, 141, 159
Los Angeles basin, index map, 142
McMurray Formation, Alberta, 240, 242, 243, 244, 245, 246
    depositional environment, 241
Macomb County, Michigan basin, 64
Magnetics
    sun-shaded, 18
    total field, 18
Maguire Islands, Alaska, 198
Main Piper Sandstone, Ivanhoe field, 170, 171
Map, form grid, 29
Mapping
    computer, 1
    contour, philosophy of, 1
    difficulties, 176
    methodology, volumetric, 220
    methods, computer, 10
Mapping-Contouring System (MCS),
    Scientific Computer Applications, 132
Maps, geophysical, methods for plotting, 11
Michigan basin, 50, 61, 63
    Antrim County, 47, 50, 64
    carbonates, 47
    evaporites, 47
    geology, 48
    Kalkaska County, 47, 50, 64
    Macomb County, 64
    Niagaran Formation, 47
    northern trend, 47, 64
    pinnacle reefs, 47
    Salina Formation, 47
    Silurian reefs, 50, 64
Mikkelsen Bay, Alaska, 186
Mineral deposits, computer-generated surfaces, 37
Minnelusa Formation
    B-Dolomite, 85
    B-Sand, 79, 80, 85
Mississippi delta, 181
Modeling
    in 3-D, 190
    3-D subsurface, 123
    anisotropy, 75
    average-oil-saturation, 105
    average oil saturation, 115
    average porosity, 112

basin simulation, 103
boundary-element, 94
evaluating error, 58
Exxon cellular system, 181
finite-difference, 94
finite-element, 94
gross pay, 112
hydrocarbon-pore-thickness (HPT), 106
hydrocarbon saturation testing, 105
net-to-gross ratio, 112
Niagaran Formation, Brown unit, 52
    Gray unit, 51
numerical, 94
oil saturation, 112
process, 3-D, 179
quality control, 58
residual method, 51
spatial framework, 99
surface, 99
techniques, 124
trend method, 51
variogram-drift, 39
water saturation, 109
Morrison Knudsen Corporation,
    EAGLES software program, 242
Morrow Formation, Kansas, 28, 29, 30, 31, 34
Multiple surfaces, computer modeling, 159
Multiple surface triangulated models, 131
Natural Neighbor Interpolant, 95
Net-feet-hydrocarbon grid (NFH), 220
Net-to-gross ratio, model, 112
Niagaran Formation, 50
    Brown unit, 49, 54, 58, 68
        modeling, 52
    Gray unit, 49, 52, 53, 54, 56, 58, 68
        modeling, 51
    Interreef carbonate, 64
    Interreef surface grid, 54
    Michigan basin, 47
        Brown unit, 47, 48
        Gray unit, 47
    Pinnacle unit, 56, 57
Nonvertical fault models, 162
Normal faulting, 124
Northern trend, Michigan basin, 47, 64
North Sea, Outer Moray Firth basin, 161, 166
North Sea reservoirs, field mapping, 162
North Slope, Alaska, 183, 196
Numerical modeling, 94
Obtuse-angled triangles, 97
Oil-in-place estimation, 105
Oil saturation, 105
    distribution, 108
    model, 112
Oregon basin field
    South Dome, 105
    Wyoming, 105, 106
Original oil-in-place (OOIP), 106
OSLO Project, 240
    location map, 242
Outer Moray Firth basin, 168
    Ivanhoe field, 161
    North Sea, 166
Oxides, spatial distribution, 43
Paleo-water
    level, 56
    thickness, 56
Parametric grids, 93, 99, 101
Park County, Wyoming, 105
Permafrost, 186

Philosophy of contour mapping, 1
Phosphate, 43
    deposit, 40
    ore bed, 41
    in situ resources, 44
Phosphoria Formation, Wyoming, 105, 106, 108, 109, 113, 115, 117, 118, 120
    erosional unconformities, 106
    hydrocarbon pore thickness, 114
    reservoir, 110
    water saturation, 111
Phosphorite
    assays, 39
    beds, 38
        chemistry, 43
    deposit, geometry, 40
    deposits, 40
Piezometric surface, 93
Pinnacle reefs, 47
    computer contouring, 279
    computer mapping, 47
    contour mapping, 3
    Michigan basin, 47
Pinocchio effects, contour mapping, 76
Piper Sandstone Formation, Ivanhoe field, 166, 168
Point Hopson, Alaska, 186
Point Sweeney, Alaska, 190
Point Thomson, Alaska, 190, 195, 196, 197
Polar fan-delta complex, 183
Porosity, capillary pressure curves, 108
Powder River basin, Wyoming, 79
Precompaction top, compaction procedure, 72
Prefault conditions, 163
Pressure curves, air-mercury capillary, 109
Projections, erroneous, contour map, 5
Prudhoe Bay, Alaska, 184, 186
Public domain data, 11
Quality control, modeling, 58
Radian Inc.
    CPS-3 Advanced Mapping Software, 161
    CPS gridding program, 125
    Full Fault Modeling System (FFMS), 127
Ranger economic zone
    Long Beach Unit, 142
    Wilmington oil field, moveable oil saturation, 149
Rectangular grid, data distribution, 4
Rectangular grids, 93, 99, 101
Red Fork Formation, Oklahoma, 32, 35
Red Fork Sandstone, Oklahoma, 283
Reefs
    compaction, 64
    pinnacle, 47
        contour mapping, 3
Regional Anisotropic Data Mapping (RADM), 78
Regionalized variable, 236
Regridding, within fault polygons, 163
Reserve estimates, calculation, 213
Reservoir attributes, 106
Reservoir continuity, 176
Reservoir properties, 144
Residual method, 53
    modeling, 51
Restored top, compaction procedure, 72
Restored Top Method, 156
Rock type characterization, 41
Sagavanirktok River, Alaska, 184
St. Clair County, Michigan basin, 64
St. Louis Formation, Kansas, 28, 31, 32
Salina Formation
    A-1 Carbonate, 47, 50

A-1 Evaporite, 47, 50
A-2 Carbonate, 47
A-2 Evaporite, 47
Michigan basin, 47, 50, 64
San Joaquin Valley, Kern County, 251
Scientific Computer Applications, Mapping-Contouring System (MCS), 132
Sedimentary basin, simulation, 93, 94
Sediment thickness, computer contouring, 279
SEDSIM model, 94
Semivariogram pair, example, 81, 87
Sericite Schist unit, Kutcho Creek, 204, 205
Shape-assist technique, 27
    computer-generated surface models, 27
Silurian reefs, Michigan basin, 50, 64
Silver, Kutcho Creek deposit, 204
Simulation modeling, 178
Single-horizon-method fault model
    Dynamic Graphics Inc., 128
    Landmark/Zycor, 128
Software, Interactive Surface Modeling, 40
Soil geochemistry, lead, 20
South Belridge field, 181
    Kern County, 251
        geology, 252
        location map, 252
        structure map, 253
South Dome, Oregon basin field, 105
Spatial framework, modeling, 99
Specific-gravity model, 3-D, 212
Stacked profile map, Vancouver Island, 10
Ste. Geneviere Formation, Kansas, 28, 31, 32
steam drive, mature, 3-D modeling, 251
Stikine Range, British Columbia, 204
Storm surge, 187
Stratamodel, Stratigraphic Geocellular Modeling system (SGM), 222
Stratigraphic correlation, 180
Stratigraphic Geocellular Modeling (SGM), 181, 225, 226, 227
    Stratamodel, 222
Stream channels, sand thickness, 283
Structural grids, building, 31
Structural irregularities, 144
Structural surface grids, 156
Structure mapping, computer contouring, 279
Subcrop, defining, 31
Sulfide deposit, Kutcho Creek, British Columbia, 203
Sumac Mines Limited, 204
Supra Piper Sandstone
    cross section, 172
    isochore map, 171
    Ivanhoe field, 169, 170, 171
Surface fitting, 101
Surface modeling, 93, 99, 100
    geologic applications, 93
Surface models, computer-generated, 27
Surfaces, computer-generated, 37
Surveys, regional, 11
Tar economic zone, Long Beach Unit, 142
Temple Avenue B-1 fault,
    Wilmington oil field, 146, 147, 148, 154, 159
Tensleep Formation,
    Wyoming, 105, 106, 108, 109, 113, 115, 117, 118, 119, 120
        average porosity, 121
        hydrocarbon pore thickness, 114
        net-to-gross ratio, 121
        reservoir, 110
        water saturation, 111, 113
Terminal economic zone, Long Beach Unit, 142

Tessellations, Voronoi polygons, 83, 84
THUMS Long Beach Company, 141
Tidelands Unit, Wilmington oil field, 141
Tigvariak Island, Alaska, 186
TIN, triangular irregular network, 4
Transition-zone oil, 108
Trend method, 53
    modeling, 51
Triangle-based finite-difference technique, 93
Triangular irregular networks (TIN), 124
Triangular meshes, 93, 95, 99
Triangular network
    data distribution, 4
    irregular, TIN, 4
Triangulated models, multiple surface, 131
True vertical thickness isochores, 162
Truncation, 180
U.S. Geological Survey, 190
Unconformity, truncated, 4
Union Pacific economic zone
    Long Beach Unit, 142
    Wilmington oil field, production map, 150
Universal kriging, 39
Vancouver Island
    aeromagnetic survey, 10
    copper concentrations, 19
    geochemical data, 24
    gold concentration, 25
    stacked profile map, 10
Variable zone modeling (VZM), 235
Variogram, linear, 39
Variogram-drift models, 39
Vertical averaging, hydrocarbon
    volumetric calculations, 219
Vertical separation, 156
    faults, 144
Volumetric mapping methodology, 220
Volumetric model, 162
Volumetrics, 113
Voronoi diagram, 95, 96, 97, 99
Voronoi mosaic, 103
Voronoi polygons, 75, 80, 81, 82, 83, 86
    tessellations, 84
Water saturation
    Phosphoria Formation, 111
    Tensleep Formation, 111
Water saturation curve, single, 111
Water saturation curves, multiple, 110
Water saturation model, 109
Waterways Formation, Alberta, 240, 241
Weighted-average algorithm, 273, 275
Weighted-average grid, 6
Wilmington anticline, 141
Wilmington oil field, 146, 150, 154
    C-608 fault, 146
    cross section, 142
    Ford economic zone, production map, 150
    horizontal trace maps, 145
    Junipero fault, 157
    Long Beach, California, 141
    Long Beach Unit, 141, 159
    Ranger economic zone, moveable oil saturation, 149
    Temple Avenue B-1 fault, 146, 147, 148, 154, 159
    Tidelands Unit, 141
    Union Pacific economic zone, production map, 150
Woronzofian transgression, 183
    sea-level maximum, 200
Zero-line projection, 276
Zinc, Kutcho Creek deposit, 204